GEOCHEMICAL TRANSPORT AND KINETICS

Papers presented at a conference at Airlie House, Warrenton, Virginia, June 1973.

Edited by

A. W. HOFMANN
B. J. GILETTI
H. S. YODER, JR.
R. A. YUND

CARNEGIE INSTITUTION OF WASHINGTON
PUBLICATION 634

Copyright © 1974 by Carnegie Institution of Washington

Library of Congress Cataloging in Publication Data

Conference on Geochemical Transport and Kinetics,
    Airlie House, 1973.
        Geochemical transport and kinetics

    Sponsored by the Carnegie Institution of Washington.
        1. Geochemistry—Congresses. 2. Diffusion—
Congresses. 3. Kinetics—Congresses. I. Hofmann,
Albrecht W., 1939-   ed. II. Carnegie Institution
of Washington. III. Title.
QE515.C7 1973   551.9   74-19286
ISBN 0-87279-644-2

PREFACE

This volume contains papers that were presented at a Conference on Geochemical Transport and Kinetics held on 4 to 6 June 1973 at Airlie House in Warrenton, Virginia. The conference was sponsored by the Carnegie Institution of Washington and was attended by 42 invited participants (see list below). The principal objective of the conference was to bring together people who are actively engaged in the study of transport and kinetic phenomena and who have interest in geochemical problems, whether they are geochemists, field geologists, ceramists, or metallurgists. One of the benefits was the opportunity to discuss the current theoretical, experimental, and observational aspects of the topic informally as well as by presentation of papers. Some of the material discussed was in preliminary form or speculative and will doubtless undergo refinement prior to publication elsewhere. The time allowed for discussion during the non-meeting hours may prove to be as valuable to the science as were the meetings, although there will be no direct evidence of that in this volume. The papers presented here were submitted by the participants about one month after the conference. It was hoped that this delay would permit authors to refine or revise their ideas in light of the conference discussions.

The proceedings are intended as a source of review papers, current research, and literature reference for those working in this field as well as those who are new to it. If the book is incomplete in its objectives, it is partly because the scope of the conference had to be limited. Many important areas of interest are not represented at all (e.g. transport and reaction kinetics in the sedimentary environment) or seriously underrepresented (e.g. geological field studies, the kinetics of solid-state reactions, and studies of ore deposits).

A serious deficiency was revealed in the trend of the discussions at the conference. The links between theory, laboratory studies, and geological field observations are still weak. It is hoped that the conference and this book will contribute toward strengthening these links.

The material has been arranged in three parts. Part I is concerned with the subject of diffusion and contains both reviews and original data where the emphasis is on laboratory measurement of diffusion coefficients. Part II, also laboratory oriented, deals with reactions such as recrystallization and exsolution (solid state precipitation) processes. The distinction between diffusion and reaction is not always easy to draw in practice. Diffusion may be thought of as atomic migration within a homogeneous phase, whereas reaction involves either a heterogeneous process or a change in grain geometry. Part III contains those contributions that are not based on new laboratory experiments (except for chemical or isotopic analyses). Some of the papers in this section are based on field observations or analyses made on rocks; others discuss how transport theory might be applied to geological problems. As might be expected, not all papers fit neatly into a single category.

We thank all participants of the conference and contributors to the proceedings, especially those whose primary interests lie with materials science rather than geochemistry. As geologists we feel fortunate to be able to take advantage of the more advanced state of transport research in the materials sciences. Special thanks are also due A. David Singer of the Geophysical Laboratory and Niels Pedersen of the Department of Terrestrial Magnetism for their invaluable help in organizing the conference. We also thank Sheila McGough, editor of the Carnegie Institution, and Janet Land, Robert Metcalf, and Stanley Weintraub of the editorial office for making a book out of a pile of manuscripts. The subject and author indexes were prepared by Mr. Weintraub.

A. W. Hofmann
B. J. Giletti
H. S. Yoder, Jr.
R. A. Yund

Bottom row (left to right): Brady, Misener, Shieh, Burt, Iiyama, Giletti, Hofmann, Yoder, Yund, McCallister, Ahrens, Birchenall, Ross. Second row: D. E. Anderson, Manning, Rye, McConnell, Goldsmith, Clayton, Heuer, Fisher, Wones, Green, Vidale. Third row: Bell, Fletcher, Chai, Taylor, Orville, Helgeson, Frey, T. F. Anderson, Sipling, Cooper, Frantz, Weisbrod, Hart, Foland, Krogh. Absent: Kushiro, Mannheim, Shaw.

PARTICIPANTS

T. J. Ahrens	California Institute of Technology
D. E. Anderson	University of Illinois
T. F. Anderson	University of Illinois
P. M. Bell	Carnegie Institution of Washington
C. E. Birchenall	University of Delaware
J. B. Brady	Harvard University
D. M. Burt	Yale University
B. H. T. Chai	Yale University
R. N. Clayton	University of Chicago
A. R. Cooper	Case Western Reserve University
G. W. Fisher	Johns Hopkins University
R. C. Fletcher	Carnegie Institution of Washington
K. A. Foland	University of Pennsylvania
J. D. Frantz	Carnegie Institution of Washington
F. A. Frey	Massachusetts Institute of Technology
B. J. Giletti	Brown University
J. R. Goldsmith	University of Chicago
H. W. Green II	University of California, Davis
S. R. Hart	Carnegie Institution of Washington
H. C. Helgeson	University of California, Berkeley
A. H. Heuer	Case Western Reserve University
A. W. Hofmann	Carnegie Institution of Washington
J. T. Iiyama	Centre de Synthèse et Chimie des Minéraux, Orléans la Source
T. E. Krogh	Carnegie Institution of Washington
I. Kushiro	Carnegie Institution of Washington
F. T. Manheim	Woods Hole Oceanographic Institution
J. R. Manning	National Bureau of Standards
R. H. McCallister	Purdue University
J. D. C. McConnell	Cambridge University
D. J. Misener	University of British Columbia
P. M. Orville	Yale University
M. Ross	U.S. Geological Survey, Washington, D.C.
D. M. Rye	Yale University
H. R. Shaw	U.S. Geological Survey, Washington, D.C.
Y. N. Shieh	Purdue University
P. J. Sipling	Brown University
H. P. Taylor, Jr.	California Institute of Technology
R. J. Vidale	State University of New York, Binghamton
A. Weisbrod	Université de Nancy
D. R. Wones	U.S. Geological Survey, Washington, D.C.
H. S. Yoder, Jr.	Carnegie Institution of Washington
R. A. Yund	Brown University

INTRODUCTORY REMARKS*

The conference was organized because of the realization that many great advances on specific problems in chemical transport and kinetics had been made and that the results are not generally known, or if known, are not appreciated by geologists. Instrumentation has become available that makes many new kinds of experimental kinetic studies feasible and rewarding. The electron microprobe has not only led to a new recognition of the limits of chemical equilibrium in rocks; it has also become an important tool for the analysis of small-scale diffusion gradients. The ion microprobe also holds great promise for the analysis of the complete range of elements and their isotopes. The central purpose of the conference was to acquaint the participants with each other's problems, generate new interactions, and delineate the most rewarding areas of research. It was also intended to disseminate the results of the conference more generally through the publication of a book integrating the field, laboratory, and theoretical aspects of transport and kinetics as presented at the conference.

There is little doubt in the minds of the organizers that transport and kinetics will become the focus of research in the geology of the future. They are already principal research topics in ceramics, metallurgy, and materials science. For sixty years the Geophysical Laboratory of the Carnegie Institution has studied systems pertinent to the major rock-forming minerals under *equilibrium* conditions. The argument was that rocks closely approached equilibrium and that if the end result were known, the nonequilibrium paths could be deduced. It is just these nonequilibrium paths with which the geologist is now most concerned. If all rocks were at perfect equilibrium, we would never know how they achieved their present state. Fortunately, rocks retain metastable states so that the rocks formed at high pressure, for example, will persist when brought to or near the surface where the geologist has access to them. Finding the thermochemical path along which a rock has traveled, deduced by its nonequilibrium features, is one of the principal tasks of the geologist.

All major types of rocks—sediments, metamorphics, and igneous rocks—abound with significant kinetics and transport problems. Sediments are excellent examples of heterogeneous nonequilibrium assemblages, which under burial will react to form a homogeneous layer in one band yet fail to react in an adjoining band. The pore fluid obviously exerts considerable influence on its behavior. The reduction of pore fluid from about 50% in the original sediment to less than 2% in the consolidated sedimentary rock is in itself a complex transport problem involving pore water, adsorbed water, interlayer water, and possibly structural water. That homogeneous layer may be quite deceiving—the electron microprobe is now telling us that only local equilibrium is established among the minerals, and that perhaps only nearest neighbors are in communication in spite of the removal of large volumes of water.

The metamorphic rocks, classified—but not without difficulty—into zones or facies, tell us that there is an approach to equilibrium of sorts. On the other hand, the sequence of isograds may be inversely related to reaction rates, especially in the lower grades. In fact, we are still struggling to learn whether a metamorphic rock goes directly to its maximum metamorphic grade or whether it must pass through all lower stages or grades *en route*. Yet there is much to understand about porphyroblasts of andalusite, hollow crystals of cordierite, Fe-Mg zoning

---

* The substance of remarks made by Dr. H. S. Yoder, Jr., in welcoming the participants to the conference.

in garnet, overgrowths on kyanite, grain size and morphological variations of quartz, and the persistence of certain phases, especially the Al_2SiO_5 minerals, through multiple metamorphic events. The preexisting zoning in a feldspar which may persist for a billion years in one rock is quickly homogenized in another rock. Many of us are still arguing whether oxygen, water, sulfur, and carbon dioxide are free and available to pass through rocks or whether these more mobile constituents are constrained within layers. Reaction rates are usually considered instantaneous, geologically speaking, and the control is predominantly the rate at which materials are brought to or away from the reaction site. Nature imposes pressure and temperature much faster than materials can be transported: that is, thermal conduction, hydrostatic loading, and stress application are always believed to be faster than diffusion. The transport of materials is therefore presumed to be the rate-determining process. However, at the site of reaction we are faced with Lindgren's so-called volume-for-volume replacement law.

Much of the great debate concerning the origins of granite has been about the role of metasomatism, partial melting, and fractionation. The questions raised run the gamut of petrology and have not been fully resolved. The debate has lost much of its impetus because there has been little new information to contribute to the solutions. A sound body of principles has evolved from a theoretical physicochemical point of view, but reliable experimental data are lacking.

Even those rocks that are unquestionably igneous in origin present an equivalently large array of problems. The accumulation of a magma from a partially melted rock under pressure is a major transport problem. The persistence of two contiguous magmas of high compositional contrast has yet to be understood. Curiously enough, when such magmas interdiffuse, the exchange is not atom-for-atom but appears to be on the basis of species or subspecies of their rock-forming minerals. The formation of phenocrysts and the depletion of the surrounding magma in one area and their absence in another area is a nucleation and transport problem still unsolved. How does one fill an amygdule completely? If a lava is quenched to a glass, what influence does its structure have on subsequent crystal nucleation? The glass formed in a cooled lava may persist for 3.5 billion years or may be quickly devitrified near a steam vent or in a fault zone. How can a lava flow exchange all its oxygen, about 60% in volume, with ground water without so much as disturbing the structure of the constituent minerals? Furthermore, why is it possible to exchange completely and selectively sodium and potassium in a feldspar without disturbing its structural state? Do the giant feldspars in a pegmatite really mean slow crystal growth under quiet conditions, or do they mean special elixirs are present, or are these large crystals grown quickly during rupture of the country rock in an expanding gas phase? Why are diamonds preserved in the corrosive kimberlite soup as it blasts its way up from 300 km in the mantle?

On a completely different scale of event, the rate of overturn in a convecting mantle requires knowledge of rheomorphic properties of rocks under extreme conditions. The explosive eruption of an andesitic volcano requires knowledge of the degassing rate of water from a vesiculating magma. In short, the entire theory of explosive volcanism depends upon the rate of diffusion of gas into and out of a liquid.

It is pointless to give more examples, for indeed every geological event one can think of depends on rate processes and involves transport of material. Some tackled these problems from an atomistic view; others dealt with more macroscopic parameters. But that was the purpose of the conference—to exchange ideas.

CONTENTS

PREFACE iii

PARTICIPANTS v

INTRODUCTORY REMARKS vi

PART I. DIFFUSION 1
    Diffusion kinetics and mechanisms in simple crystals (Manning) 3
    Vector space treatment of multicomponent diffusion (Cooper) 15
    Modeling of diffusion controlled properties of silicates (Anderson and Buckley) . . 31
    Diffusion in sulfides (Birchenall) 53
    Diffusion related to geochronology (Giletti) 61
    Alkali diffusion in orthoclase (Foland) 77
    Oxygen isotope exchange between potassium feldspar and KCl solution (Yund and Anderson) 99
    Studies in diffusion I: Argon in phlogopite mica (Giletti) 107
    Cationic diffusion in olivine to 1400°C and 35 kbar (Misener) 117
    Diffusion of tritiated water in β-quartz (Shaffer *et al.*) 131
    Diffusion of H_2O in granitic liquids: Part I. experimental data. Part II. mass transfer in magma chambers (Shaw) 139

PART II. REACTION KINETICS 171
    Coherent exsolution in the alkali feldspars (Yund) 173
    Kinetics of Al Si disordering in alkali feldspars (Sipling and Yund) 185
    Kinetics of enstatite exsolution from supersaturated diopsides (McCallister) . . 195
    Mass transfer of calcite during hydrothermal recrystallization (Chai) 205
    Oxygen isotope exchange between calcite and water under hydrothermal conditions (Anderson and Chai) 219

PART III. TRANSPORT AND REACTION IN ROCKS 229
    Criteria for quasi-steady diffusion and local equilibrium in metamorphism (Fisher and Elliott) 231
    Simple models of diffusion and combined diffusion-infiltration metasomatism (Fletcher and Hofmann) 243
    Infiltration metasomatism in the system K_2O-SiO_2-Al_2O_3-H_2O-HCl (Frantz and Weisbrod) 261
    Metamorphic differentiation layering in pelitic rocks of Dutchess County, New York (Vidale) 273
    Metasomatic zoning in Ca-Fe-Si exoskarns (Burt) 287
    A new approach to the determination of temperatures of intrusives from radiogenic argon loss in contact aureoles (Brandt) 295
    Oxygen and hydrogen isotope evidence for large-scale circulation and interaction between ground waters and igneous intrusions, with particular reference to the San Juan volcanic field, Colorado (Taylor) 299
    Mobility of oxygen isotopes during metamorphism (Shieh) 325

SUBJECT INDEX 337

AUTHOR INDEX 347

PART I. DIFFUSION

Atomic migration through crystals, aqueous fluids, melts, and glasses, or along grain boundaries is involved in many, perhaps all, chemical processes of geological interest. Diffusion controls the rates of many exsolution reactions and of most mechanisms of ductile rock deformation (diffusion creep, dislocation climb). The interpretation, and to some extent the success, of isotopic methods of determining the ages of rocks or their temperatures of formation depend heavily on the extent of diffusion of radiogenic isotopes and oxygen isotopes. These are some of the reasons why the subject of diffusion dominated the presentations and discussions at the conference.

The authors of the eleven papers of Part I of this book approach the subject of diffusion from different points of view. Manning and Cooper review the atomistic mechanisms and the phenomenology of diffusion, respectively, from a theoretical point of view. Anderson and Buckley discuss both thermodynamic and atomistic aspects of multicomponent diffusion, using specific mineral systems and boundary conditions as examples. The phenomenology of diffusion as applied to systems that model rocks is also discussed in Part III.

No attempt was made to review the solutions and methods of solution of Fick's second law, the partial differential equation $\partial c/\partial t = D(\partial^2 c/\partial x^2)$ or its more complicated extensions. The reader is referred to standard texts, e.g., J. Crank, *The Mathematics of Diffusion*, 1956. Other basic concepts are explained in Manning's paper. Among these are the activation energy, the various kinds of diffusion coefficients (tracer, intrinsic, and interdiffusion), and the equation $D = D_0 \exp(-Q/RT)$. The latter is often referred to as the "Arrhenius relation" in this book, and an additional term, "the self-diffusion coefficient," is used in some of the papers. Self diffusion is similar to tracer diffusion except that unlabeled atoms of the same chemical species as the labeled tracer atoms are present in the medium.

Birchenall and also Giletti review previous experimental work on particular systems of geological interest. The remaining papers (Foland; Yund and Anderson; Giletti; Shaffer, Sang, Cooper, and Heuer; Shaw; and Misener) present new experimental data on diffusion in minerals. They contain examples of self diffusion (Foland; Yund and Anderson), tracer diffusion or possibly intrinsic diffusion (Shaffer *et al.*), and interdiffusion (Misener). Shaw gives experimental results on the diffusion of water in silicate melts and applies these results, illustrating how water can be taken up by a convecting magma from the wall rock.

The presentations and discussions at the conference, including several contributions of experimental data not in this volume lead to the following inferences about the status of diffusion measurements on materials of geological interest:

1. Very little is known about diffusion mechanisms in complex crystalline structures such as silicates.

2. Diffusion coefficients are much more difficult to measure than has often been realized or admitted by the experimenter. Disagreements amounting to several orders of magnitude are not uncommon, and it is often the smallest value that results from the most careful measurement.

3. Experiments on metals and oxides have shown that diffusion behavior may vary drastically as a function of impurity content. Consequently, materials scientists have experimented with ever purer crystals in order to determine the intrinsic diffusion behavior of the crystal. The geologist, on the other hand, is generally not concerned with pure materials

but wishes to determine the behavior of natural materials. The extent that differences in impurity (= trace element) content influence diffusion in natural minerals is poorly known. However, with careful chemical and structural characterization of the material, an evaluation of the effect of impurities is possible. In addition, experimental techniques are available to determine diffusion coefficients that are reproducible. Some efforts are made to determine the diffusional anisotropy of crystals, for example in the papers by Shaffer *et al.*, Misener, and Giletti.

4. The geologist often wishes to extrapolate the experimentally determined diffusion coefficients to lower temperatures. This is done by use of the Arrhenius relation mentioned above. The extrapolation is useful and valid if one can assume that all relevant diffusion mechanisms are known and that they do not change over the temperature range involved. This assumption usually cannot be verified in the laboratory because diffusion at low temperatures is too small to be measured and because the material may undergo very slow changes in its atomic or defect structure at these temperatures. It is nevertheless encouraging that several of the experimental contributions presented cover a comparatively large range of temperatures and that the results are generally consistent with a simple Arrhenius relation. Ultimately, the validity of applying laboratory results to the interpretation of natural transport will depend on some verification in the "natural laboratory" (see Giletti's review paper).

5. By use of the electron microprobe it is now possible to analyze quantitatively concentration gradients on a micron scale. Such gradients can be used to determine diffusion coefficients directly. Unfortunately, the analytical sensitivity of the electron microprobe for many trace elements is inadequate. Also, the resolution of the electron microprobe may still be insufficient when transport distances are in the sub-micron range. In these cases, diffusion coefficients may be determined by bulk isotopic exchange between two phases, for example crystal and hydrothermal fluid. Examples of both experimental techniques appear in Part I. In the future, it may be possible to analyze concentration gradients with a resolution of hundreds or tens of angstroms by the use of the ion microprobe or the electron microscope.

It is clear, however, that even the most sophisticated analytical tools may yield meaningless results if the diffusion experiment itself is improperly designed. The articles presented here provide some guidance as to the design of diffusion experiments, even though not all the desirable criteria may have been met in the reported experiments.

6. Diffusion coefficients of most chemical species in minerals are so small that it is impossible to invoke volume diffusion through the solid phases as a plausible mechanism for large-scale metasomatic transport. Consequently, several authors of Part III invoke diffusion through a pore fluid or a grain boundary phase. The need for experimental data on grain boundary diffusion is clearly indicated. No such data are offered in this book, perhaps because experimentalists are out of touch with "geological reality," or, more likely, because such measurements are even more difficult to perform than those involving a single phase.

A. W. Hofmann
B. J. Giletti

DIFFUSION KINETICS AND MECHANISMS IN SIMPLE CRYSTALS[1]

John R. Manning
Metallurgy Division
National Bureau of Standards
Washington, D.C. 20234

ABSTRACT

The kinetic-atomic theory of diffusion is discussed and reviewed. Any such theory must involve a model providing a detailed picture of the paths as the atoms move through the material. Diffusion paths and diffusion mechanisms therefore are important in this theory and are discussed first. Both short-circuit diffusion mechanisms along easy diffusion paths and possible volume-diffusion mechanisms through regions of regular crystal structure are described. Simple random-walk diffusion equations are derived which yield expressions for the tracer diffusion coefficient D^* and the atom drift velocity v_F in terms of atom jump frequencies. It is noted that diffusion frequently occurs by a vacancy mechanism. When this is the case, the random-walk equations must be modified to include both correlation-factor effects and vacancy-wind effects. The origin and influence of these effects are discussed. A comparison is made between the kinetic-atomic diffusion equations and the thermodynamic-continuum diffusion equations. It is noted that cross terms which yield a dependence of the atom flux of species i on the concentration gradients or chemical potential gradients of other species k appear in these equations.

INTRODUCTION

There are two basic approaches one can take to diffusion theory in attempting to describe and understand diffusion processes: (1) the atomistic approach, in which diffusion is explained as being the result of many small discrete motions by individual atoms, and (2) the continuum approach, in which the diffusing substance is considered to be a continuum without discrete atomic structure. Most thermodynamic discussions of diffusion adopt the continuum approach, since thermodynamics is primarily concerned with initial states and final states and is not directly concerned with atomic paths between these states. Also, experimental diffusion measurements usually are done on a macroscopic scale rather than on an atomic scale and, as a result, are customarily reported in terms of continuum equations. Nevertheless, a much more complete description of diffusion processes can be obtained if one considers atomistic ideas also. Then one can treat diffusion as a more-or-less random-walk process and derive kinetic relations between measured diffusion quantities in ways which would not be possible from a strictly continuum-thermodynamic approach.

DIFFUSION MECHANISMS

Any kinetic description of diffusion must be concerned with diffusion mechanisms. One must answer the question, How does an atom or molecule move from here to there? In practice, diffusion in solids occurs both by volume diffusion through regions of good crystal structure

[1] Contribution of the National Bureau of Standards, not subject to copyright.

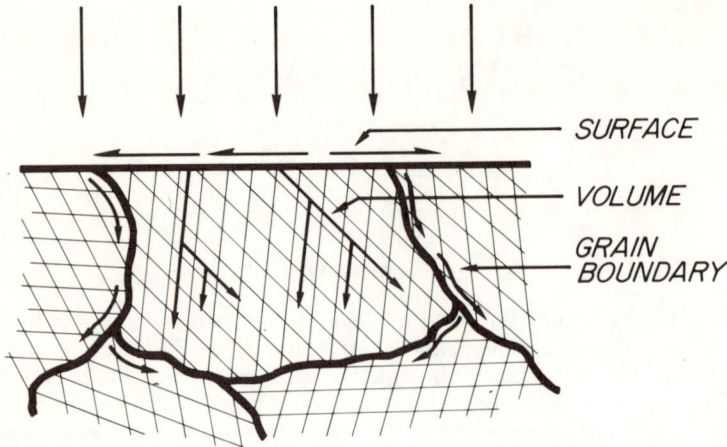

Fig. 1. Paths for volume diffusion, grain boundary diffusion, and surface diffusion. Note that volume diffusion can provide many possible paths. This multitude of possible paths allows volume diffusion to be dominant at high temperatures despite the higher mobility expected along the individual short-circuit (surface and grain boundary) paths. If diffusion and reaction throughout a grain is desired, volume diffusion is required.

and by a variety of short-circuit diffusion mechanisms where the atoms move along paths of easy diffusion. (See Fig. 1.) These easy diffusion paths usually involve surface or line defects in the crystal, such as grain boundaries, dislocation lines, or fast diffusion directions on free surfaces. Similarly, volume diffusion mechanisms in crystals usually depend on the presence of point defects, such as vacancies or interstitial atoms.

The diffusion coefficient D is usually represented as depending exponentially on an activation energy Q and the absolute temperature T according to the equation,

$$D = D_0 \exp(-Q/RT) \quad (1)$$

where R is the universal gas constant (1.987 cal mol^{-1} K^{-1}) and the pre-exponential quantity D_0 is temperature-independent. According to atomic diffusion theory, the probability of a given possible atom jump actually occurring depends exponentially on the activation energy which must be gathered locally to allow the jump, thus providing the exponential term in Equation 1, whereas D_0 is proportional to the number of possible atom paths and to the average squared jump distance. When two diffusion mechanisms operate simultaneously, each mechanism g yields a separate contribution to D, with each contribution having the form $D_{g0} \exp(-Q_g/kT)$.

Since atoms in regions of good crystal structure are more restricted and confined in their motions than are atoms on free surfaces or on grain boundaries, diffusion along surfaces or boundaries can usually proceed much more readily than can volume diffusion. Thus, if a diffusant were deposited on a surface and diffusion were allowed to occur, the diffusant normally would penetrate much farther into the crystal along the grain boundaries than through the regions of good crystal structure, where volume diffusion is necessary. In simple crystals such as many metals, the activation energy for volume diffusion typically is about twice that for grain boundary diffusion, which in turn may be about twice that for surface diffusion. This difference in activation energies might be explained qualitatively by noting that individual atom jumps along free surfaces are less constrained by neighboring atoms than are those along grain boundaries which in turn are less constrained than volume diffusion jumps.

Because of the large differences between the activation energies Q for grain boundary diffusion and those for volume diffusion (along with the strong dependence of diffusion coefficient on Q and T), grain boundary transport will dominate transport by volume diffusion in polycrystals at low temperatures. As the temperature is increased, however, volume diffusion contributions increase rapidly until finally at high temperatures the fact that there are so many more possible volume diffusion paths than grain boundary diffusion paths makes volume diffusion the prime means by which a diffusant is transported. In typical polycrystalline metal crystals (grain size 1 mm), volume diffusion usually starts to predominate at temperatures around two-thirds of the melting point. This value does not necessarily apply to more complex crystals, however. In transformation processes which require a change in chemical composition throughout an entire grain, volume diffusion will in all cases be the final limiting process regardless of how fast matter from the surface is transported into the material via grain boundaries, since volume diffusion is required to transport a reactant from the available surface or line defects into the regions of regular crystal structure within each grain.

Because of the wide variety of possible surface configurations, grain boundary orientations, and dislocation line directions, detailed models of the motions of atoms along these easy diffusion paths are difficult to establish. It appears that surface diffusion in metals occurs mainly by motion of atoms on the lattice plane immediately above the main surface plane. Such atoms, which are called adatoms, can move more freely than atoms in the main surface plane itself. Even so, the detailed paths followed by these atoms are not clear, especially for rough surfaces. Diffusion models for grain boundary diffusion are even less well established.

On the other hand, for volume diffusion, the probable diffusion paths and mechanisms can be much more readily delineated. Four general types of volume diffusion mechanisms which can be envisioned are illustrated in Fig. 2.

1. The *exchange mechanism* is merely the interchange of position of two neighboring atoms and thus requires no defect in the crystal. Such a process will usually cause a very large distortion in the crystal during the jump and consequently will require a large activation energy, making this an unlikely mechanism. A variation on the exchange mechanism is the *ring mechanism* in which three or more atoms, situated on a ring, simultaneously move to the next site around the ring. Although more atoms are involved in the elementary jump in the ring mechanism than in the exchange mechanism, there may be less distortion of the crystal lattice during the jump of the ring and hence a slightly lower activation energy. Nevertheless, this type of mechanism is still unlikely to occur.

2. A second obvious type of diffusion mechanism is the *interstitial mechanism*. Small interstitial atoms can jump directly from one interstitial site to another without greatly disturbing the other atoms. This mechanism requires that the diffusing atom first be introduced as an interstitial into the lattice. For substitutional impurities and substitutional alloys where the atoms occupy regular lattice sites, the creation of interstitials is in most cases unlikely. Nevertheless, for interstitial alloys and even for some substitutional components which can be partially excited to interstitial sites, this mechanism provides a means of rapid diffusion. Here the interstitial atom itself is the point defect which allows diffusion to occur. If there are many more possible interstitial sites than there are interstitial atoms, the intersti-

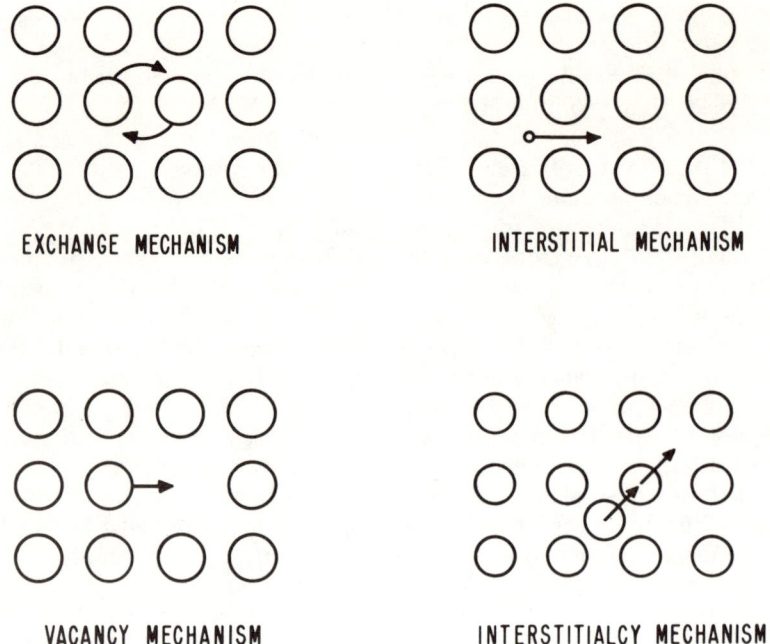

Fig. 2. Elementary atom jumps in the four main types of possible volume diffusion mechanisms.

tial atoms will not interfere appreciably with one another and the interstitials can follow paths which are essentially random walks. Motion of adatoms during surface diffusion may be very similar to this interstitial motion but with the motion on the surface restricted to two dimensions.

3. The *vacancy mechanism* is a very common type of diffusion mechanism, which depends on the presence of point defect vacancies in the crystal. In thermodynamic equilibrium, one expects a certain number of vacant lattice sites to be present in a crystal. Atoms neighboring on a vacancy then can diffuse by jumping into the vacancy. Since the jump of an atom into a neighboring vacancy causes much less distortion than the direct exchange of neighboring atoms, the vacancy mechanism is a much more likely mechanism for diffusion in substitutional alloys or simple ionic crystals than the exchange mechanism. From the viewpoint of kinetic theory, the simplest type of vacancy mechanism is that involving individual isolated vacancies, as in Fig. 2. In some instances, vacancies may be found in pairs, however, yielding a *divacancy mechanism*, or even more complex vacancy configurations may occur if vacancy concentrations are large.

4. The *interstitialcy mechanism*, also called the indirect interstitial mechanism, is a separate mechanism which differs in several important respects from the interstitial mechanism discussed above. Here the interstitial atom moves by pushing a normal lattice atom into an interstitial site and moving into the lattice site itself. The region centered on an interstitial atom can be called an interstitialcy. The location of the interstitialcy during an elementary jump in this mechanism may move twice as far as do either of the individual atoms themselves. A variation of this mechanism is the *crowdion mechanism,* where the insertion of an additional atom along a close-packed direction causes a number

of atoms along this direction to be displaced to a greater or smaller extent from normal lattice sites. The position of this imperfection then can move along the close-packed direction and cause diffusion.

Random Walk Equations for Diffusion via Independent Jumps

The illustrations of Fig. 2 show the elementary atom jumps for each of the four main types of volume diffusion mechanisms. Diffusion of any given atom takes place by a series of elementary jumps. In simple cases, as for the interstitial mechanism, the jumps are independent of one another. Then, the diffusion flux resulting from these jumps can be obtained directly from simple random-walk-type equations. For example, if diffusion occurs between two neighboring lattice planes, 1 and 2, separated by a distance λ, the number of jumps j_{12} per unit time per unit area from 1 to 2 is given by

$$j_{12} = n_1 \Gamma_{12} \tag{2a}$$

and the number of jumps j_{21} per unit time per unit area from 2 to 1 is given by

$$j_{21} = n_2 \Gamma_{21} \tag{2b}$$

where n_1 and n_2 are the number of atoms per unit area of the diffusing species on planes 1 and 2, and Γ_{12} and Γ_{21} are the jump frequencies for an atom to jump from plane 1 to plane 2 and from plane 2 to plane 1, respectively. The net atom flux J between planes 1 and 2 is given by

$$J = j_{12} - j_{21} \tag{3}$$

$$J = n_1 \Gamma_{12} - n_2 \Gamma_{21} \tag{4}$$

$$J = (n_1 - n_2)\Gamma + n(\Gamma_{12} - \Gamma_{21}) \tag{5}$$

where the average jump frequency Γ equals $\tfrac{1}{2}(\Gamma_{12} + \Gamma_{21})$ and the average concentration n per unit area on planes 1 and 2 equals $\tfrac{1}{2}(n_1 + n_2)$. Here, Equation 5 is an exact equation.

The number of atoms per unit volume c multiplied by the distance λ between planes equals the average number of atoms per unit area on the lattice planes,

$$n = c\lambda \tag{6}$$

Also,

$$(n_1 - n_2) = -\lambda\,(\partial n/\partial x) = -\lambda^2\,(\partial c/\partial x) \tag{7}$$

so,

$$J = -\lambda^2 \Gamma (\partial c/\partial x) + \lambda(\Gamma_{12} - \Gamma_{21})c \tag{8}$$

This equation can be written in the general form

$$J = -D^*\,(\partial c/\partial x) + v_F c \tag{9}$$

where D^* and v_F are quantities which can be measured in diffusion experiments, being respectively the tracer diffusion coefficient and the atom drift velocity from driving forces. In particular, D^* is the diffusion coefficient measured in tracer diffusion experiments where a very thin layer of tracer atoms diffuses into a homogeneous material in the absence of driving forces. D^* can be determined from the width of the resulting tracer concentration vs. distance profile, whereas v_F can be determined from the net drift of the profile. These profiles are illustrated in Fig. 3, where the layer of diffusing atoms is originally at $x = 0$ sandwiched between two identical homogeneous bulk specimens.

If v_F itself is proportional to the concentration gradient $\partial c/\partial x$, Equation 9 may usefully be expressed as

$$J = -D\,(\partial c/\partial x) \tag{10}$$

where

$$D = D^* - v_F c(\partial c/\partial x)^{-1} \tag{11}$$

Here D is called the intrinsic diffusion coefficient and obviously differs from the tracer diffusion coefficient D^*. If diffusion occurs on only one sublattice in a crystal, the requirement for electrical neutrality causes internal electric fields to be created which contribute to the v_F term and impose a restriction that the

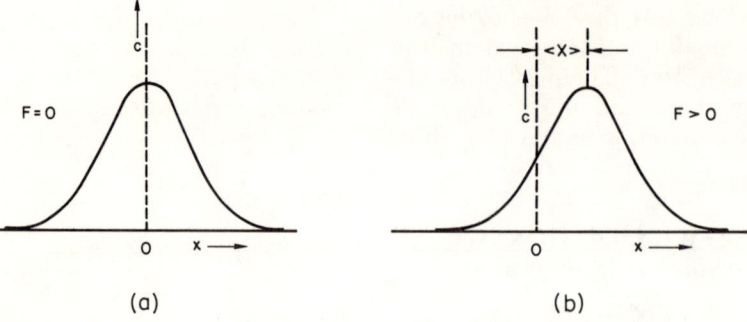

Fig. 3. Concentration-distance profile for a layer of atoms originally on the plane $x = 0$ and diffusing with constant diffusion coefficient D^*. (a) Atomic driving force F equals zero. D^* can be determined from width of profile. (b) Atomic driving force F is constant and greater than zero. Drift velocity v_F can be determined from shift of profile. Shape of profile is same as when $F = 0$. Thus, D^* still can be determined from width of profile.

total atom flux summed over all species must equal zero. In this case, the intrinsic diffusion coefficient also equals the interdiffusion coefficient. Since there are many types of diffusion coefficients which can be defined by Equation 10, depending on the v_F values provided by possible driving forces (or depending on other velocity terms, which can result from crystal lattice distortions for example), it is important when one reports a diffusion coefficient to specify the type of diffusion coefficient which was measured. When possible, specific information on the velocity terms should be provided so that the different D values can all be related to the simple D^* coefficient and hence be related to one another.

From Equations 8 and 9, it follows that

$$D^* = \lambda^2 \Gamma \qquad (12)$$

and

$$v_F = \lambda(\Gamma_{12} - \Gamma_{21}) \qquad (13)$$

According to atomic theory, the jump frequency Γ usually will depend exponentially on temperature and also will depend on crystal composition, pressure and other intensive thermodynamic variables. Thus, the atomic-kinetic approach provides an atomic explanation for the dependence of measured diffusion coefficients on temperature, composition, etc. When diffusion results are used, it is of course important that these variables be known.

In more complex crystals where jumps in a number of different directions are possible and/or the diffusing species can occupy a number of different types of lattice sites, Equations 12 and 13 generalize to the form,

$$D^* = \tfrac{1}{2} \Sigma_{j,i} \, P_j \, \Gamma_{ji} \, x_i^2 \qquad (14)$$

$$v_F = \tfrac{1}{2} \Sigma_{j,i} \, (P_j \Gamma_{ji} - P_i \Gamma_{ij}) \, x_i \qquad (15)$$

where x_i is the x-displacement for a jump of type i (starting from a site of type j and jumping to a neighboring site), Γ_{ji} is the jump frequency for this type of jump, and P_j is the probability of a given atom being in a site of type j. Also, Γ_{ij} is the jump frequency for the reverse jump from site i to site j, P_i is the probability of a given atom being in position to make this reverse jump, the sum over j is over all types of sites, and the sum over i is over all types of jumps which can be made from sites j.

When there is a driving force F, kinetic theory yields to first order

$$\Gamma_{12} = \Gamma \, (1 + \varepsilon) \qquad (16)$$

$$\Gamma_{21} = \Gamma \, (1 - \varepsilon) \qquad (17)$$

with

$$\varepsilon = \frac{\lambda F}{2kT} \qquad (18)$$

TABLE 1. Atomic driving forces in diffusion

Description of Driving Forces	Atomic Force F	Property of the Material Which Governs the Force	
Electric field (E)	$F = qE$	q	= effective charge
Temperature gradient ($\partial T/\partial x$)	$F = -\dfrac{Q^*}{T}\dfrac{\partial T}{\partial x}$	Q^*	= heat of transport
Nonideal part of chemical potential gradient ($\partial \mu'/\partial x$)	$F = -kT\dfrac{\partial \ln \gamma}{\partial x}$	γ	= activity coefficient
Centrifugal force from angular velocity (ω) and rotational radius (r)	$F = m\omega^2 r$	m	= effective molecular mass
Stress field	$F = -\dfrac{\partial U}{\partial x}$	U	= interaction energy

where k is Boltzmann's constant and T is the absolute temperature. In practice, ϵ is almost always much less than unity. Then Equation 12 and 13 yield

$$v_F = D^* F/kT \quad (19)$$

and Equation 9 becomes

$$J = -D^* \frac{\partial c}{\partial x} + \frac{cD^*F}{kT} \quad (20)$$

A list of some possible driving forces is given in Table 1.

CORRELATION AND VACANCY WIND EFFECTS

The preceding discussion of the kinetic (random walk) diffusion equations has implicitly assumed that each jump is an isolated independent event. This condition is indeed fulfilled when diffusion occurs by the simple interstitial mechanism. On the other hand, when diffusion occurs by a vacancy mechanism or an interstitialcy mechanism, the direction of one jump can be influenced by the directions of previous jumps. Instead of being independent, successive jumps are correlated. This correlation affects the diffusion equations so that these equations depend to some extent on the diffusion mechanism. To illustrate the point, kinetic diffusion equations for diffusion by a vacancy mechanism will be considered and compared to equations for an interstitial mechanism.

When diffusion occurs by a vacancy mechanism, the basic diffusion process really involves a series of jumps. A vacancy first is created at a vacancy source, moves through the crystal until it is next to a given diffusing atom, exchanges with the atom one or more times, and finally moves away through the crystal and is destroyed at a vacancy sink. The effective frequency Γ_{Re} of independent atom displacements from plane 1 to plane 2 can be written as

$$\Gamma_{Re} = N_v w_{\pi R} (P_R - P_R P_L + P_R P_L P_R - P_R^2 P_L^2 + \ldots) \quad (21)$$

Here N_v is the concentration of vacancies at vacancy sources and $w_{\pi R}$ is the frequency with which they move through the crystal to arrive at a site on plane 2 neighboring on a particular diffusing atom. After arrival at such a site, the vacancy has probability P_R of exchanging with the atom and causing an initial atom jump from plane 1 to plane 2. After such an exchange, the vacancy still neighbors on the atom and has a probability P_L of re-exchanging, causing a negative atom jump which cancels the initial jump. The total probability of such a two jump series after arrival of the vacancy on plane 2 is $P_R P_L$. This term appears in Equation 21 with a minus sign. Succeeding terms in Equation 21 result from the possibility of occurrence of series with three, four or more atom jumps.

In crystals with sufficient symmetry, all sites on a close-packed plane normal to the diffusion direction and containing the diffusing atom (chosen so that a vacancy cannot move from the $+x$ side of the tracer to the $-x$ side without stopping at a lattice site on this plane) can be regarded as effective vacancy sources and sinks, since equilibrium vacancy concentrations, unaltered by the vacancy flux, are maintained at these sites. Then, with P_R and P_L defined as including both immediate and delayed exchanges of the vacancy with the tracer, Equation 21 is an exact expression for Γ_{Re}. (In delayed exchanges, the vacancy may first wander away from the tracer but later returns to exchange with it without arrival of the vacancy at any effective vacancy sink.)

P_R can be expressed in terms of (1) the jump frequency w_R for exchange of the vacancy on plane 2 with the atom on plane 1 and (2) the competing jump frequencies $w_{R\pi}$ for the vacancy to start a path which leads it away from the site on plane 2 to an effective vacancy sink without exchanging with the atom on plane 1,

$$P_R = w_R/(w_R + w_{R\pi}) \qquad (22)$$

After suitable algebra, Equation 21 can be rewritten as

$$\Gamma_{Re} = \frac{w_{\pi R}}{w_{R\pi}} \frac{(1-P_R)(1-P_L)}{1-P_R P_L} N_v w_R \qquad (23)$$

or

$$\Gamma_{Re} = G_R f \Gamma_{Rb} \qquad (24)$$

where G_R is the vacancy wind factor, f is the correlation factor and Γ_{Rb} is the basic jump frequency for jumps from plane 1 to plane 2,

$$G_R = w_{\pi R}/w_{R\pi} \qquad (25)$$

$$f = (1-P_R)(1-P_R P_L)^{-1}(1-P_L) \qquad (26)$$

$$\Gamma_{Rb} = N_v w_R \qquad (27)$$

Here Γ_{Rb} is the jump frequency which would be obtained if all jumps were independent, being just the average probability N_v of a vacancy being at a given site multiplied by the frequency w_R of jump when the vacancy is there. Both N_v and w_R depend exponentially on temperature, N_v being proportional to $\exp(-E_f/kT)$, where E_f is the energy of formation for a vacancy, and w_R being proportional to $\exp(-E_m/kT)$, where E_m is the energy of motion, being the height of the energy barrier the atom must surmount to make the jump into the neighboring vacant site.

The paths the vacancy follows in the transition $w_{\pi R}$ are just the reverse of the paths contributing to $w_{R\pi}$. Thus, in the absence of a vacancy flux, $w_{\pi R}$ equals $w_{R\pi}$ and the vacancy wind factor then equals unity. By contrast, f is not affected by driving forces or vacancy fluxes. Since P_R and P_L are affected in equal but opposite ways by a driving force, one finds to a very good approximation (to second order in the small quantity ϵ),

$$f = \frac{1-P}{1+P} \qquad (28)$$

whether a driving force is present or not, where P equals the value of P_R or P_L in the absence of forces.

The probability P depends on the ratio of the tracer jump frequency, for example w_R, to the other competing jump frequencies, such as those for solvent jumps which contribute to $w_{R\pi}$. Since tracer and solvent jump frequencies usually have different activation energies, f in general depends on temperature. By contrast, in pure cubic crystals there often is only one type of vacancy jump frequency, and the ratio $w_R/w_{R\pi}$ is a constant. When this is the case, P and f are pure numbers independent of temperature. For example, in pure diamond structure crystals with diffusion by a vacancy mechanism f will always equal ½.

The tracer diffusion coefficient D^* will be proportional to the product $f w_2$, where w_2 is the average vacancy-tracer exchange frequency, equal to w_R in the absence of a driving force. Also,

$$f = \frac{w_\pi}{2w_2 + w_\pi} \quad (29)$$

where w_π is the value of $w_{R\pi}$ in the absence of a force and depends on jumps by atoms *other* than the diffusing tracer atom. In extreme cases, perhaps in the case of some very fast-diffusing impurities, w_2 can be much larger than w_π. Then increasing w_2 further will have very little effect on D^*. Instead, since w_2 appears in the denominator of Equation 29, f then becomes very small and D^* for the *tracer* is determined primarily by $N_v w_\pi$, where w_π is a weighted jump frequency for *non-tracer* jumps. Correlation effects therefore not only affect the measured temperature dependence (measured activation energy) of the diffusion coefficient D^* but also prevent a fast-jumping dilute impurity from diffusing much faster than the solvent (say, more than 5 times faster) unless the equilibrium vacancy concentration N_v near the impurity is appreciably larger than the N_v value when the impurity is absent.

Because the vacancy wind factor G is larger than unity for an atom jump "upstream" against the vacancy flow and less than unity for a jump "downstream," it tends to make Γ_{12} differ from Γ_{21} and hence contributes to the drift velocity v_F, as in Equation 13.

The vacancy wind effect can be important when diffusion occurs in a driving force, such as an electric field. If diffusion occurs by an interstitial mechanism, where there are no correlation effects and nothing comparable to a vacancy wind effect, one finds from Equation 20,

$$J = -D^* \frac{\partial c}{\partial x} + \frac{cD^*qE}{kT} \quad (30)$$

where the force F equals the effective charge q multiplied by the field E. For self-diffusion by a vacancy mechanism, the vacancy wind effect provides an additional force, equal to $qE(f^{-1} - 1)$,

where f is the correlation factor. Equation 20 then yields

$$J = -D^* \frac{\partial c}{\partial x} + \frac{cD^*qE}{kTf} \quad (31)$$

More generally, for impurity diffusion by a vacancy mechanism, the vacancy wind effect has a complex dependence on the solvent jump frequencies and charges. In this case

$$J = -D^* \frac{\partial c}{\partial x} + \frac{cD^*qE}{kT\psi} \quad (32)$$

where ψ may differ radically from the correlation factor f. In some instances, ψ can be much larger than unity or can even be negative.

The coefficient of cE in the second terms on the right in Equations 30, 31, and 32 equals the drift mobility μ of the diffusing species in the electric field. From Equation 30, one obtains

$$\frac{\mu}{D^*} = \frac{q}{kT} \quad (33)$$

which is the well-known Nernst-Einstein relation. As can be seen from Equations 31 and 32, the Nernst-Einstein relation often will not be valid, needing to be modified, for example, by either the simple factor f^{-1} or the more complex factor ψ^{-1}, which must be introduced on the right when, respectively, self-diffusion or impurity diffusion occurs by a vacancy mechanism.

The vacancy wind effect results from nonrandom motion of atoms other than the particular atom i whose motion is being followed, with these nonrandom motions in turn being caused by driving forces on these other atoms or from concentration gradients of the various species in the crystal. Regardless of the origin of the vacancy flux, however, the consequent vacancy wind effect operates by causing Γ_{12} to differ from Γ_{21} and thus contributing to v_F. The vacancy wind effect plays the role of an atomic driving force and might be added to the list of forces in Table 1.

Comparison of Kinetic and Thermodynamic Diffusion Equations—Cross Terms

A vacancy wind effect will arise whenever there is a flux of vacancies. Driving forces (such as those listed in Table 1) can create a vacancy flux and thus indirectly influence the motion of atom i. At the same time, these driving forces may directly act on atom i. For diffusion in a one-component system, the indirect effect will increase the net flux of species i by a factor f^{-1} as in Equation 31. For more complex systems, the indirect effect via the vacancy wind may either aid or oppose the direct effect.

A particularly interesting circumstance arises when the original influence creating the vacancy flux comes from the concentration gradients of several constituents in the crystal. Each concentration gradient gives rise to an atom flux of that species with the vacancy flux being equal in magnitude but opposite in direction to the sum of the atom fluxes. Then v_F in Equation 13 for diffusion of species i contains contributions proportional not only to $\partial c_i/\partial x$ but also terms proportional to concentration gradients $\partial c_k/\partial x$ of other species in the crystal. (A further dependence of v_F on the $\partial c_k/\partial x$ can arise from the force $\partial \mu_i'/\partial x$ from nonideal contributions to the chemical potential of species i, since the heat of mixing, and hence μ_i', may depend on the concentrations c_k of all the crystal components.) The dependence of the flux of species i on other gradients $\partial c_k/\partial x$ can considerably complicate analysis of diffusion results, especially in multicomponent systems.

The existence of cross terms where the flux J_i depends on the $\partial c_k/\partial x$ or $\partial \mu_k/\partial x$, where μ_k is the chemical potential of species k, also is an important aspect of the thermodynamic diffusion equations. These equations state

$$J_i = \Sigma_k L_{ik} X_k \qquad (34)$$

where the L_{ik} are unknown coefficients and the thermodynamic driving forces X_k for species k are given by

$$X_k = -\left[\frac{\partial \mu_k}{\partial x} - F_k\right] \qquad (35)$$

Here the F_k are the independent Newtonian forces, such as the force qE from an electric field, operating on the individual atoms of species k.

There obviously is a similarity between the basic kinetic diffusion equation, Equation 9 or Equations 30 to 32, and the expression obtained by combining Equations 34 and 35. Differences may be noted though. For example, the kinetic force $(\partial \mu_k'/\partial x)$ listed in Table 1 is not one of the forces F_k in Equation 35. Instead, μ_k' is merely one contribution to μ_k in Equation 35. Also, the kinetic driving force from a temperature gradient listed in Table 1 does not contribute to F_k. Instead there is a separate term $L_{iq} X_q$ which enters into Equation 34 to account for effects from temperature gradients with

$$X_q = -\frac{1}{T}\frac{\partial T}{\partial x} \qquad (36)$$

In some cases, the relation between the L_{ik} and the various measured diffusion coefficients can become rather complex. Nevertheless, in a number of cases it has been possible to compare the kinetic and thermodynamic equations to obtain kinetic expressions for L_{ik}. These results indicate that the vacancy wind effect can make a significant contribution to the L_{ik} even when $i \neq k$. Thus, attempts to simplify the diffusion equations by assuming that the cross terms are zero ($L_{ik} = 0$) can introduce significant errors.

Bibliography

Good basic reference books describing diffusion kinetics and mechanisms in simple crystals include:

L. A. Girifalco, *Atomic Migration in Crystals* (Blaisdell, New York, 1964) 162 pp.

J. R. Manning, *Diffusion Kinetics for Atoms in Crystals* (D. van Nostrand, Princeton, N.J., 1968) 257 pp.

P. Shewmon, *Diffusion in Solids* (McGraw-Hill, New York, 1963) 203 pp.

Conference proceedings which review these topics include:

Diffusion—Papers presented at a Seminar of the American Society for Metals, October 14 and 15, 1972 (American Society for Metals, Metals Park, Ohio, 1973).

Mass Transport in Oxides, edited by J. B. Wachtman, Jr., and A. D. Franklin, NBS Special Publication 296 (U.S. Government Printing Office, Washington, 1968).

Other pertinent review articles include:

L. W. Barr and A. B. Lidiard, "Defects in Ionic Crystals" in *Physical Chemistry, Vol. 10 Solid State*, edited by W. Jost (Academic Press, New York, 1970) pp. 151–228.

R. E. Howard and A. B. Lidiard, "Matter Transport in Solids," *Reports on Progress in Physics 27* (1964) pp. 161–240.

VECTOR SPACE TREATMENT OF MULTICOMPONENT DIFFUSION

A. R. Cooper, Jr.
Division of Metallurgy and Materials Science
Case Western Reserve University
Cleveland, Ohio 44106

ABSTRACT

Determining the eigen vectors and eigen values of the diffusion matrix for a multicomponent system allows linear diffusion processes in such systems to be easily visualized. Such phenomena as "uphill diffusion" become obvious when viewed in this geometrical way. Likewise it is straightforward to develop rules about diffusion paths in composition space. Of perhaps more importance, such a diagonalization of the diffusion matrix permits the kinetics of such transport controlled processes as precipitation, dissolution, and homogenization, to be quantitatively analyzed in multicomponent systems as readily as in binary systems. Some examples of this treatment are shown and experimental evidence on isothermal diffusion in several systems is presented. Also discussed are extension to nonisothermal diffusion and problems encountered when the diffusion matrix is composition dependent.

INTRODUCTION

Defining a multicomponent system as one in which the composition is a vector field, i.e., more than one species can vary independently, (by this definition a binary alloy with its vacancy population not at chemical equilibrium is multicomponent; a melt with three ionic species is not; a quaternary alloy constrained to remain in equilibrium with two other phases is not) we see that many systems of importance in ceramics, metallurgy and geology are multicomponent. Yet, it was not until recently when Onsager (1945) proposed a generalization of Fick's laws that multicomponent diffusion was placed on a firm phenomenological basis.

Since then a large amount of study has been given to multicomponent diffusion. At first efforts were confined to aqueous solutions (Miller 1959, 1960); often the goal was an experimental test of Onsager's Reciprocal Relations (Onsager, 1931), ORR. Studies of isothermal diffusion in multicomponent alloys (Kirkaldy, 1970) followed, and still later, attention was given to molten salts, supercooled liquids, and glasses (Varshneya and Cooper, 1972; Cooper and Varshneya, 1968; Sucov and Gorman, 1965; Oishi, 1969). Experimental efforts in such systems which can be quenched to a solid at room temperature were stimulated by the development during this period of an ideal experimental tool, the electron microprobe.

In many cases a primary interest in the later studies (Guy and Philibert, 1965; Dayanandu and Grace, 1965; Dayanandu et al., 1968; Varshneya and Cooper, 1972) was in relating the elements of the diffusion matrix to other types of diffusion coefficients such as the self-diffusion coefficients or the intrinsic diffusion coefficients. Suffice it to say here that it appears that relationships can be obtained from a suitable generalization of the Nernst-Planck equation (Nernst, 1889; Planck, 1890a, 1890b) where the coupling between the fluxes of the different ions is achieved through the effect of a compensating electric field and/or the generalization (Oishi, 1965; Ziebold and Cooper, 1965; Manning, 1970) of the Meyer-Darken equation (Darken, 1948; Meyer, 1899) where the interaction is achieved through the assumption of a compensating flow where all species move together. A method of combining both effects has been presented (Cooper, 1965).

Our interest here will be somewhat different. We want to inquire about the phenomenology of multicomponent diffusion, more specifically the relationship between the diffusion matrix and the behavior of the system during diffusion controlled processes. The approach we will adopt is identical to that previously discussed by Gupta and Cooper (1971). It is geometric, emphasizes the importance of the diffusion path and demonstrates that understanding multicomponent diffusion is no more difficult (and no easier) than understanding binary diffusion.

Vector Space for Multicomponent Systems

Typically a Gibbs triangle is used to graphically portray compositions in a ternary system. This has the advantage that it is symmetric with regard to all species. However, its generalization to higher dimensions, while possible, is not conceptually easy. Also, when we wish, right from the start, to designate the nth component as the dependent component, the symmetry of the Gibbs triangle loses its advantage. It is then convenient to use a space called (S^{n-1}) in which the pure substance of each of the $(n-1)$ *independent* components is associated with an element of a set of mutually orthogonal basis vectors ($\gamma_1, \gamma_2, \ldots \gamma_{n-1}$). This set is termed the $\{\gamma\}$ basis. A composition vector $\mathbf{c} = (c_1, c_2, \ldots c_{n-1})$, where c_i refers to the amount[1]

[1] The term amount can refer to any of several choices, for example, mass, molar, number, volume. It is important, however, that once the choice is made, it is adhered to in the description of fluxes and density.

fraction of component i, is readily visualized as seen on Fig. 1. We should note on Fig. 1 that the same vector $\mathbf{c}$ can be described in terms of a different set of basis vectors. When, for example, the vectors ($\nu_1, \nu_2, \ldots \nu_{n-1}$) are used as a set of basis vectors, the vector $\mathbf{c}$ does not change, it only gets a new representation, $\mathbf{c} = (u_1, u_2, \ldots u_{n-1})$.

If a system is isotropic and the flux and the composition gradient are always in a single direction, say y, in configuration space, then these quantities, as well, can be represented as vectors in (S^{n-1}). A difference is that the flux, $\mathbf{j}_y = (j_{1y}, j_{2y}, \ldots j_{n-1y})$, and the composition gradient,[2] $\mathbf{c}'_y = (c'_{1y}, c'_{2y}, \ldots c'_{n-1y})$ can occupy any position in (S^{n-1}), but the composition vector is always bounded by the planes $c_i = 0$ and the plane, $\sum_{i=1}^{n-1} c_i = 1$.

[2] Partial differentiation with respect to distance and time are indicated by primes and dots, respectively; underlined symbols (p. 17) denote vectors in a space of dimension $3(n-1)$.

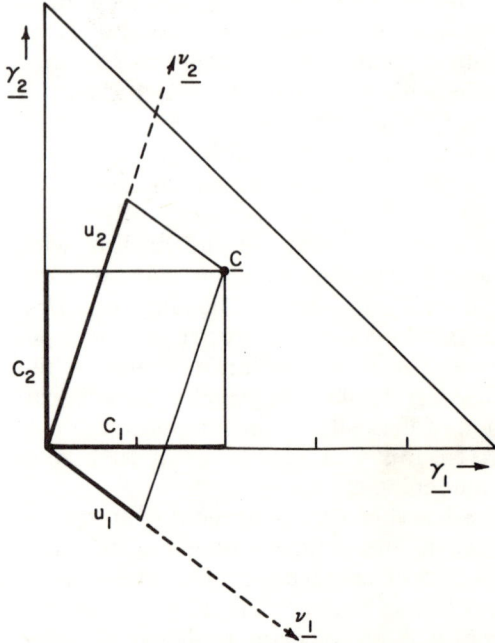

Fig. 1. An orthogonal species space (S_c^2) showing a composition vector $\mathbf{c}$ in two representations $\{\gamma\}$ and $\{\nu\}$.

We term this restricted space (S_c^{n-1}).

When a system is not isotropic the description of the flux and composition gradient vectors is more complicated, because in general there will be an element of the vector for each independent species in each of the three independent directions, x, y, z, of configuration space. Thus in general $\mathbf{j} = (\mathbf{j}_x, \mathbf{j}_y, \mathbf{j}_z)$ and $\mathbf{c}' = (\nabla c_1, \nabla c_2, \ldots \nabla c_{n-1}) = (\mathbf{c}'_x, \mathbf{c}'_y, \mathbf{c}'_z)$. Such vectors will require a space of $3(n-1)$ dimensions for their descriptions. Even though crystals of lower symmetry than cubic, and liquids in which the velocity vector does not parallel the diffusion flux may not have their diffusion behavior completely described in (S^{n-1}), we will limit our attention to diffusion that can be described in (S^{n-1}) because: (1) all the features of multicomponent diffusion can be observed, (2) descriptions in (S^2) are within the grasp of easy geometrical intuition, and (3) no experiments in multicomponent diffusion in three dimensional configurational space are available.

Diffusion Matrix and Diffusion Equations

Since we will confine our attention to phenomena that occur in a single direction in configuration space, in the following we will drop the subscript y from both the flux and gradient vectors. Then the vector equation that generalizes Fick's first law to a multicomponent system can be written

$$-(1/\rho)\,\mathbf{j} = [D]\,\mathbf{c}' \qquad (1)$$

where ρ is amount density. The diffusion matrix $[D]$ may be a function of composition. Its assumed independence from the composition gradient, $\mathbf{c}'$, makes Equation 1 linear. When it is further assumed that both $[D]$ and ρ are independent of composition,[3] then conservation of species lets us write the generalization of Fick's second law as

$$\dot{\mathbf{c}} = [D]\,\mathbf{c}'' \qquad (2)$$

Vector Equation 2 is a set of coupled simultaneous second order differential equations which can often be solved (Fujita and Gosting, 1956; Toor, 1964) for the cases of interest. However, they are typically very cumbersome since $[D]$ usually has very few zero elements. Therefore, a better insight can often be obtained by diagonalizing $[D]$ so that the simultaneous equations are uncoupled.

$[D]$ is the product of a mobility matrix, $[L]$, and a thermodynamic matrix, $[G]$, both of which are symmetric and positive definite (deGroot and Mazur 1969). The product of such a pair of matrices will itself have positive real eigen values and hence $[D]$ can always be diagonalized (Simonsen, 1973). (This property of $[D]$ is seen to depend on the ORR.)

To diagonalize $[D]$, we find its eigen values $(\lambda_1, \lambda_2, \ldots \lambda_{n-1})$ and its normalized eigen vectors $(\boldsymbol{\nu}_1, \boldsymbol{\nu}_2, \ldots \boldsymbol{\nu}_{n-1})$. Then $[\lambda] = [B]\,[D]\,[B]^{-1}$ is a diagonal matrix with the eigen values as elements, and $[B]^{-1}$ is the modal matrix formed by using the eigen vectors $\boldsymbol{\nu}_1$, $\boldsymbol{\nu}_2$, etc., as columns. As such $[B]$ is the identity transformation which takes a vector from its representation in (S^{n-1}) using the $\{\boldsymbol{\gamma}\}$ basis to its representation in the same space in which the eigen vectors $(\boldsymbol{\nu}_1, \boldsymbol{\nu}_2, \ldots \boldsymbol{\nu}_{n-1})$ are used for the set of basis vectors. For convenience of representation, henceforth we will use $\mathbf{h}$ and $\mathbf{u}$, respectively, in place of $\mathbf{j}$ and $\mathbf{c}$ for the general flux density vector and composition vector when they are represented using the $\{\nu\}$ basis. Thus, it is clear that $\mathbf{h} = [B]\,\mathbf{j}$ and $\mathbf{u} = [B]\,\mathbf{c}$. This last relation is illustrated schematically in (S^2) on Fig. 1.

Operating on both sides of Equation 1 from the left and introducing the identity matrix, $[I] = [B]^{-1}\,[B]$, into the first term on the right hand side gives

---

[3] Except when specifically mentioned to the contrary in section 7, this assumption will be in force.

$$-\frac{1}{\rho}[B]\mathbf{j} = [\lambda][B]\mathbf{c}' \quad (3)$$

Now with $[D]$ constant, $[B]$ is also constant and

$$-\frac{1}{\rho}\mathbf{h} = [\lambda]\mathbf{u}' \quad (4)$$

and instead of coupled equations we have $(n-1)$ equations each as simple as the equation for a binary system. In a like manner we can convert Equation 2 into

$$\dot{\mathbf{u}} = [\lambda]\mathbf{u}'' \quad (5)$$

a set of $(n-1)$ uncoupled binary diffusion equations, which will form the basis for much of the subsequent discussion.

Solution to Diffusion Equations and Experimental Results

Solutions to Equation 5 exist for a wide variety of initial and boundary conditions (Carslaw and Jaeger, 1959). Since the extension of these solutions to multicomponent systems has been discussed in detail previously (Gupta and Cooper, 1971), here we present only some results with a qualitative discussion and experimental confirmation where possible.

Infinite Couples

Consider two infinitely thick slabs of composition $\mathbf{p}$ and $\mathbf{q}$ to be brought together in a plane at $y = 0$ at $t = 0$ with $\mathbf{c}(+\infty) = \mathbf{p}$ and $\mathbf{c}(-\infty) = \mathbf{q}$. Looking at the problem in terms of Equation 5, we see that along the directions of the eigen vectors there will exist the simple error function solutions,

$$\frac{u_i - u_i(\mathbf{q})}{u_i(\mathbf{p}) - u_i(\mathbf{q})} = \frac{1}{2} + \frac{1}{2}\operatorname{erf}\left(\frac{y}{2\sqrt{\lambda_i t}}\right) \quad (6)$$

to the diffusion equation. Figure 2 shows this solution for the ratio $\lambda_1/\lambda_2 = 6$. Since the eigen values are positive for a thermodynamically stable single phase system, the compositions must always

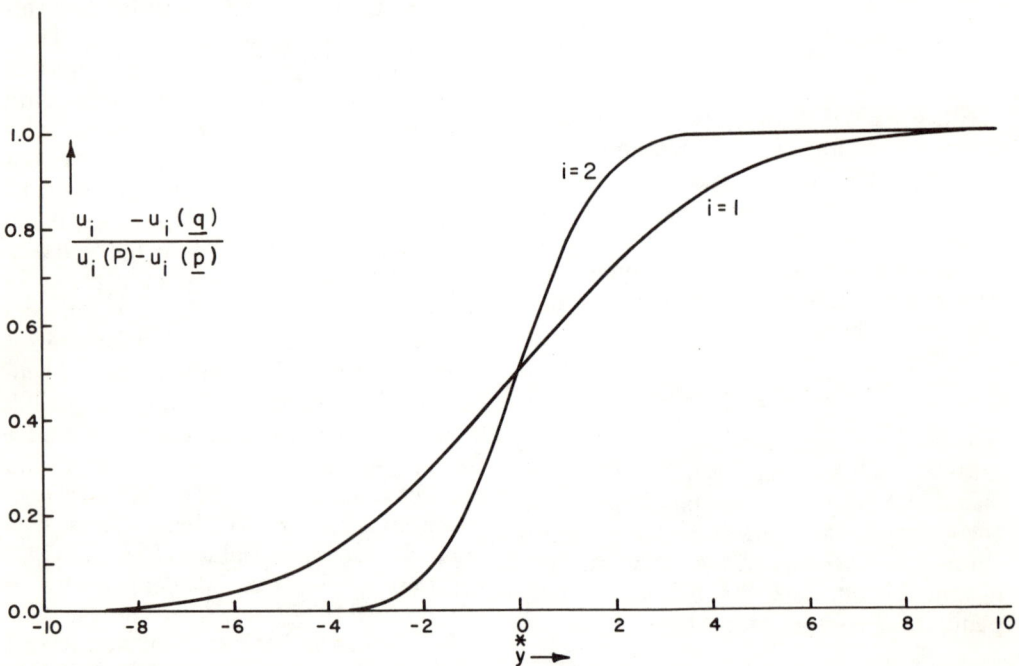

Fig. 2. Normalized composition distribution for infinite diffusion couple with $\lambda_1 = 6\lambda_2$, $y^* = y/2(D_0 t)^{1/2}$ (after Gupta and Cooper, 1971).

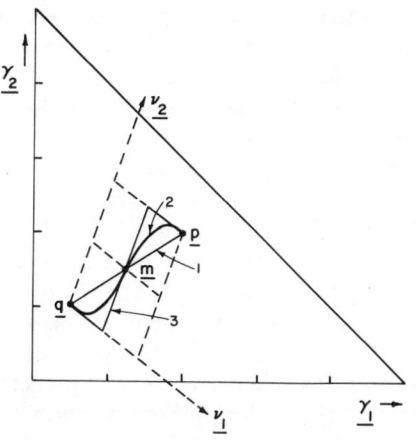

Fig. 3. Diffusion paths for an infinite couple with $\mathbf{p}_\gamma = (0.4, 0.4)$ $\mathbf{q}_\gamma = (0.1, 0.2)$ and eigen vectors ν_1 and ν_2 are indicated, as is midpoint $\mathbf{m}$. Path 1: $\lambda_1 = \lambda_2$; path 2: $\lambda_1 = 6\lambda_2$; path 3: $\lambda_1 \to \infty \lambda_2$ (after Gupta and Cooper, 1971).

For path 2, $D = D_0 \begin{bmatrix} 1.0 & -0.267 \\ -0.6 & 0.4 \end{bmatrix}$.

Fig. 3 shows (as solid lines) three diffusion paths between the compositions $\mathbf{p}$ and $\mathbf{q}$. Lines parallel to the eigen vectors are shown as broken lines and $\mathbf{m} = 1/2\ (\mathbf{p} + \mathbf{q})$ is the mid point of the couple. Path 1, the single straight line, occurs if $\lambda_1 = \lambda_2$. Path 2, the S-curved line, is consistent with results of Fig. 2 where $\lambda_1 = 6\lambda_2$. Path 3 represents the limiting path as $\lambda_1/\lambda_2 \to \infty$. The properties of the path can easily be generalized to higher (Gupta and Cooper, 1971) dimensions and simple rules can be developed such as: (1) near its extremities the diffusion path always parallels the direction of fastest eigen vector (the one associated with largest eigen value). (2) For a straight line path the path must parallel an eigen vector (when all λ's are equal, any vector is an eigen vector). (3) The path always lies within an alternate pair of parallelograms (or higher dimensional analogs) obtained by joining the midpoint $\mathbf{m}$ with the end points $\mathbf{p}$ and $\mathbf{q}$

lie between the end member compositions of the components expressed in this basis.

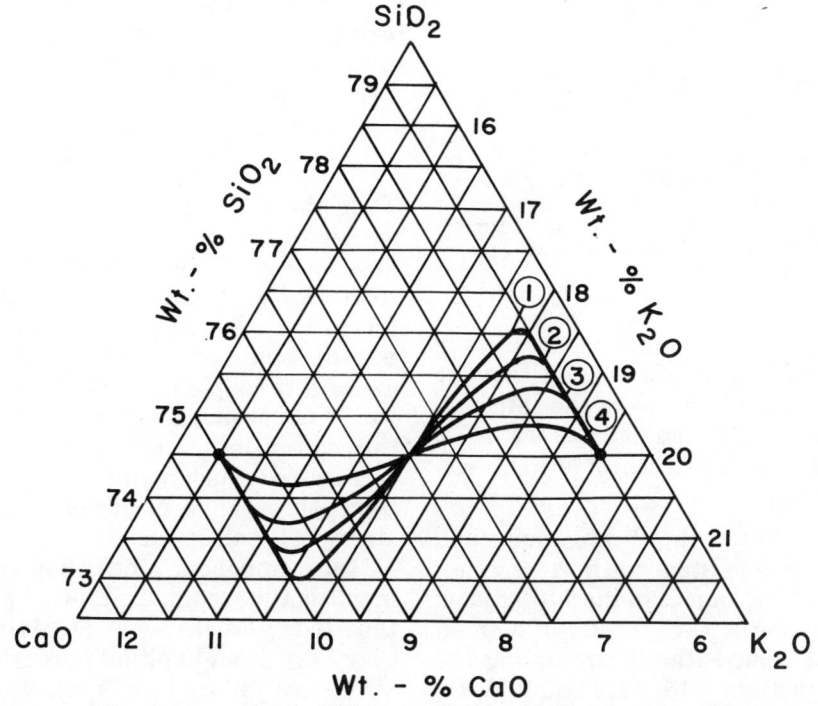

Fig. 4. Diffusion paths in system CaO-K_2O-SiO_2: 1, 652°C; 2, 750°C; 3, 800°C; 4, 920°C (after Engelke, 1972).

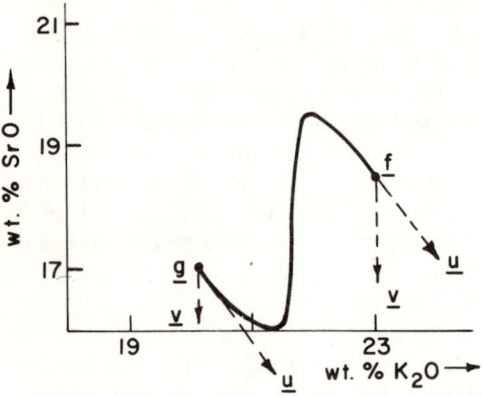

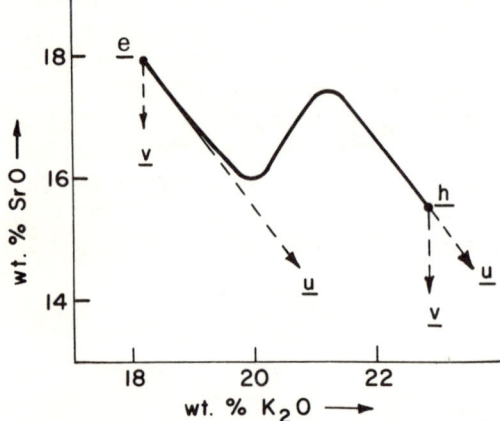

Fig. 5. Diffusion paths in the system K_2O-SrO-SiO_2 at 710°C. **u** represents fast eigen vector direction; **v** represents slow eigen vector direction; **g, f** and **e, h** represent the end member compositions of the two diffusion couples. (after Varshneya, 1972).

plotted his results on the Gibbs triangle shown as Fig. 4. They reveal that the eigen vectors for this system do not change direction as a function of temperature, but the ratio of the eigen vectors decreases as the temperature increases.

Varshneya (1970) in a study of interdiffusion in the system K_2O-SrO-SiO_2, using incremental couples in the vicinity of weight fraction 0.20, 0.18, and 0.62, also found that the eigen vectors were not greatly affected by temperature. By using couples with different directions between the end members he showed, as indicated in Fig. 5, that the eigen values and vectors and hence $[D]$ are independent of the direction (in S_c^{n-1}) between the end member compositions. He also confirmed that $[D]$ is independent of time.

Semi-infinite Slab

When an infinite slab of material lying between $y = 0$ and $y = +\infty$ of composition **p** has its surface composition at $y = 0$ instantaneously changed and maintained at a new composition, say **m**, the solution of Equation 5 has the form

$$\frac{u_i - u_{i(m)}}{u_{i(p)} - u_{i(m)}} = \text{erf}\left(\frac{y}{2\sqrt{\lambda_i t}}\right) \quad (7)$$

Thus the diffusion paths in a system with two independent species will not have an S shape but rather a single sign to the curvature, i.e., a C shape as shown for example by the paths on Fig. 3 from **p** to **m**. While the S-shaped path **pq** for an infinite couple permits global conservation of species, this is not required for the semi-infinite couple because the surface source which maintains the surface composition at **m** need not be of equal strength for all species.

Multicomponent diffusion experiments in the system Ag-Zn-Cd have been reported by Carlson *et al.* (1971) in which they have brought either pure silver or a silver-zinc alloy into contact with a mixed cadmium-zinc vapor. The surface concentration of the metal quickly

by lines parallel to eigen vectors. More important, however, is the fact that the diffusion path alone gives the information necessary (Gupta and Cooper, 1971) to determine the ratios of all the elements of $[D]$.

Recent studies in silicate melts and glasses are consistent with these rules. Using infinite couples in the supercooled liquid system K_2O-CaO-SiO_2 in a small region of composition surrounding the weight fractions 0.15, 0.10, and 0.75, respectively, Engelke (1972) studied interdiffusion at various temperatures. He

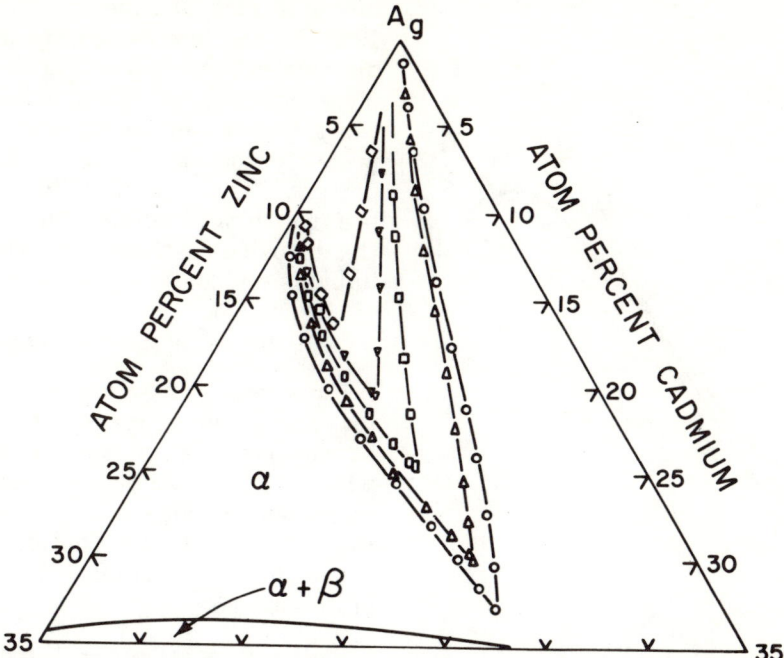

Fig. 6. Diffusion paths in the system Zn-Cd-Ag. Bulk compositions are Ag and 0.9 Ag·0.1 Zn. Surface compositions have slightly more than 10% Zn and between 10 and 35% Cd (after Carlson et al., 1971).

changed to and was maintained at the composition in equilibrium with the vapor. Figure 6 shows the diffusion paths they found. Notice that when the binary alloy 0.9 Ag · 0.1 Zn is the bulk composition, the paths show the same features as path 2 from **p** to **m** on Fig. 3. When silver is the bulk composition, however, the paths go from a straight line path 1 on Fig. 3 to a C curve, to a slight S shape as the cadmium concentration in the vapor increases. The S shape is probably a consequence of changes in $[D]$ with composition.

It is interesting to note that all paths are nearly parallel at both bulk compositions, indicating that this direction is parallel to the "fast" eigen vector.

Finite Couple

A finite couple or a multiple lamination of two compositions, say **p** and **q**, will have diffusion paths that initially appear identical to those for the same initial compositions in an infinite couple. However, as time progresses, diffusion along the faster directions approaches completion and the path becomes nearly parallel to the direction of the slowest eigen vector. This is shown in Fig. 7, which was calculated using the same initial compositions and $[D]$ as were used for path 2 in Fig. 3.

To enhance homogenization processes in multicomponent systems one should minimize the initial composition variations in the direction of the slowest eigen vector. Conversely when a system has homogenized to the extent that only a straight line path exists on (S_c^{n-1}) space it has lost almost all information regarding its initial compositions. An infinity of initial compositions different from **p** and **q** could give the path on Fig. 7 at time 3.

Dissolution-Precipitation

For dissolution or precipitation of a single phase into or from a multicomponent system, the boundary conditions

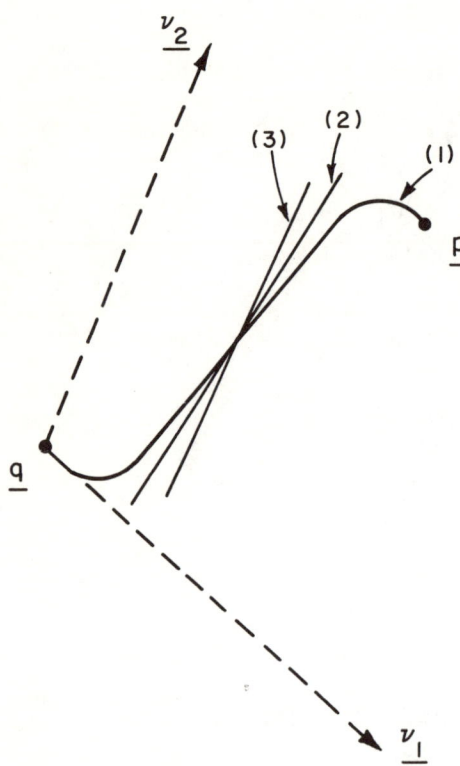

Fig. 7. Calculated diffusion paths of a finite diffusion couple with properties given in Fig. 3 and thickness b, $\frac{\lambda_1 t}{b^2}$ = 0.006 for 1, 0.606 for 2, and 1.206 for 3.

diffusion path obtained by Oishi et al., 1965, for the complementary process of dissolution as shown on Fig. 9. That the measured path has many of the general characteristics of the calculated path is probably not coincidental. What is important is that it is possible to calculate the growth rate or dissolution rate, given the diffusion coefficient matrix and the liquidus surface.

Diffusion with Convection

Equations 2 and 5 are easy to extend for the case where transport also occurs by convection. The key point is that when this is done the convection term transforms to the $\{v\}$ basis as readily as the diffusion term. Therefore, where convection is important we have, instead of Equation 5:

$$\dot{\mathbf{u}} = [\lambda]\, \mathbf{u}'' - W\, \mathbf{u}' \qquad (8)$$

where W is the velocity component parallel to $\mathbf{u}'$. Unfortunately there are no general solutions to Equation 8 even in the binary case. However, when a steady state or pseudo-steady state is

are not quite so rigid. Often the bulk composition provides one boundary condition. However, at the crystal-liquid interface the composition is free to take any point along the liquidus curve. This degree of freedom is necessary to let the rates of crystallization calculated along the different eigen directions balance. As an example Fig. 8 shows the diffusion path (calculated in this way) for the crystallization of species 3 in a hypothetical ternary system with $(\lambda_1/\lambda_2) = 4$. We see that at the bulk liquid composition $\mathbf{C}_\infty$ the diffusion path parallels the fast eigen vector v_1. Thus in order to conserve species the diffusion path crosses the dashed line of constant $c_1:c_2$ ratio before reaching the liquidus.

Let us compare this with the measured

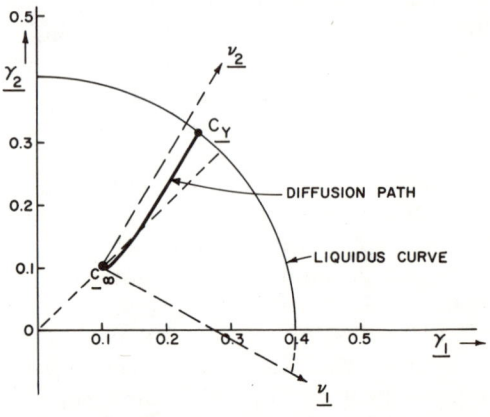

Fig. 8. A plot of the diffusion path for hypothetical planar crystallization of species 3 from supersaturated composition $c = (0.1, 0.1)$. Liquidus surface for species 3 is at $(c_1 + c_2) = 0.4$. Eigen vectors are shown.

$$D = \frac{D_0}{4}\begin{bmatrix} 13 & -3.3 \\ -3.3 & 7 \end{bmatrix},\ \frac{\lambda_1}{\lambda_2} = 4$$

(after Cooper and Gupta, 1971).

achieved, then $\dot{\mathbf{u}} = 0$ and Equation 8 can be solved (Levich, 1962). These solutions from boundary layer theory are directly applicable to dissolution or planar crystal growth in a multicomponent melt. Because the rate of solution in a binary melt depends on $D^{1/2}$ for a stationary melt (Crank, 1957), $D^{2/3}$ for a melt with force convection (Eckert, 1950), and $D^{3/4}$ for a melt with free convection (Wagner, 1949), the diffusion path should be different for each case. Conversely, therefore, the diffusion path may permit distinction among the possibilities.

When considerable shear exists in a flow process, heterogeneities tend to deform so that their maximum dimension parallels the direction of the streamlines (Cooper, 1966; Geffken, 1957). For this type of convection, in which there is a vanishing component of the velocity W in the direction of the flux density $\mathbf{u}$ in configuration space, Equation 5 is more appropriate than Equation 8 (Cooper, 1973).

UPHILL DIFFUSION

The existence of more than one independent species permits the possibility that the most efficient path toward equilibrium for a nonuniform system may involve diffusion of a species up its own concentration gradient or its own chemical potential gradient.

Diffusion Up a Concentration Gradient (Oishi, 1965; Gupta and Cooper, 1971)

Because the flux of one species is in general affected by the concentration gradients of the other species (Equation 1), it is easy to see that uphill diffusion (diffusion of a particular species up its own concentration gradient) can occur. Writing $j_i \, c'_i > 0$ as the condition for diffusion up a concentration gradient and substituting into Equation 1, we get the result that uphill diffusion occurs for species i when

$$D_{ii} < - \sum_{\substack{j=1 \\ j \neq i}}^{n-1} D_{ij} \left(\frac{\partial c_j}{\partial c_i} \right) \qquad (9)$$

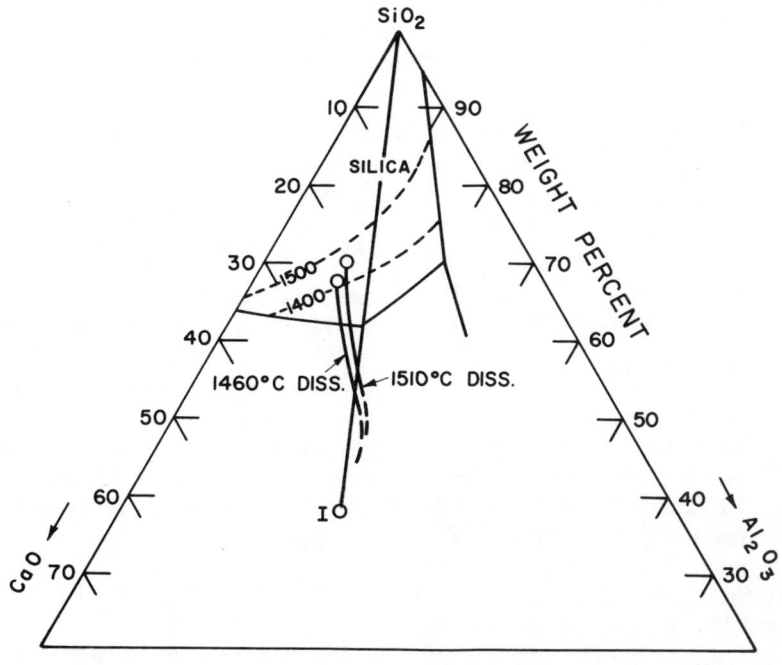

Fig. 9. Diffusion path for dissolution of SiO_2 in bulk composition I (after Oishi et al., 1965).

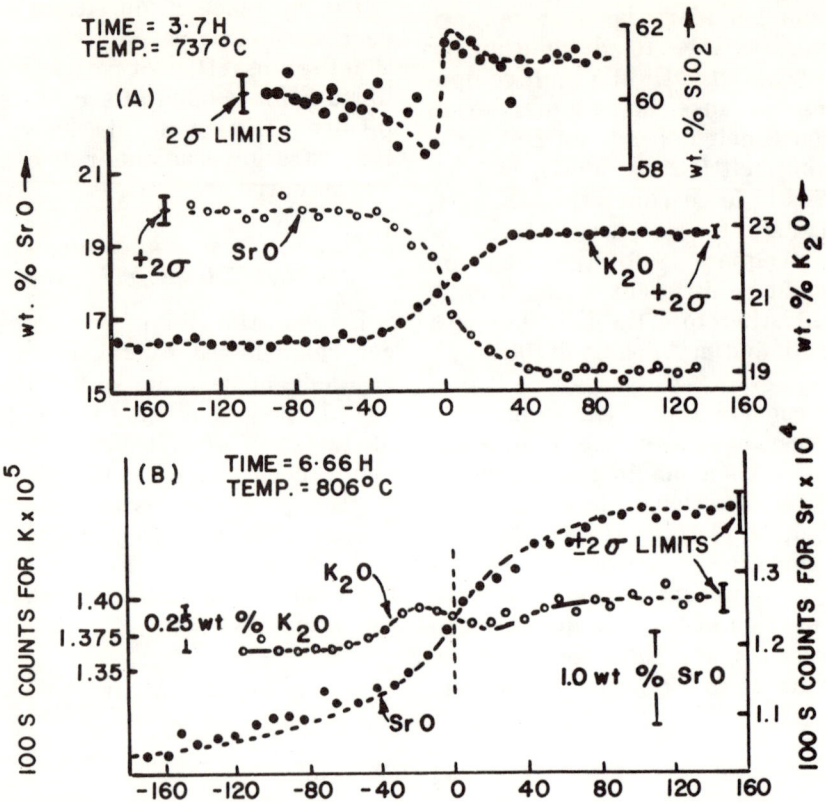

Fig. 10 (A)–(B). Various examples of uphill diffusion in the system K_2O–SrO–SiO_2 (after Varshneya and Cooper, 1968).

As was evident from Fig. 2 and its discussion, curved diffusion paths are typical in multicomponent systems. Such curved paths frequently have extremum points where the path, or its projection onto the plane given by γ_i and γ_j, parallels one of the basis vectors. When the path is parallel to γ_j, the value $\left(\dfrac{\partial c_j}{\partial c_i}\right)$ rapidly changes from extremely large positive to extremely large negative values as the extremum point is passed. As such, we expect that this extremum point separates regions of uphill diffusion from regions of normal diffusion for species i. In fact, it can be shown (Gupta and Cooper, 1971) that the region between the extremum points and the end compositions has normal diffusion, while at least part of the region between the extremum points shows uphill diffusion. This can be confirmed by substituting into Equation 9 values of $\left(\dfrac{\partial c_1}{\partial c_2}\right)$ from path 2 of Fig. 3 and D_{22} and D_{21} from the matrix given in the caption to this figure.

That uphill diffusion is not confined to one species in a ternary system is shown by the fact that in the K_2O–SrO–SiO_2 system each one of the species can be made to show uphill diffusion by the appropriate choice of end member compositions as seen in Fig. 10.

Diffusion Up a Chemical Potential Gradient

We now inquire whether a diffusion flux of a species can occur up its own

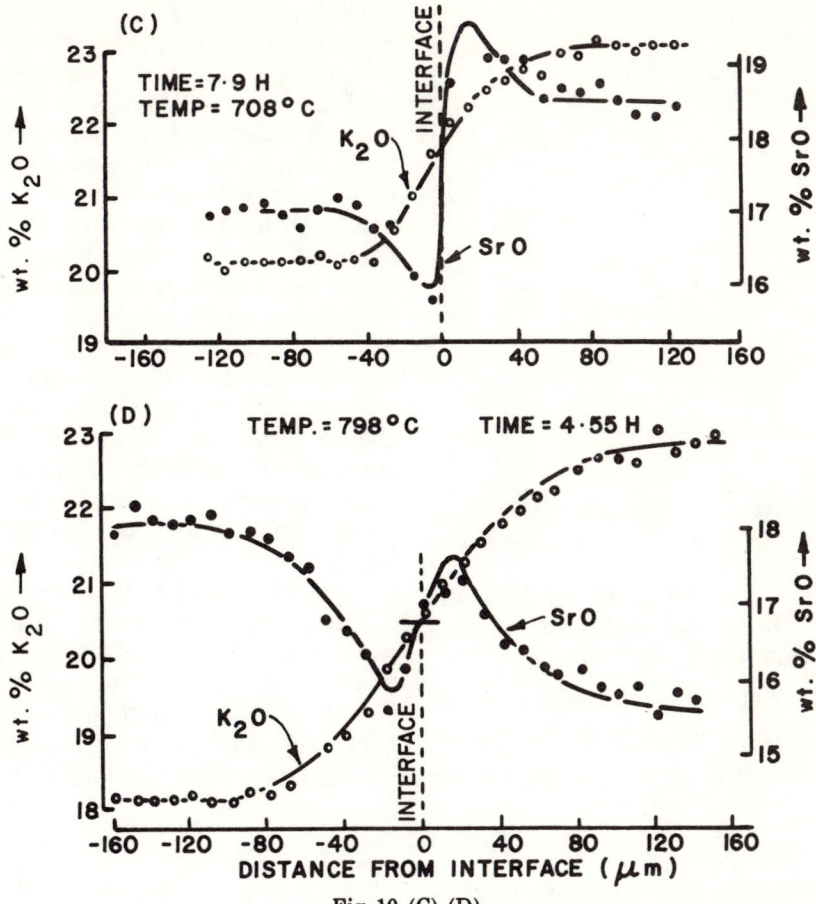

Fig. 10 (C)–(D).

chemical potential gradient; that is, can $j_i(\mu_i - \mu_n) > 0$? The fundamental equation of irreversible thermodynamics relating fluxes and chemical potential gradients is

$$-\frac{1}{\rho}\mathbf{j} = [L]\,\mu' \qquad (10)$$

where $\mu' = (\mu_1 - \mu_n), (\mu_2 - \mu_n) \ldots (\mu_{n-1} - \mu_n)$ and $[L]$ has the properties already described in section 3.

Completely analogous to the case of diffusion up a concentration gradient, the condition for diffusion up a chemical potential gradient is

$$L_{ii} < -\sum_{\substack{j=1 \\ j \neq i}}^{n-1} L_{ij}\,\frac{\partial(\mu_j - \mu_n)}{\partial(\mu_i - \mu_n)} \qquad (11)$$

There are no restrictions on $[L]$ or μ that should prevent the inequality of (11) from being satisfied. Thus we may expect that diffusion up a chemical potential gradient will be a common feature in multicomponent diffusion.

Composition Dependence of D

So far, we have considered only the case of constant $[D]$. By considering in Fig. 11 a diffusion couple near the boundary of (S_c^2) ($c_1 = 0, c_2 = 0, c_1 + c_2 = 1$), we see that it could be possible for the diffusion path to extend beyond (S_c^2). Since this is forbidden from the definition of composition, we recognize that the eigen vectors cannot have arbitrary directions, but near the boundary of (S_c^2)

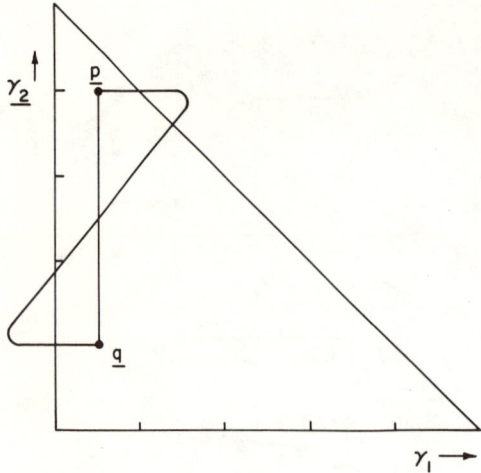

Fig. 11. A forbidden diffusion path (after Gupta and Cooper, 1971) on (S_c^2).

with the form of $[D]$ necessary to satisfy this restraint are shown. From Fig. 12 we notice particularly that some composition changes must cause rotation of the eigen vectors. It is obvious that with a larger number of independent species, equivalent restrictions occur.

Except for the trivial case where all eigen values are equal at least one of the eigen vectors is a nonconstant function of composition. It follows from this that at least one element of $[D]$ is a nonconstant function of composition.

When $[D]$ is composition dependent, the operators $' (= \partial/\partial y)$ and $[D]$ do not commute. Thus neither Equation 1 nor Equation 2 can be uncoupled by a linear transformation as was reported in earlier parts of this paper, and the partial differential equations become nonlinear.

This renders the general case of composition-dependent $[D]$ difficult to treat analytically, and numerical methods (Oishi, 1965; Gupta and Cooper, 1971) are necessary. This is not surprising, as we know that even for binary diffusion, a

an eigen vector tends to parallel the boundary. In terms of elements D_{ik}, this implies (Sundelhof, 1963)

$D_{ik} \to 0$ for all i for which $c_i \to 0$ $(i \neq k)$.

This is shown schematically in Fig. 12, where the eigen vectors of (S_0^2) along

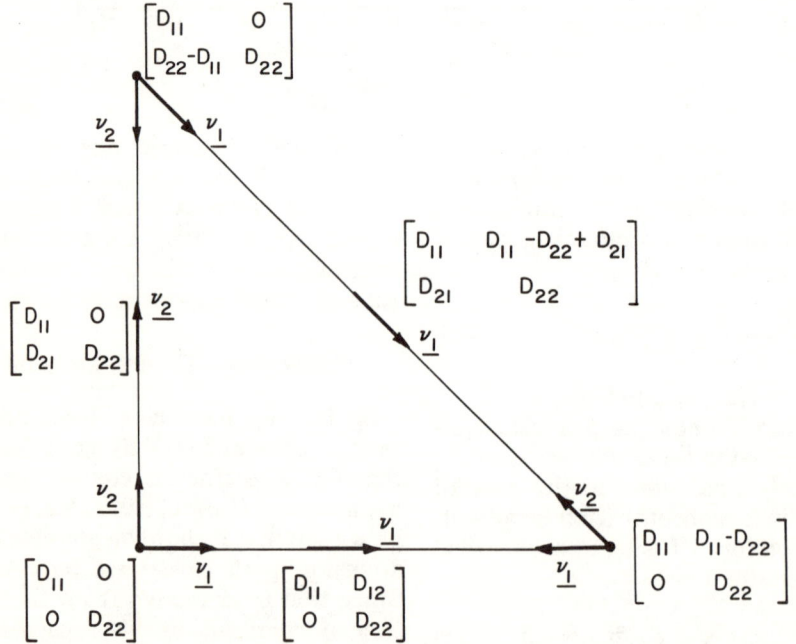

Fig. 12. (S_c^2) showing restrictions on eigen vectors and $[D]$ at the boundary (after Gupta and Cooper, 1971).

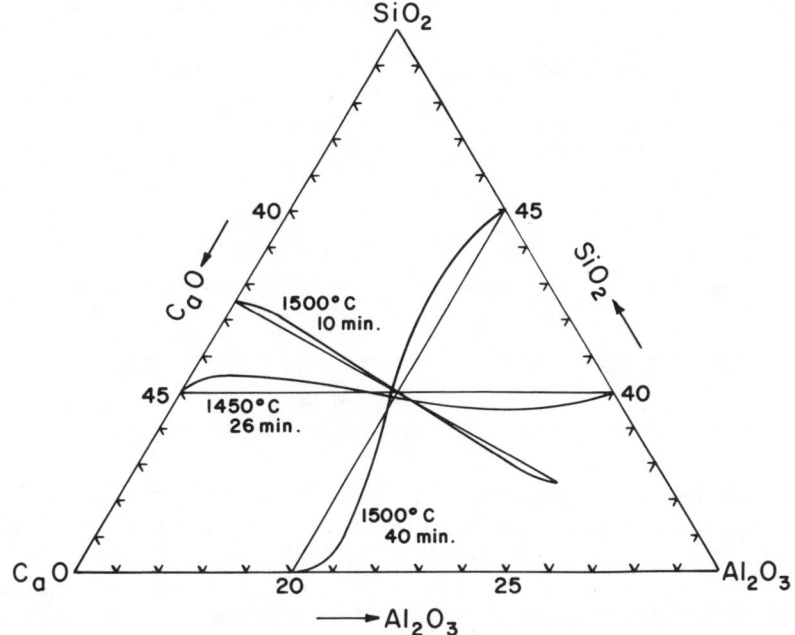

Fig. 13. Diffusion paths in the system CaO-Al$_2$O$_3$-SiO$_2$ (after Oishi, 1969).

composition-dependent $[D]$ generally requires numerical solutions of the diffusion equation.

Some rules have been deduced for the diffusion paths in the case of variable $[D]$ (Gupta and Cooper, 1971). They are, however, not nearly so concise as those for constant D. For example, no longer need an infinite couple have a diffusion path with an antisymmetric "S" shape. In a system with two independent species the path must cross the straight line joining the end members at least once but it is not limited to a single intersection. Neither need the paths at the extremities of the diffusion couple parallel the fast eigen vector because the penetration in the different eigen directions depends on the ratio of the eigen values all along the path, not just on its value at the bulk composition. Neither need the diffusion path include the middle point between the two end member compositions.

At present a general treatment is lacking that would permit the same intuitive grasp of the significance of the diffusion path as was possible when $[D]$ could be assumed constant. However, consistent with our intuition, Sauer (1973) has shown by expanding the composition in a Taylor series that as the composition differential, e.g., $\mathbf{p} - \mathbf{q}$, becomes smaller, the assumption of constant $[D]$ becomes more justifiable.

This seems confirmed by results shown on Figs. 4 and 5 where composition differentials are only about 4 wt percent and diffusion paths are consistent with constant $[D]$ In the system CaO-Al$_2$O$_3$-SiO$_2$ with a 10 mole percent composition differential, Oishi (1969) found results that also are very nearly consistent with the restrictions of a constant $[D]$ (see Fig. 13). Note for example the near parallelism of all diffusion paths at end compositions indicating direction of the fast eigen vector, and the antisymmetric S-shaped paths.

Dehoff and Guy (1969), studying diffusion in system Cu-Zn-Ni over composition differentials from 20 to 40 atomic

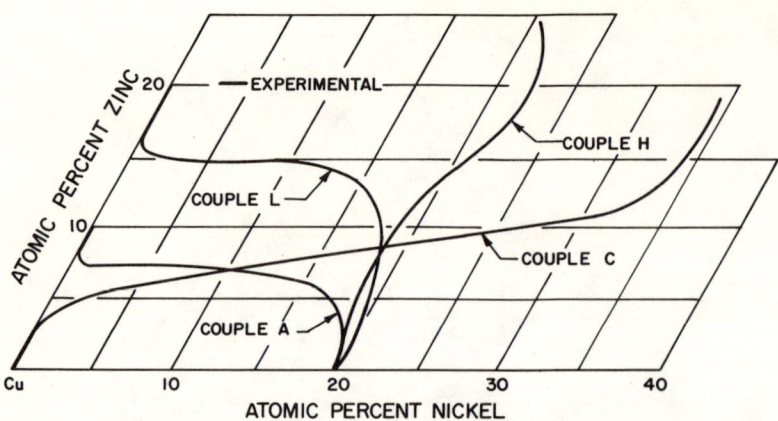

Fig. 14. Diffusion paths in the system Cu-Ni-Zn (after Dehoff and Guy, 1969).

percent, found the paths reproduced on Fig. 14 and the beginnings of significant deviations from the predictions of a constant $[D]$. The paths are not antisymmetric; they do not include the midpoint.

(Note how clearly the fact that one eigen vector must parallel the boundary is illustrated by the Cu-Zn binary.)

Finally, the work of Ziebold (1969), which extends clear across the system

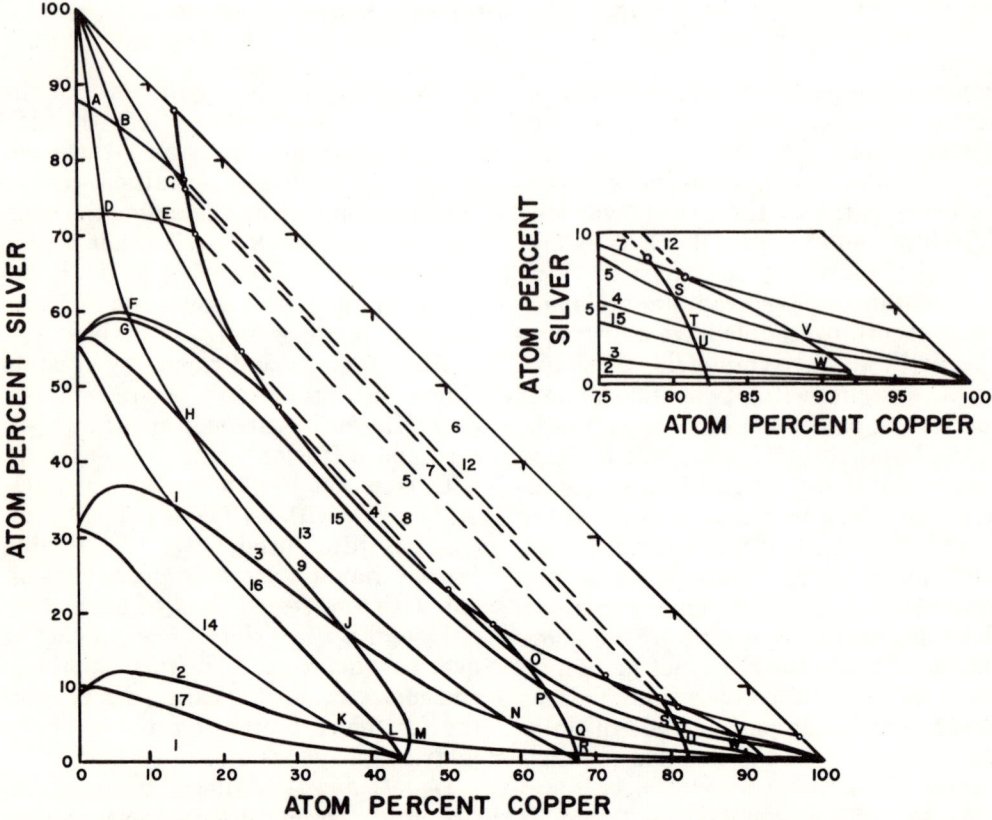

Fig. 15. Diffusion paths in the system Cu-Ag-Au at 725°C (after Ziebold, 1969).

Cu-Ag-Au, shows wider variations from the predictions of a constant $[D]$ (see Fig. 15).

Fortunately it is possible from a single diffusion experiment to formally deduce $[D]$ and its composition dependence (Gupta and Cooper, 1971) limited only by the precision of the original data and the amount of computer time available.

Nonisothermal Diffusion

When temperature, pressure, or some other intensive thermodynamic variable is not constant throughout the system in which diffusion is taking place we must add another dimension to the vector space. As an example let us consider a nonisothermal system. A new space ($S^{(n-1), T}$) is produced by adding a basis vector in the direction of increasing temperature, T. In this space let us call the "composition vector": $\mathbf{c}_T \equiv (c_1, c_2, \ldots c_{n-1}, c_q)$ where $c_q \equiv \ln T$; the flux vector: $\mathbf{j}_T \equiv (j_1, j_2, \ldots j_{n-1}, j_q)$ where j_q is the heat flux; and the gradient vector: $\mathbf{c}'_T \equiv (c'_1, c'_2, \ldots c'_{n-1}, c'_q)$. We may write as our flux equation

$$[\rho]^{-1} \mathbf{j}_T = - [D_T] \mathbf{c}'_T \qquad (12)$$

$[\rho]$ is a diagonal matrix with elements $\left(\dfrac{\partial \rho_i}{\partial c_i}\right)$, where ρ_i is the amount density of the ith quantity. As such the first $(n-1)$ terms of $[\rho]$ are just ρ, the overall amount density, and the last element,

$$\rho_{qq} = T \frac{\partial}{\partial T} \int_0^T c_P d\tau = T\rho c_P,$$

where (ρc_P) is the heat capacity per unit volume.

It remains to show that $[D_T]$ can be diagonalized. As before, $[D_T]$ is a product of two matrices which we call $[L_T]$ and $[G_T]$. $[L_T]$ is defined by the relation

$$[\rho]^{-1} \mathbf{j}T = - [L_T] \mathbf{\mu}'_T$$

where

$$\mathbf{\mu}_T = (\mu_1 - \mu_n), (\mu_2 - \mu_n), \ldots (\mu_{n-1} - \mu_n), \ln T.$$

Onsager's reciprocal relations assure the symmetry of $[L_T]$ and the fact that entropy production is always positive requires $[L_T]$ to be positive definite (de-Groot and Mazur, 1969).

$[G_T]$ is formed by adding the element $G_{qq} = 1$ to the matrix $[G]$. Thus $[G_T]$ preserves the symmetry and positive definite characteristics of $[G]$. Hence $[D_T]$, the product of two symmetric positive definite matrices, can be diagonalized; therefore we may expect that vector space treatment of nonisothermal diffusion will yield analogous benefits to those that can be achieved in isothermal systems.

Acknowledgment

An Alexander von Humboldt senior award allowed this work to be carried on. It is greatly appreciated.

References Cited

Carlson, P. T., M. A. Dayanada, and R. E. Grace, Diffusion in ternary silver-zinc-cadmium solid solutions, preprint June 14, 1971.

Carslaw, H. S., and J. C. Jaeger, *Conduction of Heat in Solids*, 2nd ed., Oxford University Press, London, 510 pp., 1959.

Cooper, A. R., Model for multicomponent diffusion, *Phys. Chem. Glasses*, 6, 55, 1965.

Cooper, A. R., Diffusive mixing in continuous laminar flow systems, *Chem. Eng. Sci.*, 21, 1095, 1966.

Cooper, A. R., Continuous glassmaking, a mixing process, *Glasteknisk Tidskrift*, 28, 27, 1973.

Cooper, A. R., and P. K. Gupta, Analysis of diffusion controlled crystal growth, in *Advances in Nucleation and Crystallization of Glasses*, L. L. Hench and S. W. Frieman, eds., Amer. Ceram. Soc., 1971.

Cooper, A. R., and A. K. Varshneya, Diffusion in K_2O-SrO-SiO_2 system effective binary diffusion coefficients, *J. Amer. Ceram. Soc.*, 51, 103, 1968.

Crank, J., *Mathematics of Diffusion*, Oxford University Press, London, 347 pp., 1957.

Darken, L. S., Diffusion, mobility and their interrelation through free energy, in binary metallic systems, *Trans. AIME*, 175, 184, 1948.

Dayanandu, M. A., and R. E. Grace, Ternary diffusion in copper-zinc-manganese alloys, *Trans. AIME*, 233, 1287, 1965.

Dayanandu, M. A., P. F. Kirsch, and R. E. Grace, Ternary diffusion in Cu–Zn–Sn solid solutions, *Trans. AIME*, 242, 885, 1968.

deGroot, S. R., and P. Mazur, *Non Equilibrium Thermodynamics*, Chap. 11, North Holland Publ. Co., Amsterdam, 1969.

Dehoff, R. T., and A. C. Guy, ONR Report, College of Engineering, University of Florida, 1969.

Eckert, E. R., *Introduction to the Transfer of Heat and Mass*, McGraw-Hill, New York, 1950.

Engelke, H., Chemische Diffusion und Brechungsindexverlauf in zusammengeschmolzenen Gläsern. Thesis, Universität Erlangen-Nürnberg, 1972.

Fujita, H., and L. J. Gosting, An exact solution of the equations for free diffusion in three-component systems with interacting flows, and its use in evaluation of the diffusion coefficients, *J. Amer. Chem. Soc.*, 78, 1099, 1956.

Geffken, W., Vorgänge der Homogenisierung in der Schmelze, Ausziehen der Schlieren, Diffusion, *Glastech. Ber.*, 30, 143, 1957.

Gupta, P. K., and A. R. Cooper, The $[D]$ matrix for multicomponent diffusion, *Physica*, 54, 39, 1971.

Guy, A. G., and J. Philibert, Determination of intrinsic diffusion coefficients in three-component solid solutions, *Z. Metallk.*, 56, 841, 1965.

Kirkaldy, J. S., Isothermal diffusion in multicomponent systems, *Advan. Mater. Res.*, 4, 1970.

Levich, Veniamin G., *Physicochemical Hydrodynamics*, Scripta Technica Inc., Washington, D.C., 1962.

Manning, J. R., Cross terms in the thermodynamic diffusion equations for multicomponent alloys, *Met. Trans.* 1, 499, 1970.

Meyer, O. E., *The Kinetic Theory of Gases*, translated by R. A. Baynes, Longman Greens and Co., London, 1899, pp. 255 ff.

Miller, D. G., Ternary isothermal diffusion and the validity of the Onsager reciprocity relations, *J. Phys. Chem.* 63, 570, 1959.

Miller, D. G., Thermodynamics of irreversible processes. The experimental verification of the Onsager reciprocal relations, *Chem. Rev.* 60, 15, 1960.

Nernst, W., Die elektromotorische Wirksamkeit der Ionen, *Z. Phys. Chem Stoechiom. Verwandschaftslehre*, 4, 129, 1889.

Oishi, Y., Analysis of ternary diffusion: Solutions of diffusion equations and calculated concentration distribution, *J. Chem. Phys.*, 43, 1611, 1965.

Oishi, Y., *Curved Diffusion Paths in Ternary Systems*, preprint of 1st U.S.-Japan Seminar on Basic Science of Ceramics, Tokyo, Japan, 1969.

Oishi, Y., A. R. Cooper, and W. D. Kingery, Dissolution kinetics in ceramic systems III, boundary layer concentration gradients, *J. Amer. Ceram. Soc.*, 48, 88, 1965.

Onsager, L., Reciprocal relations in irreversible processes. I. and II, *Phys. Rev.* 37, 405; 38, 2265, 1931.

Onsager, L., Theories and problems of liquid diffusion, *Ann. N.Y. Acad. Sci.* 46, 241, 1945.

Planck, M., Ueber die Erregung von Electrizität und Wärme in Electrolyten, *Ann. Physik. (Wiedemann)*, 39, 161, 1890a.

Planck, M., Ueber die Potentialdifferenz zwischen zwei verdünnten Lösungen binärer Electrolyte, *Ann. Physick. (Wiedemann)*, 40, 561, 1890b.

Sauer, F., Max-Planck-Institut für Biophysik, Frankfurt, private communication, 1973.

Simonsen, D., Technical University of Denmark, Lygnby, Denmark, private communication, 1973.

Sucov, E. W., and R. R. Gorman, Interdiffusion of calcium in soda-lime-silica glass at 880° to 1308°C, *J. Amer. Ceram. Soc.*, 48, 426, 1965.

Sundelhof, L. O., Isothermal diffusion in ternary systems, *Arkiv Kemi*, 20, 369, 1963.

Toor, H. L., Solution of the linearized equations of multicomponent mass transfer: I, *AIChEJ.*, 10, 448, 1964.

Varshneya, A. K., Multicomponent diffusion in glasses . . . , Thesis, Case Western Reserve University, 1970.

Varshneya, A. K., and A. R. Cooper, Diffusion in system K_2O–SrO–SiO_2 II. Cation self-diffusion coefficients, *J. Amer. Ceram. Soc.*, 55, 220, 1972.

Varshneya, A. K., and A. R. Cooper, Diffusion in system K_2O–SrO–SiO_2 III. Interdiffusion coefficients, *J. Amer. Ceram. Soc.*, 55, 312, 1972.

Wagner, C., The diffusion rate of sodium chloride with diffusion and natural convection as rate-determining factors, *J. Phys. Colloid Chem.*, 53, 1030, 1949.

Ziebold, T., *Ternary diffusion in copper-silver-gold alloys*, Thesis, Dept. of Metallurgy, Massachusetts Inst. of Technology, 1969.

Ziebold, T. O., and A. R. Cooper, Atomic mobilities and multicomponent diffusion, *Acta Met.*, 13, 465, 1965.

MODELING OF DIFFUSION CONTROLLED PROPERTIES OF SILICATES

D. E. Anderson and G. R. Buckley[1]
Department of Geology
University of Illinois at Urbana-Champaign
Urbana, Illinois 61801

Introduction

Theoretical investigation of partitioning of components among coexisting phases leads to results that may be tested directly with electron or ion microprobes. Albarede and Bottinga (1972) have proposed models for the progressive partitioning of trace elements by diffusion and growth during crystallization of melts. Anderson and Buckley (1973) have examined a model involving the exchange of major components between a crystal and an initially inhomogeneous reservoir. Although the model was developed to explore the possible role of diffusion in the zoning of metamorphic garnets, it may be used to investigate partitioning or, in conjunction with partitioning data, to analyze the formation of domains of local equilibrium during metamorphism (Blackburn, 1968; Kwak, 1970).

Unfortunately, the lack of systematic diffusion data limits the practical application of such models at this time. It is useful, however, to ask two questions at this point: (1) What kinds of diffusion data are needed to find explicit solutions? (2) How can the models be simplified? The second question is particularly important. Exchange between silicates must often involve multicomponent diffusion ($n \geq 3$).[2] The collection of data in multicomponent systems and the manipulation of multicomponent models are onerous and complicated tasks. The problem is significantly simplified if circumstances can be found in which either the number of independent diffusion coefficients is reduced, or the off-diagonal coefficients may be ignored. Notable in this respect are the efforts of Kirkaldy and co-workers to use nonequilibrium thermodynamics and kinetics to define the most economical methods of obtaining experimental data (Kirkaldy, 1957, 1958; Kirkaldy et al., 1963, 1965; Kirkaldy and Lane, 1966; Lane and Kirkaldy, 1964; Kirkaldy and Purdy, 1962). Of equal importance is the development of an atomic mobility model by Cooper (1965; see also Ziebold and Cooper, 1965); the significance of Cooper's model to diffusion in geological systems has been partially reviewed by Buckley (1973). Although more detailed and complete calculations by Manning (1961, 1967, 1968) have shown that the assumptions underlying Kirkaldy's and Cooper's models are not completely justified, the models are still useful starting points for initial calculation and investigation.

It appears that mobility models provide the most efficient method of gathering data and extrapolating the data to geological conditions beyond the immediate experimental conditions (Buckley, 1973). Nevertheless, in this paper we will focus on the application of nonequilibrium thermodynamics and kinetics. The correct formulation of flux equations, which has been clarified through the use of nonequilibrium thermodynamics, is basic to both methods. Also, this approach aids in systematizing the construction of realistic diffusion-dominated models for the interpretation of exchange and metasomatic processes.

[1] Now at Esso Production Research, Box 222, Houston, Texas 77000.
[2] The number of components refers to the number of diffusing components. This may or may not be equal to the number of thermodynamic components (Schönert, 1960; Haase, 1969).

To place the formulation of flux equations in perspective a model describing the exchange between a growing grain and an inhomogeneous reservoir is discussed in the next section. The model is discussed specifically in terms of garnet to take advantage of the growing volume of microprobe data on zoning in metamorphic garnets and of experimental and natural data on elemental partitioning between garnet and a variety of host rocks. This knowledge, when combined with general petrographic data on naturally occurring garnets may be used to place constraints on the boundary conditions required to determine the evolution of composition profiles from the appropriate flux equations.

Certain restrictions have been adhered to in all of the discussion that follows. The flux equations are written for one-dimensional diffusion in isotropic media. Well-defined methods exist for expanding the flux equations to three dimensions and for including anisotropy in them (Nye, 1957); no further insight is gained by adding these complications. Also, diffusion is assumed to take place in isothermal, isobaric conditions in the absence of applied fields.

General Models

Studies of minerals in some Grenville Province gneisses (Blackburn, 1968; Kwak, 1970) indicate that they may have equilibrated chemically in zones ranging from a few millimeters up to four centimeters in size. In addition, Blackburn suggested that the variation in zone size may be due to the change in diffusion rates with increasing grade. This idea has led some workers (Hess, 1971; Mueller and Schneider, 1971) to postulate that some zoned crystals may have formed by partial reequilibration due to changing T, P, f_{O_2}, zone size, etc. These studies immediately lead one to questions concerning the interpretation of the intercrystalline partitioning data used in geologic thermometers. Is the temperature measured a maximum, or has enough reequilibration occurred to give some lower temperature? At what temperature can it be assumed that exchange is negligible? Clearly, the answers to these questions may significantly affect geologic time and temperature interpretations; therefore, it is essential that geologists understand the exchange or diffusion processes involved in chemical and isotope equilibration and the conditions necessary for these processes to cause measurable changes in mineral composition.

The ultimate composition of any portion of a given garnet must be related to growth, original inhomogeneities in the host reservoir, and dependence of the diffusion and partition coefficients on composition and the local physical environment. The practical questions that must be answered are: (1) What is the relative importance of each of these factors under a given set of physical conditions? (2) What are the geologic constraints which can limit the choice of physical conditions available to a growing and exchanging garnet? Many of the answers to these questions are not readily available from existing theoretical, experimental, and field data. Even so, there is some advantage to be gained in setting up a general model for garnet zoning as a framework in which to discuss these known and unknown quantities.

Consider, for example, an initially homogeneous and spherical garnet (phase α) growing into and exchanging with an inhomogeneous reservoir (phase β). Assume that the system contains only two "diffusing" components. Let D^α and D^β be the mutual diffusion coefficients (Section 3) for the exchange of the two components in the garnet and in the reservoir, respectively. The initial conditions (Anderson and Buckley, 1973, Fig. 8) are

$$c_1^\alpha = c_0^\alpha, \quad 0 < R < R_1, \text{ time } t = 0 \quad (2.1)$$

$$c_1^\beta = c_0^\beta(R), \quad R_1 < R < \infty, \, t = 0 \quad (2.2)$$

where R_1 is the radius of the garnet and c_1^α and c_1^β are the concentrations of component 1 in α and β. The boundary conditions are

$$R_1 = R_1(t), t > 0, \text{(growth)}, \quad (2.3)$$

$D^\alpha (1/R(\partial R c_1^\alpha/\partial R) - c_1^\alpha/R)$
$+ \partial R_1(t)/\partial t (c_1^\alpha(R_1))$
$= D^\beta (1/R(\partial R c_1^\beta/\partial R) - c_1^\beta/R) +$
$\partial R_1(t)/\partial t (c_1^\beta(R_1))$,
$R = R_1, t > 0$, (mass balance), (2.4)

$c_1^\alpha = K_D c_1/1 - c_1^\beta(1 - K_D), R = R_1,$
$t > 0$, (partitioning),[3] (2.5)

where K_D is a distribution coefficient suitable for exchange between compounds of the type (A,B) M–(A,B) N. Finally the differential equations become (in spherical coordinates)

$\partial c_1^\alpha/\partial t =$
$\quad (1/R^2)\partial/\partial R(R^2 D^\alpha \partial c_1^\alpha/\partial R)$, (2.6)

$\partial c_1^\beta/\partial t =$
$\quad (1/R^2)\partial/\partial R(R^2 D^\beta \partial c_1^\beta/\partial R)$. (2.7)

The exact form of $R_1(t)$, $K_D(t)$ and $D(c)$ will depend on the particular growth law, composition dependence, and evolution of physical conditions used.

These equations may be applied to the class of problems dealing with the chemical redistribution that might occur when (Mn^{2+}, Fe^{2+}) garnets begin to nucleate and grow in an inhomogeneous reservoir containing (Mn^{2+}, Fe^{2+}) biotite, muscovite, and quartz as the dominant phases. Because biotite and garnet are the only two major Mn^{2+} and Fe^{2+} bearing minerals in this system, the reservoir will be treated as a single phase (β) in Equations 2.1 to 2.7. In addition, effects of the diffusion of the slower, growth controlling ions Al^{3+} and Si^{4+} will be incorporated directly into the appropriate growth law 2.3. When these assumptions apply, what constraints do existing data place on the form of the relations 2.1 to 2.7 and,

[3] For simplicity the gram-atomic volumes of α and β have been assumed equal. When this is not true, reference frames fixed in (2.4) and (2.5) will move relative to each other during growth. A solution of this problem is given by Danckwerts (1950).

ultimately, on the evolution of zoning profiles?

For the assemblage given, the partition coefficient in 2.7 may be computed from the exchange reactions

Fe-Biotite + Mn-Garnet =
    Mn-Biotite + Fe-Garnet

only when (1) the standard state for the chemical potential of each component is defined in a manner that is consistent with *all* of the coexisting phases present in the assemblage (Ramberg, 1960); (2) the form of each activity coefficient as a function of mole fractions $x_{Mn}{}^\alpha$, $x_{Mn}{}^\beta$, T and P is known; (3) the form of each chemical potential as a function of T and P is known. Although all of these required data are not available, some preliminary qualitative constraints are possible if one notes that K_D for Mn^{2+} in this and other assemblages containing garnet and biotite must be a function of temperature, pressure, f_{H_2O}, and f_{O_2}. Experimental work by Dahl (1971) indicates that this partition coefficient must decrease with increasing temperature, pressure, and f_{H_2O}. Moreover, work by Hsu (1968) suggests that the coefficient will also decrease with decreasing f_{O_2}. During burial and metamorphism one can expect the temperature and pressure to increase. The fugacity of water should decrease by progressive dehydration and, as suggested by Mueller and Schneider (1971), the fugacity of oxygen can be expected to decrease markedly at higher temperatures when natural graphite becomes an effective buffer. Thus it does not seem unreasonable to expect K_D to decrease during the first part of a thermal cycle and increase during the latter part. However, one should keep in mind that the K_D time curve may not be symmetrical, as the form of f_{H_2O} and f_{O_2} versus time curves will probably differ during prograde and retrograde metamorphism.

To obtain geologically reasonable forms of condition 2.2, one must know something about the iron and manganese distributions in biotite zone rocks just prior to the nucleation of the garnets. First approximations for this condition may be obtained by looking at compositional variations that occur in biotite zone rocks very close to the garnet isograd. Such estimates should be used with care, however, if retrograde or retrogressive effects are important. More accurate estimates will probably result from analyses of the evolution of zones of equilibration during the progressive metamorphism of an initially unaltered shale. Such studies are not available at present, possibly because of the low concentration of MnO (<0.1 wt %) in most pelites and many associated minerals, including biotite.

Geologically reasonable estimates of the magnitude and form of 2.3 are perhaps the most difficult conditions to evaluate, as they must have input from all of the other boundary conditions and thus are subject to cumulative errors. Moreover, growth theory is still in its infancy and consequently much of it has been derived for specific systems. However, assuming that many of the general macroscopic and microscopic principles governing diffusion may also be applied to growth, it is worth examining a few of the constraints that such similarities may place on the form of 2.3.

Following Shewmon (1969, p. 107), the polymorphic transformation of α into β may be written as the following flux-force equation

$$\partial R_1/\partial t \propto J^G = -M\Delta\mu/\delta \quad (2.8)$$

where J^G is the flux of phase α across the α/β boundary (in moles/cm² sec), M is a proportionality constant (mobility of interface), $\Delta\mu$ is the chemical potential difference between α and β defined for the appropriate standard state, δ is the width of the boundary, and $-\Delta\mu/\delta$ becomes the potential gradient, or driving force, for motion of the grain boundary.

Although little controlled, experimental information is available on growth in multiphase, multicomponent, charged systems, it may be possible to relate the flux of α across the α/β interface to the sum of all possible stoichiometric diffusing components of α. For example, for $(Mn,Fe)_3Al_2Si_3O_{12}$ garnet growing into biotite we may be able to write the flux equation of spessartine across the interface as

$$J^G{}_{\text{Spessartine}} = -M_{\text{Sp,Sp}} \frac{\Delta \overline{G}_{\text{Mn-bio/Mn-gar}}}{\delta}$$
$$- M_{\text{Sp,Al}} \frac{\Delta \overline{G}_{\text{Fe-bio/Fe-gar}}}{\delta}$$

giving cross effects of almandine on the growth of spessartine. The $\Delta \overline{G}$'s are the molar Gibbs free energy changes that might be expected for the transition biotite $\rightarrow$ garnet, and the G_0's must be defined for the particular standard state of interest (the standard state will depend on the phases coexisting with biotite and garnet). The net growth rate will then be proportional to

$$J^G{}_{\text{Total}} = J^G{}_{\text{Sp}} + J^G{}_{\text{Al}}$$

Assuming that such cross terms, if they exist, may be neglected, the growth of garnet into biotite can be approximated by a relationship similar to Equation 2.8. Based on evidence reviewed in Shewmon (1969), Fine (1965), and Chernov (1968), it is clear that the interface mobility must depend on the composition, temperature, and pressure of the interface so that

$$M = M(\sum_{i}^{n} c_i, T, P). \quad (2.9)$$

The exact relationship between these parameters will depend on the particular growth model and thermodynamics governing the system of interest. There are, however, certain general experimental, theoretical, and field constraints that must be met by any garnet growth model selected.

It is well known that growth behaves as an activated process and thus varies exponentially with temperature. Emphasizing the analogy between growth and diffusion, Shewmon has pointed out that M may have a form similar to D, that is

$$M = M_0 e^{-Q_G/RT} \qquad (2.10)$$

where M_0 is the intercept at $T = \infty$ on a $\ln M$ vs $1/T$ plot and Q_G is the activation energy for growth.[4] Notice, however, that although M may vary exponentially with temperature and surface composition the growth rate may still go to zero. Whether or not it does will depend on the effect of temperature and composition on the driving force for boundary migration.

For the transformation biotite → garnet several processes must be involved: (1) breakdown of biotite into ions or small units, (2) transfer of the required ions into and/or through the boundary, (3) transport of excess material along grain boundaries away from the interface or transport of additional material along grain boundaries toward the interface.[5] Execution of the first two steps must occur nearly simultaneously as the appropriate units acquire sufficient vibrational or rotational energy to break free of biotite and migrate across a relatively narrow boundary. Step three, however, must involve long distance diffusion (See discussion in Anderson and Buckley, 1973, p. 98). Moreover, regardless of the reference frame chosen, growth will involve diffusion of one or more of the slower ions Al^{3+} and Si^{4+} to or from the interface.

Whether steps one and two or step three controls the growth process is unclear at this time. It does seem fairly certain that in the garnets studied by Kretz (1966) and Jones and Galwey (1966) long-range sources of Al^{3+} and Si^{4+} were not important to garnet growth; there was no correlation between intergarnet distance and size. However, this does not rule out the possibility of a correlation between intergarnet distance and the Fe^{2+}/Mn^{2+} ratios in the garnets.

For a given mechanism of growth the boundary mobility should increase with temperature. Moreover, $\overline{\Delta G}_{\text{bio/gar}}$ should, in general, increase with increasing temperature.[6] Consequently, it is not unreasonable to expect growth to continue during an entire thermal cycle.

Most metamorphic garnets fall in a fairly narrow size range (between 0.1 and 1 cm). Because one might expect, a priori, that the mechanism controlling growth in one rock will be similar to that controlling it in another and that the growth rate will increase exponentially with temperature, it is reasonable to ask if these sizes are controlled by temperature. Although the amount of data is limited, there seems to be little correlation between garnet size and metamorphic grade. However, it appears (this must be considered tentative) that the garnet size may be more closely related to the original quartz/clay ratio. This does not seem unreasonable, as larger quartz fractions ultimately require transport of large amounts of Al^{3+} long distances through a relatively small cross sectional area of grain boundaries. It is possible, then, that garnets in pelites with relatively high quartz/clay ratios initiate growth by an interface controlled mechanism but change to a diffusion controlled mechanism as the local source of Al^{3+}, Fe^{2+}, Mn^{2+}, etc. is used up. When the quartz/clay ratio is low, garnet may proceed by an interface mechanism throughout its entire growth period. Provided the density of garnet grains is not too high, the observations of Kretz (1966) and of Jones and Galwey (1966) should be valid for either situation.

These relationships suggest that after nucleation the growth–time curve may be similar in form to the temperature–time

---

[4] By treating the boundary as a phase; Equation 2.10 may be fitted to interface controlled or diffusion controlled growth models through the composition dependence of Q.

[5] The sense and magnitude of the diffusion of the constituent ions of garnet to or from the interface will depend on the extent to which the biotite → garnet transition is a constant volume replacement.

[6] The increase in grain boundary energy (ΔG_g) with increasing garnet size will act against the driving force for the transformation (ΔG_r). However, for larger grain sizes $\Delta G_r \gg \Delta G_g$, provided the redistribution of Fe^{+2} and Mn^{+2} is not too much faster than the growth rate.

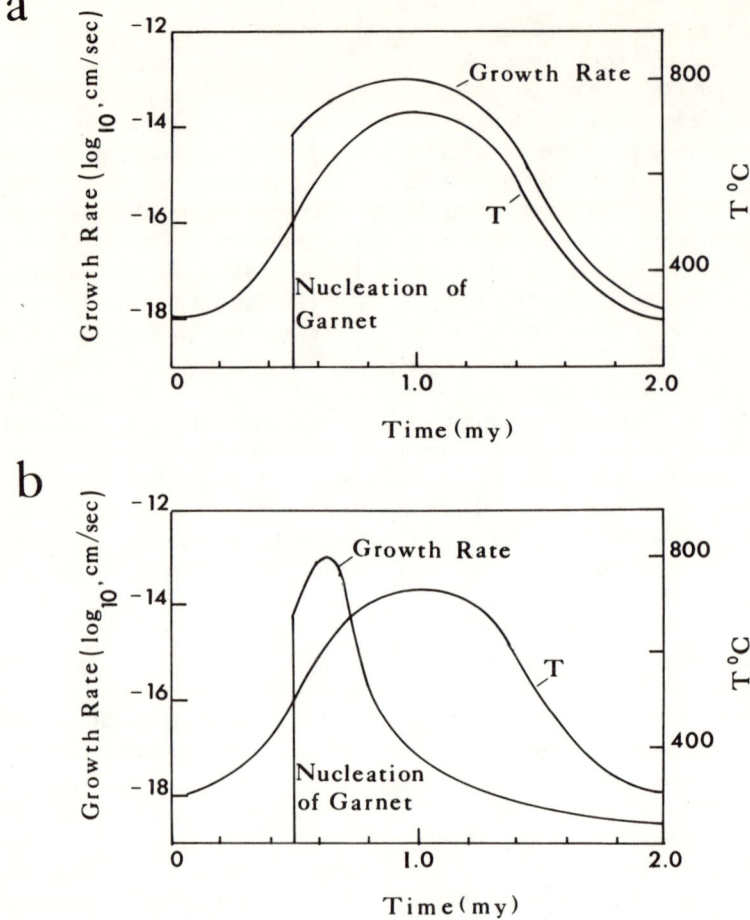

Fig. 1. Growth-time relations that might exist when garnet growth is: (a) controlled by interface processes during an entire thermal event; (b) controlled by interface processes early in a thermal event but controlled by long range diffusion later.

curve in clay-rich rocks (Fig. 1a). In other rocks the growth rate should increase with temperature until the local growth components are used up. The change from an interface controlled to a longer range diffusion controlled mechanism may cause a rapid decrease in growth rate (Fig. 1b). A quantitative evaluation of the actual growth rates and of the time required to reach the maximum growth rate requires additional information. However, assuming thermal events on the order of 1 to 10 m.y. and garnet crystals on the order of 0.1 to 1 cm the mean growth rates must have been between 10^{-13} and 10^{-15} cm/sec. Also, noting that for intervals of time when

$$R_1(t_2) - R_1(t_1)/\sqrt{D^\alpha (t_2 - t_1)} \lesssim 1 \quad (2.11)$$

and

$$D^\beta > D^\alpha$$

reequilibration will be significant. Therefore exchange coefficients averaging 10^{-15} to 10^{-16} cm^2/sec during a thermal event can significantly alter growth effects throughout most metamorphic garnets. Notice however that, although relation 2.11 may be used to estimate intervals of time when significant reequilibration may occur, it will not define the exact path of evolution toward equilibrium. To do this one must know the time dependence of the boundary conditions 2.3 to 2.5 and the appropriate diffusion coefficients.

To illustrate this approach we have applied the constraints discussed to the boundary conditions 2.3 to 2.5 and the predictions of Buckley (1973) to the Mn^{2+}–Fe^{2+} exchange coefficients. The time dependence of D^{α}, D^{β}, K_D and the growth rate $\dot{R}$ have been sketched in Figs. 2a and 2b for two forms of a time-dependent growth rate $\dot{R}(t)$. These curves were then used to sketch qualitatively the concentration profiles that might be derived from 2.6 and 2.7 (Figs. 2c and 2d). The curves in Figs. 2c and 2d are approximate solutions of 2.6 and

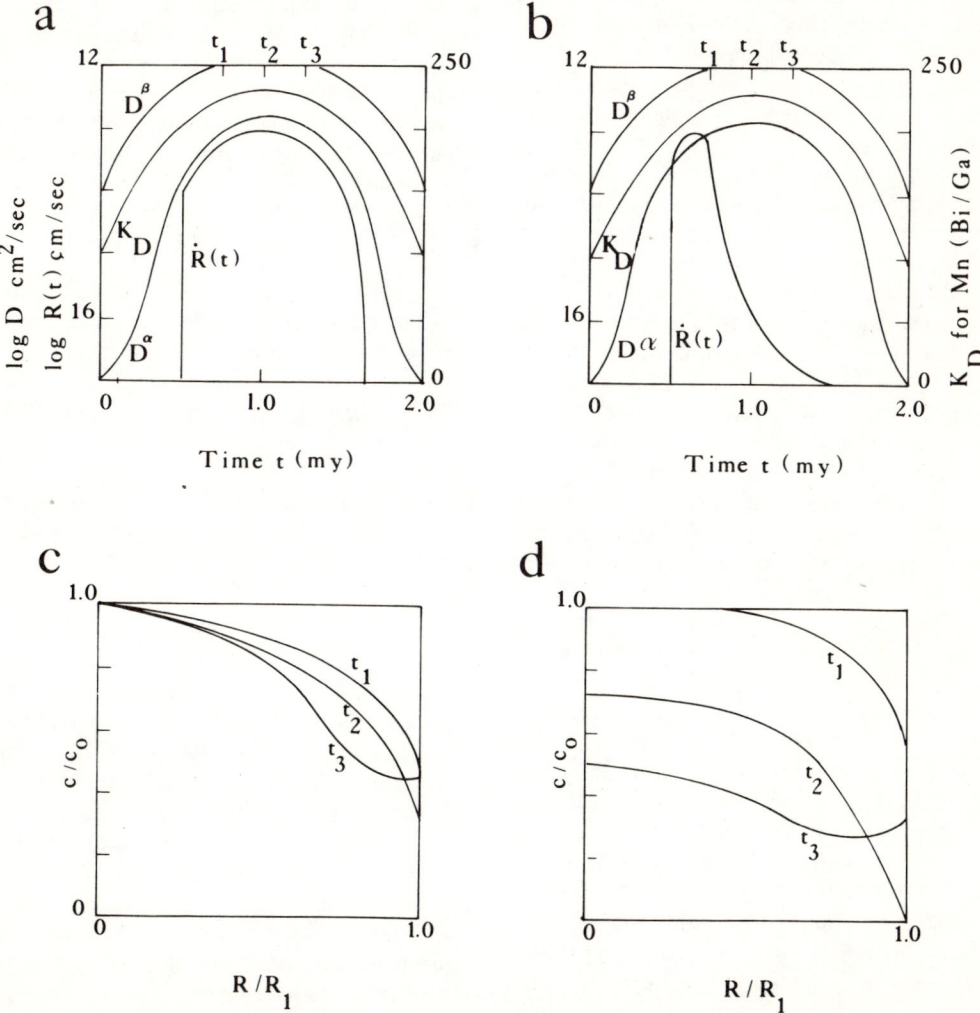

Fig. 2. Composition versus distance curves (2c, 2d) for the thermal curves depicted in Fig. 1 and the $\dot{R}(t)$, $K_D(t)$, $D^{\alpha}(t)$ and $D^{\beta}(t)$ curves of 2a and 2b. Time is given in millions of years. R/R_1 is the ratio of distance from the center of the garnet to the radius; c/c_0 is the ratio of the concentration of the diffusing component to its maximum concentration in the garnet. K_D (biotite/garnet) is assumed to be exponentially dependent on temperature. Interactions between adjacent garnets have been neglected.

2.7 based on solutions in which D^α, D^β, K_D and $\dot{R}$ are varied one at a time (Crank, 1957, gives solutions for K_D and $\dot{R}$; Anderson and Buckley, 1973). A complete solution involving the simultaneous variation of the four variables would require difficult numerical methods. Although only approximate, these curves demonstrate several points: (1) Because of the exponential decay in concentration profiles the centers of garnets resulting from growth–time curves similar to 2b will not in general be related to conditions existing at the time of nucleation of the garnet if the rock has reached sillimanite grade temperatures. (2) The general form of a zoning profile tends to be dominated by the time dependence of the partitioning condition 2.5 regardless of whether the zoning is growth dominated or diffusion dominated. (3) It does not appear possible to define a unique thermal history from a single compositional profile, as the initial conditions and boundary conditions may vary somewhat due to original inhomogeneities in the host rock. However, by simultaneously analyzing several crystals from the same outcrop it may be possible to define time–temperature paths within relatively narrow limits.

Thus far we have shown how descriptions of the evolution of certain properties of natural systems may be conveniently formulated in terms of a series of general initial and boundary conditions and differential equations. Although the exact form of the appropriate conditions and equations may vary depending on the particular system under consideration, the approach is general and offers a logical starting point for the analysis of such time-dependent systems. Moreover, these analyses must be considered prerequisites to the correct interpretation of nonequilibrium features (i.e., crystal zoning, reaction rims, etc.) found in rocks.

The model discussed was based on two relatively simple flux equations, 2.6 and 2.7, as the system was considered binary with respect to diffusion. When the number of diffusing components is greater than two, the flux equations and resulting differential equations may increase in number and complexity. Clearly, when quantitative evaluations of diffusion dependent properties are required, it is desirable to be able to formulate the flux equations in the simplest form that is consistent with available theory. In the discussion that follows we will demonstrate how nonequilibrium thermodynamics may be used to define and simplify these flux equations, and to place constraints on the type of experiments that appear to be the most useful in the measurement of the appropriate diffusion coefficients.

Diffusion Coefficients

Onsager (1945) proposed an extension of Fick's law to systems of two or more diffusing components of the form

$$J_i = -\sum_{k=1}^{n} D_{ik} \nabla c_k, \quad (i = 1, \ldots, n), \quad (3.1)$$

where J_i is the flux of component i and the D_{ik} form an $n \times n$ matrix of diffusion coefficients. The equivalent differential equations are

$$\frac{\partial c_i}{\partial t} = \sum_{k=1}^{n} \frac{\partial}{\partial x}\left[D_{ik} \nabla c_k\right],$$
$$(i = 1, \ldots, n). \quad (3.2)$$

The on-diagonal (direct) diffusion coefficients D_{ik} ($i = k$) describe the diffusion of i on its own gradient, whereas the off-diagonal (cross) coefficients D_{ik} ($i \neq k$) allow for the diffusion of i on the gradient of component k. A reference frame for defining and measuring fluxes, and hence diffusion coefficients, is usually chosen so that (Kirkwood et al., 1960; de Groot and Mazur, 1962),

$$\sum_{i=1}^{n} \alpha_i J_i = 0, \quad (i = 1, \ldots, n). \quad (3.3)$$

The α_i are normalizing or weighing factors. If, for example, the fluxes are measured across a plane located so that the volumes on either side of the plane are constant during diffusion, then

$$\alpha_i = \overline{V}_i, (i = 1, \ldots, n), \quad (3.4)$$

where $\overline{V}_i$ are partial molar volumes.

The diffusion coefficients that appear in 3.2 for binary [7] and multicomponent diffusion are chemical diffusion coefficients. Practically, coefficients of this kind may be calculated from concentration versus distance curves in annealed couples composed of two infinite or semi-infinite media, initially containing different concentrations of the diffusing components. Experience with metals, oxides, and glasses proves that chemical diffusion coefficients are usually dependent on composition as well as temperature. Or, stated differently, the activation energy for diffusion is a function of composition. Data for olivines (Misener, 1972), and the large differences in activation energies for Na and K diffusion in alkali feldspars (Bailey, 1971; Lin and Yund, 1972) reflect compositional dependence in silicates.

Because in general there are only $(n-1)$ independent concentration gradients in a system of n components,[8] 3.1 with 3.3 becomes,

$$J_i = - \sum_{k=1}^{n-1} D_{ik} \nabla c_k,$$
$$(i = 1, \ldots, n-1). \quad (3.5)$$

For a binary system, 3.5 implies that the flux of each component may be described by a single (mutual) diffusion coefficient $(D_{11} = D_{22} = \overline{D})$.

Darken (1948) derived an equation connecting the mutual diffusion coefficient to tracer diffusion coefficients D_i^* of each component of a binary system:

$$\overline{D} = x_1 D_2^* + x_2 D_1^* \left(\frac{\ln a_i}{\ln x_i} \right), \quad (3.6)$$

where x_i and a_i are, respectively, the mole fraction and activity of component i. An extension of Darken's equation (3.6) to multicomponent systems has been given by Cooper (1965) and Ziebold and Cooper (1965). The nature of Darken's equation in binary ionic systems has been examined by Cooper and Heasley (1966). Darken's equation predicts the relationship between $\overline{D}$ and D_i^* with reasonable accuracy in metal alloys. In liquids, however, the measured and calculated values of $\overline{D}$ are only in approximate agreement; the discrepancy apparently increases as the solution exhibits less ideal behavior (Trimble et al., 1965).

A kinetic analysis of Darken's equation, based on the assumption that diffusion proceeds by a vacancy mechanism, was presented by Bardeen and Herring (1951). In terms of 3.1, the vacancies are included as an additional dilute component to create, for diffusion, a ternary system. They demonstrated that kinetically Darken's equation is an approximation that is valid if (1) the total number of lattice sites is conserved; (2) vacancies are assumed to be in local thermal equilibrium, and the vacancy concentration is a unique function of composition; (3) the off-diagonal coefficients D_{ik} $(i \neq k)$ and D_{iT} (T = tracer atom) are negligibly small; (4) the jumps of tracer atoms are not coordinated. The first two assumptions have the effect of reducing the system to a binary system again. The general validity of these assumptions, especially (3) and (4), has been treated by Manning (1959, 1961, 1965, 1967, 1968). If diffusion occurs primarily by a vacancy mechanism, this validity directly affects the generality of Cooper's mobility model and the kinetic calculations of Lane and Kirkaldy (1964) reviewed in a later sec-

---

[7] Although binary diffusion ($n = 2$) is a particular example of multicomponent diffusion, the phenomenological treatment of binary systems has evolved almost independently of multicomponent systems. The qualifying terms "chemical" and "mutual" (Hartley and Crank, 1949) are used interchangeably in binary systems.

[8] This condition is not necessarily valid for multiphase systems for all choices of components.

tion. There is little evidence bearing on diffusion mechanisms in crystalline silicates. It is possible, although perhaps unlikely, that diffusion operates primarily by direct interchange or another mechanism that does not involve vacancies. Evidently, experiments designed to detect a Kirkendall shift—and hence a vacancy mechanism—or to verify Darken's equation for crystalline silicates are of vital interest.

The two independent flux equations needed to describe isothermal diffusion in a ternary system contain four diffusion coefficients:

$$J_1 = -D_{11}\nabla c_1 - D_{12}\nabla c_2, \quad (3.7)$$

and

$$J_2 = -D_{21}\nabla c_1 - D_{22}\nabla c_2, \quad (3.8)$$

if component 3 is chosen as the dependent species. At least two independent experiments (Fig. 3) are needed to determine the four diffusion coefficients at any temperature (Kirkaldy et al., 1965; Ziebold and Ogilvie, 1967). Because diffusion paths cannot usually be predicted in advance, the number of experiments required to obtain good data is indeterminate. The number of diffusion coefficients that appear in the flux equations increases rapidly as the number of diffusing components increases. Methods for simplifying the equations are discussed in the next section.

REFERENCE FRAMES

The number of independent diffusion coefficients may be reduced in one of two ways: (1) For certain combinations of fluxes and thermodynamic forces, the Onsager reciprocity theorem may be utilized (Onsager, 1931a, b, 1945). (2) By choosing a suitable reference frame, the magnitude of the off-diagonal coefficients may be diminished to zero or near zero. The second method is taken up in Section 5.

Equation 3.2 may be recast in the formalism of nonequilibrium thermodynamics as

$$J_i = -\sum_{k=1}^{n} L_{ik} X_k,$$
$$(i = 1, \ldots, n) \quad (4.1)$$

The L_{ik} and X_k are referred to as phenomenological coefficients and thermodynamic forces, respectively. For diffusion, the thermodynamic forces are gradients in chemical potentials. Onsager's theorem states that

$$L_{ik} = L_{ki}, \quad (4.2)$$

or the matrix of phenomenological coefficients is symmetric. Two distinct problems exist. First, the symmetry relations 4.2 are not true for arbitrary choices of fluxes and forces in 4.1. The derivation of Onsager's theorem is based on the microscopic analysis of the regression of fluctuations about a local equilibrium state (Onsager, 1931a, b; Yourgrau et al., 1966, pp. 23–43). The choices of fluxes and forces necessary to justify 4.2 are rigorously dictated by microscopic considerations (Yourgrau et al., 1966, p. 39). In general, however, the fluxes and forces must be chosen in macroscopic

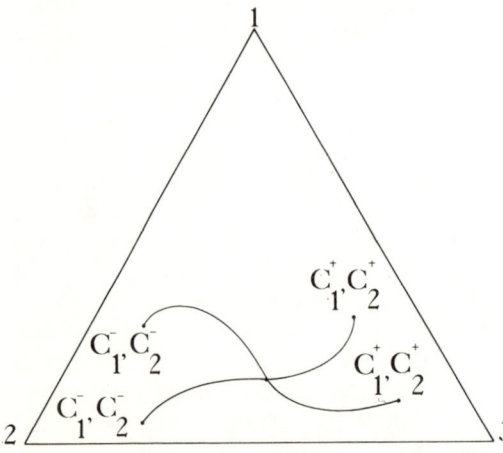

Fig. 3. Hypothetical diffusion paths in a ternary system for two separate diffusion couples. The terminal compositions of each couple are designated c_1^+, c_2^+ and c_1^-, c_2^-. The four coefficients of 3.7 and 3.8 can only be fully determined at the point of intersection of the two diffusion paths.

systems without the benefit or possibility of microscopic analysis; in these circumstances an incorrect choice of fluxes or forces is possible. There is no completely sound theoretical solution to the problem,[9] but the following macroscopic criteria appear to yield fluxes and forces consistent with 4.1 and 4.2 (Hooyman et al., 1955; Hooyman and de Groot, 1955; de Groot and Mazur, 1961; Fitts, 1962, pp. 21–50): (1) The fluxes are linear, homogeneous functions of the forces X_i. (2) The fluxes and forces are chosen so that the rate of entropy production ΔS, given by

$$T\Delta S = \sum_{i=1}^{n} J_i X_i, \qquad (4.3)$$

remains invariant under simultaneous linear transformations of the fluxes and forces. (3) At least one of the set of fluxes or the set of forces must be a mutually independent set. If neither set contains mutually independent quantities, then 4.2 is not necessarily true. It is usually possible, however, to perform a linear transformation that will produce an independent set from the dependent set (Hooyman and de Groot, 1955).

The second problem is to find diffusion fluxes and forces for 3.1 that are consistent with the preceding requirements and are connected by measurable diffusion coefficients or, at least, are connected by diffusion coefficients that may be systematically and simply computed from measured coefficients. It can be shown (Onsager, 1931a, b; Fitts, 1962) that diffusion coefficients for a mass-fixed frame may be defined in such a way that they obey Onsager's theorem. Fluxes in this frame are measured with respect to a plane across which there is no net flow of mass. Volume changes during diffusion commonly lead to mass flow and, unless special precautions are taken,[10] this frame is not easily accessible to experiment. Moreover, solutions of Fick's law (3.2) do not give diffusion coefficients in a mass-fixed frame except in the very unusual circumstance that there is no mass flow during diffusion. A mass-fixed frame is defined by (Kirkwood et al., 1960)

$$\sum_{i=1}^{n} J_i^M = 0, \qquad (4.4)$$

and the flux equations are

$$J_i^M = -\sum_{k=1}^{n} D_{ik}^M \nabla \mu_k,$$

$$(i = 1, \ldots n). \qquad (4.5)$$

Neither the fluxes nor the forces in 4.5 form independent sets and, as noted by Kirkwood et al. (1960), the matrix of coefficients may not be symmetric. An independent set of forces may be obtained by rewriting 4.5 as

$$J_i^M = -\sum_{k=1}^{n-1} \overline{D}_{ik}^M \nabla(\mu_k - \mu_n),$$

$$(i = 1, \ldots, n-1), \qquad (4.6)$$

where 4.4 has also been used. Then

$$\overline{D}_{ik}^M = \overline{D}_{ki}^M. \qquad (4.7)$$

A simpler procedure is to establish a reference frame moving with the local velocity of one of the components and appeal to the Gibbs-Duhem relations,

$$\sum_{i=1}^{n} (c_i \nabla \mu_i)_{T,P} = 0 \qquad (4.8)$$

to form a mutually independent set of forces. In liquids, the solvent is often chosen as the reference component, but it is immaterial which component is chosen as the "solvent."

For the reasons outlined in the next paragraph, diffusion coefficients are usually measured in a volume-fixed or

---

[9] Hence the suggestion of Fitts (1962) that 4.2 be regarded as a postulate to be verified by experiment for a particular class of fluxes and forces.

[10] Mass-flow may be detected by embedding inert markers in the diffusion couple (Darken, 1948; Hartley and Crank, 1949). Movement of the markers during diffusion is commonly called the Kirkendall effect in metals.

molecular reference frame. Also, the coefficients are usually calculated from measurable concentration gradients rather than chemical potential gradients. However, diffusion coefficients in volume-fixed or molecular frames derived from concentration gradients are not, except fortuitously, subject to the Onsager relations. For ease of manipulation, it is convenient to transform the experimental coefficients to theoretical coefficients in a solvent-fixed frame. This involves not only transformation of the diffusion coefficients, but also the thermodynamic factors (Kirkwood *et al.*, 1960; Fitts, 1962),

$$\nabla \mu_i = \sum_{k=1}^{n-1} (\partial \mu_i / \partial c_k) \nabla c_k,$$
$$(i = 1, \ldots, n - 1). \quad (4.9)$$

required for the conversion of concentration gradients to chemical potential gradients.

Since Darken's paper in 1948, the nature of the diffusion coefficient calculated from Fick's law (3.1, 3.2), and its relationship to $\overline{D}$ (3.6) in binary systems has been the subject of numerous papers. Summaries are given by Crank (1957, pp. 219–229, 236–240), Trimble *et al.* (1965), and Wagner (1969). For binary diffusion, the mutual diffusion coefficient may be determined from a concentration versus distance curve for either component. For a solution with constant partial molal volumes, the integration of 3.1 or 3.2 yields a diffusion coefficient in a volume-fixed reference frame identical to $\overline{D}$ (Crank, 1957; Kirkwood *et al.*, 1960; Trimble *et al.*, 1965). Alternatively, if concentrations are expressed in mole fractions and fluxes in moles cm^{-2} sec^{-1}, and if the same restrictions apply to the partial molal volumes, the diffusion coefficient found on integration is consistent with a molecular (mean-molar, number-fixed) reference frame (Trimble *et al.*, 1965). Because the on-diagonal and off-diagonal coefficients in multicomponent solutions are calculated from penetration curves by a similar method, the thermodynamic conditions imposed on the integration of Fick's law give coefficients in either a volume-fixed or molecular reference frame. If the mean molar volumes are not a linear function of composition, the diffusion coefficient obtained cannot be identified theoretically with $\overline{D}$ (Crank, 1957; it is not clear which, if any, reference frame has been specified during integration). Practically, even for moderately large departures from linear behavior, the two coefficients may be numerically identical. This will be especially true in silicates if $\overline{D}$ is to be determined by graphical integration under penetration curves measured with an electron microprobe. Large errors, which may be reduced to some extent by using probability plots (Hall, 1953; Crank, 1957), are associated with this technique.

It is convenient to illustrate the following discussion by reference to the system Mg_2SiO_4-Fe_2SiO_4-Mn_2SiO_4. For the present, it is assumed that the system behaves as an ideal solution for all compositions. The crucial problem is the proper formulation of equations in terms of chemical potential gradients. This problem has been treated for aqueous electrolyte solutions by Kirkwood *et al.* (1960) for glasses by Cooper (1965) and in general terms by Sundheim (1957) and Schönert (1960).

Although neutral (molecular) components are more easily dealt with in equations containing chemical potentials, elemental (ionic) components have a simpler intuitive appeal and are less remote from measurable concentration gradients. Also, diffusion mechanisms are usually visualized as the migration of single ions relative to a lattice-fixed frame. The relationship of the solvent-fixed frame to the diffusion of individual cations and anions may be examined by constructing a special lattice-fixed frame for silicates. Advantage may be taken at this point of two conditions that appear to be approximately valid for natural exchange processes. Exchange commonly occurs between cations of the same charge and between phases with a fixed silicate or aluminosilicate framework that does not appear to participate in the process.

A lattice-fixed frame may be instituted by measuring fluxes by the number of atoms per unit area per second, distances in lattice spacings and concentrations in the number of atoms per unit cell. For silicates, this reference frame has one significant property: If a particular site is stoichiometrically invariant, there can be no concentration gradient or flux of the component occupying that site. For example, the tetrahedral site in pure olivines is filled solely by silicon in the stable configuration. If the total number of tetrahedral sites is conserved and there is no other ion that can substitute stably for silicon, there can be no macroscopic flux of silicon during diffusion. The controlling factor is stoichiometry; the argument still holds if two ions share the same structural site in a fixed ratio (e.g., ideally Al and Si in alkali feldspars). It should be emphasized that the flux of silicon disappears because a concentration gradient is absent. A tracer experiment may well record a finite diffusion coefficient at any given temperature, reflecting the mixing of isotopes. Also, a flux of silica will appear in other reference frames.[11]

The ionic components Fe^{2+}, Mg^{2+}, Mn^{2+}, Si^{4+} and O^{2-} are not independent; to preserve electrical neutrality they must be combined locally in stoichiometric proportions. This does not imply the compounds are perfect; if, for example, vacancies are present there must be an equal number of cation and anion vacancies. In natural or impure compounds, neutrality may be preserved in part by the creation or presence of ions of different charge (e.g., Fe^{3+} in olivine) in combination with vacancies. Following Schönert (1960), the stoichiometric equations may be applied to obtain neutral components (Mg_2SiO_4, etc.). For exchange between different phases (e.g., olivine-clinopyroxene) oxides are the only neutral components available and there is a corresponding increase in the number of diffusing components.[12]

The conditions imposed by the definition of the reference frame are

$$J_0^L = 0, \qquad (4.10)$$

$$c_0 = \text{constant}, \nabla c_0 = 0 \qquad (4.11)$$

and

$$\sum_{i=1}^{n} J_i^L = 0, \qquad (4.12)$$

$$\sum_{i=1}^{n} c_i = 0, \ \sum_{i=1}^{n} \nabla c_i = 0, \qquad (4.13)$$

where the subscripts 0, 1, 2 and 3 denote the species SiO_4, Mg, Fe, and Mn, respectively. In combining Si and O to define a fixed lattice, we have assumed that the oxygen site is also significant. Small deviations in stoichiometry will not significantly alter the conditions 4.10 to 4.13. By virtue of these conditions, the general flux equations

$$J_i = - \sum_{k=0}^{n} D_{ik} \nabla c_k,$$

$$(i = 0, 1, 2, 3) \qquad (4.14)$$

---

[11] The magnitude of this flux will vary. For the common metamorphic garnets the transformation from a lattice-fixed to a volume-fixed frame will result in a negligible flux of Al or Si (assuming a fixed Al/Si ratio in the tetrahedral site), since the mean molar volumes of common garnets vary little with composition. A much more significant change occurs in the system forsterite-fayalite. For many natural exchange reactions, the total range of compositions involved is relatively small, and the volume change may not be very large.

[12] Whereas Fe-Mg exchange between two olivines is binary diffusion, exchange between Fe-Mg olivine and Fe-Mg clinopyroxene is ternary diffusion. The presence of Ca in the clinopyroxene in finite amounts, even if it does not participate in the exchange, strictly creates a quarternary system with respect to diffusion. For small concentrations of CaO, however, the system may closely approximate ternary behavior.

reduce to, if $i = 1$ is chosen as the dependent component in 4.13,

$$J_i = - \sum_{k=2}^{n} \overline{D}_{ik}{}^L \nabla c_k,$$

$$(i = 2, 3) \quad (4.15)$$

with

$$\overline{D}_{ik}{}^L = D_{ik}{}^L - D_{i1}{}^L. \quad (4.16)$$

There is no reason to suppose symmetry relations among the coefficients $D_{ik}{}^L$ or $\overline{D}_{ik}{}^L$. The only possible further simplification of 4.15 is to assume, or determine experimentally, that the off-diagonal coefficients are much smaller than the on-diagonal coefficients and may be effectively ignored.

The flux equations may be written in terms of chemical potential gradients. In that case small deviations from exact stoichiometry may have a marked effect. The condition 4.10,

$$J_0{}^L = 0,$$

defines a solvent-fixed frame in which component $i = 0$ is the solvent; and symmetry relations may be deduced immediately from the derivations of Kirkwood et al. (1960). Using the additional condition (Cooper, 1965)

$$\nabla \mu_0 = 0 \quad (4.17)$$

and the Gibbs-Duhem relation, the flux equation

$$J_i{}^L = - \sum_{k=0}^{n} L_{ik}{}^L \nabla \mu_k,$$

$$(i = 0, 1, 2, 3), \quad (4.18)$$

transforms to

$$J_i{}^L = - \sum_{k=1}^{n} \overline{L}_{ik}{}^L \nabla \mu_k,$$

$$(i = 1, 2, 3), \quad (4.19)$$

with

$$\overline{L}_{ik}{}^L = L_{ik}{}^L - L_{i0}{}^L (c_k/c_0) \quad (4.20)$$

and

$$\overline{L}_{ik}{}^L = \overline{L}_{ki}{}^L. \quad (4.21)$$

Equation 4.19 may be further simplified through the use of 4.13, but unless great care is exercised, the symmetry relations may be lost. Even with the use of 4.13, however, the flux equations still contain two independent components. Although it is always possible to arrange for condition 4.13 to be true, it is not easy or often possible to assign chemical potentials to ionic components unambiguously. On the other hand, if we had chosen neutral components such as Mg_2SiO_4 and arbitrarily referenced diffusion to one of these components (as a "solvent"), we would have also obtained flux equations with two independent components. There is no particular difficulty attached to the definition of the chemical potential of molecular components. Molecular or oxide components may be simply correlated to the thermodynamic analysis of metasomatic systems offered by Thompson (1959) to define local equilibrium conditions.

It is possible to connect the flux equations in the lattice-fixed frame to the mobilities of individual ions and gradients in molecular chemical potentials by a method parallel to that of Cooper (1965). However, because of the concentration units in 4.15 for the lattice-fixed frame, the resultant equations are not very practical. And, as noted previously, in view of the questions raised by Manning about the assumptions made by Bardeen and Herring, the general validity of this method for crystalline silicates is open to some doubt in the absence of experimentation. For a volume-fixed frame, the treatment of Cooper (1965) may be adapted immediately by assuming that the volume of a phase is determined largely by the packing of SiO_4 (and/or AlO_4) tetrahedra. This assumption (or the same assumption made in terms of oxygen packing) appears to have a more restricted range of validity in crystalline silicates than in silicate glasses.

Kinetic Considerations

Lane and Kirkaldy (1964) attempted to estimate the relative magnitude of on-diagonal and off-diagonal coefficients in a general substitutional solution through the application of transition-state theory. A complete quantitative analysis demands statistical mechanics either to calculate the magnitude of some of the terms that appear in the kinetic equations or to relate these terms to measurable thermodynamic quantities. But the origins of the off-diagonal coefficients and the circumstances in which their magnitudes are diminished may be investigated qualitatively without detailed calculations. It is worthwhile, even in a preliminary fashion, to examine the implications of kinetic models for diffusion in silicates.

The kinetic calculations are made with respect to a molecular reference frame. Relative to a fixed, external coordinate system, the molecular frame moves with the average particle velocity (Haase, 1969). To illustrate the kinetic approach we shall again refer to diffusion in olivines. The discussion of diffusion mechanisms in olivines is not, and is not intended to be, either comprehensive or complete.

Oxygen atoms in olivines are arranged approximately in a hexagonal, close-packed pattern. Following Azároff (1961a, b), the structure may be visualized as occupied and unoccupied octahedra and tetrahedra. The occupied polyhedra correspond to the familiar oxygen-coordination polyhedra; the unoccupied polyhedra have similar dimensions but contain no central cation in the stable configuration. The unoccupied cation site is not a vacancy in the normal sense. In olivine, each occupied octahedron normally contains Fe, Mg, or Mn and shares two faces with unoccupied octahedra. The remaining faces are shared with unoccupied tetrahedra. Occupied octahedra are joined only along edges. The occupied tetrahedra normally contain silicon ions and one face is shared with an unoccupied tetrahedron. The remaining faces are shared with unoccupied octahedra.

Consider now diffusion of Mg parallel to the a axis. There are chains in this direction composed successively of occupied and unoccupied octahedra (Fig. 4). As suggested by Azároff, we shall assume that diffusion paths through shared polyhedral faces are energetically more favorable than paths that pass through shared edges.[13] Direct interchange of Mg (or Mg-Fe) atoms in olivine always involves an edge path between occupied octahedra. The intermediate step corre-

---

[13] A note of caution should be injected. For example, the experimental work of Austerman and Wagner (1966) on diffusion of Be in BeO strongly implies that the more immediately obvious paths are not utilized. Their results do not necessarily invalidate Azároff's approach but rather emphasize the difficulty of intuitively evaluating all factors in a qualitative analysis of this kind.

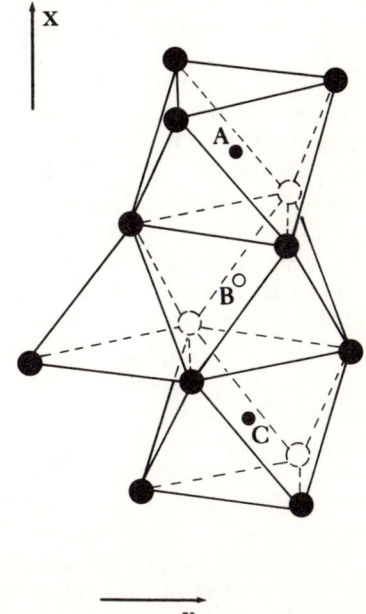

Fig. 4. Part of a chain of occupied and unoccupied octahedra parallel to the x-crystallographic axis of olivine. The solid circles (A, C) represent the centers of occupied octahedra, and the open circle (B) marks the center of an unoccupied octahedron.

sponds to the formation of an activated complex that brings two cations close together and displaces a significant number of oxygen atoms from their stable configuration. The activation energy required appears to be very large relative to some other diffusion paths.

Let us assume for the moment that the transfer of an atom from A to C (Fig. 4) begins with a jump, through the shared octahedral face, from A to B. A number of paths are available to the atom at B (if the most probable event, the immediate jump of the atom from B to A, does not occur):

1. The ion may exchange directly with the ion at C. Although the interchange takes place through a shared octahedral face, the activation energy for this process must still be relatively large (the activated complex is composed of two cations in close proximity).

2. If site C is unoccupied, the ion may pass directly from B to C. A true vacancy may exist at C, if the number of octahedral ions is less than the total number of octahedral sites that might be stably occupied. Alternatively the site at C may be temporarily vacant because the atom from that site has jumped to an adjacent polyhedron. No diffusion can result from this process unless a complex, and thus improbable, ring mechanism returns the atom formerly at C to A. The existence of vacancies is well documented in metals and simple oxides; similar data are lacking for silicates, but to the extent that silicates resemble simple oxides, their presence may be suspected.

Other paths may be constructed between A and C. It is not intended here to enumerate or debate the nature of these various paths. The important point is that with detailed calculations, a definite probability could be assigned to each path, derived from the individual probabilities for each step in the path. For path 2, assuming a true vacancy is located at C, there are two steps. The probability that an ion will jump from A to B in any instant includes these factors (Shewmon, 1963): the number of unoccupied nearest-neighbor sites; the probability that any of these neighboring sites is vacant and not filled momentarily by another migrating ion; the probability that the ion will jump into the site. The same factors apply to the jump from B to C, although the probability of a vacancy at C is much less than the probability that B is vacant. The probability that an atom will jump is also influenced by the immediate environment and is thus dependent on composition.

It seems likely that among the numerous possible paths only a few will be reasonably probable and, as in metals, one path may emerge as the most probable. Verification of the probable path (or paths) must await comparison between calculation and tracer diffusion data (especially the anisotropy of the data, e.g., Austerman and Wagner, 1966). Independent proof of the existence and nature of defect structures, in particular cation and anion vacancies, would be invaluable.

Lane and Kirkaldy (1964) consider the relative magnitude of the off-diagonal coefficients from a kinetic point of view for diffusion by two different mechanisms. The first pertains to diffusion by direct interchange, with the formation of an activated complex composed of species i and k (i.e., Fe-Mn, Fe-Mg and Mg-Mn complexes). According to Lane and Kirkaldy, the off-diagonal coefficients will disappear in a solution in which

$$\Delta G^{\dagger}_{12} = \Delta G^{\dagger}_{13} = \Delta G^{\dagger}_{23} \quad (5.1)$$

and

$$\nu_{12} = \nu_{13} = \nu_{23} \quad (5.2)$$

Here $\Delta G_{ik}^{\dagger}$ is the Gibbs free energy associated with the formation of an ik activated complex and ν_{ik} is the rate of forward transition through the activated state. When these conditions are met, then (in a molecular reference frame)

$$D_{11}{}^m = D_{22}{}^m \quad (5.3)$$

and

$$D_{12}{}^m = D_{21}{}^m = 0 \quad (5.4)$$

Diffusion by a vacancy mechanism is also considered. For this it is necessary to assume that direct interchanges of atoms are very infrequent and that the vacancies are everywhere in local equilibrium. In this situation the off-diagonal coefficients (D_{ik}) will disappear if

$$P_1 = P_2 = P_3 \qquad (5.5)$$

where P_i is the jump probability for atom i.

For vacancy diffusion, the jump probabilities are given by

$$P_i = \nu_{iv} \exp(-\Delta G^\dagger_{iv}/kt)\, \gamma_i/\gamma^\dagger_{iv} \qquad (5.6)$$

The subscript v in 5.6 denotes a vacancy and γ_i and $\gamma^\dagger_{iv}$ are respectively the activity coefficient of i and an iv activated complex. Approximate magnitudes of the jump probabilities may be estimated from tracer and binary diffusion data (Lane and Kirkaldy, 1964). The utilization of binary diffusion data has a special relevance to the treatment of geologic problems; many natural silicates form solutions with compositions that fall close to binary joins in the multicomponent system. For example, in the system Mg_2SiO_4-Fe_2SiO_4-Mn_2SiO_4, naturally occurring olivines are restricted to compositions close to the forsterite-fayalite and fayalite-tephroite joins. Similarly, although in a different form, coexisting clinopyroxenes and orthopyroxenes in metamorphic rocks tend to be Fe-Mg solutions at a nearly constant Ca content. In this case, Ca (CaO, $CaSiO_3$) may be treated in the flux equations, to a first approximation, as an invariant component. The same kinds of arguments may be extended to many natural silicates.

By neglecting the compositional dependence of the P_i (or equivalently the D_i^{*m}), we may relate the jump probabilities to self-diffusion coefficients in the molecular frame (D_i^m) by (Lane and Kirkaldy, 1964),

$$\lim_{x_i \to 0} P_i = \lim_{x_i \to 0} D_i^{*m}/N\lambda^2 \qquad (5.7)$$

where N is the total number of moles per unit volume of atoms on a particular lattice plane and λ is a lattice spacing. The use of limits in 5.7 indicates values taken at infinite dilution (a more exact estimate may be gained from Equation 59 in Lane and Kirkaldy's paper). Theoretically, it is probably not realistic to neglect the compositional dependence of the diffusion coefficients in silicates. However, the effect may be mitigated to some extent in applying the results to natural systems by the presence of small gradients. In pelitic rocks, the extent of substitution of Mg or Mn for Fe in minerals such as garnet and staurolite is limited by low initial concentrations of these elements. The concentration dependence of the diffusion coefficients may be markedly diminished within these narrower limits.

At 1000°K, Buckley (1973) has proposed the following limiting values of D_i^{*m} for diffusion in olivines:

$$D_{Fe}^{*m} = 10^{-11}\ (cm^2/sec)$$
$$D_{Mn}^{*m} = 10^{-14}\ (cm^2/sec) \qquad (5.8)$$
$$D_{Mg}^{*m} = 10^{-18}\ (cm^2/sec)$$

These values may be substituted in the equations (Lane and Kirkaldy, 1964)

$$L_{ik} = -x_1 x_2 \{P_1 - P_2 - \Sigma(x_i P_i)\} N\lambda^2/RT \qquad (5.9)$$

and

$$L_{ii} = \{x_i P_i (1 - 2x_i) + x_i^2 \sum_i (x_i P_i)\} N\lambda^2/RT \qquad (5.10)$$

to obtain the ratio

$$L_{ii}/L_{ik}. \qquad (5.11)$$

It must be emphasized that the use of 5.9 and 5.10 is only justified in certain circumstances. There is no condition in these equations that provides for the maintenance of electrical neutrality during diffusion; more general equations containing this restriction have been formulated by Lane and Kirkaldy (1965, 1966). In applying the unrestricted equations here, we envisage diffusion of Fe, Mn, and Mg through a framework of

SiO$_4$ tetrahedra that is in the limit essentially immobile. This is tantamount to disregarding any interaction between the mobile cations (Fe, Mn, and Mg) and the framework except to balance charges. The fact that the mobile cations and Si occupy distinctly different sites, combined with other arguments advanced previously in Section 4, provide considerable justification for this approach. As the diffusing cations have the same charge, the condition that determines electrical neutrality reduces in this case to the same condition that defines the reference frame.

The ratios L_{ii}/L_{ik} for Fe-Mg-Mn olivines, derived from 5.7, 5.8, 5.9, and 5.10, are given in Fig. 5. The distribution of contours in Fig. 5 is determined mainly by the differences in the D_i^{*m}. For Fe, Mg, and Mn diffusion in olivines the differences are large and a moderate error in the estimated values will not displace the contours very much. Similarly, the compositional dependence of the diffusion coefficients, unless very large, will not markedly affect the diagrams. The analysis of Buckley (1973) suggests that garnet will behave much like olivine with respect to multicomponent diffusion, and Fig. 5 is a fair representation of the distribution of contours for garnet. Moreover, at any temperature, D_{Ca}^{*m} is about an order of magnitude greater than D_{Mg}^{*m} but is still significantly less than the diffusion coefficients of Fe or Mn. Only minimal changes will occur if Ca is put in the place of Mg in Fig. 5.

To the extent that they are important in silicates, vacancy-wind and correlated-jump effects will modify the results in Fig. 5 (Manning, 1968).

As an illustration of the use of the calculations, we may consider diffusion in

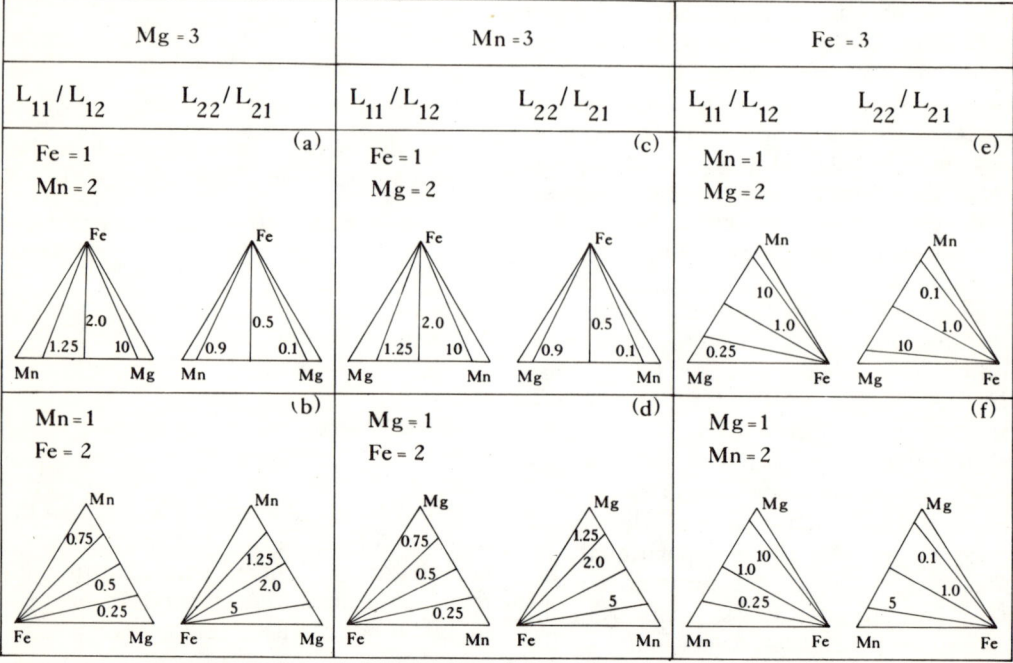

Fig. 5. Calculated ratios L_{ii}/L_{ik} for diffusion in olivines assuming tracer diffusion coefficients $D_{Fe}^* = 10^{-11}$, $D_{Mn}^* = 10^{-14}$ and $D_{Mg}^* = 10^{-18}$ cm^2/sec. Results for all possible choices of dependent components (3) and independent components (1, 2) are shown, although some of the diagrams are equivalent. Concentrations are expressed as mole fractions, and the contours are based on solutions of 5.9 and 5.10 at 0.05 intervals. Note that diagrams for L_{ii}/L_{ik} when $D_1^* = D_2^* = D_3^*$ are identical to those of (a). The contours do not apply along the binary joins.

olivines with compositions near the forsterite-fayalite join. The flux equations are

$$J_1 = -L_{11} \nabla \mu_1 - L_{12} \nabla \mu_2$$

and

$$J_2 = -L_{21} \nabla \mu_1 - L_{22} \nabla \mu_2,$$

when component 3 is the dependent component. The manner in which Fe, Mg, and Mn are designated components 1, 2, and 3 determines the relative magnitudes of the off-diagonal coefficients. Some of the possibilities are:

1. Fe = 1, Mn = 2, and Mg = 3. The ratios L_{11}/L_{12} are greater than 10 to 1 except for very fayalitic solutions (Fig. 5a) and increase with decreasing Mn content. As $\nabla \mu_2$ is also small, the off-diagonal term may be ignored in the flux equation for Fe. On the other hand, the ratios L_{22}/L_{21} are small and the coefficients L_{21} are correspondingly large. Unless $\nabla \mu_{Fe}$ is small, the off-diagonal term ($L_{MnFe} \nabla \mu_{Fe}$) must be retained in the flux equation for Mn. Indeed, the calculations suggest that the diffusion of Mn is largely dependent on the gradient of Fe; significant diffusion of Mn may occur even though $\nabla \mu_{Mn}$ is small or nonexistent.

2. Mn = 1, Fe = 2, and Mg = 3. The results are reversed. L_{12} are large and L_{21} are relatively small (Fig. 5b). Again a large off-diagonal term appears in the flux equation for Mn.

3. Fe = 1, Mg = 2, and Mn = 3. The off-diagonal coefficients are the same size as the on-diagonal coefficients in both flux equations (Fig. 5c). Unless one of the independent gradients is small, all four terms must be retained in the flux equations. This remains true for Mg = 1, Fe = 2, and Mn = 3.

It is evident from Fig. 5 that it is impossible to minimize both off-diagonal coefficients simultaneously.

The flux equations may also be written in terms of concentration (mole fractions) gradients:

$$J_1 = -D_{11} \nabla x_1 - D_{12} \nabla x_2$$

and

$$J_2 = -D_{21} \nabla x_1 - D_{22} \nabla x_2.$$

Conversion of the chemical potential gradients to concentration gradients may be accomplished for an ideal solution by the relations (Kirkwood et al., 1960):

$$\nabla \mu_i = \sum_{k=1}^{2} (\partial \mu_i / \partial x_k) \nabla x_k$$

and

$$\partial \mu_i = RT \, \partial \ln x_i \ (i = 1, \ldots, n-1).$$

For Fe = 1, Mn = 2, Mg = 3, the ratios D_{11}/D_{12} are very large for all ternary compositions (>100:1). Conversely, the ratios D_{22}/D_{21} are very small ($\simeq 0.02$) for all compositions. The magnitudes of the ratios are reversed for Mn = 1 and Fe = 2. Again it is not possible to minimize both off-diagonal terms simultaneously. The results for other combinations of components 1, 2, and 3 may be summarized briefly as follows:

1. Mn = 1, Mg = 2, and Fe = 3 or Mg = 1, Mn = 2, and Fe = 3. The ratios D_{ii}/D_{ik} are all approximately equal to one for all compositions.

2. Fe = 1, Mg = 2, Mn = 3. The D_{11}/D_{12} ratios are very large (>100:1) and the ratios D_{22}/D_{21} very small for all compositions. The results are reversed for Mg = 1 and Fe = 2.

The preceding discussion assumes that diffusion occurs via a vacancy mechanism. The ratios L_{ii}/L_{ik} for direct interchange of atoms are zero only when conditions 5.1 and 5.2 are true. Calculations by Buckley (1973) suggest that the factors ν_{ik} have approximately the same magnitudes for diffusion of Fe, Mn, Mg and Ca in silicates. The sizes of the factors $\Delta G_{ik}^\dagger$ are dependent on the particular structures considered, but it is unlikely that they will ever be all approximately equal. Although the calculations have not been attempted, it appears that the ratios L_{ii}/L_{ik} for diffusion by direct interchange in olivines will have a form very similar to those illustrated in Fig. 5 for a vacancy mechanism. If this is true, then kinetic calculations of the off-diagonal coefficients will not be of much assistance in distinguishing diffusion mechanisms. This appears to be true of diffusion in aqueous electrolyte solutions (Lane and Kirkaldy, 1965).

Conclusions

When the nonequilibrium properties of natural systems are to be used to estimate the average values of time-dependent parameters such as T, P, or f_{O_2}, order of magnitude estimates of diffusion rates will often be sufficient; errors arising from an improper choice of reference frames and the resulting flux equations may usually be neglected. However, when details of the time dependences of these parameters are desired, it is necessary to analyze observable properties with much greater care. We have proposed a general analytical approach that may ultimately be used to place constraints on the thermal histories of metamorphic rocks. A correct interpretation of such histories relies to a great extent on the correct formulation of the appropriate boundary conditions and differential equations.

We have attempted to show that rigorous formalisms exist for the proper selection of flux equations and diffusion coefficients necessary to describe exchange diffusion. It should be evident, however, that extrapolation of such data from experimental to metamorphic conditions is at best difficult. Models relating tracer diffusion and exchange diffusion presently offer the best chance for success in this area, although they are, as yet, untested in silicates. Even though many natural systems may be reduced to dilute ternary systems with respect to diffusion, kinetic calculations suggest that off-diagonal coefficients may be important in the flux of some components. At least the calculations provide a useful guide for choosing dependent components and formulating the flux equations in their simplest form. Obviously, experimental corroboration of Kirkaldy's or Cooper's model in a silicate system would be most useful.

Acknowledgments

This paper includes work supported by grants from the National Science Foundation (GA 1676) and the Research Board of the University of Illinois.

References Cited

Albarede, F., and Y. Bottinga, Kinetic disequilibrium in trace element partitioning between phenocrysts and host lava, *Geochim. Cosmochim. Acta, 36,* 141–156, 1972.

Anderson, D. E., and G. R. Buckley, Zoning in garnets—diffusion models, *Contrib. Mineral. Petrol., 40,* 87, 1973.

Austerman, S. B., and J. W. Wagner, Cation diffusion in single-crystal and polycrystalline BeO, *J. Amer. Ceram. Soc., 49(2),* 94–99, 1966.

Azároff, L. V., Role of crystal structure in diffusion. I. Diffusion paths in closest-packed crystals, *J. Appl. Phys., 32,* 1658–1662, 1961a.

Azároff, L. V., Role of crystal structure in diffusion. II. Activation energies for diffusion in closest-packed structures, *J. Appl. Phys., 32,* 1663–1667, 1961b.

Bailey, A., Comparison of low-temperature with high-temperature diffusion of sodium in albite, *Geochim. Cosmochim. Acta, 35,* 1073–1081, 1971.

Bardeen, J., and C. Herring, Diffusion in alloys and the Kirkendall effect, in *Atom Movements,* ASM Cleveland, 87–111, 1951.

Blackburn, W. H., The spatial extent of chemical equilibrium in some high grade metamorphic rocks from the Grenville of southeastern Ontario, *Contrib. Mineral. Petrol., 19,* 72–92, 1968.

Buckley, G. R., Application of experimental oxide diffusion data to geologic problems, submitted to *Amer. Mineral.,* 1973.

Chernov, A. A., Effects of trace components on the growth rates of a crystal, *Growth Cryst. (USSR), 3,* 31–34, 1968.

Cooper, A. R., Jr., Model for multicomponent diffusion, *Phys. Chem. Glasses, 6,* 55–61, 1965.

Cooper, A. R., and J. H. Heasley, Extension of Darken's equation to binary diffusion in ceramics, *J. Amer. Ceram. Soc., 49,* 280–283, 1966.

Crank, J., *The Mathematics of Diffusion,* Oxford University Press, London, 347 pp., 1957.

Dahl, O., Hydrothermal studies of garnet-mica equilibria in the system $3(FeO,MnO)\text{-}2Al_2O_3\text{-}$

12SiO$_2$-K$_2$O-H$_2$O, *Geol. Fören. Stockholm Förh., 90,* 331–348, 1971.

Danckwerts, P. V., Unsteady-state diffusion or heat conduction with moving boundary, *Trans. Faraday Soc., 46,* 701–712, 1950.

Darken, D. S., Diffusion, mobility and their interrelation through free energy in binary metallic systems, *Trans. AIME, 175,* 189–201, 1948.

de Groot, S. R., and P. Mazur, *Non-equilibrium Thermodynamics,* North-Holland Pub. Co., Amsterdam, 510 pp., 1962.

Fine, M. E., *Introduction to Phase Transformations in Condensed Systems,* Macmillan, New York, p. 133, 1965.

Fitts, P. D., *Nonequilibrium Thermodynamics,* McGraw-Hill, New York, 173 pp., 1962.

Haase, R., *Thermodynamics of Irreversible Processes,* Addison Wesley, Reading, Mass., 509 pp., 1969.

Hall, L. D., An analytical method of calculating variable diffusion coefficients, *J. Chem. Phys., 21,* 87–89, 1953.

Hartley, G. S., and J. Crank, Some fundamental definitions and concepts in diffusion processes, *Trans. Faraday Soc., 45,* 801–818, 1949.

Hess, P. C., Prograde and retrograde equilibria in garnet-cordierite gneisses in south-central Massachusetts, *Contrib. Mineral. Petrol., 30,* 177–195, 1971.

Hooyman, G. J., and S. R. de Groot, Phenomenological equations and Onsager relations, *Physica, 21,* 73–76, 1955.

Hooyman, G. J., S. R. de Groot, and P. Mazur, Transformation properties of the Onsager relations, *Physica, 21,* 360–366, 1955.

Hsu, L. C., Selected phase relationships in the system Al-Mn-Fe-Si-O-H: A model for garnet equilibria, *J. Petrology, 9,* 40–83, 1968.

Jones, K. A., and K. A. Galwey, Size distribution, composition and growth kinetics of garnet crystals in some metamorphic rocks from the West of Ireland, *Quart. J. Geol. Soc. London, 122,* 29–44, 1966.

Kirkaldy, J. S., Diffusion in multicomponent metallic systems, *Can. J. Phys., 35,* 435–440, 1957.

Kirkaldy, J. S., Diffusion in multicomponent metallic systems. I. Phenomenological theory for substitutional solid solution alloys, *Can. J. Phys., 36,* 899–906, 1958.

Kirkaldy, J. S., R. J. Brigham, and D. H. Weichert, Diffusion interactions in Cu-Zn-Sn as determined from infinite and finite couples, *Acta Met., 13,* 907–915, 1965.

Kirkaldy, J. S., and J. E. Lane, Diffusion in multicomponent metallic systems. IX. Intrinsic diffusion behaviour and the Kirkendall effect in ternary substitutional solutions, *Can. J. Phys., 44,* 2059–2072, 1966.

Kirkaldy, J. S., J. E. Lane, and G. R. Mason, Diffusion in multicomponent metallic systems. VII. Solutions of the multicomponent diffusion equations with variable coefficients, *Can. J. Phys., 41,* 2174–2186, 1963.

Kirkaldy, J. S., and G. R. Purdy, Diffusion in multicomponent metallic systems. V. Interstitial diffusion in dilute ternary austenites, *Can. J. Phys., 40,* 208–217, 1962.

Kirkwood, John G., R. L. Baldwin, P. J. Dunlop, L. J. Gosting, and G. Kegeles, Flow equations and frames of reference for isothermal diffusion in liquids, *J. Chem. Phys., 33,* 1505–1513, 1960.

Kretz, R., Grain-size distribution for certain metamorphic minerals in relation to nucleation and growth, *J. Geol., 74,* 147–173, 1966.

Kwak, T. A. P., An attempt to correlate nonpredicted variations of distribution coefficients with mineral grain internal inhomogeneity using a field example studied near Sudbury, Ontario, *Contrib. Mineral. Petrol., 26,* 199–224, 1970.

Lane, J. E., and J. S. Kirkaldy, Diffusion in mulicomponent systems. VIII. A kinetic calculation of the Onsager L coefficients in substitutional solid solutions, *Can. J. Phys., 42,* 1643–1657, 1964.

Lane, J. E., and J. S. Kirkaldy, A quasi-crystalline model of diffusion in ternary liquid systems, *Can. J. Chem., 43,* 1812–1828, 1965.

Lane, J. E., and J. S. Kirkaldy, Diffusion in multicomponent aqueous systems, *Can. J. Chem., 44,* 477–485, 1966.

Lin, T. H., and R. A. Yund, Potassium and sodium self-diffusion in alkali feldspar, *Contrib. Mineral. Petrol., 34,* 177–184, 1972.

Manning, J. R., Tracer diffusion in a chemical concentration gradient in silver-cadmium, *Phys. Rev., 116*(1), 69–79, 1959.

Manning, J. R., Diffusion in a chemical concentration gradient, *Phys. Rev. 124*(2), 470–482, 1961.

Manning, J. R., Correlated walk and diffusion equations in a driving force, *Phys. Rev., 139* (1A), A126–A135, 1965.

Manning, J. R., Diffusion and the Kirkendall shift in binary alloys, *Acta Met., 15,* 817–825, 1967.

Manning, J. R., Vacancy-wind effect in diffusion and deviation from thermodynamic equilibrium conditions, *Can. J. Phys., 46*(23), 2633–2643, 1968.

Misener, D. J., Interdiffusion studies in the system Fe$_2$SiO$_4$-Mg$_2$SiO$_4$, *Carnegie Inst. Washington Yearb, 71,* 516–520, 1972.

Mueller, G., and A. Schneider, Chemistry and genesis of garnets in metamorphic rocks, *Contrib. Mineral. Petrol., 31,* 178–200, 1971.

Nye, J. F., *Physical Properties of Crystals,* Oxford University Press, London, p. 322, 1957.

Onsager, L., Reciprocal relations in irreversible processes. I, *Phys. Rev., 37,* 405, 1931a.

Onsager, L., Reciprocal relations in irreversible processes. II, *Phys. Rev., 38,* 2265–2278, 1931b.

Onsager, L., Theories and problems of liquid diffusion, *Ann. N.Y. Acad. Sci., 46,* 241–265, 1945.

Ramberg, H., Energy transfer from differentiation in a differential pressure system under non-equilibrium conditions: A discussion of "Partial Quantities," *J. Geol., 68,* 110–113, 1960.

Schönert, H., Diffusion and sedimentation of electrolytes and nonelectrolytes in multicomponent systems, *J. Phys. Chem., 64,* 733–737, 1960.

Shewmon, P. G., *Diffusion in Solids,* McGraw-Hill, New York, 203 pp., 1963.

Shewmon, P. G., *Transformations in Metals,* McGraw-Hill, New York, p. 394, 1969.

Sundheim, B. R., Transport processes in multicomponent liquids, *J. Chem. Phys., 27*(3), 791–795, 1957.

Thompson, J. B., Local equilibrium in metasomatic processes, in *Researches in Geochemistry,* Vol. 1, P. H. Abelson, ed., Wiley, New York, pp. 427–457, 1959.

Trimble, L. E., D. Finn, and A. Cos Garea, Jr., A mathematical analysis of diffusion coefficients in binary systems, *Acta Met., 13,* 501–507, 1965.

Wagner, C., The evaluation of data obtained with diffusion couples of binary single-phase and multiphase systems, *Acta Met., 17,* 99–107, 1969.

Yourgrau, W., A. van der Merwe, and G. Raw, *Treatise on Irreversible and Statistical Thermophysics,* Macmillan, New York, 268 pp., 1966.

Ziebold, T. O., and A. R. Cooper, Jr., Atomic mobilities and multicomponent diffusion, *Acta Met., 13,* 465–470, 1965.

Ziebold, T. O., and R. E. Ogilvie, Ternary diffusion in copper-silver-gold alloys, *Trans. AIME, 239,* 942–953, 1967.

DIFFUSION IN SULFIDES

C. Ernest Birchenall
Professor of Metallurgy
Department of Chemical Engineering
University of Delaware
Newark, Delaware 19711

Introduction

The sulfides are widely distributed in nature and are used in many different ways in commercial applications. There are correspondingly diverse reasons for trying to understand their properties. Ferrous sulfide is a good example of a substance for which effort from several disciplines yielded a wealth of related information. Geologists investigated the phase boundaries intensively when there was hope that the pyrite-pyrrhotite equilibrium, preserved by natural quenching, might serve as a geological barometer and thermometer. Solid-state physicists, especially in Japan, studied the unusual electronic and magnetic properties of ferrous sulfide. Crystallographers found the low-temperature transitions in electron-spin orientations and point-defect structures worth considerable effort. Metallurgists have been concerned for decades with the unfortunate effects of ferrous sulfide inclusions on the ductility of steels. The fast growth of the sulfide on iron in sulfur-containing environments at elevated temperatures, a problem of some magnitude in the petroleum industry, kindled my interest in this substance.

In place of a detailed review of the work of my students on ferrous sulfide, I shall survey more generally diffusion in sulfides, treating ferrous sulfide as one of several examples. The reader is also referred to the recent review of diffusion in the chalcogenides of Zn, Cd, and Pb by Stevenson (1973).

Diffusion in Sulfides Compared with Diffusion in Oxides

Although diffusion in oxides is still poorly understood, much more is known about oxides than sulfides. A general comparison has some value mainly because several aspects of the influence of defect structure on diffusion seem to be defined more clearly in the sulfides than in the oxides.

One difference affecting ion mobilities in these compounds is that the sulfide ion is much larger than the oxide ion. Moreover, the larger sulfide ion has a polarizability more than three times as large as the oxide ion. Therefore, the smaller cations occupy larger and softer interstices in the sulfides. Identifying the interstices as octahedral and tetrahedral sites if bounded by six or four anions, respectively (regardless of small distortions from regular packing), the constrictions in the channels connecting these sites also are wider and softer in the sulfides. The strains necessary to squeeze cations through these barriers into empty sites determine the activation energies for diffusion. Hence, lower activation energies can be expected for sulfides than for the corresponding oxides. Table 1 lists

TABLE 1. Ionic radii in angstrom units for selected ions

	Fe^{2+} 0.74	Fe^{3+} 0.64	O^{2-} 1.40	F^- 1.36
Cu^+ 0.96	Cu^{2+} 0.72		S^{2-} 1.84	Cl^- 1.81
Zn^+ 0.88	Zn^{2+} 0.74		Se^{2-} 1.98	Br^- 1.95
Ag^+ 1.26	Ag^{2+} 0.89		Te^{2-} 2.21	I^- 2.16
Cd^+ 1.14	Cd^{2+} 0.97			
Au^+ 1.37		Au^{3+} 0.85		

ionic radii for a few anions and cations considered in this account plus a few extras for comparison.

Diffusion in crystals depends on the motion of point defects. The sulfides show two related types of disordering more clearly than the oxides. Bertaut (1956) determined the structure of the vacancy superlattice in the monoclinic phase that has a composition near Fe_7S_8. At higher temperatures the vacant cation sites are more randomly distributed, but a strong short-range interaction or clustering must persist to account for the lower mobility of these vacancies compared with the mobility of the vacancies near the FeS composition. A similar dependence of vacancy mobility is found in highly nonstoichiometric ferrous oxide where x-ray measurements have demonstrated the presence of vacancy clusters (Roth, 1960, and Koch and Cohen, 1969).

SILVER AND COPPER SULFIDES

The crystal structure usually defines either the octahedral (e.g., NaCl structure) or tetrahedral sites (e.g., zincblende structure) as the structure sites. The other sites then become interstitial sites through which the ions may jump, or sometimes must jump, in the course of a unit diffusion process. If the ions enter the interstitial sites and jump from one such site to another without returning to structure sites often, the mechanism is called interstitial. Cation diffusion by interstitial mechanisms probably is more frequent in sulfides than in oxides because of the size effect. If the interstitial ion displaces an ion from a structure site into a neighboring interstitial site, the mechanism is interstitialcy. If the motion largely consists of an ion jumping from a structure site into a neighboring vacant structure site, perhaps passing quickly through an interstitial site, it is a vacancy mechanism. However, αAg_2S, the cubic high-temperature form, has silver ions distributed almost at random over the octahedral and interstitial interstices (Rahlfs, 1953). In such a case there is no distinction between interstitial and vacancy mechanisms. The jump from one site to another requires a very small activation energy. In a sense, the cation part of the structure has melted while the anion part remains nearly intact.

Figure 1 shows the cation diffusivities for silver tracers in αAg_2S (Allen and Moore, 1959 and Okazaki, 1967). The activation energy reported by Allen and Moore is only 3.45 kcal per gram mole, somewhat lower than the activation energy required for the silver ion to diffuse in aqueous solution at room temperature. Below 177°C, the monoclinic βAg_2S has lower diffusivities and an activation energy of about 10.5 kcal per gram mole. The mechanism appears to be predominantly vacancies on silver ion sites.

This sequence of structures, disordered cations at high temperature, cations on characteristic sites with vacancies nearly randomly distributed at intermediate temperatures, and ordered cation sites at low temperatures, as in Fe_7S_8, seems to be a consistent pattern. Although the sequence probably represents a universal tendency, it is obscured by phase transformations, including melting, that intrude at different levels of cation disorder in the different sulfides and oxides.

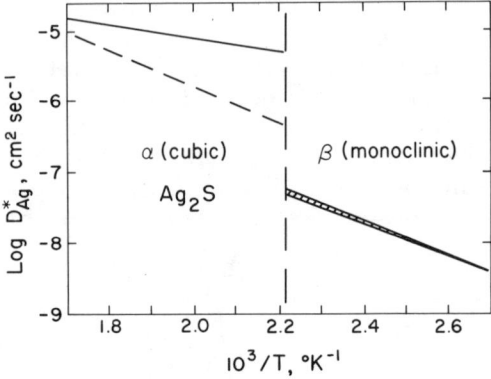

Fig. 1. Self-diffusion of silver in cubic (α) and monoclinic (β) Ag_2S. Solid lines, Allen and Moore (1959). Dashed line, Okazaki (1967).

At low temperatures the phase equilibria in the cuprous sulfides are complex (Mathieu and Rickert, 1972). Diffusion data are sparse and obtained mainly from sulfide growth measurements. It is interesting that the results of Etienne (1970) for low chalcocite and low digenite between 30° and 73°C differ from each other by only about a factor of two and lie about one order of magnitude below the extrapolation of the βAg_2S line. The results of Bartkowicz et al. (1969) for 350° to 550°C lie less than an order of magnitude above the extrapolation of the αAg_2S line. This striking similarity in diffusion behavior must reflect similar defect structures. The copper diffusion results are summarized in Table 2.

Sulfide ion diffusivities in silver sulfide have been shown to be much slower and to require much higher activation energies than cation diffusion. Ishiguro et al. (1953) reported [1]

$$D_S = 2.4 \times 10^{-4} \exp(-24.0/RT) \text{ cm}^2/\text{sec}$$

for the α (cubic form, while Peschanski (1950) reported

$$D_S = 0.24 \exp(-26.3/RT) \text{ cm}^2/\text{sec}$$

for the β (monoclinic) form of Ag_2S.

[1] All activation energies in the exponential factors are in kcal per gm mole.

Cadmium Sulfide

CdS has useful photoconductive properties. In combination with Cu_2S it also may form useful photovoltaic devices. Pure CdS has n-type electronic behavior because of interstitial cadmium ions. The cadmium self-diffusivity has been measured several times, most recently by Jones (1972). He observed qualitatively the same features that others found: (1) Self-diffusivity of cadmium increases with cadmium pressure or decreases with increasing sulfur pressure, consistent with a cadmium interstitial defect model. (2) Cadmium self-diffusivity can be increased by doping with acceptor impurities. (3) Although the sulfide is hexagonal (wurtzite structure) there is little or no variation in diffusivity parallel versus perpendicular to the c axis. (4) A small proportion of fast-diffusing tracer ions runs well ahead of the remainder. The slower tracer ions are presumed to exchange with ions from structure sites to become temporarily immobilized. Excluding the small, fast-diffusing component, Jones found between 602° and 1255°C in excess cadimum vapor

$$D^*_{Cd} = 1.2 \exp(-53 \pm 5/RT) \text{ cm}^2/\text{sec}$$

Figure 2 summarizes the published self-diffusion data for cadmium and sulfur in cadmium sulfide that is not doped

TABLE 2. Diffusion coefficients for copper in cuprous sulfides

T °C	D cm²/sec		Reference
	low temperature chalcocite	low temperature digenite	
30	3.1×10^{-11}	1.5×10^{-11}	
73	2.8×10^{-10}	1.2×10^{-10}	Etienne
	$D_0 = 8.1 \times 10^{-3}$	3.6×10^{-2}	(1970)
	$Q = 5.87*$	$6.10*$	
	high-temperature chalcocite		
350	8.7×10^{-5}		
450	2.7×10^{-4}		Bartkowicz et al.
550	4.0×10^{-4}		(1969)
	$D_0 = 3.3 \times 10^{-2}$		
	$Q = 7.30*$		

*kcal/g mole.

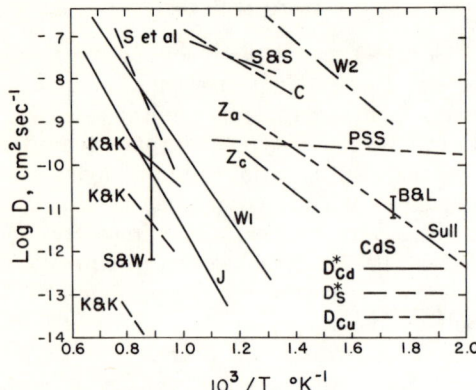

Fig. 2. Self-diffusion of cadmium and sulfur and diffusion of copper in CdS.

(*W1*) Woodbury (1964). Under saturated cadmium vapor. At 800°C using sulfur vapor as well, self-diffusion of cadmium was linearly dependent on the donor impurity concentration.

(*S&W*) Shaw and Whelan (1969). The upper limit of the range (vertical line) at 850°C is for saturated cadmium vapor, the lower limit is extrapolated to saturated sulfur vapor. The self-diffusion coefficient, which was measured along the hexagonal axis had a plateau from P_{Cd} of 0.1 to 10^2 torr.

(*K&K*) Kumar and Kroger (1971). Self-diffusion of cadmium was measured under 1 atmosphere pressure of cadmium vapor. Self-diffusion of sulfur was measured under 2 atmospheres of sulfur pressure (upper dashed line) and 4 atmospheres of cadmium pressure (lower dashed line).

(*J*) Jones (1972). Self-diffusion of cadmium, under saturated cadmium vapor (to 1280°K) and excess cadmium beyond, showed no dependence on crystallographic orientation. In excess sulfur vapor the cadmium diffusivities were about 10-fold lower.

(*S et al.*) Sysoev et al. (1969). Sulfur self-diffusion in CdS. Crystals grown from the melt under inert gas pressure.

(*C*) Clarke (1959).

(*W2*) Woodbury (1965).

(*S&S*) Szeto and Somorjai (1966). Observed a fast and a slow component and postulated interstitial-substitutional equilibrium distribution.

(*PSS*) Purohit et al. (1969). Used CdS crystals dipped into cuprous chloride solution to form a Cu_2S layer on the surface.

(*Sull*) Sullivan (1969).

(Z_a, Z_c) Zmija (1971), and Zmija and Demianiuk (1971). The subscripts refer to diffusion in the a and c directions, respectively.

(*B&L*) Birchenall and Lu, unpublished data.

intentionally. Some of the variations in the results reflect the sensitivity to the sulfur or cadmium activity. (A similar dependence on sulfur activity was not observed in the more highly disordered Ag_2S.) However some of the variations, especially in the slopes, reflecting activation energies are undoubtedly experimental disagreement which may be due in some measure to impurities in the single crystals that were used. The references and some additional information about the measurements are given in the extended caption.

CdS with a layer of Cu_2S forms a p-n photovoltaic junction that is potentially useful as a solar energy converter. Diffusion of Cu into CdS appears to be essential in forming the p-n junction within the CdS. Further diffusion of Cu in CdS may lead to a decrease in photocell properties. Numerous measurements of bulk diffusivity have been made. They are included in Fig. 2 for direct comparison with Cd and S self-diffusion. The most precise measurements to the lowest temperature seem to be those of Sullivan (1969) by an indirect capacitance method that detected the width of the insulating intrinsic layer bounded by the p-n junction and the CdS-Cu_2S interface. His method does not detect the fast-interstitial tail observed by others (Zmija, 1971 and Zmija and Demianiuk, 1971).

We have been concerned that copper might diffuse rapidly along CdS grain boundaries to form an insulating layer, isolating whole grains from the solar cell. Copper was vapor deposited on coarse-grained polycrystals[2] of CdS and diffused for 24 hours at 300°C. A slightly beveled surface exposed the near-surface region to depths of 0 to about 30 microns. Microprobe scans across isolated grain boundaries revealed no excess copper in those regions. The most useful results

[2] I am grateful to Mr. Ludewig Vanderberg for growing these crystals and to Mr. Tien-Lien Lu for carrying out the diffusion measurements. Microprobe measurements of the copper in CdS were made by Mr. James Ficca at Micron, Inc. Details of this work will be reported when work in progress is completed.

came from the regions around pores, filled with vapor-deposited copper, from which grain boundaries emerged. Microprobe traces normal to the surfaces of such pores permitted calculation of a bulk diffusion coefficient for copper in cadmium sulfide, ignoring the fast-diffusing tail, that falls on Sullivan's line at 300°C. What our method lacks in precision it makes up in directness. Microprobe traces parallel to pore surfaces intersecting grain boundaries between 5 and 10 microns from the pore again showed no excess copper at the boundaries.

At 400°C Zmija found that for impurity diffusion in CdS, $D_{Cu} \gg D_{Ag} > D_{Au}$, which is in the order expected, the smallest ion diffusing most easily, presumably by an interstitial mechanism.

Ferrous Sulfide

Ferrous sulfide ($Fe_{1-\delta}S$) has the hexagonal nickel arsenide structure above about 300°C. Both iron and sulfur diffuse about twice as fast parallel to the c axis as perpendicular to it. Nickel sulfide shows a smaller enhancement in the c direction, and its diffusion coefficients have nearly the same magnitudes as those in ferrous sulfide (Klotsman, et al., 1966).

It is possible to use the magnitude of the anisotropy and its virtual independence of composition and temperature to investigate the relative ease of atom jumps along the various channels in the structure (Condit, 1969). In the case of ferrous sulfide it appears that jumps seldom start and end in neighboring sites on the same basal plane and take place between planes by jumping from a normal (octahedral) site into an interstitial (tetrahedral) site instead of jumping directly along the c axis between two octahedral sites. The notion that diffusion both parallel and perpendicular to the c axis takes place at rates limited by the same sort of ion jump, from an octahedral site into a tetrahedral site which has a vacancy in the adjoining basal

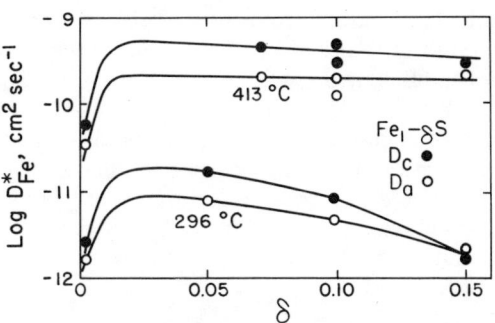

Fig. 3. Self-diffusion of iron in ferrous sulfide single crystals parallel (filled points) and perpendicular (open points) to the hexagonal (c) axis. Unpublished data by R. H. Condit, R. R. Hobbins, Jr., and C. E. Birchenall (submitted to *Oxidation of Metals*).

plane, is consistent with the observation that the activation energy is the same for both orientations.

Figure 3 shows that at 413°C, as at all temperatures investigated between 352° and 697°C, the iron ion diffusivity rises rapidly with the first increase in δ, to about 0.02, then practically levels off as δ goes to 0.15. This leveling of D corresponds to a decrease of vacancy mobility because the vacancy concentration increases as δ increases. At 296°C (and down to at least 250°C) the diffusivity drops again for large δ, particularly along the c direction. These conditions correspond to the monoclinic structure and Bertaut's vacancy superlattice. The vacancies segregate preferentially to alternate basal planes and tend to avoid being nearest neighbors. On the basis of these observations it is concluded that short-range ordering (clustering) of the vacancies reduces their mobilities and long-range ordering reduces the mobilities still more.

Bertaut (1953) also has observed structural modifications in FeS at low temperatures that lead to supercell reflections. Fasiska (1972) investigated FeS and $Fe_{0.94}S$ and concluded that at high temperatures the iron vacancies are nearly random but that close-packed layers become pleated by small displacements of sulfur atoms and that twins and stacking faults are common consequences of quenching.

With respect to geological problems that depend on stoichiometric ratios, it is important that composition changes can be effected by the motion of vacant sites which are much more mobile than the ions; that is,

$$C_V D_V = C_{Fe} D_{Fe}$$

where C's are concentrations and $C_{Fe} \gg C_V$. When D_V is 10^{-8} cm²/sec, distances of about 1 cm can be homogenized in times of about 1.5 years. Thus a compositional gradient would have to be present on a coarse scale or the pyrrhotite cooled rapidly on a geological scale to prevent homogenization above a few hundred degrees centigrade. Very large pressures are required for moderate decreases in diffusivities.

Summary

Mechanisms of diffusion in several sulfides have been reviewed in order to show that diffusive transport in crystals depends upon the crystal structure and especially upon their point defects, their concentrations, and their arrangements.

Acknowledgment

The preliminary work on copper diffusion in cadmium sulfide cited here received support under contract no. NSF/RANN/SE/GI-34872 through the Institute of Energy Conversion, University of Delaware.

References Cited

Allen, R. L., and W. J. Moore, Diffusion of silver in silver sulfide, *J. Phys. Chem., 63*, 223–226, 1959.

Bartkowicz, I., E. Fryt, and S. Mrowec, Self-diffusion and concentration of defects in cuprous sulfide, *Zesz. Nauk. Akad. Gorn-Hutn. Krakowie Ceram, 14*, 19–34, 1969.

Bertaut, E. F., Contribution à l'étude des structures lacunaires: La pyrrhotine, *Acta Crystalogr., 6*, 557–561, 1953.

Bertaut, E. F., Structure de FeS stoechiométrique, *Bull. Soc. Fr. Minéral. Cristallogr., 79*, 276–292, 1956.

Clarke, R. L., Diffusion of copper in CdS crystals, *J. Appl. Phys., 30*, 957–960, 1959.

Condit, R. H., Diffusion path networks in oxides, *Mater. Sci. Res., 4*, 284–303, 1969.

Etienne, A., Electrochemical method to measure the copper ionic diffusivity in a copper sulfide scale, *J. Electrochem. Soc., 117*, 870–874, 1970.

Fasiska, E. J., Some defect structures of iron sulfide, *Phys. Status Solidi, A10*, 169–173, 1972.

Ishiguro, M., F. Oda, and T. Fujino, On the self-diffusion of sulphur ion in the ionic crystal α-Ag₂S and its structure sensitive properties (I), *Mem. Inst. Sci. Ind. Res. Osaka Univ., 10*, 1–6, 1953.

Jones, E. D., Measurement of the self-diffusion of cadmium into cadmium sulphide using radiotracer techniques, *J. Phys. Chem. Solids, 33*, 2063–2069, 1972.

Klotsman, S. M., A. N. Timofeev, and I. S. Trakhtenberg, Mechanisms of the self-diffusion of nickel in nickel monosulfide and the reactional diffusion of nickel in the nickel-sulfur system, in *Surface Interactions between Metals and Gases*, V. I. Arkharov and K. Gorbunova, eds., Consultants Bureau Translation, Plenum, New York, pp. 90–97, 1966.

Koch, F., and J. B. Cohen, The defect structure of Fe₁₋ₓO, *Acta Crystallogr., B25*, 275–287, 1969.

Kumar, V., and F. A. Kröger, Self-diffusion and the defect structure of CdS, *J. Solid State Chem., 3*, 387–400, 1971.

Mathieu, H. J., and H. Rickert, Elektrochemisch-thermodynamische Untersuchungen am System Kupfer-Schwefel bei Temperaturen T = 15–90°C, *Z. Phys. Chem. N. F., 79*, 315–330, 1972.

Okazaki, H., Deviation from the Einstein relation in average crystals. Self-diffusion of Ag⁺ ions in α Ag₂S and α Ag₂Se, *J. Phys. Soc. Jap., 23*, 355–360, 1967.

Peschanski, D., Détermination de coefficients d'autodiffusion par la méthode des échanges isotopiques II.—Autodiffusion des ions S⁻⁻ et Ag⁺ dans le sulfure d'argent β, *J. Chim. Phys. Physicochim. Biol., 47*, 933–941, 1950.

Purohit, R. K., B. L. Sharma, and A. K. Sreedhar, Diffusion of copper in CdS crystal from Cu₂S layer, *J. Appl. Phys., 40*, 4677–4678, 1969.

Rahlfs, P., Ueber die kubischen, hochtemperatur-Modifikationen der Sulfide, Selenide, und

Telluride des Silbers und des einwertigen Kupfers, *Z. Phys. Chem.*, *B31*, 157–194, 1935.

Roth, W. L., Defects in the crystal and magnetic structures of ferrous oxide, *Acta Crystallogr.*, *13*, 140–149, 1960.

Shaw, D., and R. C. Whelan, The dependence of Cd diffusion and electrical conductivity in CdS on Cd partial pressure and temperature, *Phys. Status Solidi*, *36*, 705–716, 1969.

Stevenson, D. A., Diffusion in the chalcogenides of Zn, Cd, and Pb, in *Atomic Diffusion in Semiconductors*, D. Shaw, ed., Plenum, New York, p. 7, 1973.

Sullivan, G. A., Diffusion and solubility of Cu in CdS single crystals, *Phys. Rev.* *184*, 796–805, 1969.

Sysoev, L. A., A. Y. Gel'fman, A. D. Kovaleva, and N. G. Kravchenko, Measurement of the coefficients of self-diffusion of sulfur in CdS monocrystals, *Izv. Akad. Nauk SSSR, Neorg. Mater.*, *5*, 2208–2209, 1969.

Szeto, W., and G. A. Somorjai, Optical study of copper diffusion in CdS single crystals, *J. Chem. Phys.*, *44*, 3490–3495, 1966.

Woodbury, H. H., Diffusion of Cd in CdS, *Phys. Rev.*, *134A*, 492–498, 1964.

Woodbury, H. H., Diffusion and solubility of Ag in CdS, *J. Appl. Phys.*, *36*, 2287–2293, 1965.

Zmija, J., Studies on the diffusion of gold, silver, and copper in CdS and CdSe single crystals by a method based on measurements of the resonant capacitance of piezoelectric transducers, *Acta Phys. Pol.*, *39A*, 531–538, 1971.

Zmija, J., and M. Demianiuk, Studies on the diffusion of copper, silver, and gold in single-crystalline cadmium sulfide, *Acta Phys. Pol.*, *39A*, 539–553, 1971.

DIFFUSION RELATED TO GEOCHRONOLOGY

B. J. Giletti
Department of Geological Sciences and
Materials Research Laboratory
Brown University
Providence, Rhode Island 02912

ABSTRACT

Diffusion of geochronologically important nuclides has been studied in a wide variety of minerals and by three different approaches. The first approach was to examine field situations where age discordances were found. These studies yielded the basic systematics of discordance patterns in various minerals for the different dating systems. The second approach was to employ a known geological case where one clearly defined process caused the age discordances and to make detailed measurements of ages and concentrations as functions of some critical distance parameter. Relative diffusivities could be obtained from this type of study as well as crude estimates of their temperature dependence. The third approach was to try to determine the diffusivities of the different species in particular minerals in the laboratory. The result of the third approach has been much data but little useful knowledge, owing to the large differences in the results of experiments. This discouraging picture of the experimental approach is rapidly changing. Some good data are appearing, with the promise of quite a bit more.

Criteria for the design of appropriate experimental determinations are given with specific references to minerals of geochronological importance. The criteria include the following requirements for the experiments: Minerals must be stable; there must be only one solid phase; a significant fraction of the migrating species must actually be transported; the relation between particle size and effective grain size for diffusion must be determined; the shape of the particles must be known; diffusional anisotropy in the solid must be known.

Particularly careful attention must be paid to the possibility that other diffusion mechanisms than the one(s) observed might become important at temperatures so low that transport is too slow to be measured in the laboratory. The correlation of laboratory measurements with carefully analyzed field situations appears to be the best hope here.

The different isotopic dating methods are treated separately. Examples are chosen to illustrate cases of much-quoted data where faulty experimental design led to incorrect results, as well as cases in which the transport measurements are thought to give valid volume diffusion data.

Aspects of spontaneous fission track observations and of electrical conductivity are discussed in relation to their possible influence on our knowledge of volume diffusion and geochronology. It is concluded that the fission track observations may be important, while the electrical conductivity does not appear to be a significant parameter.

Finally, the applicability of the diffusion results to geological settings is considered. It is tentatively concluded that it will soon be possible to estimate the diffusion constants for particular systems and temperatures. Knowing the constants will aid in determining which of several processes, including diffusion, was operative in a given geological setting. For cases in which diffusion was the primary factor, the data should permit useful new conclusions to be drawn.

Introduction

One of the basic problems in geochronology is how to determine if a rock or mineral has remained a closed system for precisely the time interval between the event which affected it, and which we wish to date, and the present. Various processes can open the system to the gain or loss of nuclides. They include chemical reaction, recrystallization, and diffusion. This paper will focus on diffusion, primarily volume diffusion in solids.

It is sometimes difficult to determine whether diffusion or one of the other processes was the dominant, or only, one which operated. A possible means to re-resolve this is to see if the analytical results could be due to dissolution of all or some of the crystals so that all the constituents were involved to the same degree. In contrast, a diffusion mechanism would permit the selective gain or loss of some nuclides while leaving others in the crystal relatively undisturbed. An example of the latter would be the biotite Rb-Sr vs. K-Ar data in Fig. 1 (from Hart, 1964). Cases where diffusion is believed to have been the controlling mechanism in changing isotopic ages can be found in the literature for all three major dating systems: K-Ar, Rb-Sr, and U-Th-Pb.

Geochronologists have devised methods to determine some of the dates of events isotopically recorded in the rocks of complex terranes. This is a significant achievement; however, only a partial chronology can be deduced. For example, minerals in rocks undergoing plastic deformation and metamorphism are not likely to remain closed to argon loss. Consequently, events that occurred before that deformation are probably not observable as part of the K-Ar record of those rocks today.

It should be possible to go beyond this sort of analysis to learn more about events in the past, based on the record of reset or partially reset ages in rocks. For example, if the diffusional behavior of K, Ar, Rb, and Sr were known as a function of temperature, pressure, and mineral species and if the diffusivities were different and had different temperature dependences, it should be possible to take data from a carefully studied terrane and derive a time-temperature history for the event that caused the resetting. In all such cases, the transport of the different nuclides might be fairly accurately determined, so that a set of conditions described by the product of the diffusion constant, D, and time could then be deduced. Since D is a function of temperature, we would arrive at a set of time-temperature pairs which could be fairly narrowly constrained. Efforts to determine the rates of diffusion of these critical nuclides have been under way for some time.

Three types of study have been carried out to gain a better understanding of the complexities of the age resetting problem. These were: the dating of rocks or terranes which yielded discordant isotopic ages; the systematic dating of rocks in which a single process controlled the discordance effects to be observed; and laboratory studies of diffusion.

Discordance Observations

The earliest method of studying open systems in rocks and minerals used in dating was isotopic age determination of a rock or terrane for which the resulting dates were discordant. These studies yielded isotopic age patterns which suggested that there was a systematic relationship among the various age methods and mineral systems. A hierarchy of stability of particular minerals to age resetting was proposed. Schemes for the interpretation of the discordant ages were devised. They included the now standard concordia plots discussed by Wetherill (1956), and then by Tilton (1960), that dealt with the U-Pb system, and the strontium development lines, or isochrons, proposed by Nicolaysen (1961).

In most cases, the mechanism by which

nuclides entered or left a rock or mineral was not critical to the model. All models could be applied to cases in which diffusion into or out of the solid phase was the sole or primary exchange mechanism, although only the model by Tilton required this mechanism.

Age Discordances Controlled by One Geological Process

The second type of study grew out of the information obtained in the early work just described. This type was based on the choice of a field setting where only one process had affected the rocks. A systematic sampling of the area was then carried out. The classic example of this approach was reported by Hart (1964) on the effect of the intrusion of the Eldora stock into the Precambrian in the Colorado Rockies. In a traverse starting within the stock and moving out into the country rock, clear age patterns were obtained (Fig. 1). The response of a particular mineral to an episode of heating could be followed by means of the reset age from the contact with the intrusive outward to where the effect of the contact metamorphism could no longer be detected clearly.

Comparable studies have been made adjacent to mafic dikes (Hanson and Gast, 1967; and Westcott, 1966). In all of these studies, it is possible to estimate the geometry and thermal history of the intrusive rocks and therefore of the surrounding rocks. From this, an estimate of the diffusivities of the nuclides may be obtained. Unfortunately, the geometry of the intrusive, the intrusion temperature, the ambient temperature of the country rock, and the location of the contact at depth relative to the host rock are among the parameters which are poorly known and which lead to large uncertainties in determining the correct diffusivities and their temperature dependence.

A similar type of study was attempted by Hurley et al. (1962) based on uplift of rocks along the Alpine fault in New Zealand. Here, the premise was that argon was being lost by the micas when they were at depth, and the change to a

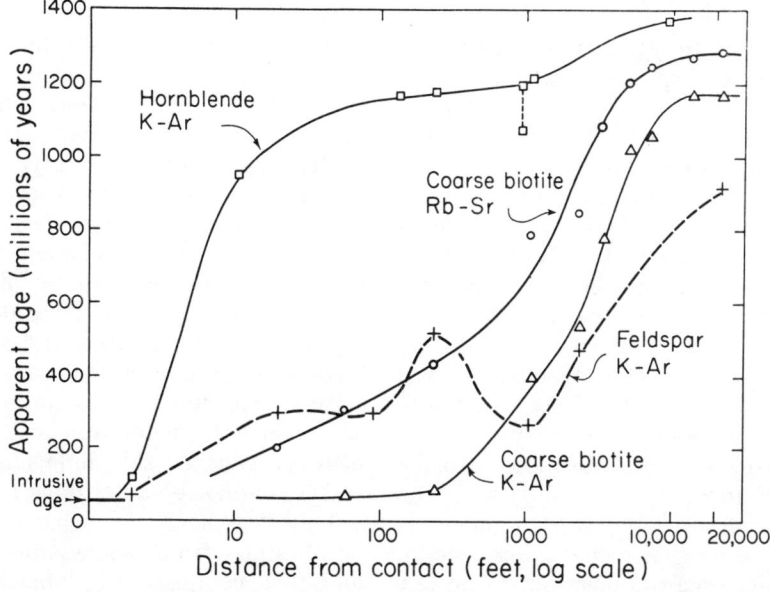

Fig. 1. Radiometric ages determined in Precambrian rocks adjacent to the intrusive Eldora stock, Colorado. (From Hart, 1964).

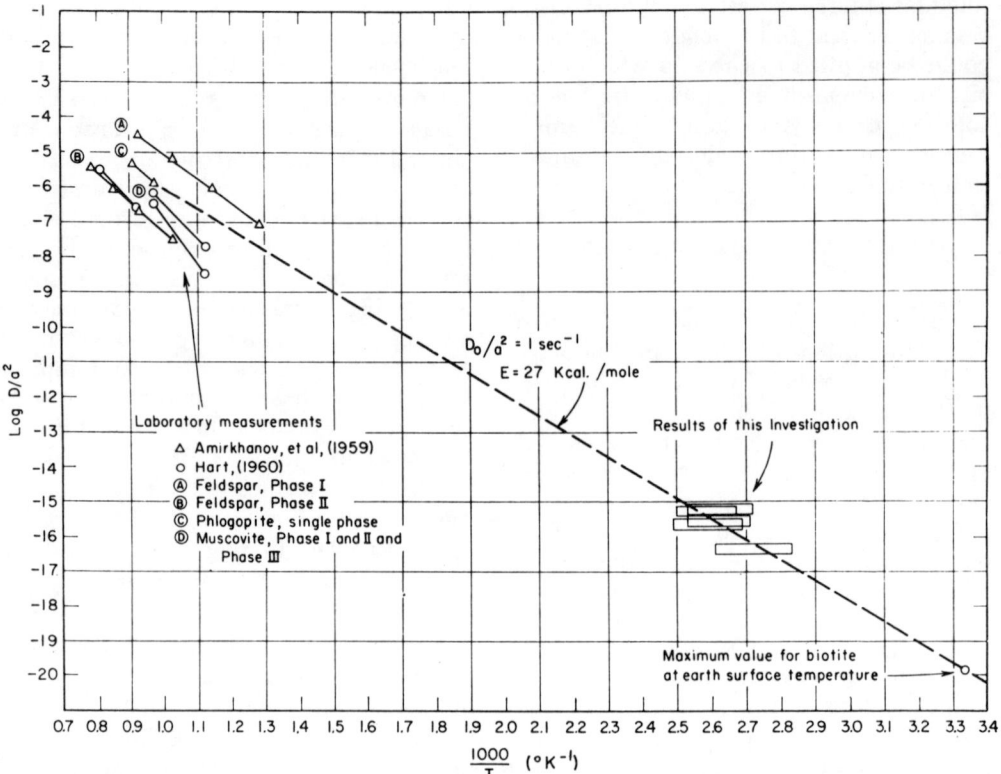

Fig. 2. Arrhenius plot connecting high temperature, experimentally determined values of the diffusion constant with those inferred from the data on the Alpine fault, New Zealand (from Hurley et al., 1962).

closed system resulted from uplift and subsequent cooling. Using an Arrhenius plot (see Fig. 2) based on some reported argon diffusivities experimentally determined at moderately high temperatures, the diffusivities calculated from the New Zealand data were also plotted, and activation energies were then obtained by connecting the two sets of data. Inherent in this approach was the need to know the thermal history of the rocks during their uplift. A geothermal gradient was assumed and an uplift of 9000 feet was taken to represent the displacement. There is now a serious question as to the amount of throw on the fault, with the possibility that it may be as much as 40,000 feet (Suggate, 1963). This would, of course, change the position of the low temperature points on the Arrhenius plot and lead to greater activation energies.

All of these efforts at the interpretation of field data to determine diffusivities are theoretically sound. The primary difficulty is the usual one of knowing the geological history sufficiently well to be able to compute a precise thermal history for the rocks in question. Further efforts along these lines are to be encouraged. It is particularly important to obtain data on diffusion which occurred at low temperatures for long times, and this cannot be done under laboratory conditions. The critical criterion for this sort of study is the precise determination of the thermal history. Without this, independent diffusivities cannot be determined.

Very important information can still be obtained from field studies, however, and this, coupled with laboratory data, should provide the essential control on low temperature behavior. For example, it should be possible to match field and laboratory results when there is some time–temperature overlap. This can be done because different diffusing species such as Ar, Sr, and Pb have different diffusivities and activation energies, making control of the actual time–temperature conditions possible. Analysis of rocks in the lower temperature regimes should then indicate if the mechanisms for diffusion remain the same, since the time–temperature profile should be calculable. This approach may prove to be the only way to determine the temperature dependence of diffusion in the low temperature range. This is a very promising avenue.

Laboratory Diffusion Studies

Much of the work just reviewed can be described as the use of geologically controlled settings to determine diffusion parameters. It is hoped, of course, to turn this approach around and to use diffusion data to reconstruct geological history. Owing to the difficulties encountered in making field-based diffusion measurements, attention was turned to the laboratory measurement of diffusivities. This work has taken several directions but until very recently has not had much impact on the interpretation of geochronological data. This lack of impact is largely the result of a voluminous and conflicting literature.

Criteria for Experimental Design

It is important in the measurement of diffusion in minerals that certain experimental criteria be met. Owing to the trace abundances of some of the diffusing species and the need to have data on isotopic compositions, it is often necessary to carry out the experiment with an aggregate of grains rather than with a single crystal. Until very recently, analyses were done on a material balance basis, rather than by tracing concentration profiles through a single crystal. Ion microprobe analysis will greatly facilitate this sort of study because it will permit just such profiles to be measured on a very fine scale. Other experimental approaches are used and are discussed in other papers in this volume.

The primary focus of this set of criteria will be for the self-diffusion of a species in a solid. Diffusion characterized by exchange of one ionic species by another, interdiffusion, leads to somewhat different experimental design criteria. Although the truth embodied in these criteria may appear self-evident, in many reports it is clear that the criteria were ignored. Some of these criteria are:

1. To measure volume diffusion in a solid phase, that phase must be stable during the experiment. If the phase breaks down or reacts chemically, other transport mechanisms will obtain, or else the diffusion will occur through a different substance. The stability in question must also include nonrecrystallization. For the determination of self-diffusion, it is best if no chemical change at all occurs. This requirement often makes a hydrothermal experiment necessary.

2. The presence of more than one solid phase plus the hydrothermal fluid complicates the diffusion rate study markedly, and requires careful control of transport. Most diffusivity determinations which involve more than one solid phase and employ only bulk analysis cannot yield correct D values for the solid phases because the net transport rates become either interdependent or additive. These ambiguities are eliminated if a single solid phase is present with the fluid.

3. Diffusive transport of a component in a solid must be shown to involve all of that component, that is, all the atoms of that species move by the same mechanism even though only a fraction may exchange with the surroundings. Alternatively, it must be demonstrated that more than one population of that component is present and that they move by different

mechanisms. The latter is difficult to demonstrate unless all the present experimental criteria are met in this type of experiment. Failure to meet some of these criteria can lead to results that resemble the effects of diffusive transport of two populations by two mechanisms, whereas the properly designed experiment would demonstrate volume diffusion of a single population by a single mechanism.

The remaining criteria refer to those cases where particle aggregates of one solid phase are used. It is then necessary to determine:

4. The relation of the particle size to the effective grain size for diffusion.

5. The shape of the particles and its relation to the crystallographic axes of the material. The shape of the particles is critical in choosing the geometric computational model to be used.

6. The diffusional anisotropy in the particles and its relation to the crystallographic axes and particle shape. For anisotropic minerals, this effect may dominate the shape effect.

The choice of geometric computational model depends largely on the mathematical difficulty of the solution of the diffusion equation for the particular boundary conditions. It has been possible to obtain comparatively simple solutions to such geometries as transport across an infinite plane sheet of finite thickness, radial transport in an infinite circular cylinder of finite radius, and isotropic transport in a sphere. For details, see Crank (1957) or Carslaw and Jaeger (1959).

Thus far, the process by which transport takes place in a mineral has been ignored. In any laboratory measurement of diffusional transport, a choice must be made between determination of self-diffusion and interdiffusion. As both may be of importance to geochronology, it is necessary to distinguish which process is being studied. Self-diffusion will be taken to mean the migration of a single atomic species. Its transport may be followed by using different isotopic tracers of that element, all of which are assumed to behave identically.

In contrast, interdiffusion involves the exchange of one elemental species for another. The result is a net chemical change for the solid phase, for example, the exchange of Fe for Mg or of Na for K. In the first process, self-diffusion, migration of the species need not depend on the rate of transport of any other species. In the second, the migrations are coupled to maintain electrical neutrality, and the faster moving species will have a net transport rate influenced by the slower rate-controlling one.

Of critical importance to this discussion is the actual mechanism by which species migrate in different solids at different temperatures, or even the mechanism by which one species migrates under one set of conditions. Other contributors to this Conference have dealt with this topic. Of particular importance to the geochronologist is the identification of the important diffusion mechanism under one set of conditions and the determination that this is the operative mechanism under the relevant geological conditions, where he cannot make a direct diffusion measurement. There is always the possibility that there is a mechanism that contributes negligible transport at high temperatures but has such a small dependence on temperature that it dominates the diffusion process at low temperatures, where the other mechanisms become very slow. Extrapolation of the rates from high to low temperatures would then obviously lead to predicted transport rates that are too low.

One such competing mechanism might be found when transport occurs through vacancies in the structure of a crystal, and the vacancies are caused by two different processes. Some of the vacancies will be produced by the thermal motions of the atoms when, statistically, some atoms have enough energy to jump out of their lattice positions (see Manning, this volume). However, particularly in geological materials, the crystal is also likely to include many impurities of different electrical charge from that which would normally exist in that position. In such cases, there may be a vacancy in some nearby position in order to maintain electrical neutrality. Clearly, vacancies of the first kind depend for their existence on the total energy in the system. This means that they are most abundant at high temperatures and their concentration can be computed as a function of temperature. Vacancies of the second type exist because of the impurities and since the impurities remain at low temperatures, the vacancies also remain.

Laboratory Measurements of Diffusion in the K-Ar System

There have been numerous attempts to measure the diffusion of argon in minerals. A very useful review and critique of the literature is given in Mussett (1969). In this paper, Mussett is severely critical of most of the diffusion measurements made. On the whole, his criticisms are justified. It is worth noting, however, that there is considerably more cause for optimism than the somewhat pessimistic tone of his conclusions four years ago. It is possible to make reproducible argon diffusion measurements of sufficient accuracy on geologically important minerals to permit useful geological interpretation.

The present discussion will be limited to those cases in which the data are still referred to, the results are recent, or the results give every indication of being valid. The discussion will not include cases in which the diffusion measured was the result of fast neutron or other heavy particle irradiation. Such irradiation induces many new defects in the structure which affect the argon transport measured. The experiments will then be partly the result of normal argon diffusion and partly the result of transport as the new defects are annealed out.

A summary of the published feldspar argon-diffusion data was shown by Mussett (1969) and is reproduced here (Fig. 3). This illustrates the problems faced by the geochronologist in attempting to use diffusion data to interpret ages. In the temperature region 400° to 1000°C, argon diffusivities in K-feldspar have values at any given temperature which range over six orders of magnitude. In addition, the temperature dependence varies both between different results and as a function of temperature.

The main problem with these data is that the criteria listed above have nearly all been violated in many of these experiments. The requirement that only one solid phase be present is violated most consistently in perthites. This indicates that not only are two phases possibly contributing argon but that with long-term heating at high temperatures the feldspar will homogenize to a single phase. During the reaction a totally different transport process can occur. Subsequently, markedly different conditions arise because a new phase has been produced. In addition, it is difficult to determine the actual grain size which should be used for the boundary condition in the diffusion computation. The particle size is clearly not the correct one, as the exsolution lamellae are often far smaller than the particles. Since perthites vary in size from submicroscopic to large patches, it is not possible to choose some common size to compute the D values. The dimensions of the lamellae are not known for the feldspars represented in Fig. 3.

The work on feldspars points out the great need to characterize the material on which the study is made. We must take the experience of the materials scientists much more seriously. They have found this careful characterization essential in their work. Often, their materials are simpler than those used by geologists. The careful characterization of mineral specimens is absolutely necessary in diffusion studies.

An alternative way to look at these feldspar data is to assume that the primary problem is the multiphase aspect

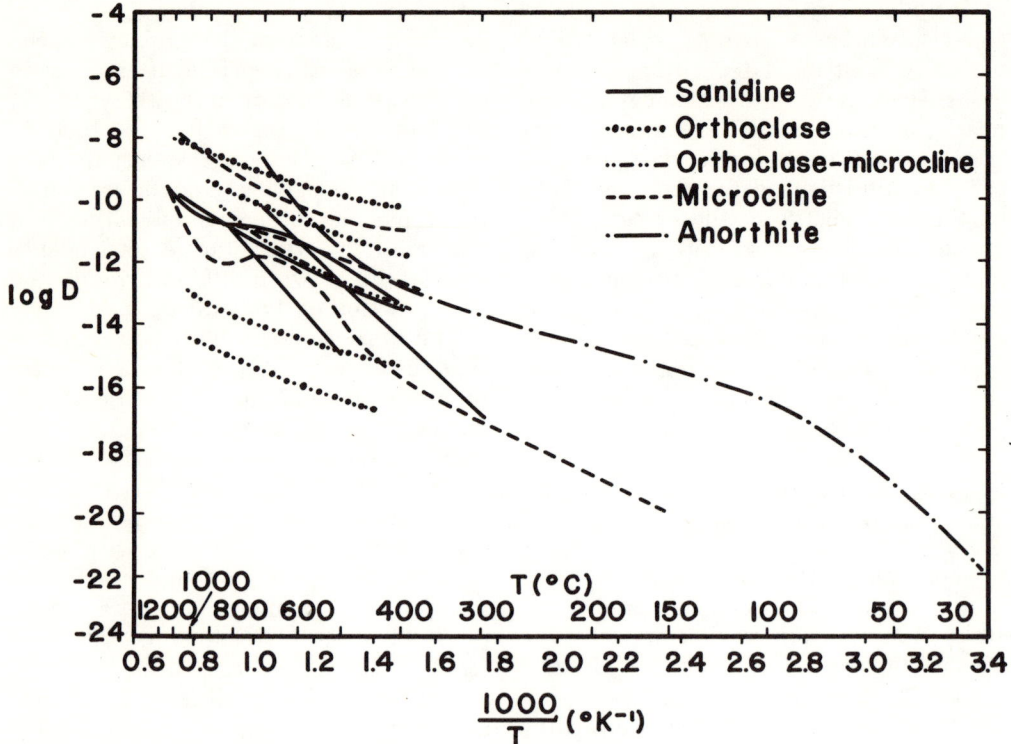

Fig. 3. Argon diffusion results by various investigators shown as an Arrhenius plot. Note that the difference for any given temperature may exceed six orders of magnitude (After Mussett, 1969).

and that no homogenization occurs during the experiment. This would suggest that the D values differ by a factor of a million because the effective grain sizes differ. The square root relationship for the dimensional parameter would then suggest that the effective grain sizes differ by three orders of magnitude. This is possible. It means that the reporting of K-Ar ages is no longer a complete process unless the particle size of the mineral is given as well as the relation of the particle size in the rock to that in the analysis; if there is exsolution, the particle size and the size and spacing of the lamellae should be given. However, improved computational methods will be needed to deal with the complex geometry involved.

Recently, Foland (1974) has shown that a *single phase, homogeneous* K-feldspar yields reliable argon diffusion data, including a linear Arrhenius plot over a three hundred degree temperature range. By taking carefully sized aliquots of the orthoclase and heating them for different lengths of time under identical conditions, he obtained the curve shown in Fig. 4. There is a direct functional relationship between f, the fraction of the total argon which is lost during the experiment, and the square root of time. Any one of the points on the curve shown is sufficient to define the entire curve uniquely. If the same diffusion mechanism obtains, and there is only one "population" of argon in the solid, all the points should fall on the same curve. The five points do. Therefore, at least 71% of the argon is behaving normally in terms of simple diffusional transport. There is no evidence from these results that any of the argon would do otherwise

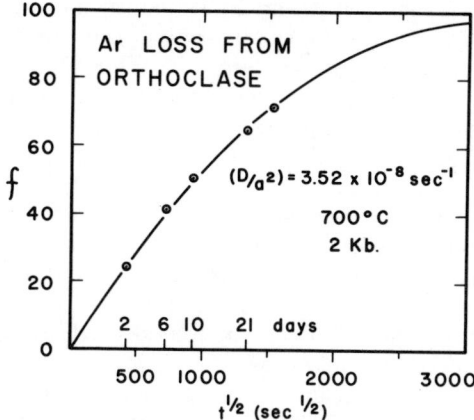

Fig. 4. Fraction of total radiogenic argon lost vs. time$^{1/2}$. Diffusion model is a sphere. Any one point suffices to define the line. Other points thus confirm that the same diffusion mechanism(s) operates for at least three quarters of the argon present. (From Foland, 1972.)

if runs of longer duration were to be carried out.

Where sanidine samples consisting of a single phase were studied (Baadsgaard et al., 1961; and Newland, 1963), there is good agreement both in activation energy and the actual value of the diffusivity with Foland's (1974, Fig. 4) data.

There is a smaller body of data on the micas, and it too fails some of the criteria for a valid measurement. The primary constraint on use of the data arises because criterion 1, mineral stability, was violated. In most cases, the mica was heated in a vacuum, resulting in the loss of its structural water near 600°C.

In a much-quoted paper, Evernden et al. (1960) made numerous diffusion measurements on various minerals including several micas. They noted a difference of several orders of magnitude in measured Ar loss upon heating, depending on whether the mineral glauconite was in vacuo or at 1,000 or 10,000 psi water pressure. Further, the hydrothermal data gave a reasonably linear Arrhenius plot, while the vacuum experiments did not (see Evernden et al., 1960, Fig. 2). It is curious that, despite the glauconite data and the specific reference to and description of water loss in micas, they also reported in vacuo data for other micas. A more detailed description of their phlogopite results appears in another paper in this volume (Giletti, Argon in phlogopite mica).

This dehydration problem in micas has been referred to by numerous workers. Brandt et al. (1967) showed the effect of a supporting water pressure on biotite and phlogopite during argon diffusion experiments. Despite the clear evidence of difficulty in the dry heating of micas, Hanson (1971) reported argon diffusion data for biotite heated dry. He describes the change in d-spacing for the (001) plane as a result of heating and shows that the argon loss correlated with this change. At some point in the experiment the material is no longer biotite but a layered, dehydrated silicate which, in some cases, is ruby red. This is not the material being dated in rocks. Further, this dehydrated mineral, which is not found in rocks, would have to be reconstituted to replace the water and make it resemble again the biotite that is actually dated. This rehydration would probably be accompanied by further argon loss. Consequently, the applicability of these dry heating results is limited.

Other minerals have been studied with regard to argon diffusion. Among them have been the pyroxenes, with particular attention paid to the excess argon (that radiogenic argon in the mineral which is in excess of the amount that resulted from decay of the potassium during the history of the mineral), which is thought to have been incorporated in the mineral at its time of formation. References to the earlier work can be found in Mussett (1969) and in Schwartzman (1971).

As Kistler et al. (1969) and Schwartzman (1971) point out, there is a serious exsolution problem with the pyroxenes. The results resemble the feldspar results and probably for the same reason. The presence of two phases, orthopyroxene and clinopyroxene, presents the same geometric problems as do the two feldspars in perthites. One potential difference could be the site of the potassium. In the perthites, the K-feldspar is the host phase and, of course, is the K-rich phase. Schwartzman (1971) notes that the reverse situation may exist in the orthopyroxenes in that the clinopyroxene lamellae are likely to have the greater K content. No single-phase argon diffusion measurements have been reported for pyroxenes.

A second aspect of the K-Ar dating system which should be discussed is the possibility of K diffusion. A critical consideration here is that for the majority of cases the material being dated is a potassium mineral, which means that its K concentration is fixed within broad limits. Since there is usually negligible K isotope fractionation during geological processes (for possible exceptions see Letolle, 1963; Schreiner and Verbeeks, 1965; and Schreiner and Welke, 1971) the isotopic composition of potassium at any given time in the past may be assumed uniform in all phases. Thus, while there may be isotope exchange among the different mineral phases in a rock, the effect on the potassium will be nil. Of much greater importance is the possibility of ion exchange owing to different chemical equilibrium conditions. If the rock is a closed system, the dominant effect would be the exchange of Na for K and *vice versa* among the minerals present. If the rock is open to alkali exchange, then the effect could be one of K or Na metasomatism.

Where the mineral is grossly out of equilibrium with its environment, it is possible for the effect to be solution-precipitation (O'Neil and Taylor, 1967). Where the departure from equilibrium is not great, diffusive exchange may occur instead. There are no good data to determine the conditions under which one process yields to the other in importance. In the case of orthoclase, Foland shows in this volume that the diffusive exchange is controlled by the slower-moving species.

His reported K and Na self-diffusion constants differ markedly, but in the case of ion exchange the rate is similar to that which might be predicted from the K self-diffusion alone. Preliminary experiments on phlogopite by me show the same relationship.

At high temperatures, the difference in diffusivities between Ar and K is large for orthoclase and phlogopite, and the activation energies in orthoclase differ so that the diffusivities diverge further at lower temperatures.

The diffusion of K in biotite was reported by Hofmann and Giletti (1970). The comparison of these results with preliminary argon diffusivities again suggests a lower K diffusivity.

Results in the case of K diffusivity are still too sparse to permit quantitative discussion. The qualitative conclusion is that the primary focus has been placed correctly on argon rather than potassium in the numerous diffusion studies.

Laboratory Measurements of Diffusion in the Rb-Sr System

The Rb-Sr decay system has been used for dating for some time now, but problems persist in the correct interpretation of the ages obtained. When the system has been open at some time, and for unknown duration, after the formation of a mineral, various effects can occur. These include net gain or loss of Rb or Sr, isotope exchange of the Sr, or any combination of these. Because neither Rb nor Sr is an essential constituent of most minerals, particularly those used for Rb-Sr dating, the abundances of Rb and Sr may change in a mineral that remains chemically stable to significant major element changes.

It has been recognized for some time that ion exchange can alter Rb-Sr ages. Experiments in which a mineral was not in chemical equilibrium with a solution were attempted (Kulp and Bassett, 1961; Gerling and Ovchinnikova, 1962; and Kulp and Engels, 1963). These were all low-temperature measurements (approximately 100°C) and generally resulted in chemical reaction rather than diffusion.

Two studies were carried out in which high-temperature diffusion measurements were attempted (Deuser, 1963; and McNutt, 1964). The Deuser investigation was on biotite in hydrothermal solution. Unfortunately, the water contained only Rb and Sr isotopic tracers, so the resulting isotope exchange with the mica must have involved net transport of Rb and Sr into the mica with resulting extraction of other cations, presumably dominated by K. The reported analysis, assuming the presence of the Rb and Sr tracers in the fluid in unchanged amount (as though this were a typical isotope dilution experiment), cannot be interpreted correctly because of the change in the Rb and Sr material balance in the fluid.

The McNutt experiments were complex owing to the presence of more than one solid phase in the initial charge. They were further complicated by the observation of breakdown of the biotite to opaque materials, perhaps magnetite. The process was too complex to permit any direct quantitative diffusion interpretation.

A different approach to the Rb-Sr transport problem was that by Baadsgaard and van Breemen (1970). They took a well-dated quartz monzonite which appears to have had a simple geological history and placed several kilograms, consisting of three or four pieces of the rock, into a muffle furnace in air and heated the sample isothermally for 100 hours. They then separated the minerals and analyzed them for their Rb, Sr, and Sr isotope compositions. This experiment was carried out at 800°, 860°, 920°, 985°, and 1025°C.

Their study has much in common with that of Hanson (1971) described above. In both cases the mineral is heated while still in the rock and then extracted. The heating in air has several effects which complicate any conclusions that might be drawn about diffusional exchange. Both the muscovite and the biotite in the rock break down at any of the run temperatures reported. The amount of the breakdown would be a function of the run duration and the tightness of the rock to water loss. At the higher temperatures, the authors report that the rock partially fused. It is probably safe to say that the minerals were not in chemical equilibrium at any time. The exchanges were, no doubt, influenced by the presence of phases not normally found in rocks (the mica breakdown products, for example).

It is possible that the experiment of Baadsgaard and van Breemen may be useful in interpreting a very high level intrusive or a volcanic contact metamorphic effect, but there is little possibility that this study will be applicable to the kinetics of diffusional exchange.

Data on Rb diffusion in a biotite have been reported by Hofmann and Giletti (1970). The diffusion rate appears to be approximately one third that of potassium and the activation energy is fairly low.

In their paper, Hofmann and Giletti discuss a method for the determination of the fraction of equilibrium approached by any species. It was stated that a system which is both chemically and isotopically out of equilibrium can be studied in a special way. If such a system consists of a single solid phase and a water solution, for example, the rate at which a chemical equilibration will occur between two species, such as Na and K, will be the same as the rate at which either of those species will achieve isotopic equilibrium. Thus, when a certain fraction of Na is transported from one phase to the other (relative to the net amount which has been transported when equilibrium is reached), the same fraction of the isotopes of K would also be transported. But this statement is not necessarily true: This author has observed that for phlogopite, Na isotopes will equilibrate long before the K isotopes do and also long before there is Na-K chemical equilibrium. Foland (this volume) reports the same result for orthoclase. It is clear that the Na diffuses by a mechanism that is not coupled to the K diffusion.

An important question involving Sr transport in minerals is whether or not the radiogenic Sr^{87} behaves like the common Sr. Preliminary studies by Hofmann (1969) and Foland (1972) indicate that when the normal hydrothermal experiments under controlled sample conditions are employed, all the Sr behaves in the same way within experimental error. That is to say, the initial location of the Sr and its origin do not seem to have as much effect as the many jumps it must make before it reaches the grain edge.

It should be noted in passing that in the preceding discussion we distinguish between behavior of radiogenic Sr^{87} and the common Sr. This has nothing to do with the isotopic fractionation that might occur as a result of diffusion. Such fractionation is not likely to be observed because the usual treatment of the Sr data is to normalize them to an Sr^{86}/Sr^{88} ratio of 0.1194.

Clearly, there is a great lack of transport data that can be used in the interpretation of discordant Rb-Sr dates. On the other hand, there is considerable interest in this subject now, and it is likely that in a few years this problem will be largely resolved. It should then be possible to determine the extent of the diffusion effect relative to other transport processes in the geochronologically complex terranes.

Laboratory Measurements of Diffusion in the U-Pb System

One of the first age discordances to be discovered was in the U-Pb system. Here, Pb^{206}/U^{238} ages were found to be low relative to Pb^{207}/U^{235} ages, which in turn were low relative to the Pb^{207}/Pb^{206} ages. One suggestion to explain this pattern is the loss, perhaps by diffusion, of one of the decay intermediates in the U^{238}-Pb^{206} chain, such as radon gas (Rn^{222}). Radon leakage measurements made by Giletti and Kulp (1955) and Adams *et al.* (1972) suggest that this effect is small in most well-crystallized minerals. However, the effect is important for very fine grained materials and for secondary alteration products of uranium minerals; and it may be important in interpreting concordia plots in which other data suggest that data points have been displaced vertically downward.

The primary reason for this minor effect is the short half-life of Rn^{222} (3.825 days). Any instantaneous loss of all the radon in a mineral will have a negligible effect on the measured age because there is so little radon present at any one time. Diffusional loss must take place over a significant portion of the mineral's history.

Although the U-Th-Pb system is most frequently used in the dating of zircons, few experimental efforts have been made to study diffusional behavior of U, Th, or Pb. One set of experiments on this topic (which appears not to have been followed up) was reported by Pidgeon *et al.* (1966). They subjected a metamict Ceylon zircon to a 2 molal NaCl water solution at 500°C and 1000 bars for times varying from 1 to 312 hours. The lead lost by this process was as much as approximately 63%, while the uranium loss was only about 5%. The primary question derives from the fact that the starting material was so metamict that it gave no x-ray diffraction pattern. It was not reported if any of the run products was sufficiently annealed to give a pattern. In any case, it would be most interesting to measure Pb loss from non-metamict zircons.

RELATED CONSIDERATIONS

In any effort to understand the ramifications of the diffusion data, it is worth considering the relation of the experimental design to the "experiments" in nature which we hope to interpret. In particular, do we modify the minerals in some way during their preparation or by the way we do the experiments that introduces uncertainties or errors in experimental results? A case in point is the possibility that cracks or defects in the crystals may be introduced when the minerals are comminuted. The usual process of grinding or chopping with a Waring blender may introduce fractures and defects. The data presented elsewhere in this volume by Foland and by me suggest that fracturing does occur but that it is restricted to only a few cracks, since the size effect for different particle sizes is approximately what would be expected for perfect grains of that particle size. It may be noted in passing that one advantage of the whole rock heating experiments of Hanson (1971) and Baadsgaard and van Breemen (1970) is that the minerals are not comminuted until after the diffusion experiment is complete.

The introduction of lattice defects during comminution is much more difficult to evaluate. Numerous defects doubtless are produced during this preparation process, but whether they are significant to the observed transport is not known.

Another experimental difficulty lies in the possible presence of damaged portions of the crystal structure owing to the formation of spontaneous fission tracks of uranium 238. These are used in their own right to measure rock ages.

Such tracks are tunnels for transport whose length, in micas for example, is on the order of 10 to 20 μm (Maurette et al., 1964). Some studies have been carried out to determine the behavior of such tracks upon annealing of the mineral (Fleischer et al., 1964, and Fleischer et al., 1965). Annealing temperatures are rather low for micas. Track densities in phlogopite are reduced to less than half their original value upon heating at 400°C for 16 hours.

It was observed, however, that these are the tracks which can be seen clearly and measured with a microscope after standard acid etching is applied. It was shown by Price and Walker (1962), using electron microscopy, that the damaged tracks persist apparently unchanged even when heated to 800°C for one hour. It is likely that some portions of the tracks become isolated from others by short annealed segments. These segments block the acid and so inhibit the etching but do not fill in the entire track. The result is that it is not clear to what extent these tracks affect the rapid transit of atomic species through a crystal. The importance of such a process affecting diffusion in a mineral during a mild heating would depend not only on its actual effect as determined in laboratory experiments but also on the age of the mineral at the time of heating. An old mineral would have had more tracks than a young one and so could have permitted more transport.

Finally, it should be noted that measurements of electrical conductivity have been performed on some minerals, particularly micas, owing to their use as insulators. Tolland and Strens (1972) showed that in the phlogopite-annite series the conductivity is ten to one hundred times greater parallel to the layers than perpendicular to them. These authors also showed that the conductivity increases with iron content, with a marked break occurring as the Fe content rises to a value corresponding to about 50% annite content. The difficulty in comparing these low-temperature results with the diffusion results, however, is that the conductivity is probably occurring by electron hopping, particularly among octahedral-site iron ions.

Diffusion and Geochronology

One of the questions which remain after a discussion of the diffusion process studies is whether they can be applied to specific geological settings. It can be expected that in the next few years self-diffusion and interdiffusion constants will be known for various elements in several of the important rock-forming minerals over a significant range of temperatures and, perhaps, pressures. It remains to be seen how successfully these data can be used to unravel complex geology. Because the application will not always be direct, an interpretative approach will be needed. Some examples of the route to be taken will illustrate the nature of the approach.

A prime consideration in the application of diffusion data is that diffusion data will have been obtained for a mineral specimen other than that in question. Can the D values determined on biotite A be applied to biotite B? If the diffusivity must be measured for each specimen dated, the task is a tedious one indeed. If the D value obtained for biotite A were to apply to all biotite specimens, then only a few diffusivities would be needed to establish this value.

On the basis of limited data from this laboratory on argon diffusion in biotite and muscovite, there appears to be no significant difference in argon diffusion between micas, provided that the mica compositions are similar. Two muscovites of different age but similar composition gave argon diffusivities which were similar within experimental uncertainties (Robbins, 1972). There is a difference, however, between micas in the phlogopite-annite series if the iron to magnesium ratios differ (Norwood, 1973). This is currently being explored further.

In the feldspars, there are more possible differences, including composition, structural state (Al-Si ordering), and amount of exsolved second feldspar phase. Foland's (1974 and this volume) and Lin and Yund's (1972) data indicate that diffusivities can now be estimated to within an order of magnitude for homogeneous K-feldspars for Ar and K, and activation energies to approximately 5 kcal/g-atom.

While it is too early to make a judgment, it appears that the differences on an atomic scale, such as the number and types of defects, may play a role in the precise value of D, but that they may not dominate the value such as to produce orders of magnitude differences in D. This is a subject which requires much more investigation, but the preliminary results are promising as a step toward the generalization of mineral properties.

In most plutonic rocks that have K-feldspar, the mineral particle is really a perthite. Diffusion of different species in microcline and orthoclase has been studied and will doubtless continue. Some work on albite is in progress. Once the single phases have been studied, a logical extension is to the two-phase lamellar perthites. This is much more complex, but should be a solvable problem.

From these examples, it should be clear that the actual value of D for a given element in a mineral at a given temperature may in the near future be known to a fair degree of accuracy. This accuracy is not likely to be very high, but that need not be a serious handicap to the interpretation of the geology. There are other factors, however, which may play as great, or greater, a role in deciphering a particular geological puzzle.

These factors relate to the external influences on a mineral. One of these is the presence or absence of sufficient fluid in the rock to provide a suitable transport medium, either by flow or by diffusion in the fluid, to permit elements to reach other minerals rapidly once they reach the surface of their original host. Another factor is the always present possibility that some other process may have occurred, such as chemical reaction or recrystallization. There may be clearcut cases in which it can be shown that diffusion was dominant, others in which another process operated, and some in which the dominant factor cannot be identified with certainty. Knowledge of the diffusivities will aid in determining which effect is important.

I wish to thank Drs. K. A. Foland, S. R. Hart, and A. W. Hofmann for critical reading of the manuscript and numerous useful suggestions.

References Cited

Adams, J. A. S., P. M. C. Barretto, and R. B. Clark, Radon 222 loss from zircons and sphenes; stochastic considerations of discordant lead isotopic ages (abstract), *Geol. Soc. Amer. Abstr. with Programs, 4,* 430, 1972.

Baadsgaard, H., J. Lipson, and R. E. Folinsbee, The leakage of radiogenic argon from sanidine, *Geochim. Cosmochim. Acta, 25,* 147–157, 1961.

Baadsgaard, H., and O. van Breemen, Thermally-induced migration of Rb and Sr in an adamellite, *Eclogae Geol. Helv., 63,* 31–44, 1970.

Brandt, S. B., V. N. Smirnov, I. L. Lapides, N. V. Volkova, and V. I. Kovalenko, Radiogenic argon as geochemical indicator of hydrothermal stability of some minerals, *Geokhimiya, no. 8,* 1010–1012, 1967.

Carslaw, H. S., and J. C. Jaeger, *Conduction of Heat in Solids,* 2nd ed., Oxford Univ. Press, London, 510 pp., 1959.

Crank, J., *The Mathematics of Diffusion,* Oxford Univ. Press, London, 347 pp., 1957.

Deuser, W. G., The effects of temperature and water pressure on the apparent Rb-Sr age of micas, Ph.D. thesis, Pennsylvania State Univ., 107 pp., 1963.

Evernden, J. F., G. H. Curtis, R. W. Kistler, and J. Obradovich, Argon diffusion in glauconite, microcline, sanidine, leucite, and phlogopite, *Amer. J. Sci., 258,* 583–604, 1960.

Fleischer, R. L., P. B. Price, E. M. Symes, and D. S. Miller, Fission-track ages and track-annealing behavior of some micas, *Science, 143,* 349–351, 1964.

Fleischer, R. L., P. B. Price, and R. M. Walker, Effects of temperature, pressure, and ionization on the formation and stability of fission tracks in minerals and glasses, *J. Geophys. Res., 70,* 1497–1502, 1965.

Foland, K. A., Cation and Ar^{40} diffusion in orthoclase, Ph.D. thesis, Brown Univ., Providence, R.I., 143 pp., 1972.

Foland, K. A., Ar^{40} diffusion in homogeneous orthoclase and an interpretation of Ar diffusion in K-feldspars, *Geochim. Cosmochim. Acta, 38,* 151–166, 1974.

Foland, K. A., Alkali diffusion in orthoclase, this volume.

Gerling, E. K., and G. V. Ovchinnikova, Causes of low age values determined by the Rb-Sr method, *Geokhimiya, no. 9,* 755–762, 1962.

Giletti, B. J., and J. L. Kulp, Radon leakage from radioactive minerals, *Amer. Mineral., 40,* 481–496, 1955.

Giletti, B. J., Studies in diffusion I: Argon in phlogopite mica, this volume.

Hanson, G. N., Radiogenic argon loss from biotites in whole rock heating experiments, *Geochim. Cosmochim. Acta, 35,* 101–107, 1971.

Hanson, G. N., and P. W. Gast, Kinetic studies in contact metamorphic zones, *Geochim. Cosmochim. Acta, 31,* 1119–1153, 1967.

Hart, S. R., The petrology and isotopic mineral age relations of a contact zone in the Front Range, Colorado, *J. Geol., 72,* 493–525, 1964.

Hofmann, A. W., Hydrothermal experiments on equilibrium partitioning and diffusion kinetics of Rb, Sr, and Na in biotite-alkali chloride solution systems, Ph.D. thesis, Brown Univ., Providence, R.I., 105 pp., 1969.

Hofmann, A. W., and B. J. Giletti, Diffusion of geochronologically important minerals under hydrothermal conditions, *Eclogae Geol. Helv., 63,* 141–150, 1970.

Hurley, P. M., H. Hughes, W. H. Pinson, Jr., and H. W. Fairbairn, Radiogenic argon and strontium diffusion parameters in biotite at low temperatures obtained from Alpine Fault uplift in New Zealand, *Geochim. Cosmochim. Acta, 26,* 67–80, 1962.

Kistler, R. W., J. D. Obradovich, and E. D. Jackson, Isotopic ages of rocks and minerals from the Stillwater Complex, Montana, *J. Geophys. Res., 74,* 3226–3237, 1969.

Kulp, J. L., and W. H. Bassett, The base-exchange effects on potassium-argon and rubidium-strontium isotopic ages, *Ann. N.Y. Acad. Sci., 91,* 225–226, 1961.

Kulp, J. L., and J. Engels, Discordances in K-Ar and Rb-Sr isotopic ages, in *Radioactive Dating,* International Atomic Energy Agency, Vienna, p. 219, 1963.

Letolle, R., Sur l'abondance relative de l'isotope 41 du potassium suivant son origine geologique, *C. R. Acad. Sci., Paris, 257,* 3996–3998, 1963.

Lin, T. H., and R. A. Yund, Potassium and sodium self-diffusion in alkali feldspar, *Contrib. Mineral. Petrol., 34,* 177–184, 1972.

Manning, J. R., Diffusion kinetics and mechanisms in simple crystals, this volume.

Maurette, M., P. Pellas, and R. M. Walker, Etude des traces de fission fossiles dans le mica, *Bull. Soc. Fr. Mineral. Cristallogr., 87,* 6–17, 1964.

McNutt, R. H., A study of strontium redistribution under controlled conditions of temperature and pressure, *Mass. Inst. Technol. Dep. Geol. Geophys. U.S.A.E.C. 12th Annu. Rep. AT(30-1) 1381,* pp. 125–157, 1964.

Mussett, A. E., Diffusion measurements and the potassium-argon method of dating, *Geophys. J. Roy. Astron. Soc., 18,* 257–303, 1969.

Newland, B. T., On the diffusion of radiogenic argon from potassium feldspars, unpublished

M.S. thesis, Univ. of Alberta, Edmonton, 119 pp., 1963.

Nicolaysen, L. O., Graphic interpretation of discordant age measurements on metamorphic rocks, *Ann. N.Y. Acad. Sci., 91,* 198–206, 1961.

Norwood, Curtis B., Radiogenic argon diffusion in the biotite micas, unpublished M.S. thesis, Brown Univ., Providence, R.I.

O'Neal, J. R., and H. P. Taylor, The oxygen isotope and cation exchange chemistry of feldspars, *Amer. Mineral., 52,* 1414–1437, 1967.

Pidgeon, R. T., J. R. O'Neil, and L. T. Silver, Uranium and lead isotopic stability in a metamict zircon under experimental hydrothermal conditions, *Science, 154,* 1538–1540, 1966.

Price, P. B., and R. M. Walker, Observations of charged-particle tracks in solids, *J. Appl. Phys., 33,* 3400–3406, 1962.

Robbins, Gary A., Radiogenic argon diffusion in muscovite under hydrothermal conditions, unpublished M.S. thesis, Brown Univ., Providence, R.I.

Schreiner, G. D. L., and A. A. Verbeek, Variations in K^{39}/K^{41} and the movement of potassium in a granite-shale contact region, *Proc. Roy. Soc., Ser. A, 285,* 423–429, 1965.

Schreiner, G. D. L., and H.-J. Welke, Variations in K^{39}/K^{41} ratio and movement of potassium in heated and stress xenoliths, *Geochim. Cosmochim. Acta, 35,* 719–726, 1971.

Schwartzman, D. W., Excess argon in the Stillwater complex and the problem of mantle-crustal degassing, Ph.D. thesis, Brown Univ., Providence, R.I., 120 pp., 1971.

Suggate, R. P., The Alpine fault, *Trans. Roy. Soc. N.Z., Geol., 2,* 105, 1963.

Tilton, G. R., Volume diffusion as a mechanism for discordant lead ages, *J. Geophys. Res., 65,* 2933–2945, 1960.

Tolland, H. G., and R. G. J. Strens, Electrical conduction in physical and chemical mixtures. Application to planetary mantles, *Phys. Earth Planet. Interiors, 5,* 380–386, 1972.

Westcott, M. R., Loss of argon from biotite in a thermal metamorphism, *Nature (London), 210,* 83, 1966.

Wetherill, G. W., Discordant uranium-lead ages, *Trans. Amer. Geophys. Union, 37,* 320–326, 1956.

ALKALI DIFFUSION IN ORTHOCLASE

K. A. Foland
Department of Geological Sciences and Materials Research Laboratory
Brown University
Providence, Rhode Island 02912 [1]

ABSTRACT

The diffusion of alkalies in a natural, homogeneous orthoclase (Or_{94}) has been studied in hydrothermal experiments at 2 kbar from 500° to 800°C. Diffusivities were determined by measuring isotope exchange between feldspar grains and approximately 2-molar alkali chloride solutions, where these had compositions such that the crystals did not change during an experiment. Alkali equilibrium partitioning determinations for the natural orthoclase with trace amounts of Rb are similar to literature values for synthetic samples. Diffusion coefficients were calculated using both the ideal cylindrical model, which appears to be the best representation of alkali diffusion in the anisotropic structure, and an isotropic model for spherical grains. The choice of models does not affect the systematics of the results; the main effect of using the cylindrical model in preference to the spherical is to raise D values by a factor of about 2.5. That is, diffusivities are changed by a relatively small amount but the activation energies change hardly at all. Results on three different grain sizes which vary by a factor of 4 demonstrate that the particle size is the effective dimension for diffusion. For the temperature intervals studied self-diffusion coefficients give simple Arrhenius relations which are based on the cylindrical model and are described by:

	D_0 (cm²/sec)	Q (kcal/mole)
Na	8.9	52.7
K	16.1	68.2
Rb	38.	73.

At 800°C, $D_{Na} \sim 1000\ D_K \sim 3000\ D_{Rb}$. Such a high D value for Na, relative to K and Rb, suggests a direct interstitial mechanism. However, the larger ionic sizes of K and Rb may prohibit direct interstitial transport so that a different mechanism, possibly a vacancy or interstitialcy one, may predominate in the larger alkali cations. The diffusion coefficients for Na and K in this orthoclase are in general agreement with those reported for other K-feldspars.

INTRODUCTION

The process of diffusion in minerals, especially silicates, has recently become a subject of increasing interest. Diffusional processes and rates have been recognized as having important implications for many aspects of petrology and geochemistry. Feldspars are of obvious interest because they comprise such a large portion of the crust.

In the context of this volume it is not necessary to detail applications of kinetic data, so only a few possible ones are noted here. Helgeson (1971) has shown

[1] Present address: Department of Geology, University of Pennsylvania, Philadelphia, Pa. 19174.

that the rates of mass transfer among silicates and aqueous solutions are in some cases diffusion controlled. The application of the infiltration metasomatism theory discussed by Hofmann (1972) is dependent on the rate of solid-fluid equilibration relative to the rate of fluid flow. Albarede and Bottinga (1972), in a discussion of disequilibrium partitioning between phenocrysts and host lava, have shown the importance of diffusional kinetics in both solid and melt. The application of the phase relations and ion diffusivities in the Fe-Ni system to concentration gradients in meteorite metal phases by Wood (1964) and others has added much to our knowledge of the cooling rates and origin of meteorites. Some detailed microprobe studies, such as Grant and Weiblen's (1971), have shown concentration gradients within minerals. Alkali feldspar exsolution and the formation of perthites have long been subjects of great interest. The importance of diffusion to exsolution rates has been discussed by Yund and McCallister (1970).

One of the most interesting geologic applications of diffusion data is to geochronology when interpreting discordant ages. Jäger et al. (1967) have used discordant ages to estimate Alpine cooling rates and, by extension, rates of Alpine uplift. Direct knowledge of the temperatures at which various minerals close, with respect to gain or loss of the nuclides of interest, is desirable for such applications.

Investigations of diffusion in feldspar include many argon studies but only a few on cation mobility. Discussions of and references to Ar diffusion may be found in Mussett (1969) and Foland (1974). Mérigoux (1968) and Yund and Anderson (this volume) have studied oxygen mobility in alkali feldspars. Studies of alkali diffusion include: Jensen (1952), Na in microcline-perthite; Jagitsch and Olsson (1954), interdiffusion of K and Na in albite-perthite couples; Sippel (1963), Na in orthoclase-perthite, microcline-perthite, and albite; Bailey (1971), Na in albite; Lin and Yund (1972), Na in albite and K in microcline; and Petrović (1974), K-Na interdiffusion in adularia and albite.

This report presents data on the diffusivities of K, Na, and Rb in orthoclase. No attempt is made to apply the data to specific examples at this time.

Experimental Method

The sample used in this study is a natural orthoclase from Benson Mines, near Star Lake, N.Y. Analyses and descriptions of this feldspar are given by Foland (1974); only a very brief description is given here. Chemical, x-ray, and microprobe analyses have shown that the Benson Mines feldspar is homogeneous and nonperthitic. It has a composition corresponding to Or_{94} and an intermediate degree of Al-Si tetrahedral ordering as it plots near the "orthoclase line" on a $b-c$ diagram (Wright and Stewart, 1968). The specimen is clear and colorless with well-developed cleavage and shows no alteration, zoning, or twinning. Optical and x-ray examination and unit cell refinements of the feldspar after hydrothermal treatment show it to be stable under the experimental conditions. Carefully sized grains with average diameters of 125 ± 6, 232 ± 10, and 480 ± 17 microns were used. The errors given represent two standard deviations of the mean for a population of approximately 60 individual grains.

The important aspects of suitability and stability of natural samples used for kinetic studies, especially feldspars which are commonly perthitic, have been discussed by Foland (1974) and Giletti (this volume). A one-phase K-feldspar has been used in order to obtain the simplest behavior and to provide an end-member upon which to build. The method used here, isotopic exchange in a system which is already in equilibrium except for isotopic distribution, is similar to that described by Hofmann and Giletti

(1970). Diffusion coefficients obtained this way are tracer or self-diffusion coefficients.

The actual experiments to measure diffusivities were those of isotope exchange between orthoclase (Or_{94}) and a $2M$ aqueous alkali chloride solution. With this approach, equilibrium cation partitioning must first be determined for the orthoclase-fluid system, particularly K-Na partitioning, so that the sample does not change during an experiment. Changes in major element chemistry will likely cause structural changes and the observed transport may be modified by other kinetic effects. Petrović (1973) suggests that the ion exchange process in feldspars under hydrothermal conditions takes place by dissolution and reprecipitation rather than diffusion, when the Na/K ratio of the fluid departs markedly from equilibrium. Since the diffusion coefficient may depend on composition and structure, changes in major element chemistry may produce the additional complication of diffusion coefficient gradients.

An ideal experiment is one in which the chemical compositions of fluid and solid phases remain constant and only isotope exchange occurs. For such an experiment, the amount of potassium exchange may be expressed as:

$$f = \frac{C41^{t}_{fl} - C41^{0}_{fl}}{C41^{e}_{fl} - C41^{0}_{fl}} \quad (1)$$

where:

$f =$ the fractional approach to isotopic equilibrium for K;
$C41^{0}_{fl} =$ the initial atom fraction of K^{41} in K, in the fluid;
$C41^{t}_{fl} =$ the atom fraction of K^{41} in K, in the fluid after time t; and,
$C41^{e}_{fl} =$ the atom fraction of K^{41} in K, at equilibrium.

Analogous expressions may be written for other isotopes in both fluid and solid phases. The above formulation is valid as long as the kinetic isotope effect and the equilibrium isotopic fractionation are negligible. Once the fractional approach has been determined, diffusion coefficients may be calculated for diffusion from a well-stirred reservoir of limited volume.

The choice of diffusion model is not clear cut and the literature contains conflicting evidence. In a study of Na diffusion in albite, Bailey (1971) indicated that Na diffusion perpendicular to (001), (010), and (110) cleavage faces was the same within experimental error. However, this apparent isotropic behavior, based on two shallow profiles determined by a few points, would apply to Na-feldspar and not necessarily K-feldspar. Experiments by Lin (1971) suggested approximately isotropic diffusion in K-feldspar. These observations are based on gross ion exchange of microcline 1 cm³ cubes. With such exchange, feldspar crystals are expected to fracture (Petrović, 1973) and transport may be primarily along grain boundaries. On the other hand, while Petrović (1974) indicates nearly isotropic behavior in the (a, c) plane, he finds that between 800° and 1000°C the value of D for the (010)* direction is about one hundredth that of D in the (110)* direction. Such an anisotropy is consistent with the crystal structure of K-feldspar. Consequently, for the purpose of the present study, the diffusion process could be approximated by the ideal cylindrical model. That is, diffusion in the (a, c) plane is isotropic, but diffusion perpendicular to it (in the b direction) is negligible.

On the basis of these considerations, the cylindrical model has been adopted here. However, D values are also calculated using an isotropic model for spherical particles, hereafter referred to as the spherical model. Deviations due to the shape of feldspar grains and to a slight variation of grain size within any collection of grains have been discussed by Lin and Yund (1972) and by Foland (1974) on the basis of the work of Jain (1958) and Gallagher (1965). Suffice it to say here that these factors will not introduce significant errors ($>20\%$) unless the fractional approach to equilibrium is greater than 80%. Of course, it is possible that for different species different geometric models apply. Values calculated with the spherical model will permit a more meaningful comparison with other studies in which the spherical model has been employed.

Values of D have been calculated from approximate solutions given by Carman and Haul (1954) which are applicable to diffusion from a well-stirred reservoir of limited volume when the diffusion time is small. For both spherical and cylindrical models:

$$\left(\frac{M_t}{M_\infty}\right) = (1 + \alpha)$$

$$\left[1 - \frac{\gamma_1}{\gamma_1 + \gamma_2} \text{eerfc} \left\{\frac{C\gamma_1}{\alpha}\left(\frac{Dt}{a^2}\right)^{1/2}\right\}\right.$$

$$\left. - \frac{\gamma_2}{\gamma_1 + \gamma_2} \text{eerfc} \left\{-\frac{C\gamma_2}{\alpha}\left(\frac{Dt}{a^2}\right)^{1/2}\right\}\right]$$

$$+ \text{ higher terms} \qquad (2)$$

where:

$\gamma_1 = \frac{1}{2}[(1 + 4\alpha/3)^{1/2} + 1]$ (spherical model); or:
$\gamma_1 = \frac{1}{2}[(1 + \alpha)^{1/2} + 1]$ (cylindrical model);
$\gamma_2 = \gamma_1 - 1$;
eerfc $z = \exp z^2$ erfc z;
$a = $ particle radius;
$D = $ diffusion coefficient;
$t = $ time;
$\left(\frac{M_t}{M_\infty}\right)(\equiv f) = $ the fractional approach to equilibrium;
$C = 3$ (spherical model); or
$C = 2$ (cylindrical model); and
$\alpha = $ equilibrium amount of diffusing species in fluid/ that in solid.

The higher terms in these approximate solutions may be neglected.

D values were obtained by determining the fractional approach to equilibrium with expressions like (1) and then using expression (2) to obtain (D/a^2), except in some cases where the corresponding exact solutions given by Crank (1956) are more appropriate because large values of f were attained in the experiment. The actual particle radius was used to calculate D values.

EXPERIMENTAL METHODS

Experiments described here were performed under hydrothermal conditions in cold-seal pressure vessels using H_2O as a pressure medium (2 kbar, 500° to 800°C). Temperature was maintained to less than ±2°C and pressure to within ±50 bars. Run duration ranged from a few hours to about seven weeks. The run-up and cool-down times were negligible compared to total heating time.

Hydrothermal Technique

Charges, sealed into gold tubes, consisted of weighed amounts of orthoclase and hydrothermal solution. The actual fluid to solid ratio of charges ranged from about 2 to 0.2 by weight. A typical charge might contain 100 mg each of orthoclase and hydrothermal solution. After a run, the charges were opened into a Millipore filter apparatus permitting a quantitative recovery and separation of feldspar and filtrate liquid.

Starting solutions were prepared using stock salt solutions which were pipetted and weighed in polyethylene bottles. The total concentration of the hydrothermal solutions, mainly KCl and NaCl, was approximately $2M$ in Cl$^-$.

Analytical Determinations

Concentrations of Na and most K were determined by flame photometry using Li internal standards. All Rb and some K concentrations were determined mass spectrometrically. All isotope abundance and isotope dilution analyses were performed on a 6-inch-radius mass spectrometer using a single Ta filament. Na^{22} activity was measured by γ-ray scintillation counting.

DIFFUSION

The fractional approaches to equilibrium for K and Rb were determined by measuring the fluid isotopic compositions, while those for Na were determined by gamma counting of the solid and liquid. The tracers used were Na^{22}, K^{41}, and Rb^{87}.

Equilibrium Determinations

Throughout this report, (%K) is used to denote the quantity $100 \times \{[K]/([K]+[Na])\}$ (in moles) in the fluid phase; the superscripts 0, e, and t are used for initial, equilibrium, and end of run values, respectively.

Equilibrium alkali partitioning was determined by a series of experiments at each temperature. Runs were designed so that the orthoclase would buffer the solution (i.e., low ratio of alkalies in fluid to alkalies in solid). $(\%K)^t$ and $(K/Rb)^t$ values were determined after each experiment, and a number of successive experiments were performed, each having a starting composition closer to the equilibrium value. The procedure was carried out for both directions, that is, $(\%K)^e$ values were approached from both higher and lower initial values. Equilibrium partition coefficients for Or_{94} and $2M$ (K, Na)Cl aqueous solutions at 2 kbar are given in Table 1. The equilibrium fluid is enriched in both Na and Rb relative to K over the solid and this enrichment increases with decreasing temperatures. For reference, the Benson Mines orthoclase has the molar ratios:

$(Rb/K) = 1.59 \times 10^{-3}$, and
$(Na/K) = 6.59 \times 10^{-2}$.

TABLE 1. Partition coefficients

T (°C)	P_{Na}*	P_{Rb}*
800	0.139 ± 0.002	0.48 ± 0.02
745	0.117 ± 0.002	0.44 ± 0.02
700	0.0974 ± 0.0015	0.38 ± 0.02
600	0.0634 ± 0.0025	

* $P_X = \dfrac{(X/K) \text{ (orthoclase)}}{(X/K) \text{ (fluid)}}$

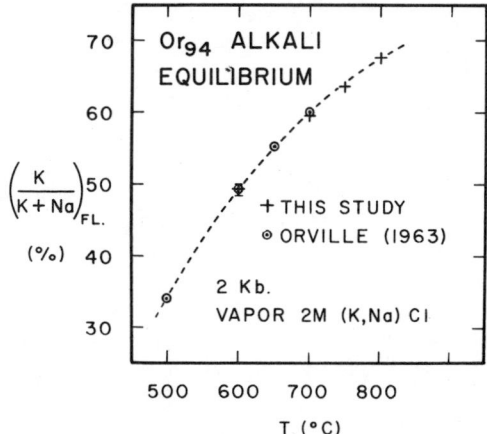

Fig. 1. Equilibrium alkali fluid compositions (in mole per cent) for Or_{94} as a function of temperature.

$(\%K)^e$ values are shown as a function of temperature in Fig. 1. Data from Orville (1963), in some cases interpolated from his graphs, are shown for comparison. The dashed curve in Fig. 1 is drawn only to show the consistency between the two sets of data. Rb partition coefficients are similar to those recently reported by Beswick (1973). Thus, the alkali partitioning for this natural orthoclase is very similar to that reported for synthetic alkali feldspars.

Kinetic Results

Alkali diffusivities are given below in Tables 2, 3, and 5. The measured average particle radius has been used for the a value. Uncertainties for D are obtained by considering only the uncertainty in the fractional approach to equilibrium. Thus when comparing D values of different grain sizes an additional small uncertainty, arising due to the uncertainty in a, should be included. Depending on the values for α and f, the uncertainty in D will not always be symmetrical about the best value for D. For this reason maximum and minimum D values are quoted. Log D's are simply the base 10 logarithms of D in cm²/sec. When denoting cations, charge signs have been omitted.

TABLE 2. Na self-diffusion data.

Run	T (°C)	t (day)	a (μm)	α	f	D_{Na} (cm²/sec)*	(D/a^2) (sec^{-1})	D_{Na} (cm²/sec)
78	800.	0.551	116.	1.982	0.614	8.4×10^{-11}	1.3×10^{-6}	$1.7 \pm 0.1 \times 10^{-10}$
74	800.	1.92	240.	2.505	0.571	9.2×10^{-11}	3.3×10^{-7}	$1.9 \pm 0.1 \times 10^{-10}$
84X	745.	18.83	240.	3.706	0.748	2.4×10^{-11}	7.9×10^{-8}	$4.6 \pm 0.2 \times 10^{-11}$
61D	700.	23.11	116.	44.29	0.728	5.5×10^{-12}	7.4×10^{-8}	$1.0 \pm 0.1 \times 10^{-11}$
32G	700.	20.83	116.	9.174	0.787	7.2×10^{-12}	9.4×10^{-8}	$1.26 \pm 0.05 \times 10^{-11}$
48	600.	7.00	116.	5.865	0.150	3.5×10^{-13}	5.8×10^{-9}	$7.8 + 1.7 - 1.5 \times 10^{-13}$
50	600.	40.80	116.	10.08	0.309	3.3×10^{-13}	5.2×10^{-9}	$7.0 \pm 0.8 \times 10^{-13}$
62	600.	19.91	63.5	0.723	0.584	3.0×10^{-13}	1.6×10^{-8}	$6.3 \pm 0.5 \times 10^{-13}$
63	600.	19.91	63.5	0.957	0.524	2.7×10^{-13}	1.4×10^{-8}	$5.8 \pm 0.5 \times 10^{-13}$
118Y	500.	47.0	63.5	6.086	0.120	1.0×10^{-14}	5.5×10^{-10}	$2.2 + 0.8 - 0.7 \times 10^{-14}$

*Based on the spherical model. Other diffusivities according to cylindrical model.

Na Diffusion

Na diffusion results are given in Table 2. Maximum and minimum D_{Na} values are obtained by assigning an uncertainty of ± 0.015 in f.

The fractional approach to equilibrium for Na is based on the initial specific activity of Na^{22} in the solution and final activity in the solid. Weighed orthoclase samples were counted and were assumed to have the original Na concentration. Due to the large α's, the relative change of initial to final activity is less for the solution than for the solid. Because of the increase in sensitivity obtainable, it is advantageous to calculate f based on final solid activity.

D_{Na} values were calculated from fractional approaches to equilibrium which ranged from 12% to 79%. Na self-diffusion is very fast compared to other cations. D_{Na} changes by a factor of $\sim 10^4$ from 500° to 800°. Multiple determinations at 600° and at 800° show good agreement and, within experimental error, different grain sizes give the same diffusion coefficient.

In some runs, specifically those at 600° and 700°, neither initial nor final solutions had the exact value of $(\%K)^e$. However, the rate of Na isotopic equilibration is very fast compared to the rate of Na chemical equilibration (see below), and the results will not be affected because for the duration of the experiment the orthoclase Na concentration will not change significantly. This is borne out by the agreement of D_{Na}'s from runs 62 and 63 where fluid compositional change is in different directions. Figure 2 is an Arrhenius plot of the D_{Na} data. Both spherical and cylindrical models are shown for comparison. Lines are least-square best-fit lines determined by the method of York (1966).

K Diffusion

Table 3 gives the potassium diffusion data. Values of f, based on measurements for K isotopic composition of initial and final solutions, vary from 1 to 55%. Their uncertainties are derived by considering the uncertainties of mass spectrometer measurements. It is emphasized that D_K's presented in this section are calculated from isotope exchange data for experiments at chemical equilibrium; that is, $(\%K)^o$ values for these runs were

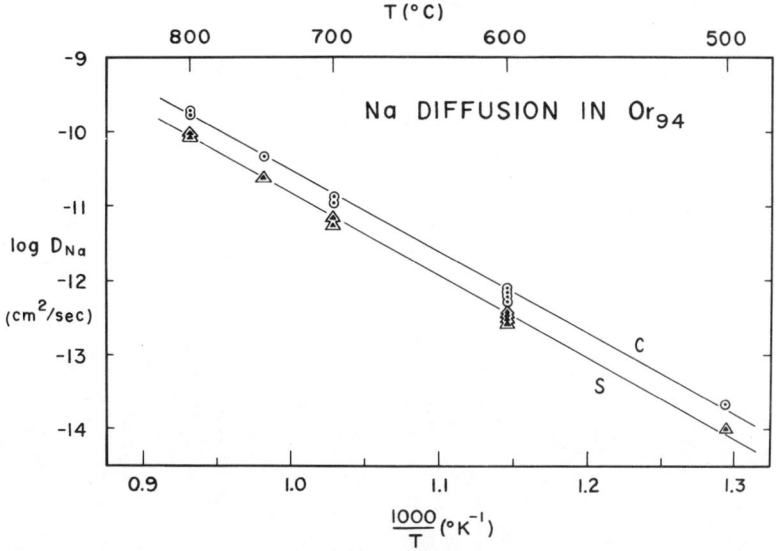

Fig. 2. Arrhenius plot for Na diffusion in Or_{94}. Open circles and upper line correspond to the cylindrical model, triangles and lower line apply to the spherical model.

TABLE 3. K self-diffusion data

Run	T (°C)	t (day)	a (μm)	α	f	D_K (cm²/sec)*	(D/a^2) (sec⁻¹)	D_K (cm²/sec)
46	800.	4.28	116.	0.470	0.154	9.5×10^{-14}	1.6×10^{-9}	$2.1 \pm 0.3 \times 10^{-13}$
80	800.	4.48	63.5	0.114	0.553	9.2×10^{-14}	5.0×10^{-9}	$2.0 \pm 0.1 \times 10^{-13}$
83	800.	10.42	240.	0.406	0.137	1.1×10^{-13}	4.1×10^{-10}	$2.4 + 0.5 \times 10^{-13}$ $- 0.4$
96	700.	6.95	63.5	0.0772	0.270	3.5×10^{-15}	1.9×10^{-10}	$7.8 \pm 0.4 \times 10^{-15}$
48	600.	7.00	116.	0.319	0.011	1.4×10^{-16}	2.3×10^{-12}	$3.1 + 5.3 \times 10^{-16}$ $- 2.7$
50	600.	40.80	116.	0.542	0.013	6.8×10^{-17}	1.1×10^{-12}	$1.5 + 1.5 \times 10^{-16}$ $- 1.0$

*Based on the spherical model. All other diffusivities according to cylindrical model.

the values shown as crosses in Fig. 1. The alkali salt solutions can be prepared very precisely; therefore the actual deviations from chemical equilibrium are very small and depend upon the accuracy of $(\%K)^e$ values. Except at 600°C these ratios are reproducible to better than 1% in $(K/K + Na)$.

From Table 3 it is evident that D_K is much smaller than D_{Na}. The difference, three orders of magnitude at 800°, increases with decreasing temperature. D_K values at 800° for three experiments using different sizes of orthoclase show good agreement.

Except for 600° runs, the data in Table 3 derive from experiments in which the initial fluid composition was that at equilibrium. The small initial departure from equilibrium of the 600°C runs could cause significant errors. However, D_K for runs 48 and 50 have large uncertainties anyway because of the very small amount of exchange.

Figure 3 shows that the data probably obey the Arrhenius relation with a straight line. The line is primarily defined by data at only two temperatures, 700° and 800°. The 600° data have large uncertainties but their consistency justifies extension of the line to that temperature. (In the log D vs. $1/T$ plots, unless indicated by error bars, the uncertainties are within the open circles.) Both cylindrical and spherical models are shown with least-squares lines for all the D_K data.

Ion Exchange Rates

Thus far, results obtained from isotope exchange data in experiments at chemical equilibrium have been considered. However it is possible, once $(\%K)^e$ is known, to calculate diffusion coefficients based on chemical ion exchange between nonequilibrated orthoclase and solution. In this case:

$$f = \frac{(\%\ K)^t - (\%\ K)^0}{(\%\ K)^e - (\%\ K)^0} \quad (3)$$

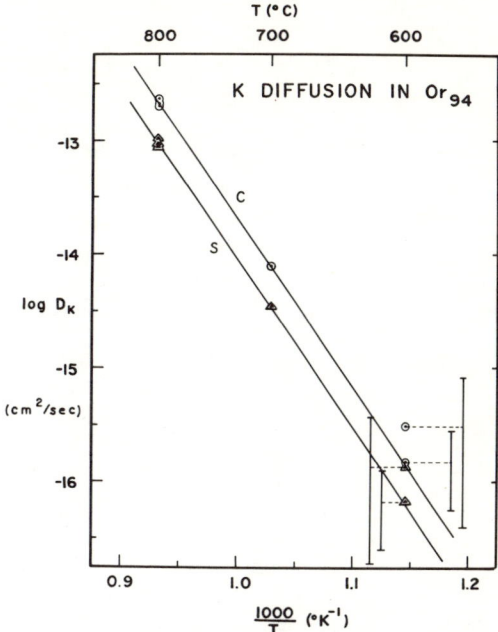

Fig. 3. Arrhenius plot for K diffusion in Or_{94}. Same representation as Fig. 2. For clarity, error bars for 600° data are displaced.

Thus f for this purpose is the fractional approach to chemical equilibrium for the fluid phase. D values, which might be termed apparent K diffusion coefficients, are based on ion exchange and have been calculated for partitioning runs which were neither initially nor finally at equilibrium. There is a net flow of K in and Na out of the feldspar, or vice versa, in such experiments. For these calculations the proper α is the equilibrium value, which may be calculated. Diffusion coefficients ($\overline{D}_K$) calculated in this manner are summarized in Table 4.

The equilibrium values in Fig. 1 and Table 1 have been used. Only runs for which $(\%K)^0$, $(\%K)^t$, and $(\%K)^e$ may be distinguished by more than the analytical error have been considered. The f's have large uncertainties due to the small differences between initial, final, and equilibrium compositions and due to the relatively large uncertainties of the flame photometric measurements. Resulting

TABLE 4. Data for calculation of apparent K diffusion coefficients based on ion-exchange rates*

Run	T (°C)	t (day)	α	(% K)⁰	(% K)^t	f	$\bar{D}_K$ (cm²/sec)	$\bar{D}_K$ (MAX)	$\bar{D}_K$ (MIN)
33	800.	4.67	0.137	60.70	64.55	0.54 ± 0.08	2.5 × 10⁻¹³	4.2 × 10⁻¹³	1.5 × 10⁻¹³
34	800.	4.88	0.129	52.50	61.51	0.57 ± 0.05	2.5 × 10⁻¹³	3.4 × 10⁻¹³	1.8 × 10⁻¹³
39	800.	5.98	0.262	64.56	65.95	0.42 ± 0.10	2.4 × 10⁻¹³	4.6 × 10⁻¹³	1.1 × 10⁻¹³
51E	747.	14.02	0.0996	60.46	61.83	0.40 ± 0.10	1.8 × 10⁻¹⁴	6.2 × 10⁻¹⁴	3.6 × 10⁻¹⁵
51F	747.	14.02	0.0982	56.23	60.00	0.51 ± 0.09	3.7 × 10⁻¹⁴	6.9 × 10⁻¹⁴	2.0 × 10⁻¹⁴
51G	747.	14.02	0.339	57.86	59.11	0.21 ± 0.09	2.4 × 10⁻¹⁴	5.7 × 10⁻¹⁴	6.7 × 10⁻¹⁵
30D	700.	9.81	0.0802	50.20	52.71	0.26 ± 0.06	5.6 × 10⁻¹⁵	9.6 × 10⁻¹⁵	3.0 × 10⁻¹⁵
32D	700.	20.83	0.0687	52.50	53.93	0.20 ± 0.10	3.5 × 10⁻¹⁵	9.2 × 10⁻¹⁵	7.5 × 10⁻¹⁵
53	700.	7.00	0.0541	55.33	58.36	0.70 ± 0.12	1.0 × 10⁻¹³	3.0 × 10⁻¹³	4.3 × 10⁻¹⁴
60	700.	10.52	0.0462	61.72	60.64	0.52 ± 0.23	1.4 × 10⁻¹⁴	7.9 × 10⁻¹⁴	2.4 × 10⁻¹⁵
56	600.	8.17	0.0433	46.87	47.21	0.16 ± 0.10	6.4 × 10⁻¹⁵	2.1 × 10⁻¹⁴	7.6 × 10⁻¹⁷
57	600.	6.08	0.0692	52.50	51.99	0.15 ± 0.10	1.7 × 10⁻¹⁵	5.8 × 10⁻¹⁵	1.4 × 10⁻¹⁶
62	600.	19.91	0.0493	52.79	51.25	0.27 ± 0.14	1.2 × 10⁻¹⁵	3.9 × 10⁻¹⁵	2.3 × 10⁻¹⁶
63	600.	19.91	0.0572	45.80	46.33	0.22 ± 0.14	4.9 × 10⁻¹⁶	2.2 × 10⁻¹⁵	9.5 × 10⁻¹⁸
35	600.	13.70	0.121	32.16	34.16	0.12 ± 0.05	1.3 × 10⁻¹⁵	2.9 × 10⁻¹⁵	4.1 × 10⁻¹⁶
41	600.	16.72	0.0736	41.25	43.10	0.24 ± 0.08	2.2 × 10⁻¹⁵	4.6 × 10⁻¹⁵	8.3 × 10⁻¹⁶
42	600.	17.76	0.119	38.62	40.01	0.13 ± 0.08	1.3 × 10⁻¹⁵	3.8 × 10⁻¹⁵	1.8 × 10⁻¹⁶
43	600.	14.20	0.120	44.40	45.10	0.15 ± 0.09	2.2 × 10⁻¹⁵	6.4 × 10⁻¹⁵	3.1 × 10⁻¹⁶

*For all experiments, $a = 63.5$ μm. Cylindrical model. (% K)^e values in Table 1 and Fig. 1.

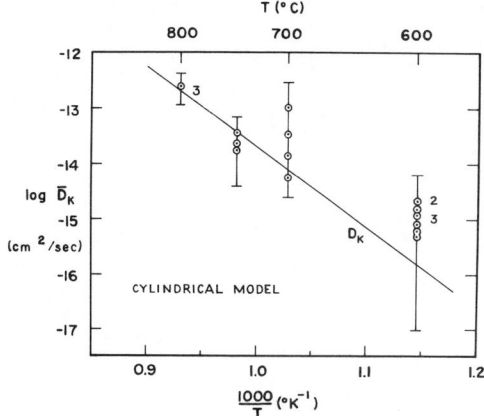

Fig. 4. Arrhenius plot of $\overline{D}_K$ (apparent K diffusion calculated from ion exchange data). Error bars give the total uncertainty for all values at the same temperature; those for individual points appear in Table 4. Solid line represents K self-diffusion as shown in Fig. 3. All results are based on the cylindrical model.

$\overline{D}_K$ values have correspondingly large uncertainties.

The striking feature in these experiments is that the rate of ion exchange is so slow. The calculated values reflect this. Figure 4 is a Arrhenius plot for the $\overline{D}_K$ data using the cylindrical model. Circles represent individual determinations which have maximum and minimum values given in Table 4. Error bars represent maximum and minimum values of all individual analyses at one temperature.

The individual measurements at the same temperature were not treated by normal statistical methods because they are not strictly members of the same population; that is, each value derives from an experiment of somewhat different conditions (see $(\%K)^0$ and $(\%K)^t$, Table 4). The line, which is shown for comparison, represents D_K as established in the preceding section.

Values of $\overline{D}_K$ at 800° and 745° are very similar to those for K diffusion (solid line in Fig. 4) measured under conditions of chemical equilibrium. Though not inconsistent, the 700° and 600° data are less convincing. At the lower temperatures, the value of D_K does not fall within the uncertainties of all $\overline{D}_K$'s. Probable reasons for these departures, very low α's and mechanical crushing of particles, will be discussed later.

In most of these runs the net K transfer was solid to fluid, that is, (%K) increased during the run. Runs in which the solid gained K gave the same results as those in which the solution gained K.

In these partitioning runs, the rate of approach to Na isotopic equilibration was much greater than the rate of approach to Na (and therefore K) chemical equilibration. This is shown especially in some 700° experiments where full Na isotopic equilibrium was achieved but the chemical equilibration had only gone about 10% of the way.

Rb Diffusion

Rb diffusion data appear in Table 5. Rb fractional approach to equilibrium was calculated in the normal manner, i.e., based on initial and final isotopic composition of the fluid phase.

Because of the very slow rates, most of the f values are small. The relative uncertainties for low f are large and the resulting D_{Rb} values have large uncertainties. Further, for a small f especially large errors may result if the alkali composition of the fluid is not in equilibrium with the orthoclase.

The three 800° experiments agree within the analytical error. These runs were made with different size particles. D_{Rb} at 800° appears to be well established. D_{Rb} at lower temperatures, however, is not known as accurately. A large number of 700° experiments were conducted, but unfortunately most of them had initial compositions which were far from chemical equilibrium. Run 96 represents a run with the proper initial fluid (equilibrium composition) and is considered to give the best measure of D_{Rb} at 700°. D_{Rb} values for all 700° data in

TABLE 5. Rb diffusion data

Run	T (°C)	t (day)	a (μm)	α	f	D_{Rb} (cm²/sec)*	(D/a^2) (sec⁻¹)	D_{Rb} (cm²/sec)
46†	800.	4.28	116.	1.151	0.052	2.6×10^{-14}	4.3×10^{-10}	$5.8 {+1.7 \atop -1.5} \times 10^{-14}$
80†	800.	4.48	63.5	0.239	0.226	2.4×10^{-14}	1.4×10^{-9}	$5.5 \pm 0.3 \times 10^{-14}$
83†	800.	10.42	240.	0.853	0.054	3.4×10^{-14}	1.4×10^{-10}	$8.2 {+2.3 \atop -2.1} \times 10^{-14}$
84X	745.	18.83	240.	0.955	0.026	5.1×10^{-15}	2.0×10^{-11}	$1.2 {+0.8 \atop -0.7} \times 10^{-14}$
51G	747.	14.02	63.5	0.813	0.066	2.7×10^{-15}	1.5×10^{-10}	$6.1 {+1.6 \atop -1.4} \times 10^{-15}$
96†	700.	6.95	63.5	0.203	0.064	7.4×10^{-16}	4.2×10^{-11}	$1.7 {+0.5 \atop -0.4} \times 10^{-15}$
30D	700.	9.81	63.5	0.353	0.075	1.8×10^{-15}	9.8×10^{-11}	$4.0 \pm 0.9 \times 10^{-15}$
61A	700.	23.11	63.5	0.224	0.236	4.7×10^{-15}	2.6×10^{-10}	$1.0 \pm 0.2 \times 10^{-14}$
61C	700.	23.11	63.5	0.162	0.222	2.4×10^{-15}	1.3×10^{-10}	$5.4 \pm 0.5 \times 10^{-15}$
61D	700.	23.11	116.	11.12	0.054	1.5×10^{-14}	2.5×10^{-10}	$3.3 {+2.2 \atop -1.6} \times 10^{-14}$
50	600.	40.80	116.	0.696	0.029	9.9×10^{-16}	1.6×10^{-11}	$2.2 {+1.4 \atop -1.1} \times 10^{-15}$
63	600.	19.91	63.5	0.169	0.048	1.1×10^{-16}	5.9×10^{-12}	$2.4 {+1.1 \atop -0.9} \times 10^{-16}$

*Based on spherical model. Other diffusivities according to cylindrical model.
†Runs with initial equilibrium partitioning.

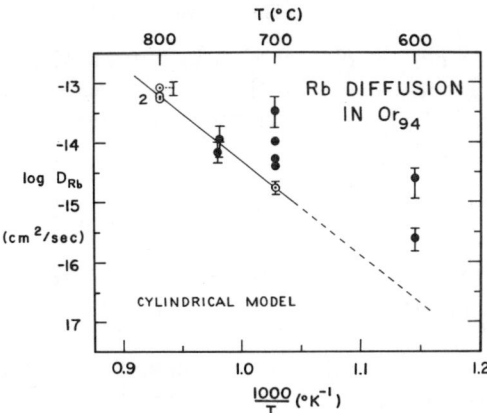

Fig. 5. Arrhenius plot for D_{Rb} according to the cylindrical model. Open circles represent runs with initial equilibrium partitioning; solid circles, nonequilibrium runs.

Table 5 show a range of approximately one order of magnitude. Those runs not at equilibrium show the effects which probably result from not adhering to the equilibrium requirement discussed above.

P_{Rb} at 600° is not known accurately, and runs 50 and 63 have similar departures. These nonequilibrium runs had compositions such that there was a net addition of Rb to the fluid. The D values are probably too high, as are the 700° nonequilibrium results. D_{Rb} at 600° is not accurately known but is probably less than the values given in Table 5.

Data from Table 5 are plotted in Fig. 5, where the solid circles indicate nonequilibrium experiments and therefore less reliable data. A straight line drawn through the reliable data points gives the best values for the Rb Arrhenius relation. Primarily because of the small temperature range, the slope of the line has a comparatively large uncertainty corresponding to ±5 kcal/g-atom. The 745° data are consistent with this, but the nonequilibrium points at lower temperatures clearly fall above it.

Discussion

It can be seen from Tables 2, 3, and 5 and from Figs. 2 and 3 that the choice of geometric models does not alter the systematics of the results. The net effect of selecting a cylindrical model with a strong anisotropy in preference to an isotropic model is that the D values are higher by a factor of 2 to 3. The data presented here do not, therefore, confirm the cylindrical model but would be consistent with it. While the isotope exchange technique used is very sensitive, it has the disadvantage that a model must be known or assumed unless the grain shape can be varied.

Diffusivities in the form (D/a^2) are determined so that a value for the effective dimension of diffusion, a, must be established in order to arrive at a D value. The parameter a is normally taken to be the actual particle size of the material in question, but the two are not always equivalent. It is possible that a mosaic exists, for example in fractured or perthitic samples, within the particles so that the effective grain size is reduced. To evaluate this factor for the orthoclase sample, several sizes have been used. Pertinent results for K, Na, and Rb are those at 800°C (see Tables 2, 3, and 5). For the same species at 800°C, (D/a^2) values show the expected differences, while D values for sizes with radii which vary by a factor of 4 show agreement. This relationship is exactly what is expected for the case where particle size is equivalent to the effective size, and it indicates that the particles behave as single crystals. Ar^{40} diffusion results for the Benson orthoclase show the same relationship, which applies regardless of the geometric model (Foland, 1974).

The agreement of D values for different particle sizes has additional implications. Ion and isotope exchange between fluid and surface layer(s) is probably very rapid. If this effect is significant with regard to total transport, large perturbations could result. Smaller particles would have a relatively larger surface area and therefore should be more seriously affected. Assuming that any surface effects are proportional to surface area, then the effects on the 63.5 μm (radius) grains would be nearly 15 times as great as those for the 240 μm grains. The agreement in D for experiments using the largest and the smallest sizes indicates that surface effects are not significant so long as the fluid/solid ratio is moderate and the fractional attainment of equilibrium is not too small. Lin and Yund (1972), using the same general technique, have also demonstrated that surface effects are negligible.

Within the range of experimental temperatures, the Arrhenius relation appears to be obeyed with a single activation energy for each species. Table 6 summarizes the frequency factors and activation energies for the diffusivities according to both cylindrical and spherical models. Ar^{40} diffusion parameters for this orthoclase (Foland, 1974) are included for comparison.

In the spherical model, the parameters would describe isotropic transport. However, for the cylindrical model, they apply to diffusion in the directions of major transport, that is, in the (a, c) plane. Throughout the remainder of this report, discussion will apply to the cylindrical model and to diffusion in the major transport directions.

Na Diffusion

Sodium self-diffusion in Benson Mines orthoclase is well defined between 500° and 800°C. Diffusivities are described by:

$$D_{Na} = (8.92^{+6.68}_{-3.83}) \exp - (52700 \pm 1100/RT).$$

Above 500°C, D values for Na are higher than those for other alkali cations and Ar^{40}.

It is of interest to compare the Or_{94} sodium data to the data for other feldspars as shown in Fig. 6. These data are for a variety of feldspars but concern here is only with data for potassium feldspars. The data of Jensen (1952) and Sippel (1963) are subject to question since in both cases perthitic samples were used. As Lin and Yund (1972) point out, these measurements must be influenced by grain boundary diffusion and the presence of two feldspar phases. Petrović (1974) has calculated Na tracer diffusion coefficients from interdiffusion measurements using an Or_{86} adularia and KCl. His measurements (see Fig. 6) show a range in D but are consistently lower, approximately by a factor of 4, than the values indicated by a higher temperature extrapolation of the Or_{94} data. A similar

TABLE 6. Summary of frequency factors and activation energies for the Arrhenius relation
$D = D_0 \exp(-Q/RT)$

	Na	K	Rb	Ar^{40}*
Experimental temperature range:	500°–800°C	600°–800°C	700°–800°C	500°–800°C
Cylindrical model:				
Q (kcal/mole)	52.7 ± 1.1	68.2 ± 0.9	73. ± 5.	43.1 ± 1.1
D_0 (cm²/sec)	8.92 +6.68 −3.83	16.1 +8.9 −5.7	38.0 +589. −35.7	0.0140 +0.0114 −0.0063
Spherical model:				
Q (kcal/mole)	52.8 ± 0.9	68.2 ± 1.0	72. ± 5.	43.8 ± 1.0
D_0 (cm² sec)	4.97 +2.62 −1.72	7.19 +3.9 −2.51	13.9 +142. −13.8	0.00982 +0.00660 −0.00371

Uncertainties are one sigma. All data are for Benson Mines orthoclase.
*The argon data are taken from Foland (1974) and have been recalculated using the cylindrical model.

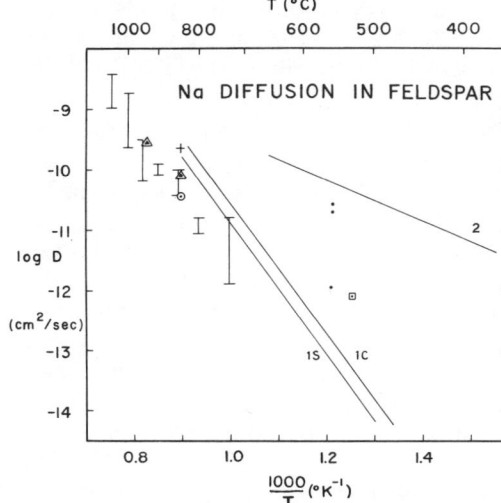

Fig. 6. Comparison of feldspar Na diffusion coefficients. Representation is (straight lines are Arrhenius relations): (1) (This study) Or_{94}, according to the cylindrical model (C) and the spherical model (S); (2) (Lin and Yund, 1972) Ab_{100}; brackets—approximate range of values by Petrović (1974) from interdiffusion between Or_{86} (adularia) and KCl; solid circles—Jensen (1952), perthite; open circle—Sippel (1963), orthoclase perthite; triangles—Sippel (1963), albite; square —Bailey (1971), albite; and cross—Sippel (1963) microcline perthite.

activation energy is indicated. It is not clear at this time whether the differences in D are experimental or the result of differences in the samples, e.g., composition with respect to Na/K. Nonetheless, there is general agreement, showing that two different specimens have Na diffusivities determined by different techniques which differ by less than one order of magnitude.

K Diffusion

K self-diffusion in Or_{94} is precisely defined at 800° (by three determinations) and at 700°C (by one determination). The 600° data with large uncertainties are consistent with a linear Arrhenius relation which is principally defined by the 800° and 700°C data. That relation is:

$$D_K = (16.1^{+8.9}_{-5.7}) \exp - (68200 \pm 900/RT).$$

The self-diffusion behavior of K in the 600°–800°C range is distinctly different from that of Na in both magnitude and activation energy.

The K data for Or_{94} are compared with similar data for other K-feldspars in Fig. 7. There is striking similarity between the Or_{94} data and those for Or_{100} (Lin and Yund, 1972). The Lin and Yund data derive from essentially the same technique except that microcline (exchanged to pure K end-member) and KCl, with K^{40} as a tracer, were used. Their data, using both molten KCl and 2-M aqueous KCl solutions in the interval 600°–800°C, yielded a linear Arrhenius relation with little scatter. Since Lin and Yund's D values are calculated for an isotropic model, line 2 (Lin and Yund) must be compared with line 1S (present results) in Fig. 7. (Note that use of the cylindrical model would likewise displace the Or_{100} data to higher values by the same amount as the Or_{94} data.) The activation energies (68 kcal for Or_{94} vs. 70 for Or_{100}) are virtually identical. D values differ somewhat in magnitude where D_K's for Or_{100} are higher, by fac-

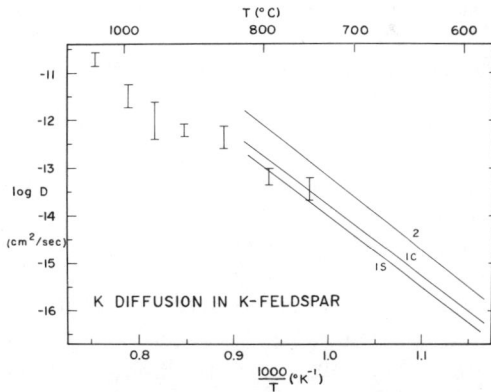

Fig. 7. Comparison of K-feldspar K diffusion coefficients. Straight lines are Arrhenius relations. (1) (This study) Or_{94} orthoclase according to the cylindrical (C) and spherical (S) models; (2) (Lin and Yund, 1972) Or_{100} microcline. Brackets: approximate range of values by Petrović (1974) from interdiffusion between Or_{86} (adularia) and KCl.

tors of about 7.5 to 5.5 at 800° to 600°C, than D_K's from this study.

The differences in D could be attributed to differences in the feldspar specimens. These might be in structural state; composition, including the Al/Si ratio; impurity or defect concentrations; or the effective dimension for diffusion. It is not possible, without further study and experimentation, to ascribe the observed relative behavior to any single factor. One simple explanation is that for Or_{100} the effective dimension is less than the grain size used by Lin and Yund. This is conceivable since their sample, originally twinned and perthitic, was exchanged to pure Or_{100} and such exchange may produce microfractures. Such behavior, the presence of a mosaic within grains, is exhibited in some studies of Ar diffusion in perthitic samples. If this were the case, the Or_{100} D_K's could be identical with those for Or_{94}.

There is also reasonable agreement between the Or_{94} data and the lower temperature K diffusion coefficients given by Petrović (1974) as shown in Fig. 7. Again, his K tracer coefficients were calculated from adularia K-Na interdiffusion profiles measured on the microprobe. At temperatures above about 850° extrapolated D_K's for Or_{94} would be somewhat higher but not unreasonably so. However, Petrović's data indicate a significantly lower activation energy. As discussed above, the reason for the differences is not clear. The tracer diffusion coefficients as calculated by Petrović from interdiffusion experiments are consistently a little lower for both K and Na than the values determined in this study.

In summary, Fig. 7 indicates very general agreement among three sets of data for K diffusion in K-feldspar.

Ion Exchange Kinetics

The sluggish approach to equilibrium in ion exchange reactions is striking. The experiments in Table 4 are true ion exchange, as there is no evidence for a dissolution-reprecipitation phenomenon and the (K, Na)Cl solutions never departed drastically from alkali equilibrium. The apparent K diffusion coefficients, denoted $\overline{D}_K$, are calculated from ion exchange rates in the same manner as for self-diffusion coefficients. These experiments were designed so that the amount of alkalies in orthoclase was greater than the amount in the fluid. Therefore, the Na/K ratio in the orthoclase may remain nearly constant in experiments in which the Na/K ratio of the fluid undergoes significant changes.

In these ion exchange experiments, it is possible to calculate an interdiffusion coefficient in the same way but using $\alpha =$ alkalies (fluid)/alkalies (orthoclase). If this is done, α is somewhat larger and therefore the D values will be higher. However, this procedure would yield an interdiffusion coefficient which is 2 to 3 times the value of $\overline{D}_K$ (Table 4), with the exact amount depending on the individual experiment. Petrović (1974) indicates that K-Na interdiffusion coefficients can be described by a fixed-framework model, that is, they are determined by molar concentrations and tracer diffusion coefficients which are approximately independent of composition. Using the Or_{94} composition and the Na and K tracer diffusion coefficient data, an interdiffusion coefficient 15 times larger than D_K would be predicted. The observed rates are much slower than one would predict on the fixed-framework model.

Noting this behavior, diffusion coefficients $\overline{D}_K$ were calculated on the basis of ion exchange data. So doing, there is first order agreement between $\overline{D}_K$ and D_K which *a priori* might not have been predicted. All data at 800° and 750°C yield values of $\overline{D}_K$ which are consistent with K self-diffusion behavior. Some lower temperature $\overline{D}_K$ values (one datum at 700°C and ½ the data at 600°C) are somewhat higher than D_K. Several possibilities may be advanced to explain these deviations. The object of the experiments was to obtain equilibrium data, so low fluid/solid ratios were used and the orthoclase was sometimes crushed when pressure was first applied to the charge. When crushing occurs, the grain size will usually be smaller and if the original size is used, the $\overline{D}_K$ value will be too high. This effect would be most important at 600° where low ratios are necessary because the reaction is so slow. Also, surface effects may be important when α values are very low. Any initial rapid surface exchange will be significant if the value for f is low. Surface exchange may be responsible for a significant amount of the observed transport at 600°. Additionally, the value of $(\%K)^e$ at 600° is not determined as precisely as at other temperatures. These factors can account for high values, and there is apparent first order equivalence between D_K and $\overline{D}_K$. Further, this equivalence applies to ion exchange in both directions of change in the fluid Na/K ratio.

The reason for this behavior is unclear, but although electroneutrality must be maintained, Na and K are apparently not coupled in the normal sense. The system may represent more complicated, perhaps ternary, diffusion behavior involving an unknown species. Since these experiments were hydrothermal, the proton is one possibility.

As a result, it appears that in such experiments the rate of fluid approach to Na/K equilibrium is controlled by orthoclase K diffusion. B. Giletti (personal communication) has found the same type of behavior in experiments on phlogopite, and Yao and Kummer (1967) found anomalously low interdiffusion coefficients for beta-aluminas.

Rb Diffusion

Rb diffusion values are less well determined. There is much scatter at lower temperatures because many experiments were nonequilibrium runs in which initial (K/Rb), and in some cases $(\%K)^0$, were not the equilibrium values. For those runs with appropriate fluid compositions (where $(K/Rb)^0 = (K/Rb)^e$) agreement is good. At 700° and 800°, the data best define Rb self-diffusion in Or_{94}. Based on data at these temperatures, the Arrhenius relation is:

$$D_{Rb} = (38.0^{+589.}_{-35.7}) \exp - (73000 \pm 5000/RT).$$

Runs not initially at equilibrium with respect to (K/Rb) will yield incorrect D_{Rb} values which are too high if the fractional approach to equilibrium is based only on isotope exchange; that is, uptake calculated from Equation 1 alone is no longer an accurate measure of the self-diffusion rate. Any net addition of Rb in the fluid will constitute a large portion of the apparent isotope exchange for any experiment where f is small. Further, nonequilibrium Rb runs may represent fairly complex ternary diffusion behavior. Additional experiments, with the proper fluid compositions, must be performed below 700° in order to obtain accurate D_{Rb} values at low temperatures.

There are no other data for Rb diffusion in feldspars.

General

The diffusivities obtained for the Benson Mines orthoclase are summarized for comparison in Fig. 8. While the behavior illustrated in that figure may apply generally to K-feldspars, it is based on only one natural sample. However, the K, Na, and Ar^{40} results for other homogeneous K-feldspars show general agreement. Petrović (1974) suggests that alkali diffusion in alkali feldspar is intrinsically controlled at temperatures above 600°C, that is, the diffusion coefficients are intrinsic properties which are independent of both the impurity concentrations and the thermal history of the specimen. Examination of Figs. 6 and 7 fails to establish this conclusively, as there are differences between different samples and the reason for these is not clear. Clearly, more data are desirable; however, avail-

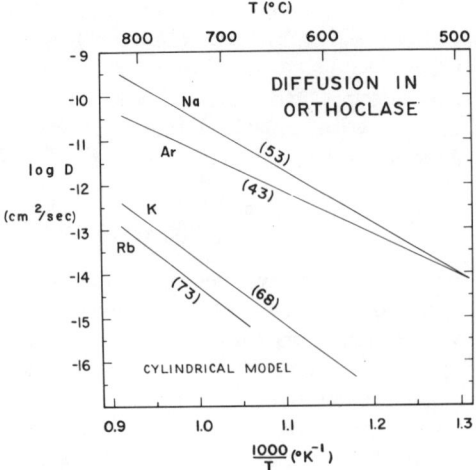

Fig. 8. Summary of diffusivities in the Benson Mines orthoclase (Or_{94}) as given in Table 6. Activation energies, in kcal/mole, given in parentheses. Based on the cylindrical diffusion model.

able data suggest that the diffusivities for different samples do not differ greatly; that is, for three K-feldspar specimens which must have some differences in impurity concentration, for example, divalent cations, and thermal history, the measured K diffusivities are within an order of magnitude of the Or_{94} values.

The behavior of K relative to Ar^{40} as shown in Fig. 8 has implications for K-Ar dating. Lower temperature extrapolation of the K and Ar^{40} Arrhenius relations indicates, as has usually been assumed, that low K-Ar ages for K-feldspar are due to Ar^{40} loss rather than gain of K.

For the Benson orthoclase, at 500° to 800°C diffusivities decrease in the order $Na > Ar^{40} > K > Rb$. Activation energies which, except for Rb, are well defined decrease in the order $Rb \geq K > Na > Ar^{40}$.

For alkali diffusion, the rates decrease and the activation energies increase with increasing ionic size. This apparently reflects the greater difficulty of moving the larger cations past their neighbors. The significant differences between behavior of the smaller sodium ion and the larger alkalies appears to be greater than that typically observed in ionic solids as well as organic ion exchangers. Manning (1968) points out that if the jump frequency for an impurity (Na) is much greater than that for the solvent species (K), then the jump frequency of a fast moving impurity is limited by that for the slower moving, more abundant cation. Thus, the relative magnitudes of D's would be inconsistent with a simple vacancy or interstitialcy mechanism unless the vacancy concentrations and K jump frequencies are considerably enhanced near an impurity (Na). An alternative is to have a direct interstitial mechanism for Na or all alkalies, since such correlation effects are not exhibited by an interstitial mechanism unless the concentration of interstitials is very high (Manning, 1968).

Diffusion in alkali feldspars is not easily amenable to simple treatment owing to the complexity and irregularity of the crystal structure. Alkali ions are in irregular 9-fold oxygen coordination, and tetrahedral cation bond lengths vary by nearly 10%, depending on whether the cation is Al or Si. Complications due to the domain structure are also introduced. A conventional approach would indicate either vacancy or interstitial transport in the principally ionic structure.

Lin and Yund (1972), on the basis of very different rates and activation energies, have suggested different mechanisms for Na and K self-diffusion in the respective end-member feldspars. They postulate that a direct or indirect interstitial mechanism may be more important for sodium, but potassium migrates by a vacancy mechanism. However, since their albite was exchanged microcline, Petrović (1974) concludes that Lin and Yund have overestimated the Na volume diffusion coefficient in albite.

Petrović (1974) has elaborated upon a plausible model of alkali diffusion based on kinetic and crystal chemical data. Frenkel defects, consisting of an alkali vacancy and an interstitial alkali, would control cation migration that is primarily by a vacancy mechanism. His analysis of the relatively open, idealized structure shows large (0, 0, ½) and equivalent interstitial sites. Principally, cations would jump from one regular site to another. The supposition that cation-anion vacancy pairs, Schottky defects, do not predominate is key to the model and is based on the observation of a high Si-O bond energy (about 90 kcal/mole) relative to a lower activation energy for alkali diffusion.

There seems to be no compelling reason why oxygen vacancies could not play an important role, even though there is no detectable Al-Si disordering during the experiments. The possibility of a catalytic effect of H_2O or an H_2O species on the framework bonds cannot be excluded. To a first approximation, diffusion rates of alkalies appear to be the same whether measured with molten salts or under hydrothermal conditions; also, Ar^{40} diffusion coefficients are the same whether measured by vacuum or hydrothermal heating (Foland, 1974). However, Yund and Anderson (this volume) have unambiguously shown that under hydrothermal conditions the rate of oxygen diffusion in K-feldspar is greatly enhanced over that for "dry" conditions. Their D values for hydrothermal experiments are several orders of magnitude higher than those for K (at about 500°C) and are described by an activation energy of about 30 kcal/mole, an activation energy which seems to be characteristic of silicate O diffusion under hydrothermal conditions. H_2O obviously has some effect on the feldspar framework, but the exact mechanism is unclear. Note further, even trace amounts of H_2O can cause hydrolytic weakening in quartz (Griggs, 1967). Perhaps one should ask: Is a molten salt experiment really "dry" and, in this connection, what is a truly "dry" experiment? If H_2O does have an effect, the diffusion coefficients may depend upon H_2O fugacity or may be different under truly anhydrous conditions.

It is suggested that Na transport is primarily by a direct interstitial mechanism as suggested by the very large difference between D_{Na} and D_K values. Petrović (1974) shows that it is difficult to move large cations from one interstitial site to another in the idealized structure. However, movement of the smaller, 0.99 Å radius, Na^+ should be possible with minor accommodation by the framework structure. Noting the irregularities in the actual structure and the possible effects of an H_2O species on the framework, this does not seem unreasonable. The mechanism for the larger alkalies is uncertain, but interstitial transport would require substantially more structural accommodation. The increase in ionic size, approximately 35% in K^+ relative to Na^+, may prohibit direct interstitial movement of K and Rb. Accordingly, if this were the case, migration of the larger alkalies could be predominantly via vacancy or interstitialcy mechanisms. Here the vacancy mechanism description of Petrović may generally apply and we note that interstitial Na ions would produce vacancies for the transport of larger alkalies. Kinetic data would be consistent with this interpretation.

Ar^{40} diffusion proceeds at relatively high rates with an activation energy distinctly lower (by 10 and 25 kcal/mole) than those for Na and K. Uncharged Ar^{40} is significantly larger than the cations (crystal ionic radius, 1.54 Å for Ar^+, Weast and Selby, 1967; covalent radius, 1.91 Å, Gould, 1962) but may not be as restricted by repulsive electrostatic forces. By Petrović's analysis of Ar diffusion, Ar occupies interstitial sites and jumps from one interstitial site to another across a vacancy. If Ar can migrate interstitially and is found in interstitial sites, then its activation energy would represent the enthalpy of migration. This could be larger than the enthalpy for Na migration, since the Na activation energy must contain terms for interstitial production as well as migration. If size limits interstitial transport, then Ar^{40} interstitial transport would require additional sites not available for large cations. While the direct interstitial mechanism would therefore seem unlikely, the same correlation effects discussed above would apply. K^{40} decays to Ar^{40} which emits a 1.46 MeV gamma ray and this gives the Ar^{40} about a 28 eV recoil which may be sufficient to displace the atom from the normal lattice position (Mussett, 1969). Apart from any charge imbalance, Ar^{40} must considerably distort the normal structure, and there may be radiation damage due to the Ar^{40} recoil. A characteristic defect could be associated with each argon atom. Therefore, an alternative is that Ar^{40} is transported with the defect and the 43 kcal activation energy is that for migration.

Until additional research is performed, it is not possible to arrive at more definitive conclusions regarding mechanisms. Further work on the details of anisotropies for each species and on the lower temperature diffusivities is also needed.

Conclusions

1. Alkali partitioning between a natural orthoclase with trace concentrations of Rb and aqueous chloride solutions is similar to that reported for synthetic samples.
2. For the Benson Mines orthoclase, the actual particle size is the effective dimension for diffusion. For this specimen, the isotope exchange technique yields results showing good internal agreement.
3. Even though the cylindrical model is preferred, the choice of geometric models does not greatly affect the resulting diffusivities and activation energies. The effects are systematic.
4. Within their experimental temperature intervals, all diffusing species appear to conform to the Arrhenius relation, with one activation energy for each. Parameters to describe these are given in Table 6.
5. There are marked differences in diffusivities for the orthoclase. At 800°C, $D_{Na} \sim 10\ D_{Ar} \sim 1000\ D_K \sim 3000\ D_{Rb}$. Activation energies increase in the order $Ar^{40} < Na < K \leq Rb$. Therefore, for the alkalies, diffusivities decrease and activation energies increase with increasing ionic size.
6. Large differences between the diffusivity of Na and that of K and Rb suggest different mechanisms. It is tentatively suggested that Na is transported primarily by a direct interstitial mechanism and the larger alkalies by a vacancy or perhaps an interstitialcy mechanism.
7. The mechanism of Ar^{40} diffusion could be similar to that for Na or the larger alkalies, but the possibility that it involves migration of a characteristic defect should be considered. The relative diffusivities of K and Ar^{40} indicate that discordant, low K-Ar ages for K-feldspar are due to Ar^{40} loss rather than K gain.
8. Data for Na and K diffusion in different K-feldspar specimens show general agreement. The kinetic results presented here, especially the relative behavior in one sample shown in Fig. 8, seem well enough established to be applied at least on a preliminary basis to various problems in petrology and geochemistry.

Acknowledgments

This research was performed as part of a dissertation at Brown University. Professor B. J. Giletti, thesis supervisor, is thanked for his contributions to many phases of the study. The author also thanks R. A. Yund for helpful discussions and the use of laboratory facilities, Y. Isachsen for the specimen of Benson Mines orthoclase, T.-H. Lin for discussions and a program to calculate D values, and R. Petrović for a preprint of his feldspar manuscript and permission to include his data in this paper. H. Faul, B. Giletti, A. Hofmann, and R. Petrović contributed suggestions for improvement of the manuscript. Contributions in the form of computer time and technical assistence from the Department of Geology at the University of Pennsylvania are acknowledged. Financial support was provided in part by NSF Grant GA-1649 and by Advanced Projects Research Agency Contract SD-86 awarded to B. J. Giletti.

REFERENCES CITED

Albarede, F., and Y. Bottinga, Kinetic disequilibrium in trace element partitioning between phenocrysts and host lava, *Geochim. Cosmochim. Acta, 36,* 141–156, 1972.

Bailey, A., Comparison of low-temperature with high-temperature diffusion of sodium in albite, *Geochim. Cosmochim. Acta, 35,* 1073–1081, 1971.

Beswick, A. E., An experimental study of alkali metal distributions in feldspars and micas, *Geochim. Cosmochim. Acta, 37,* 183–208, 1973.

Carman, P. C., and R. A. W. Haul, Measurement of diffusion coefficients, *Proc. Roy. Soc., Ser. A, 222,* 109–119, 1954.

Crank, J., *The Mathematics of Diffusion,* Oxford Univ. Press, London, 1956.

Foland, K. A., Ar^{40} diffusion in homogeneous orthoclase and an interpretation of Ar diffusion in K-feldspars, *Geochim. Cosmochim. Acta, 38,* 151–166, 1974.

Gallagher, K. J., The effect of particle size distribution on the kinetics of diffusion reactions in powders, *Reactiv. Solids, Int. Symp., 5th, 1964,* 192–203, 1965.

Giletti, B. J., Diffusion related to geochronology, this volume.

Gould, E. S., *Inorganic Reactions and Structure,* revised edition, Holt, Rinehart and Winston, New York, 1962.

Grant, J. A., and P. W. Weiblen, Retrograde zoning in garnet near the second sillimanite isograd, *Amer. J. Sci., 270,* 281–296, 1971.

Griggs, D., Hydrolytic weakening in quartz and other silicates, *Geophys. J. Roy. Astron. Soc., 14,* 19–31, 1967.

Helgeson, H. C., Kinetics of mass transfer among silicates and aqueous solutions, *Geochim. Cosmochim. Acta, 35,* 421–470, 1971.

Hofmann, A., Chromatographic theory of infiltration metasomatism and its application to feldspar, *Amer. J. Sci., 272,* 69–90, 1972.

Hofmann, A. W., and B. J. Giletti, Diffusion of geochronologically important nuclides in minerals under hydrothermal conditions, *Eclogae Geol. Helv., 63,* 141–150, 1970.

Jäger, E., R. Niggli, and E. Wenk, Rb-Sr Altersbestimmungen an Glimmern der Zentralalpen, *Beitr. Geol. Karte Schweiz, N. F.* 134, 1967.

Jagitsch, R., and M. G. Olsson, Geologische Diffusionen in kristallierten Phasen: Mischkristallbildung in Na-K Feldspaten, *Proc. Int. Symp. Reactiv. Solids, 3rd, 1952,* 463–470, 1954.

Jain, S. C., Simple solutions of the partial differential equation for diffusion (or heat conduction), *Proc. Roy. Soc., Ser. A, 243,* 359–374, 1958.

Jensen, M., Solid diffusion of radioactive sodium in perthite, *Amer. J. Sci., 250,* 808–821, 1952.

Lin, T.-H.; Potassium self-diffusion in microcline, unpublished Ph.D. thesis, Brown University, 1971.

Lin, T.-H., and R. A. Yund, Potassium and sodium self-diffusion in alkali feldspar, *Contrib. Mineral. Petrol., 34,* 177–184, 1972.

Manning, J. R., *Diffusion Kinetics for Atoms in Crystals,* Van Nostrand, Princeton, 1968.

Mérigoux, H., Étude de la mobilité de l'oxygène dans les feldspaths alcalins, *Bull. Soc. Fr. Minéral. Cristallogr., 91,* 51–64, 1968.

Mussett, A. E., Diffusion measurements and the potassium-argon method of dating, *Geophys. J. Roy. Astron. Soc., 18,* 257–303, 1969.

Orville, P. M., Alkali ion exchange between vapor and feldspar phases, *Amer. J. Sci., 261,* 201–237, 1963.

Petrović, R., The effect of coherency stress on the mechanism of the reaction albite + K^+ = K-feldspar + Na^+ and on the mechanical state of the resulting feldspar, *Contrib. Mineral. Petrol., 41,* 151–170, 1973.

Petrović, R., Diffusion of alkali ions in alkali feldspars, in *The Feldspars,* W. S. MacKenzie and J. Zussman, eds., pp. 174–182, Manchester Univ. Press, Manchester, 1974.

Sippel, R. F., Sodium self diffusion in natural minerals, *Geochim. Cosmochim. Acta, 27,* 107–120, 1963.

Weast, R. C., and S. M. Selby, *Handbook of Chemistry and Physics,* Chemical Rubber Co., Cleveland, 1967.

Wood, J. A., The cooling rates and parent planets of several iron meteorites, *Icarus, 3,* 429–459, 1964.

Wright, T. L., and D. B. Stewart, X-ray and optical study of alkali feldspar: I. Determination of composition and structural state from refined unit-cell parameters 2V, *Amer. Mineral., 53,* 38–87, 1968.

Yao, Y-F. Y., and J. T. Kummer, Ion exchange properties of and rates of ionic diffusion in beta-aluminas, *J. Inorg. Nucl. Chem.*, *29*, 2453–2475, 1967.

York, D., Least squares fitting of a straight line, *Can. J. Phys.*, *44*, 1079–1086, 1966.

Yund, R. A., and T. F. Anderson, Oxygen isotope exchange between potassium feldspar and KCl solutions, this volume.

Yund, R. A., and R. H. McCallister, Kinetics and mechanisms of exsolution, *Chem. Geol.*, *6*, 5–30, 1970.

OXYGEN ISOTOPE EXCHANGE BETWEEN POTASSIUM FELDSPAR AND KCl SOLUTION

R. A. Yund
Department of Geological Sciences
Brown University
Providence, Rhode Island 02912

and

T. F. Anderson
Department of Geology
University of Illinois
Urbana, Illinois 61801

ABSTRACT

Rates of oxygen isotope exchange between a $2M$ KCl solution and either microcline (Or_{100}) or adularia (Or_{100}) were measured between 400° and 700°C at 2 kbar. In the absence of net sodium and potassium exchange, the oxygen exchange occurs by diffusion of an oxygen bearing species, probably hydroxyl ions or water. Dissolution and redeposition of the feldspar did not contribute to the observed oxygen isotope exchange, and no change in the aluminum-silicon ordering occurred. The diffusion data for microcline are given by $D = (2.8 \pm 4.7) \times 10^{-6} \exp[(-29{,}600 \pm 900)/RT]$ cm²/sec and for adularia by $D = (5.3 \pm 1.0) \times 10^{-7} \exp[(-29{,}600 \pm 1100)/RT]$ cm²/sec. Oxygen exchange between feldspar and either CO_2 or air is much slower because this presumably occurs by the intrinsic self-diffusion of oxygen ions in feldspar.

INTRODUCTION

O'Neil and Taylor (1967) have shown that the extent of oxygen isotope exchange is equal to the extent of net alkali ion exchange when the alkalies in the feldspar and chloride solution are far from equilibrium. They identified the mechanism of this exchange as a fine-scale recrystallization involving a reaction front moving through the crystal with local dissolution and redeposition in a fluid film at the interface between the exchanged and unexchanged feldspar.

Mérigoux (1968) measured the rate of oxygen isotope exchange under conditions of gross alkali disequilibrium between feldspar and chloride solution. He also measured the rate of oxygen exchange between adularia ($Or_{86.5}$) or albite (Ab_{99}) and distilled water. In the latter experiments, the rate of oxygen exchange was slower and appeared to be diffusion controlled, but he did not demonstrate the absence of fine-scale recrystallization of his samples. This is especially important since he used fine-grained material, and the extent of isotopic exchange below 660° did not exceed 10% of the oxygen in the feldspar.

We initiated this study to determine the oxygen isotope exchange rate under conditions of alkali equilibrium between feldspar and solution and to determine or place closer limits on the mechanism of the exchange under these conditions. In addition to determining the structural state of the feldspars before and after our hydrothermal experiments, we can compare our results with the rate of potassium isotope exchange in the same microcline under identical conditions (Lin and Yund, 1972). A few preliminary experiments were also done to determine the mobility of oxygen in feldspar under anhydrous conditions for comparison with the hydrothermal experiments.

Experimental Methods

The alkalies in an adularia crystal (~Or_{98}) from Kristallina, Switzerland (American Museum of Natural History no. 26545) and a maximum microcline (Or_{89}) from Amelia, Virginia, were exchanged twice in molten KCl at 900°C to produce essentially pure potassium feldspars. These materials were used for subsequent oxygen isotope exchange experiments with a $2M$ KCl solution. The O^{18}/O^{16} ratio of the solution was approximately $+56$ per mil relative to either feldspar.[1] Approximately equal weights of feldspar and solution were sealed in a gold tube and the experiments conducted at 2 kbar in standard cold-seal pressure vessels with a temperature uncertainty of less than ±5°C. One experiment was carried out using distilled water with the same oxygen isotope composition as the KCl solution.

In addition to those described above, several experiments were done using either CO_2, unpurified air, or dried air as the exchange medium rather than an aqueous solution. The CO_2, which was about $+143$ per mil relative to the feldspars, was circulated in a closed system, but because of the low exchange rate in these experiments and the temperature limitation of the apparatus, additional experiments were carried out in air at 1100° or 1107°C in a platinum crucible. However, in these experiments the difference in oxygen isotope composition between atmospheric oxygen and the feldspars was only about 13 per mil. One experiment consisted of heating the sample in the atmosphere. In the other four experiments the air was first dried in a dry ice–acetone cold trap before it came in contact with the sample.

After each hydrothermal experiment, the sample was carefully washed and dried. Most samples were then resieved (125–149 micron size fraction) to remove any grains which were broken during the experiment. There was no difference in the results for samples which were resieved and those which were not. The products were examined optically and by x-ray powder diffraction for sanidine and other evidence of recrystallization. Least-squares cell refinements were carried out for several samples using the computer program written by Evans et al. (1963).

All feldspars were reacted with bromine pentafluoride at 450°C for 12 hours (Clayton and Mayeda, 1963). The resulting oxygen gas was converted to carbon dioxide by reaction with hot graphite. Duplicate analyses were performed on all samples. The average reproducibility of the duplicate analyses was better than ±0.2 per mil. The isotope compositions of the solutions were determined by equilibration with CO_2 at 25°C (Epstein and Mayeda, 1953).

Discussion of Results

The percent oxygen isotope exchange in these experiments was calculated from the difference in the O^{18}/O^{16} ratio of the feldspar before and after the experiment. A correction was made for feldspar-water fractionation using the data of O'Neil and Taylor (1967). The temperature, duration, and exchange data for the hydrothermal experiments are given in Table 1.

Assuming that the oxygen isotope exchange is due to the diffusion of an oxygen-bearing species, the data in Table 1 can be used to calculate diffusion coefficients. The validity of this assumption will be considered below. Making the further approximation that the diffusion is isotropic and that the grains are spherical, the diffusion coefficients can be calculated using the solution for the diffusion equation for a well-stirred reservoir given by Crank (1956, p. 89). The approximation of spherical geometry is good except for greater than 75% exchange (Jain, 1958). The radius of this hypo-

---

[1] Oxygen isotope abundance ratios are normally expressed in the δ or per mil notation where $\delta = \dfrac{R - R_{std}}{R_{std}} \times 10^3$ per mil, and R and R_{std} = O^{18}/O^{16} of the sample and standard, respectively.

TABLE 1. Experimental results for oxygen isotope exchange between microcline (M) or adularia (A) and a $2M$ KCl solution at 2 kbar

Sample	Temp. (°C)	Time (hrs.)	Isotopic Equilibration (%)	D cm^2/sec
*M	700	337	53.3	$(7.4 \pm 1.5) \times 10^{-13}$
M	700	340	49.3	$(5.5 \pm 1.1) \times 10^{-13}$
M	700	60	26.4	$(7.3 \pm 1.6) \times 10^{-13}$
†M	600	672	31.1	$(9.0 \pm 1.9) \times 10^{-14}$
M	550	1056	25.7	$(3.4 \pm 0.7) \times 10^{-14}$
M	500	840	15.0	$(1.3 \pm 0.3) \times 10^{-14}$
M	450	2354	12.4	$(3.3 \pm 1.1) \times 10^{-15}$
M	400	2352	6.1	$(7.6 \pm 3.0) \times 10^{-16}$
†A	700	672	35.3	$(1.3 \pm 0.3) \times 10^{-13}$
A	700	340	27.8	$(1.3 \pm 0.3) \times 10^{-13}$
A	700	60	11.7	$(1.1 \pm 0.2) \times 10^{-13}$
A	600	672	16.7	$(2.1 \pm 0.4) \times 10^{-14}$
A	550	1058	11.6	$(5.5 \pm 1.2) \times 10^{-15}$
A	500	840	6.0	$(2.1 \pm 1.4) \times 10^{-15}$
A	450	2355	5.7	$(6.7 \pm 3.0) \times 10^{-16}$
A	400	2352	3.2	$(2.1 \pm 1.4) \times 10^{-16}$

*Distilled water rather than $2M$ KCl solution was used for this one experiment.
†See Table 2 for cell parameters of these samples.

thetical sphere is taken as half of the mean of the size fraction employed (125–149 microns). This value of 68.5 ± 6 microns is consistent with the average size of the grains calculated from the density and weight of a known number of grains. This method gives a value of 126 microns for the edge of the grains assuming perfect cubes, and a diameter of 156 microns assuming spheres. Details of this method for calculating diffusion coefficients and a discussion of the errors involved are given by Lin and Yund (1972). The calculated diffusion coefficients are given in Table 1 and are shown on an Arrhenius plot in Fig. 1. The reported uncertainty in the D values includes the estimated uncertainties in the isotope compositions of the microcline, adularia, and KCl solution of ±0.2, ±0.4, and ±0.15 per mil, respectively, and an uncertainty of ±6 microns in the radius of the grains.

The difference in the rate of oxygen exchange between the adularia and the microcline is easily accounted for. A similar difference has been observed for potassium diffusion in this microcline (Lin and Yund, 1972) and in an orthoclase (Foland, this volume). The higher apparent diffusion rate in the microcline is probably due to microcracks in the microcline (along twin planes or cracks created by the exchange in molten KCl) which reduce the effective diffusion distance.

The data for our adularia sample are similar to Mérigoux's (1968) data for an adularia of Or$_{86.5}$. His data are shown on Fig. 1 for comparison. The difference between his and our results may be attributed to several causes, including the uncertainty in the effective radius of both his and our grains. (He used grains whose mean radius varied from 19 to 39 microns.) In addition, his experiments were done at lower water pressure, 325 to 600 bars.

A least-squares fit of the data in Fig. 1 yields the following values for the activation energies and preexponential factors in the relation $D = D_0 \exp(-Q/RT)$. For microcline, $Q = 29.6 \pm 0.9$ kcal/mole and $D_0 = 2.8 (\pm 4.7) \times 10^{-6}$ cm^2/sec. For adularia, $Q = 29.6 \pm 1.1$ kcal/mole and $D_0 = 5.3 (\pm 1.0) \times 10^{-7}$ cm^2/sec. Identical activation energies are expected if the only difference be-

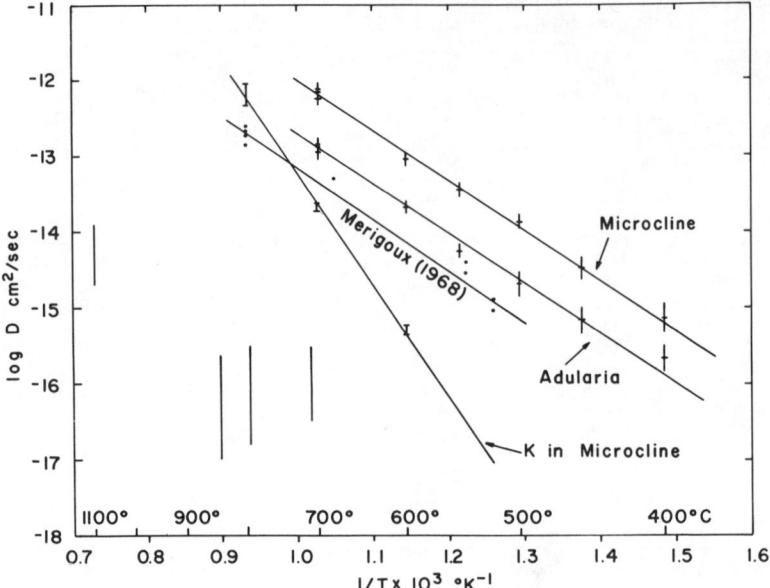

Fig. 1. The diffusivity of oxygen-bearing species as a function of temperature for microcline and adularia is shown by the upper two curves. Diffusivities were calculated from hydrothermal oxygen isotope exchange experiments. Mérigoux's (1968) data for adularia and Lin and Yund's (1972) data for potassium diffusion in microcline are shown for comparison. The vertical bars show the range of values calculated from the feldspar-air (1100°C) and feldspar-CO_2 (712–843°C) oxygen isotope exchange experiments (ignoring the "negative" values) in Table 3.

tween the two samples is the effective diffusion radius. For comparison, Mérigoux (1968) reported activation energies of 32 kcal/mole for adularia and 37 kcal/mole for albite.

To demonstrate that the data shown in Fig. 1 have any significance as true volume diffusion coefficients, it must be shown that the exchange is not, at least in part, due to another process. The fine-scale dissolution and reprecipitation mechanism which O'Neil and Taylor (1967) observed in their experiments involving net alkali ion exchange appears to be the most likely competing mechanism. If this process occurred in our experiments, the newly precipitated phase would be a disordered feldspar (sanidine), at least at 600°C or above.

Approximately 10% sanidine was observed optically and by x-ray diffraction in the two experiments with microcline at 700°C for 337 and 350 hours. The percent oxygen isotope exchange in these experiments was 53.3 and 49.3, respectively. There was either no sanidine or less than one percent sanidine in the two experiments with adularia at 700°C for 340 and 672 hours (27.8% and 35.3% oxygen exchange, respectively). No sanidine was observed in any of the other experiments even though they exhibited up to 31% oxygen isotope exchange.

Clearly, sanidine is absent or the amount is much less than the percent oxygen exchange in these experiments. Even though some recrystallization does occur in the microcline experiments after several days at 700°C, this process apparently does not contribute significantly to the observed oxygen isotope exchange. The outer portion which recrystallizes has already exchanged essentially all of its oxygen by diffusion.

In addition to the development of sanidine, the Al/Si distribution of an ordered feldspar may change continuously and this can be determined by a least-squares refinement of the x-ray powder data. Unit cell dimensions for the initial potassium-exchanged adularia and microcline and for the products from two of the exchange experiments are given in Table 2. Although appreciable oxygen isotope exchange had occurred in both of these experiments, 31.1% for the microcline and 35.3% for the adularia, there appears to be no significant change in their cell dimensions and hence in their structural states (Orville, 1967; Wright and Stewart, 1968). Even if the small changes in the cell dimensions were significant, they do not correspond to an increase in the Al/Si disorder.

A final argument against dissolution and redeposition can be made by comparing the data for oxygen isotope and potassium isotope exchange in this microcline. The experiments reported here are identical to those of Lin and Yund (1972) except for the substitution of the O^{18} tracer for the K^{40} tracer. The same microcline was used for both sets of experiments. The calculated diffusion coefficients for potassium and oxygen in this microcline are compared in Fig. 1. At 600°C and below, the calculated diffusion coefficients for potassium are much lower than those for oxygen. Such a relation is impossible if the oxygen isotope exchange was due to dissolution and redeposition because the potassium isotope exchange rate would be at least as rapid as the oxygen exchange rate.

We originally considered that the oxygen isotope exchange rates which Mérigoux (1968) observed might be high because of recrystallization. After some initial dissolution of the $Or_{86.5}$ adularia in the distilled water of his experiments, minor net alkali exchange between the solid and the fluid would be necessary to establish alkali equilibrium. If this slight departure from alkali equilibrium had increased the rate of recrystallization significantly, Mérigoux's calculated diffusion coefficients would be too large. Comparison with our results show that recrystallization was not significant in his experiments.

From the preceding arguments we conclude that the oxygen isotope exchange observed in the hydrothermal experiments is due to diffusion of an oxygen-bearing species through the feldspar structure. From the rate data alone there is no way to identify the diffusing species; it could be oxygen ions, hydroxyl ions, water, etc. Comparison of the exchange rates for hydrothermal and nonaqueous experiments provides some information concerning this question.

The rate of oxygen isotope exchange between CO_2 or air and feldspar is much slower than that between feldspar and an aqueous solution. Data for the CO_2 and air experiments are given in Table 3. Diffusion coefficients calculated from these exchange experiments are only very approximate because of the limited oxygen isotope exchange and because of the small difference in the isotopic compositions of the feldspar and air. Because the amount of exchange is very small and a correlation between percent exchange and length of the experiment cannot be dem-

TABLE 2. Cell parameters for microcline and adularia before and after hydrothermal oxygen isotope exchange

Material	a, Å	b, Å	c, Å
Microcline starting material	8.590(5)*	12.965(6)	7.213(3)
Microcline from line 4, Table 1, after run	8.588(4)	12.958(4)	7.217(2)
Adularia starting material	8.597(1)	12.994(2)	7.191(1)
Adularia from line 9, Table 1, after run	8.604(3)	13.002(3)	7.193(1)

*Numbers in parentheses are one standard deviation in the last digit as calculated by the cell refinement program.

TABLE 3. Experimental results for oxygen isotope exchange between microcline (M) or adularia (A) and either a CO_2, air, or dried air reservoir at approximately one atmosphere

Sample	Grain Size (microns)	Reservoir	Temp. (°C)	Time (hrs.)	Isotopic Equil.* (%)
M	88–105	CO_2	712	20	negative
M	88–105	CO_2	712	138	0.2
M	88–105	CO_2	712	280	0.7
M	88–105	CO_2	800	121	0.5
M	88–105	CO_2	800	313	0.2
M	88–105	CO_2	843	496	0.2
A	88–105	CO_2	712	20	0.3
A	88–105	CO_2	712	138	0.6
A	88–105	CO_2	712	280	0.9
A	88–105	CO_2	800	121	0.9
A	88–105	CO_2	800	313	0.8
A	88–105	CO_2	843	165	0.9
A	88–105	CO_2	843	496	0.5
A	88–105	air†	1100	473	4.0
A	105–125	air†	1100	473	negative
A	125–149	air‡	1107	672	5.0
M	125–149	air‡	1107	672	5.2
A	125–149	air‡	1107	1218	negative
A	125–149	air‡	1107	1218	12.1

*The uncertainty in these values is large and varies from about 30% to 100% for the CO_2 experiments and about 100% for the air experiments.
†Unpurified air.
‡Air dried in acetone–dry ice cold trap.

onstrated, the exchange may be largely at or near the surface of the feldspar grains where the structure is disturbed. Therefore D values calculated from the data in Table 3 would represent maximum values for the diffusion of oxygen ions in feldspar. In this connection it should be mentioned that we are assuming that the rate of exchange of an oxygen atom from a CO_2 or O_2 molecule to the feldspar surface is not a rate controlling step. This assumption appears valid for many oxides which have been investigated, including $MgAl_2O_4$, α-Al_2O_3, MgO. (See, for example, Ando and Oishi, 1972.)

If the data in Table 3 are used to calculate diffusion coefficients, the values for the CO_2 experiments are on the order of 10^{-16} cm²/sec and those for the 1107°C air experiments are on the order of 10^{-15} cm²/sec. The range of values is shown by vertical bars on Fig. 1. The only other data which we are aware of for comparison are the CO_2-albite exchange experiments of Dontsova (1955). We have estimated the D value from her 1100°C exchange data, and it is on the order of 10^{-12} cm²/sec. This value appears high compared to our results; however, Dontsova's experiments were performed with albite grains of 6 micron radius and the structurally disturbed outer layer may have been large compared to the radius. No details are given concerning her starting material, and it is probably inappropriate to estimate D values from her data.

Provided that the pressure difference of 2 kbar between the air or CO_2 experiments and the hydrothermal experiments is not significant, a majority of the available data suggests that the rate of oxygen diffusion in feldspar is very low under nonhydrothermal conditions. If intrinsic volume diffusion occurs in the CO_2 and air experiments, the diffusing species is presumably oxygen ions. The higher exchange rates in the hydrothermal experiments represent the greater mobility of another species which we tentatively suggest is either water or hydroxyl ions.

Conclusions

At or near alkali ion equilibrium, the exchange of oxygen isotopes between feldspar and a hydrothermal solution occurs by diffusion of an oxygen-bearing species such as hydroxyl ions or water molecules. This oxygen isotope exchange does not affect the distribution of the tetrahedral ions or the mobility of the potassium ion. However, the mechanism may be similar to that proposed by Donnay et al. (1959). They envisaged the rapid diffusion of hydrogen ions and the hydrolyzation of the tetrahedral ion–oxygen bond. The slower diffusing hydroxyl ions or water could then exchange with these hydroxyl groups containing an oxygen from the original feldspar structure. We now know, however, that water does not affect the tetrahedral site ordering. Therefore, the fraction of hydrolized bonds per tetrahedron must be low in order to prevent this disordering. Additional support for this general mechanism comes from the observed hydrolytic weakening of quartz, and the Frank-Griggs hypothesis that slip can occur more easily when Si-O-Si bridges adjacent to a dislocation are hydrolyzed by the migration of water (Griggs, 1967).

In addition to clarification of this mechanism, the dependence of the oxygen exchange rate on both the partial and total pressure of water should be considered. Another question concerns how far the alkali ratio in the feldspar has to be out of equilibrium with the hydrothermal solution before the dissolution and redeposition mechanism identified by O'Neil and Taylor (1967) becomes significant. Answers to these and other questions are important for understanding the kinetics of mineralogical and geochemical processes involving feldspars.

References Cited

Ando, K., and Y. Oishi, Self-diffusion of oxygen ion in $MgAl_2O_4$ single crystal, *J. Ceram. Ass. Jap.*, *80*, 324–326, 1972.

Clayton, R. N., and T. K. Mayeda, The use of bromine pentafluoride in the extraction of oxygen from oxides and silicates for isotopic analysis, *Geochim. Cosmochim. Acta*, *27*, 43–52, 1963.

Crank, J., *The Mathematics of Diffusion*, Oxford Univ. Press, London, 1956.

Donnay, G., J. Wyart, and G. Sabatier, Structural mechanism of thermal and compositional transformations in silicates, *Z. Kristallogr.*, *112*, 161–168, 1959.

Dontsova, E. I., Investigation of the mobility of oxygen in silicates and aluminosilicates by the isotope tracer method, *Dokl. Akad. Nauk SSSR*, *103*, 1065–1067, 1955.

Epstein, S., and T. K. Mayeda, Variations in the O^{18}/O^{16} ratio of natural waters, *Geochim. Cosmochim. Acta*, *4*, 213–224, 1953.

Evans, H. T., Jr., D. E. Appleman, and D. S. Handwerker, The least squares refinement of crystal unit cells with powder diffraction data by an automatic computer indexing method (abstr.), Amer. Crystallogr. Ass. Meeting, Cambridge, Mass., March 28, 1963, 42–43.

Foland, K. A., Alkali diffusion in orthoclase, this volume.

Griggs, D. T., Hydrolytic weakening of quartz and other silicates, *Geophys. J. Roy. Astron. Soc.*, *14*, 19–31, 1967.

Jain, S. C., Simple solutions of the partial differential equation for diffusion (or heat conduction), *Proc. Roy. Soc. (London), Ser. A.*, *243*, 359–374, 1958.

Lin, T.-H., and R. A. Yund, Potassium and sodium self-diffusion in alkali feldspar, *Contrib. Mineral. Petrol.*, *34*, 177–184, 1972.

Mérigoux, H., Étude de la mobilité de l'oxygène dans les feldspaths alcalins, *Bull. Soc. Fr. Mineral. Cristallogr.*, *91*, 51–64, 1968.

O'Neil, J. R., and H. P. Taylor, The oxygen isotope and cation exchange chemistry of feldspars, *Amer. Mineral.*, *52*, 1414–1437, 1967.

Orville, P., Unit cell parameters of the microcline–low albite and the sanidine–high albite solid solution series, *Amer. Mineral.*, *52*, 55–86, 1967.

Wright, T. L., and D. B. Stewart, X-ray and optical study of alkali feldspar: I. Determination of composition and structural state from refined unit-cell parameters and 2V, *Amer. Mineral.*, *53*, 38–87, 1968.

STUDIES IN DIFFUSION I: ARGON IN PHLOGOPITE MICA

Bruno J. Giletti
Department of Geological Sciences
and Materials Research Laboratory
Brown University
Providence, Rhode Island 02912

ABSTRACT

Diffusion of radiogenic argon 40 has been measured from a natural phlogopite (annite = 4%) under isothermal conditions in the range 600° to 900°C in hydrothermal pressure vessels at 2000 bars. The data indicate that only one mechanism of diffusion is dominant and one argon population exists in this temperature–pressure range. The particle sizes used are shown to be very nearly the effective grain sizes for diffusion. The argon transport occurs preferentially parallel to the mica layers, although significant amounts may also migrate across them. The results of sixteen observations yield a linear Arrhenius plot whose equation is:

$$D = (0.75^{+1.7}_{-0.52}) \exp - [(57.9 \pm 2.6) \times 10^3/RT] \text{ cm}^2 \text{ sec}^{-1}.$$

These results are several orders of magnitude lower than those in the literature, and reflect better experimental design. Experiments have been reported in which unexpectedly large amounts of argon were released in the initial stages of heating. No such large release is found here, and the maximum is probably much less than 2% of the argon in the mica.

Extrapolation of behavior to lower temperatures is well controlled by the data and, if the same diffusion process is dominant at the lower temperatures, argon loss will be less than 5% if a phlogopite is heated to 300°C for 100 m.y. With a normal geothermal gradient, this suggests that the thermal regime in the upper 10 km of continental crust and the uppermost 5 km of oceanic mantle would permit the phologopite to retain its argon. This assumes mica flakes of at least 1 mm diameter.

Introduction

One of the major problems in K-Ar geochronology is the determination of whether or not a mineral has been an open system to the gain or loss of potassium or radiogenic argon. It has become an accepted premise that if a mineral is heated during a metamorphic episode, its K-Ar age is reset to some extent and frequently is reset to yield the age of the metamorphism. If the K concentration in the mineral is not changed very much, as is often the case with potassium minerals, the major age resetting must derive from gain or loss of radiogenic argon.

The resetting process will be treated in this discussion from the point of view of volume diffusion of argon. Those cases which involve recrystallization or significant chemical change in the mineral, while they may be important, are beyond the scope of this paper.

Phlogopite was chosen for the first study in argon diffusion even though micas such as muscovite and biotite are used far more extensively for dating. The primary reason for the choice stems from the practical need to obtain significant amounts of argon transport during the experiment. Phlogopite is stable at significantly higher temperatures than are the other micas, provided that a modest water pressure is maintained, and diffusion rates increase with temperature.

Experimental determination of argon loss from phlogopite was first made in heating experiments carried out by Evernden et al. (1960) (See Fig. 2). These were incremental heating experiments in vacuo, where the sample was taken to a given temperature and held there for a period of from 2 to 48 hours. The gas released in this time was then collected and the sample temperature raised, usually by approximately 75°C, and a new gas collection made. This process was continued to the melting point.

The experimental design suffers from two difficulties. At some temperature (Evernden et al. suggested near 600°C) the mica breaks down and loses hydroxyl water. This implies that any argon loss measurements made above that temperature are no longer being made on the phlogopite itself. Rather, the Ar transport occurs during the mica breakdown and then from the breakdown products. Any diffusion constants measured above the 600° point need not bear any relation to argon diffusion in phlogopite.

The second difficulty lies in the very small amounts of argon Evernden et al. obtained at temperatures below 600°C. The two sets of runs released 0.32% and 0.22% of the total sample argon for the five temperature intervals studied below 600°C. It is not necessary that this small amount was lost by the mineral by true volume diffusion. It is possible that this was surface-adsorbed argon. In light of these two difficulties, it is not surprising that the data do not yield a linear Arrhenius plot and that the D values obtained differ from those to be presented here by several orders of magnitude.

The present study employed long term isothermal heating runs in which the mineral was in its stability field so that transport was measured for phlogopite itself throughout any experiment. This was possible by use of hydrothermal equipment in which water pressure was normally 2 kbar.

Procedures

The specimen chosen for study was a single phlogopite book from south Burgess County, Ontario, donated by Dr. V. Manson of the American Museum of Natural History, New York City (AMNH no. 13528). Based on chemical analysis, its formula is: $(K_{1.83}Na_{0.43}Ca_{0.01})$ $(Mg_{5.45}Fe^{II}_{0.16}Fe^{III}_{0.07}Ti_{0.13}Mn_{0.08}Al_{0.21})$ $(Al_{2.19}Si_{5.81}O_{20})((OH)_{0.64}F_{1.38})$ (Shapiro and Botts, 1970). If the solid solution phlogopite-annite is assumed to be based on just the Mg and total Fe as Fe^{II}, the mica is Ann_4 (out of 100). It has a K-Ar age of 1120 m.y.

The specimen, approximately $18 \times 11 \times 2$ cm, was cut up with scissors and a razor blade after the outer portions were trimmed off. The material was then comminuted by cutting it up, dry, in a Waring blendor. Careful sizing was carried out using wet and dry sieves. Flake dimensions were then measured under a microscope.

All the data reported are for hydrothermal runs in which the phlogopite and either distilled water or an alkali chloride solution were sealed into a gold tube and placed in a cold-seal bomb with water as the external pressure medium. The only exception to this procedure was one run at 900°C in which Ar was the external medium. All runs were at 2 kbar pressure except at 900°C, which was at 1 kbar. The fluid to solid weight ratio in each charge was approximately one.

Alteration or reaction products observed at the end of the runs were minor. While it is difficult to assess the amount of such material, a reasonable estimate is approximately one half of one percent.

The amount of argon lost during the run was determined by taking the difference between the radiogenic argon present in an aliquot of the starting material and the radiogenic argon present in the mica after the hydrothermal run. Fresh mica was used in making up all charges.

The argon analysis was carried out using methods standard to K-Ar dating.

TABLE 1. Dimensions of phlogopite flakes

Nominal Size	Mesh Range	Average Radius (μm)	Average Half Thickness (μm)	Aspect Ratio
40	40–45	323	3.8	85
100	100–120	110	1.91	58
200	200–230	67	1.67	40

The argon was extracted by vacuum fusion, mixed with an Ar^{38} tracer and analyzed by isotope dilution on a mass spectrometer. The procedure has been described elsewhere (Giletti, 1966). Argon analyses have a standard deviation of ± 1 sigma = 1%.

Results and Diffusion Models

The phlogopite mica used for all the experiments was carefully sized and then measured under a microscope. The dimensions are shown in Table 1.

The average radius is computed by taking a best estimate of the length and width of the flakes, finding the area of the resulting rectangle, and then finding the radius of the circle with the same area. The actual average length to width ratio ranges from 1.31 for the 40 mesh to 1.57 for the 200 mesh mica.

Six determinations were made for the radiogenic argon content of the mica. The results are given in Table 2 and average 5.38×10^{-4} scc Ar_{rad}/g phlogopite.

Nineteen charges were run isothermally at temperatures in the range 600° to 900°C. All but three were for the 200 mesh size. The fraction of the initial argon content lost in the course of the experiment, f, is listed in Table 3.

In order to compute diffusivities, it is necessary to assume an appropriate diffusion model (Crank, 1967). As these flakes are very thin sheets (see aspect ratios above), two models were tested. One, the infinite cylindrical model, assumes that diffusional transport is only parallel to the layers. The other, the infinite plate model, assumes transport occurs only across the layers.

Two factors enter into the geometrical considerations. One is whether the particle size used is the effective grain size for diffusion, or whether there is a mosaic of smaller grains in each particle. The use of different mesh sizes with the same basic geometrical computation model permits the determination of this parameter. It can be seen from Table 4 that the uncorrected values (D/a^2 or D/l^2) differ considerably for the different sizes, while the values (D) corrected for size are quite similar. From this it can be inferred that the effective grain size for diffusion cannot be very much smaller than the particle size. The particle size will be used hereafter as though it were the true effective grain size for diffusion.

The second parameter governing the geometrical interpretation is the diffusional anisotropy. This relates to the choice between the cylindrical and plate models and, ultimately, some combined refinement of these. In this paper, only the two simple models will be considered.

The choice between the two models could ideally be made by a microanalyt-

TABLE 2. Radiogenic argon content of phlogopite starting material

Mesh	$\dfrac{Ar^{40}_{atm}}{Ar^{40}_{total}}$	Ar^{40}_{rad} ($\times 10^4$ scc/g)
40	0.062	5.39
40	0.069	5.34
200	0.076	5.38
200	0.097	5.34
200	0.083	5.40
200	0.077	5.44
	Mean Value	5.38

TABLE 3. Argon diffusion in phlogopite (infinite cylinder diffusion model)

Run Number	T °C	Run Duration (Days)	(Seconds $\times 10^{-6}$)	Mesh Size	f^*	$\dfrac{D}{a^2}$† (sec^{-1} $\times 10^9$)	D (cm² sec^{-1} $\times 10^{12}$)
1	900	1	0.0703	200	0.362	434.	19.5
2	900	6	0.505	200	0.668	253.	11.4
3	799	9	0.7645	200	0.247	17.5	0.787
4	800	6	0.4762	200	0.283	37.8	1.70
5	800	2	0.1718	200	0.145	25.1	1.13
6	800	11	0.9856	40	0.129	3.50	3.65
7	800	0.4	0.0338	200	0.094	53.3	2.39
8	750	5	0.4039	100	0.103	5.40	0.653
9	750	5	0.4039	200	0.131	8.79	0.389
10	750	14	1.191	200	0.197	6.88	0.309
11	750	12	1.010	200	0.172	6.24	0.280
12	750	14	1.207	200	0.169	4.97	0.223
13	700	6	0.4950	40	0.024	0.232	0.242
14	700	6	0.4950	200	0.073	2.16	0.097
15	650	23	1.966	200	0.070	0.51	0.023
16	650	43	3.715	200	0.091	0.46	0.021
17	650	43	3.715	200	0.064	0.215	0.0097
18	600	72	6.202	200	0.022	0.0158	0.00071
19	600	72	6.202	200	0.025	0.0193	0.00087

$*f = \dfrac{(Ar^{40}{}_{rad} \text{ in mica initially}) - (Ar^{40}{}_{rad} \text{ in mica after heating})}{Ar^{40}{}_{rad} \text{ in mica initially}}$

†Equation 5.41, p. 71, Crank, 1967.

ical technique in which a grain is systematically traversed and the Ar content measured inward from the edge toward the center. Such traverses would be carried out parallel to the different crystallographic axes. It was not possible to determine the primary transport direction by this direct method. Instead, the difference in aspect ratio for the different mesh sizes was employed to select the appropriate model. The cylindrical model calculation yields a term D/a^2 where a is the radius of the flake. The plate model calculation yields a term D/l^2, where l is the half-thickness of the flake. The calculated values for both the cylindrical and plate models are given in Table 3. The critical data are the D values for the two models. In any given model, all the D values should be the same for the same

TABLE 4. Comparison of particle size effects and diffusion model choice

Run Number	Mesh	Cylinder Model			Plate Model		
		$\dfrac{D}{a^2}$ ($\times 10^8$)	D_{cyl} ($\times 10^{13}$)	$\dfrac{D_{40}}{D_{200}}$	$\dfrac{D}{l^2}$ ($\times 10^8$)	D_{plate} ($\times 10^{15}$)	$\dfrac{D_{40}}{D_{200}}$
800°C							
6	40	0.350	36.5	2.4	1.25	1.81	0.49
3	200	1.75	7.87		6.10	1.70	
4	200	3.78	17.0		13.2	3.69	
5	200	2.52	11.3		9.58	2.66	
7	200	5.33	23.9		24.0	6.69	
	200	Ave.	15.0			3.69	
750°C							
13	40	0.0232	2.42	2.5	0.0978	0.141	0.57
14	200	0.216	0.97		0.880	0.246	

temperature, since the actual particle dimensions are employed in going from D/a^2 or D/l^2 to the D. Table 3 shows the ratios of the D values for 40 mesh vs. 200 mesh for each model. The D_{40}/D_{200} is approximately 2.5 for the cylindrical model and is approximately 0.5 for the plate model.

If the particles were not perfect, but had some cracks, the average grain size would be somewhat smaller than the particle size. This would yield a D_{40}/D_{200} greater than 1 if, on the average, the cracks were equally spaced in the two mesh sizes. (The calculation involves multiplication by the grain size, and if it is taken too large, the D is too large.) This result is suggested by the data for the cylindrical model. The analogous interpretation for the plate model data would require the actual grain size to be larger than the particle size.

The results given above led to the choice of the cylindrical model. This choice is not as clear-cut as might be desired, and the actual behavior probably includes a significant amount of transport parallel to and perpendicular to the c axis, with somewhat greater net transport along the mica layers.

The choice of cylindrical versus plate model has considerable importance in that, owing to the large aspect ratio, the D values computed will differ by three orders of magnitude. On the other hand, the fact that the data in Table 4 suggest the cylindrical model is particularly significant just because the flakes are such extremely thin sheets. Transport would be expected to occur across the layers in substantial amounts even if the primary transport direction were along the layers just because of the extremely small c-axis dimension.

An Arrhenius plot showing log D vs. $1/T$ is given in Fig. 1. All 19 points are shown. Owing to the uncertainty in the data conversion for different mesh sizes, the least squares fit to the data using the York (1966) program employed only the sixteen 200 mesh data points. The calculated slope yields an activation energy of 57.9 ± 2.6 kcal/g-atom^{40} Ar$_{rad}$ and the value for $D_0 = 0.75^{+1.7}_{-0.52}$ cm^2 sec^{-1}. The uncertainties quoted are for one standard deviation and derive from the scatter in the data. Other experimental factors leading to uncertainties are not included.

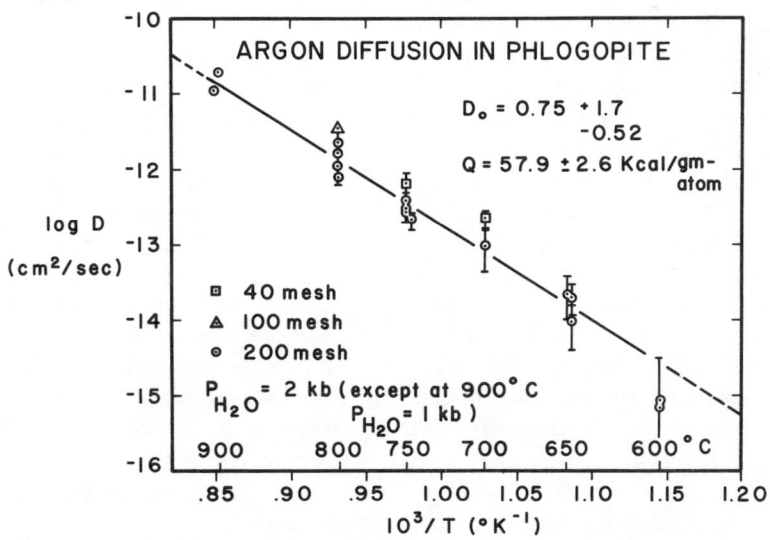

Fig. 1. Arrhenius plot for radiogenic argon diffusion in phlogopite mica (annite 4%).

Discussion

From Fig. 1 it can be seen that the radiogenic argon loss from the phlogopite conforms to a single mechanism of diffusion insofar as the Arrhenius relation is concerned. Of particular interest are the end points of the curve at 900° and 600°C. The 900° data points were obtained from runs made at 1 kbar instead of 2 kbar for all the other data. If there is a pressure effect, it is either small or it is counterbalanced by some other effect, such as a new diffusion mechanism with a high activation energy which comes into prominence at the higher temperatures. It will be assumed that there is no marked effect in going from 2 to 1 kbar. It should be noted that the stability field of phlogopite extends to approximately 1100°C at 1 kbar (Wones, 1967, and Wones and Eugster, 1965), so that breakdown is not occurring.

At the low end of the temperature range in these measurements, the two data points fall below the curve, although the uncertainties are large. The large uncertainty derives from computing the Ar loss by difference; that is, the Ar remaining after the run is subtracted from the initial Ar measured in the raw material. In the two 600°C runs, the loss was 2%, so that we are taking the small difference between two large numbers.

It is possible to suggest from the 600°C data in Fig. 1 that a reduction in the diffusion rate of Ar relative to that predicted by the Arrhenius relation might be occurring. It is not yet reasonable to make such a statement owing to the large uncertainty in the 600°C data points. Further work to extend the data to different pressures and to higher and lower temperatures is in progress. It is just this question of linearity of the Arrhenius plot and its implications of mechanism that will be tested.

There is considerable scatter in the data which is not explained. A given temperature will yield differences of as much as a factor of 3 between the highest and lowest D values. No correlation could be found with any of the possible variants in such an experiment.

Any attempt to define a variety of populations of argon in these data runs into the difficulty of this data scatter. The largest amount of argon lost was 67% in one of the 900°C runs. If there is a fraction of the argon that behaves differently, it will have to be one third or less of the total.

Other workers have found that there is a fraction of the argon in some minerals which will come off almost at once during an isothermal heating experiment (Brandt and Bartnitsky, 1964; Hanson, 1971). The data for the 600°C runs show an argon loss of 2% after ten weeks and that this value is on the low side for the best fit Arrhenius plot of Fig. 1. If there were some initial burst of argon loss during an isothermal run, it would have to be less than 2% of the argon present. Further, if the 600° argon loss is approximately that which should occur, then the initial burst effect must be quite small indeed.

The data of Evernden *et al.* (1960) can now be compared with those presently collected (see Fig. 2). The lower straight line represents the data of the present study. This again confirms the results Evernden *et al.* (1960) obtained for glauconite: that a different diffusion rate is measured for micas under hydrothermal conditions as opposed to dry heating. The phlogopite in the present study was stable throughout the heating.

It is suggested that the diffusivities measured in the present study represent the correct values of argon diffusion out of a phlogopite under a 2 kbar hydrostatic pressure between 600° and 900°C. Further, it is suggested that these data apply to any natural phlogopite in a rock where the total pressure is not much more than 2 kbar. A further condition for the latter statement is that the partial pressure of water did not fall below that

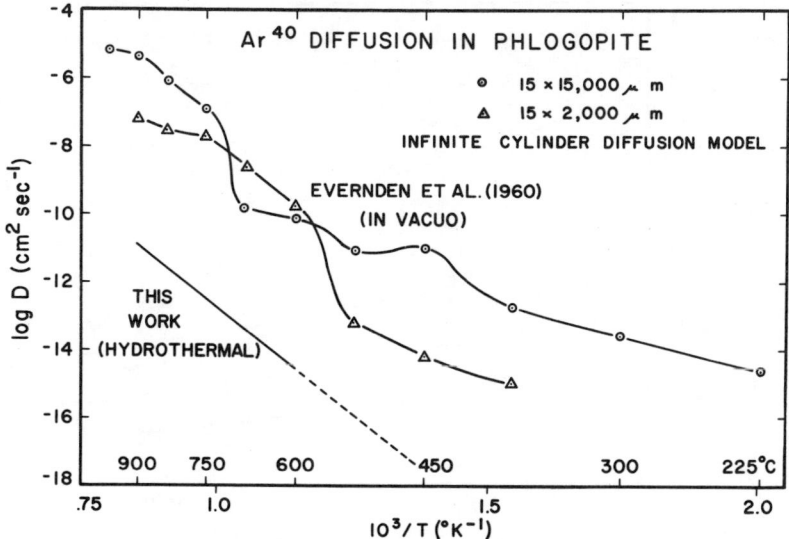

Fig. 2. Comparison of published argon diffusivities in phlogopite with those in this report. Note differences in both absolute values and linearity. See text for discussion.

required to maintain the stability of the phlogopite.

Other than for local contact metamorphic settings, it is unlikely that a phlogopite will exist naturally at temperatures in excess of 600°C and pressures of only approximately 2 kbar. Consequently, it is necessary to extrapolate the present data to lower temperatures in order to attain geologically meaningful conditions (the geothermal gradient would require much higher pressures for a 600° regime). Any extrapolation introduces several uncertainties, however, and it is useful to consider these in the case of phlogopite.

Owing to the high activation energy, there is only a comparatively short extrapolation to temperatures below which little or no significant transport of argon would occur for geologically significant times and on crystals of modest dimensions (for example, $D = 1.3 \times 10^{-19}$ at 400°, which is four orders of magnitude below the 600°C point, and the data collected also span four orders of magnitude). Given the control of the Arrhenius plot from 900° to 600°C on a $1/T$ plot, this extrapolation, of no more than twice the length of the line for which there are data points, is still rather well defined.

If a different diffusion mechanism is also occurring, which has a much lower activation energy and which becomes significant at the lower temperatures, there is no evidence for it down to 600°C. Such a phenomenon would have the effect of increasing argon transport at the lower temperatures so that estimates of maximal time-temperature effects would have to be revised downward. It is difficult to see how this type of effect, if it exists at still lower temperatures, can be determined in laboratory experiments.

Using the cylindrical diffusion model, it is possible to compute the temperature at which a phlogopite grain must be held for given times in order that particular fractions of its argon be lost. Such curves are given in Fig. 3. They suggest that it is unlikely that any but very fine grained phlogopite will be significantly affected at temperatures below 300°C unless they are heated for several hundred million years. A crustal phlogopite would have to be buried at least 10 km, assuming a normal geothermal gradient, in order to suffer appreciable argon loss. This as-

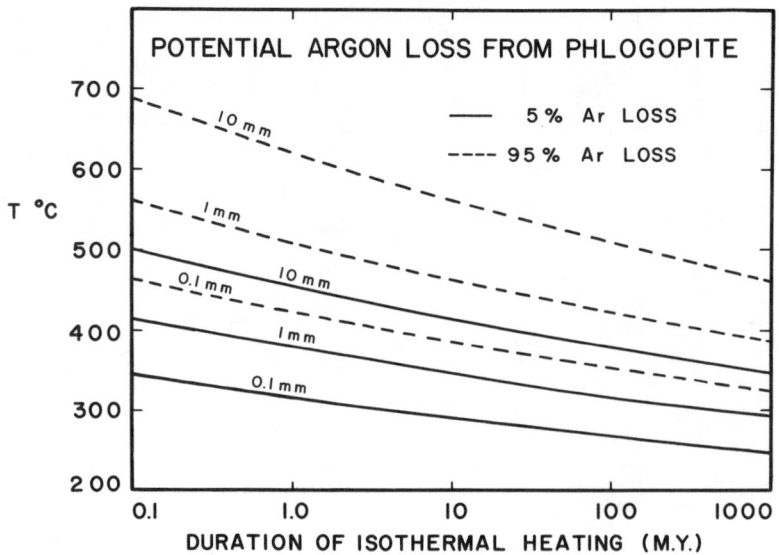

Fig. 3. Fractions of radiogenic argon lost by volume diffusion from phlogopite as functions of temperature and duration of heating. Dimensions are for flake diameter and assume argon transport parallel to layers (infinite cylinder diffusion model).

sumes that the pressure effect of such burial does not affect the diffusion of Ar.

It has been suggested that phlogopite is one possible water-bearing mineral which can exist in the upper mantle (Yoder and Kushiro, 1969). Based on the present diffusion data, it should be possible for phlogopite to retain its argon if it is in the uppermost 5 km of sub-oceanic mantle and the grains are at least one mm in diameter. This is worth noting in connection with possible drilling into the mantle and localities where it is thought oceanic mantle is now at the surface owing to faulting.

Another consequence of the data in Fig. 3 is that a phlogopite will not lose all of its argon readily. It will take 100 m.y. for a 1 mm diameter flake to lose 95% of its argon at 425°C. This suggests that unless the terrane is strongly heated, there is the possibility of incomplete resetting of the argon age.

It should be noted that this work concerns phlogopite which is 4% annite. Preliminary work by Norwood and me at this laboratory has shown that increase in the Fe content of the trioctahedral mica lowers the activation energy and increases the diffusivity. Although this work is still in its early stages, it is already clear that use of the phlogopite data to predict biotite argon diffusion would lead to serious underestimates at low temperatures.

Regardless of the diffusion model used, the mathematical functions yield a crude linear relationship between fraction of argon lost and the linear dimension of the grain size up to losses of over 50%. In any case where there is observed age discordance, it is necessary to be able to estimate the diffusive effects. To this end, the reporting of K-Ar data is no longer complete without a measure of the size of the mineral particles in the rock which were dated. The mesh size of a mineral separate used in the analysis is also of interest, but it need bear no relation to the original size if the rock was crushed and ground. Consequently, mesh size is not an adequate substitute for actual size in the rock as determined, for example, in thin sections.

Conclusions

1. Volume diffusion in phlogopite was measured and the results are consistent with a single argon diffusion population and mechanism.

2. The particle size of the phlogopite was shown to be very nearly the effective grain size for diffusion.

3. There is a preferential transport of argon parallel to the mica layers (infinite cylinder diffusion model), although the amount of this preference is difficult to determine.

4. The diffusivity is given by

$$D = (0.75^{+1.7}_{-0.52}) \exp - [(57.9 \pm 2.6)/RT] \text{ cm}^2 \text{ sec}^{-1}$$

using the cylindrical model for 2 kbar water pressure and between 600° and 900°C.

5. There is no evidence of Ar loss significantly in excess of that predicted by the diffusion model early in the isothermal heating experiments. Such loss is probably much less than 2% of the argon initially in the mineral.

6. There is no significant evidence that other diffusion mechanisms with lower activation energies exist, at least down to 600°C.

7. Previously reported phlogopite argon diffusion data were high by at least four orders of magnitude owing to Ar loss primarily by processes other than volume diffusion.

8. Extrapolation of the data to lower temperatures, assuming the same diffusion mechanism, is possible with relatively small uncertainties owing to the reasonably tight control on the Arrhenius plot and to the high activation energy which results in negligible D values below 300°C in most geologically meaningful settings.

9. Neglecting pressure effects, phlogopite should retain its argon under normal geothermal gradient conditions to depths of approximately ten kilometers in the continental crust and, if present, in the uppermost five kilometers of oceanic mantle.

Acknowledgments

The phlogopite specimen used for this work was kindly supplied by Dr. V. Manson of the American Museum of Natural History, New York, for which I am most grateful. Drs. L. Shapiro and S. Botts of the U.S. Geological Survey supplied the chemical analysis of the mica. Ms. S. Sachs and Mr. C. Norwood aided in various aspects of the computations and laboratory work.

This research was supported in part by National Science Foundation grant GA-1649 and by an Advanced Research Projects Agency contract to Brown University for the study of materials.

I thank A. R. Cooper, A. Heuer, and A. E. Mussett for several useful comments on the manuscript.

References Cited

Brandt, S. B., and E. N. Bartnitsky, Losses of radiogenic argon in potassium–sodium feldspars on heat activation, *Int. Geol. Rev., 6,* 1483, 1964.

Crank, J., *The Mathematics of Diffusion,* Oxford Univ. Press, London, 347 pp., 1967.

Evernden, J. F., G. H. Curtis, R. W. Kistler, and J. Obradovich, Argon diffusion in glauconite, microcline, sanidine, leucite, and phlogopite, *Amer. J. Sci., 258,* 583–604, 1960.

Giletti, B. J., Isotopic ages from southwestern Montana, *J. Geophys. Res., 71,* 4029–4036, 1966.

Hanson, G. N., Radiogenic argon loss from biotites in whole rock heating experiments, *Geochim. Cosmochim. Acta, 35,* 101–107, 1971.

Shapiro, L., and S. Botts, Chemical analysis supplied by U.S. Geological Survey, 1970.

Wones, D. R., A low pressure investigation of the stability of phlogopite, *Geochim. Cosmochim. Acta, 31,* 2248–2253, 1967.

Wones, D. R., and H. P. Eugster, Stability of biotite: experiment, theory, and application, *Amer. Mineral., 50,* 1228–1272, 1965.

Yoder, H. S., and I. Kushiro, Melting of a hydrous phase: phlogopite, *Amer. J. Sci., 267-A,* 558–582, 1969.

York, D., Least squares fitting of a straight line, *Can. J. Phys., 44,* 1079–1086, 1966.

CATIONIC DIFFUSION IN OLIVINE TO 1400°C AND 35 KBAR

D. J. Misener
Department of Geological Sciences
University of British Columbia
Vancouver, B.C.

ABSTRACT

Thirty diffusion experiments were performed on crystalline samples of Fe-Mg olivine. Interdiffusion coefficients of Fe and Mg have been determined between 900° and 1100°C and from "one atmosphere" to 35 kbar using diffusion couples of fayalite from Rockport, Mass., and $Fo_{91}Fa_9$ olivine from St. John Island, Red Sea, Egypt. The diffusion of cations is strongly dependent on olivine composition and crystallographic orientation. The diffusion coefficient varies with temperature and pressure according to an empirical Arrhenius relationship, with an activation enthalpy parallel to the c axis of:

$$\Delta H^* = 49.83 + 9.05\, N_2 \text{ kcal/mole}$$

where

$$N_2 = \text{Mg cation mole fraction}$$

An average value of 5.50 cm³/mole was calculated for the activation volume of diffusion.

Diffusion couples of Red Sea olivine–MgO powder and couples of $Fo_{93}Fa_7$ olivine–synthetic forsterite (Fo_{100}) were used to determine the interdiffusion coefficient between 1200° and 1400°C. The interdiffusion coefficient increases with increasing Fe content and with temperature. Diffusion is faster parallel to the c, [001], axis than parallel to either a, [100], or b, [010]. At 7 cation mole percent in the olivine, the [001] activation enthalpy is 65.6 ± 3.6 kcal/mole.

Calculations of ionic electrical conductivity in olivine, using results of this investigation agree with observed conductivity measurements. The results of the present study indicate that at depths greater than 100 km in the mantle ionic conduction is the dominant mechanism of electrical conduction. Estimates of temperature versus depth are made using the derived conductivities in conjunction with conductivity-depth profiles calculated from published electromagnetic depth sounding results.

INTRODUCTION

Recent experimental determinations of electrical conductivity in olivine (Duba, 1972; Duba and Nichols, 1973; Shankland, 1969), taken in conjunction with profiles of conductivity versus depth in the mantle (McDonald, 1957), have been used to calculate temperatures in the mantle (Duba et al., 1973). The temperature and pressure dependence of the electrical conductivity of single crystals of olivine has been determined up to 1300°C (Duba et al., 1973) and up to 10 kbar (Hughes, 1955). Measurements of the electrical conductivity have been made on olivine crystals assumed to approximate upper mantle composition (fayalite content = 10%). In the work reported here, the interdiffusion coefficients of Mg and Fe in olivine are reported and coupled with the above work to refine estimates of the conduction mechanism and geothermal gradients.

Theory

The conduction mechanism has been found to change from impurity conduction at low temperature ($T < 800°C$) to one of intrinsic semiconduction ($800°C \leq T \leq 1200°C$) and possibly to one of ionic conduction at high temperature ($T > 1200°C$), (Shankland, 1969; Duba, 1972). At depths greater than 100 kilometers, the dominant conduction process seems likely to be ionic if one assumes the temperature profile corresponding to the oceanic geotherm of Ringwood (1966) and Clark and Ringwood (1964).

If the process of ionic electrical conduction and that of ionic diffusion are the same, the relationship between the electrical conductivity and the diffusion coefficient may be expressed by the Nernst-Einstein equation

$$\frac{\sigma}{D} = \frac{e^2 z^2 c}{kT} \quad (1)$$

where ez = charge of the migrating species; c = concentration of the migrating species; k = Boltzmann's constant; and T = absolute temperature. The few reported studies of cationic diffusion in olivine (Clark and Long, 1971; Buening and Buseck, 1973; Misener, 1972) suggest that vacancy diffusion governs the rate of ionic migration in a chemical potential gradient; consequently, the processes may operate by the same mechanism and be strictly coupled.

The interdiffusion coefficient of Fe and Mg has been determined in olivine using a diffusion couple technique (Misener, 1972). Interdiffusion coefficients were obtained for the olivine solid solution up to 1100°C and 35 kbar. Between 1200° and 1400°C, diffusion couples of MgO powder and single crystals of forsterite were used. Two high-temperature runs are also reported for a crystal-crystal couple consisting of synthetic forsterite and 7% Fa olivine.

The oxidation state of the Fe ions in the olivine changes the electrical conductivity by as much as a factor of 10^3 (Duba et al., 1973) and thus when comparison of diffusion and conductivity are made care must be taken to assure that the experimental samples have a similar Fe^{+++}/Fe^{++} ratio. The choice of Red Sea olivine crystals for the diffusion experiments permits reliable comparison with the work of Hughes (1955), Duba (1972), and Duba and Nichols (1973).

The equation necessary for the evaluation of the interdiffusion coefficient is

$$\tilde{D}(N_2^*) = \frac{(N_2^+ - N_2^-)/V_m(N_2^*)}{2t \, (\partial N_2/\partial x)_{x=x^*}}$$
$$\left[(1-Y^*) \int_{-\infty}^{x^*} \frac{Y}{V_m} dx + Y^* \int_{x^*}^{\infty} \frac{1-Y}{V_m} dx \right] \quad (2)$$

where $\tilde{D}(N_2^*)$ = interdiffusion coefficient evaluated at a distance on the concentration profile $x = x^*$; $V_m(N_2^*)$ = molar volume of cations at $N_2 = N_2^*$; Y = auxiliary variable = $\frac{N_2 - N_2^-}{N_2^+ - N_2^-}$; $Y = Y^*$ at $x = x^*$; N_2^+, N_2^- = initial concentrations of component 2 in the two diffusion couple members; and t = time interval.

A complete derivation of Equation 2 is given by Wagner, (1969). Figure 1 illustrates the necessary integrals and derivatives for the calculation of the interdiffusion coefficient.

Equation 2 has been used successfully in determining interdiffusion coefficients in the systems $MgO-Cr_2O_3$ (Greskovich and Stubican, 1969), and $TiO_2-Cr_2O_3$ (O'Keefe and Ribble, 1972), and in the oxide-spinel system $MgO-MgAl_2O_4$ (Whitney and Stubican, 1971).

There are certain advantages in using Wagner's formulation for diffusion studies in silicate systems. It is not necessary to determine accurately the location of the Matano interface (Matano, 1933) or the original diffusion couple interface. Variation in the molar volume across the olivine solid solution can be included in the determination of $\tilde{D}$ using the values for V_m obtained by Fisher and Medaris

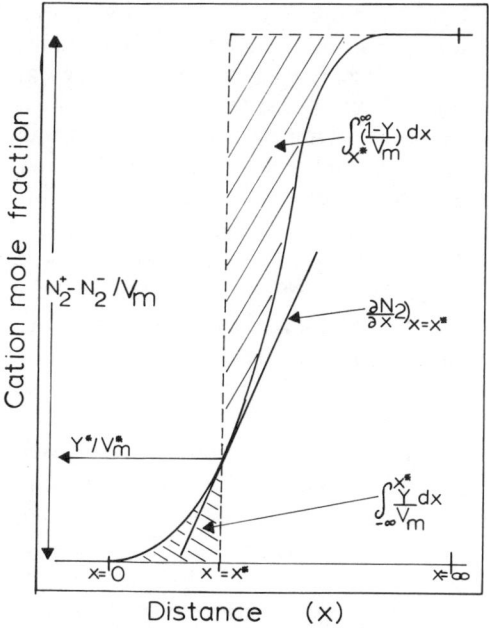

Fig. 1. Graphical representation of the quantities necessary for the evaluation of the interdiffusion coefficient.

where $\Delta H^* =$ activation enthalpy for diffusion; $\Delta V^* =$ activation volume for diffusion; $R =$ gas constant; $T =$ absolute temperature; and $P =$ ambient hydrostatic pressure.

Experimental Details and Procedures

Table 1 lists the electron microprobe analyses of the olivine samples used in this study. Individual crystals were mounted on an x-ray goniometer head and the standard precession method was used to align the crystals. Discs 0.070 in. thick were cut from the oriented crystals, and cores of 0.125 in. diameter were subsequently cut from the individual discs. Five cores were remounted on the goniometer head and precession photographs taken to confirm orientation. The cores were all within 3° of arc of their intended orientation.

Two types of "one atmosphere" diffusion experiments were performed. In the first, crystallographically oriented discs were placed in contact and wrapped in an inert metal foil. In experiments at temperatures over 1000°C samples were wrapped in Pt foil and those at lower temperatures in $Ag_{70}Pd_{30}$. The foil-wrapped diffusion couple was tightly wrapped in Pt wire and placed in a silica glass tube which was evacuated. In the second type of experiment, an oriented disc of olivine

(1969) and Yoder and Sahama (1957). An arbitrary reference point may be selected for analysis of the profile.

The temperature and pressure dependence of diffusion coefficients has been confirmed by numerous experiments (Lazarus and Nachtrieb, 1963), and may be written in the form:

$$\tilde{D}(T_0 P) = \tilde{D}_0 \exp(-\Delta H^*/RT) \exp(-P\Delta V^*/RT) \qquad (3)$$

TABLE 1. Electron microprobe analysis of the olivine samples

Oxide	Olivine Sample (Wt %)			
	Synthetic	Fayalite	Fo 1	Fo 2
FeO	0.00	67.30	9.05	7.18
MnO	0.00	2.38	0.16	N.D.
NiO	0.00	0.19	0.42	0.39
MgO	57.55	0.07	50.27	51.08
SiO_2	42.62	29.13	40.83	41.70
Total	100.17	99.07	100.73	100.35

N.D. = not determined. Synthetic forsterite grown by flame fusion process (Shankland, 1969). Other samples are natural crystals.

was pressed against a pellet of MgO powder, inserted into a Pt tube, followed by more MgO powder, and the tube tightly crimped at both ends. The Pt tube was placed in a silica glass tube as in the other type of experiment.

Platinum-wound furnaces were employed in all high temperature experiments. Platinum–platinum 90% rhodium 10% thermocouples were used to monitor temperature. Total temperature uncertainty due to positioning in the thermal gradient, thermocouple error, and temperature controller error was ±5°C over the temperature range studied.

At the completion of each experiment the samples were slowly cooled (over a period of two to three hours). If the tubes appeared cloudy (due to devitrification) or a "pop" was not heard when the tubes were opened, it was assumed that the vacuum was not maintained and the experiment was rejected. Between 1200° and 1400°C a nitrogen atmosphere was used to prolong the life of the silica glass tubes, allowing experiments of up to two weeks.

Immediately upon extraction from the silica glass tube, the diffusion couple was impregnated with epoxy under vacuum and prepared for microprobe analysis. The original interface was oriented perpendicular to the polished surface.

The high pressure experiments were carried out on a 0.75 in. diameter single-stage, solid-media pressure apparatus developed and modified at the Geophysical Laboratory (Boyd and England, 1960). The pressure was known to an accuracy of ±2 kbar.

Temperature was controlled using a solid state controller (Hadidiacos, 1972), with a Pt–Pt 90% Rh 10% thermocouple. The thermal gradient in the sample cavity was determined, and over the diffusion zone, approximately 200 μm, temperature variation was approximately 2°C. Total temperature uncertainty was ±10°C.

Graphite capsules were used as sample holders in order to control the ambient P_{O_2} and, even for experiments of several hundred hours duration, olivine remained stable. Experiments using fayalite in graphite capsules have produced similar results (Akimoto et al., 1967 and Akimoto and Fujisawa, 1968).

A uniform experimental procedure was adopted for all high pressure experiments. Initially, pressure was increased to within 5 kbar of the desired pressure, temperature was increased to the desired value, then the pressure was increased to the final value. At the completion of each experiment, temperature and pressure were simultaneously lowered over a period of two to three hours. The graphite capsule containing the diffusion couple was impregnated with epoxy and prepared for electron microprobe analysis in

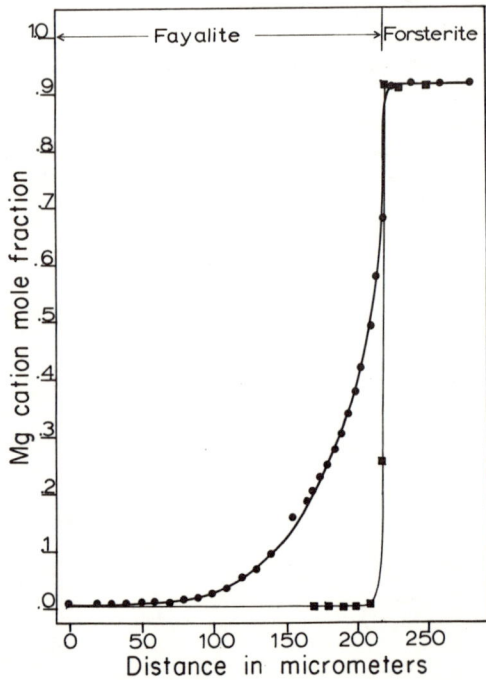

Fig. 2. Mg concentration profile for experiment no. 6 (1000°C, 311 hr, 1 atm). Plotted circles are corrected microprobe analyses along the diffusion zone. The continuous fluorescence effect (squares) is also plotted.

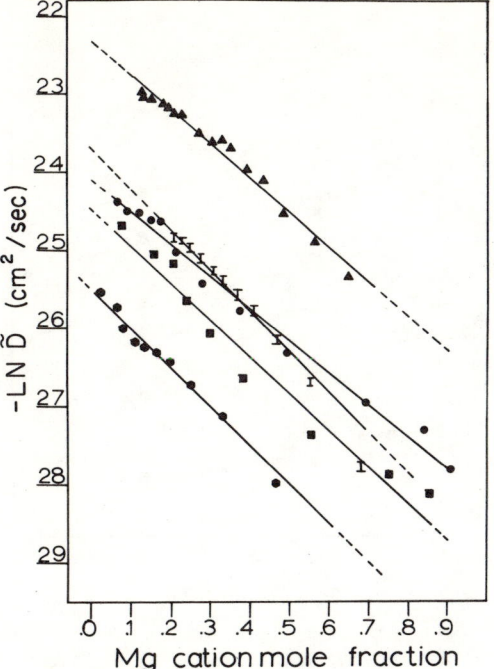

Fig. 3. Calculation of ln $\tilde{D}$ versus Mg cation mole fraction at 1 atm and parallel to [001] Experiment 4 (900°C) (hexagons), Experiment 14 (950°C) (squares), Experiment 1 (1000°C) (circles), Experiment 6 (1000°C) (perpendicular lines), Experiment 40 (1100°C) (triangles).

and Albee, 1968) with the coefficients given by Albee and Ray (1970).

The continuous fluorescence effect (Reed and Long, 1963) may become important when measuring small amounts of one element in a phase near a boundary with another phase which is rich in that element. A calibration couple of forsterite-fayalite was prepared in a manner similar to that for the one atmosphere experiments and analyzed on the electron microprobe. The resulting curve is plotted in Fig. 2. The correction is of minor importance within 10 μm of the interface and is negligible farther away.

RESULTS AND ANALYSIS

A list of the one atmosphere and high pressure experiments is given in the Appendix. Figure 2 is a representative Mg

the same manner as the one atmosphere experiments.

Concentration profiles were obtained for each diffusion experiment using the M.A.C. model 400 electron microprobe at the Geophysical Laboratory. The concentration data were collected at intervals of 2, 5, 10, or 20 μm, depending on the rate of change of composition with distance.

The microprobe was calibrated using standards of 7.42% and 31.14% Mg, 5.62%, 17.18%, and 31.86% Fe, 4.19% Mn, and 19.10% Si. The raw data, in counts per second, were reduced to oxide weight percent at the time of analysis using a computer program developed by Finger and Hadidiacos (1972). The Bence-Albee correction procedure was used for the matrix corrections (Bence

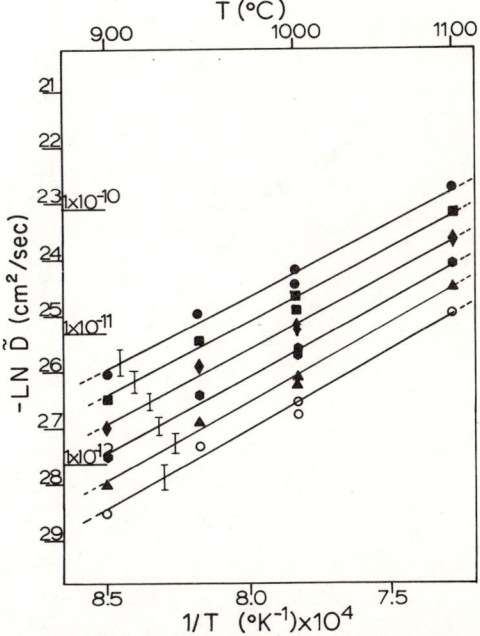

Fig. 4. Ln $\tilde{D}$ versus $1/T$ (°K)$^{-1}$ × 10^4 for D calculated at 10 (solid circles), 20 (squares), 30 (diamonds), 40 (hexagons), 50 (triangles), and 60 (open circles) Mg cation mole percent. All data at 1 atm and parallel to [001]. Error bars indicate one standard deviation calculated using regression analysis.

concentration profile for a forsterite-fayalite couple at 1000°C. Figure 3 illustrates the results of calculations of $\tilde{D}$ versus concentration using Equation 2. The resulting profiles show a logarithmic dependence of the interdiffusion coefficient on concentration in the range $(0.1 \leq \mathrm{Mg/Mg + Fe} \leq 0.8)$.

Values of $\tilde{D}$ as a function of $1/T$ (°K^{-1}), at selected concentrations, are plotted in Fig. 4. Regression analysis of $\ln \tilde{D}$ parallel to [001] versus $1/T$ results in a general Arrhenius relationship:

$$\tilde{D} = [1.53 \pm 0.25 - 1.12 N_2] \times 10^{-2} \exp \left[\frac{49.83 \pm 4.5 + 9.05 N_2}{RT} \right] \quad (4)$$

where N_2 = cationic mole fraction Mg $(0.1 \leq \mathrm{Mg} \leq 0.8)$. In all figures using

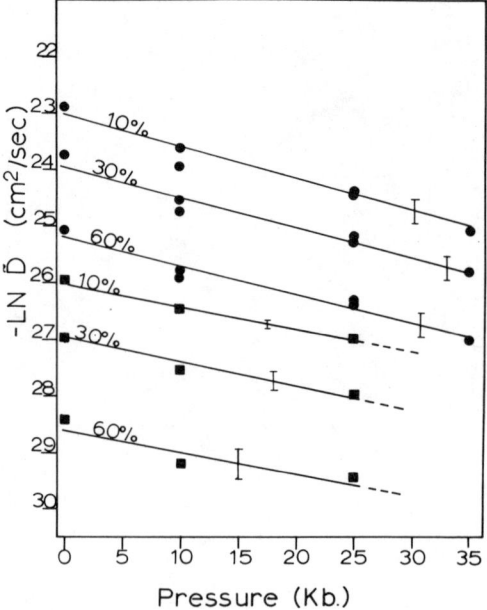

Fig. 6. Ln $\tilde{D}$ versus pressure calculated at 10, 30, and 60 Mg cation mole percent. Experiments at 900°C (squares), at 1100°C (circles).

regression techniques, the standard deviation in the dependent variable is indicated by the large error bars.

The results of the diffusion experiments parallel to [001] at pressures up to 35 kbar, at 1100°C, are shown in Fig. 5. Ln $\tilde{D}$ shows a linear decrease with increasing mole percent Mg as in the one atmosphere experiments.

Figure 6 shows the effect of hydrostatic pressure on the interdiffusion coefficient. Regression analysis of the data for 900° and 1100°C yield values of the activation volume, ΔV^*, plotted in Fig. 7. The average value of ΔV^* is approximately 5.50 cm^3/mole with a slight decrease in ΔV^* between 1100° and 900°C. Calculated values of ΔV^* using experiments 3-P and 21-P yield values consistent with the average ΔV^* of 5.50 cm^3/mole.

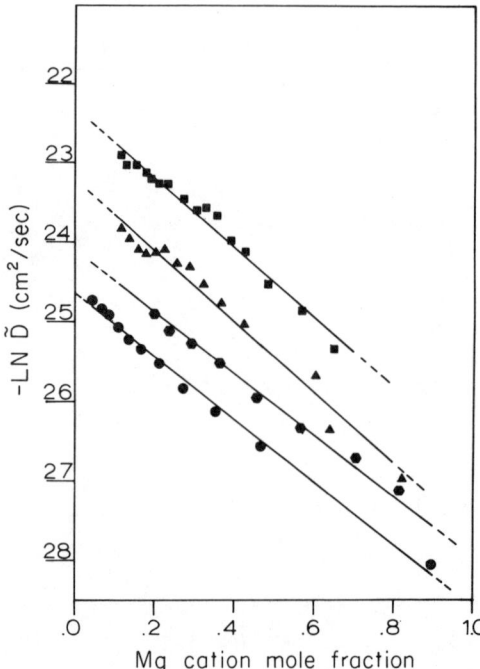

Fig. 5. Ln $\tilde{D}$ versus Mg cation mole fraction at various pressures. Experiment 40 (1 atm) (squares), Experiment 1-P (10 kbar) (triangles), Experiment 18-P (35 kbar) (circles), Experiment 14-P (25 kbar) (hexagons). All experiments at 1100°C and parallel to [001].

A diffusion couple of MgO powder–fayalite crystal was run at 1100°C and 25 kbar (experiment no. 12-P) to observe the effect of the crystal-powder interface on the diffusion coefficient. Calculated

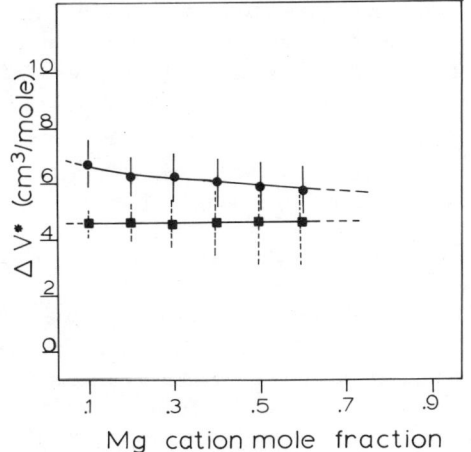

Fig. 7. ΔV^* versus concentration (Mg cation mole fraction). (Circles) 1100°C, (squares) 900°C.

values of $\tilde{D}$ agree with the crystal-crystal experiments within experimental error.

Figure 8 shows profiles parallel to the a, b, and c crystallographic axes at 1250°C and one atmosphere. Figure 9 represents calculations of $\tilde{D}$ versus Mg concentration for the profiles in Fig. 8. As in the lower temperature experiments, ln $\tilde{D}$ decreases with increasing Mg concentration.

The plot of ln $\tilde{D}$ versus concentration departs from linearity at Mg concentrations greater than 98%. This is probably due to large errors in the calculation of the derivative (dN^*_2/dX) in this portion of the profile. A similar but less pronounced effect is observed when the Mg concentration is less than 92%.

Fig. 10 illustrates the regression analysis of ln $\tilde{D}$ versus $1/T$ (°K^{-1}) for the interdiffusion of Fe-Mg in olivine between 1200° and 1400°C parallel to the c axis. The standard error in ln $\tilde{D}$, as calculated using the regression analysis, is indicated

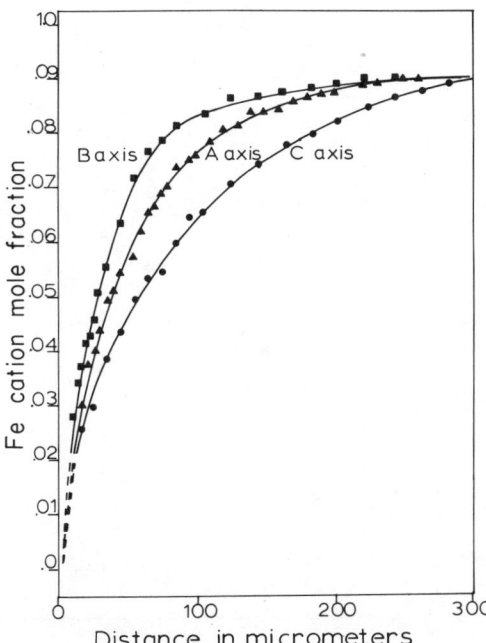

Fig. 8. Fe concentration profiles for Experiment 29, parallel to a axis (triangles); Experiment 30, parallel to b axis (squares); Experiment 31, parallel to c axis (circles). Experiments at 1 atm, 1250°C, and 408 hr.

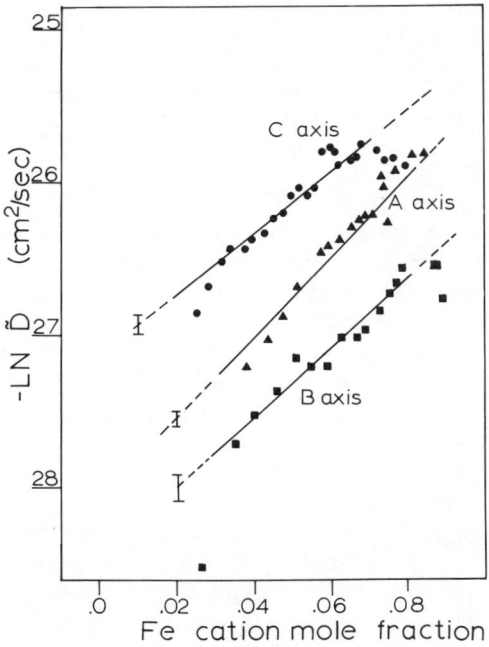

Fig. 9. Ln $\tilde{D}$ versus Fe cation mole fraction. Experiment 29, parallel to a axis (triangles); Experiment 30 parallel to b axis (squares); Experiment 31, parallel to c axis (circles).

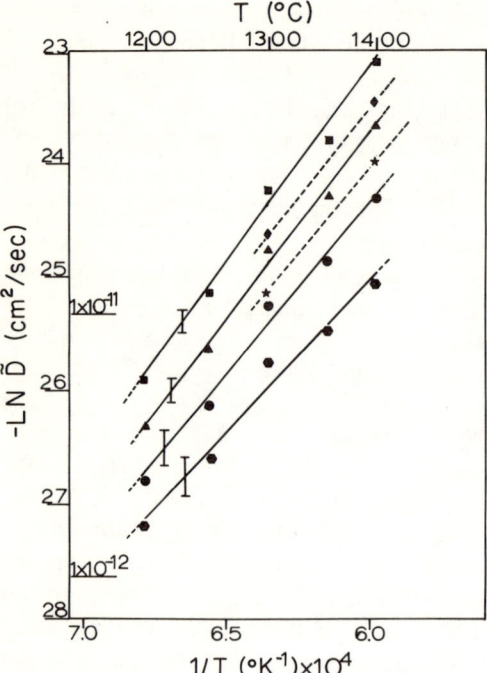

Fig. 10. Ln $\tilde{D}$ versus $1/T$ (°K^{-1}) × 10⁴ for $\tilde{D}$ calculated at 3 (hexagons), 5 (circles), 7 (triangles), and 9 (squares) Fe cation mole fraction. Calculations for synthetic forsterite-Fo₉₃Fa₇ olivine diffusion couple, $\tilde{D}$ calculated at 5 (stars) and 7 (diamonds) Fe cation mole fraction. Error bars indicate one standard deviation.

by the large error bars. The slight increase in the values of $\tilde{D}$ obtained from the crystal-crystal experiments, relative to crystal-powder experiments, may indicate a decrease in the contact resistance in the crystal-crystal couple relative to the crystal-powder couple.

The dependence of $\tilde{D}$ on crystallographic orientation between 1200° and 1400°C is illustrated in Fig. 11. Diffusion parallel to the c axis is approximately 4.5 times faster than diffusion parallel to the b axis at 1200°C. At 1400°C this ratio has decreased to a value of approximately 2.6.

Figure 12 shows literature values of previous studies of 1 atm diffusion rates of cations in olivine along with the results from the present study. The results of Naughton and Fujikawa (1959) and Jander and Stamm (1932) were obtained for rates in powdered samples; the other results are for studies on single crystals.

Geophysical Application of the Diffusion Coefficients

The interdiffusion coefficient has been used to calculate the conductivity as a function of temperature. This calculation is based on the assumption that $\tilde{D}$ measured at low Fe concentrations approximates the self-diffusion coefficient for vacancies. Results of this calculation using Equation 1 are shown in Fig. 13 along with the previously determined values of electrical conductivity in olivine. Interdiffusion data from the experiments between 1200° and 1400°C were used in the calculations. Agreement between calculated and experimentally determined values of the electrical conduc-

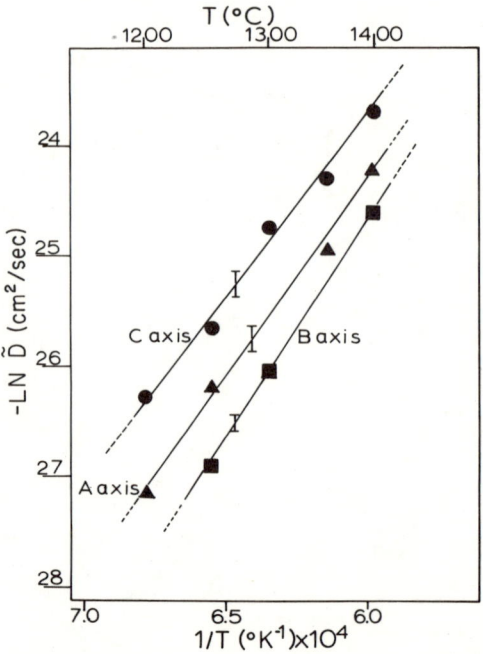

Fig. 11. Ln $\tilde{D}$ versus $1/T$ (°K^{-1}) × 10⁴ for diffusion parallel to a axis (triangles), b axis (squares), and c axis (circles). All calculations for Fe cation mole fraction = 0.07.

tivity and the enthalpy of activation for conductivity indicates that ionic migration is probably the dominant mechanism of electrical conduction in olivine when the temperature is greater than 1200°C.

Estimates of temperatures in the mantle may also be made using the P-T dependence of $\tilde{D}$. From Equation 1

$$\sigma_m = \frac{\tilde{D}(T,P)K_1}{T} \qquad (5)$$

where σ_m = independent determination of conductivity in the mantle; $\tilde{D}(T, P)$ = calculated value of the interdiffusion coefficient at T and P (Fo$_{91}$Fa$_9$ olivine);

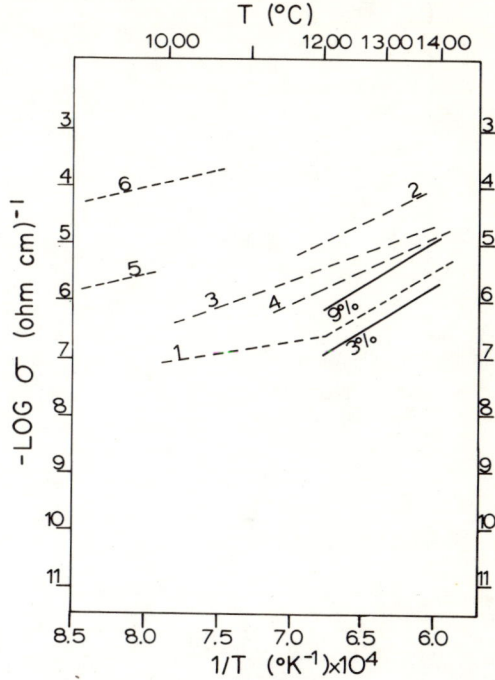

Fig. 13. Experimental determinations of log σ versus $1/T$ (°K^{-1}) $\times 10^4$. Solid line, results of log σ calculated using Equation 2 for Fe cation mole fraction of 0.03 and 0.09. Results from other authors: (1) Synthetic Forsterite, Shankland (1969), (2) Red Sea Olivine, Hughes (1959), (3) and (4) San Carlos Olivine, Duba and Nichols (1973), (5) Fo$_{92.6}$ olivine, Kobayashi and Maruyama (1971), (6) Fo$_{82}$ olivine, Mizutani and Kanamori (1967).

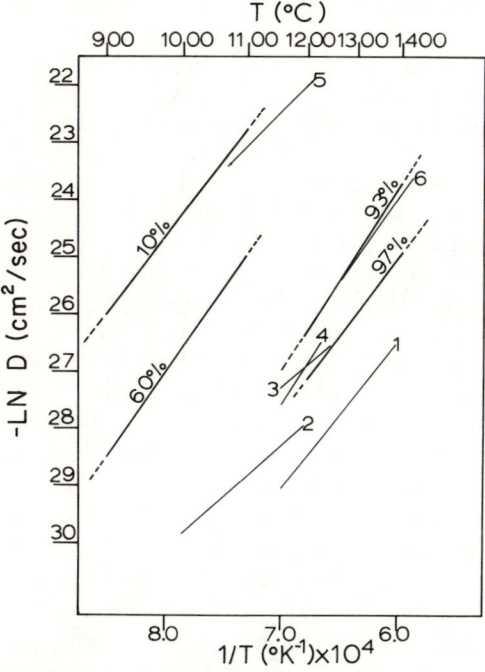

Fig. 12. Experimental determination of ln $\tilde{D}$ versus $1/T$ (°K^{-1}). Results of the present study (dashed lines). Percentages correspond to Mg cation mole percents. Results of other authors: (1) Co^{++} in Co$_2$SiO$_4$, Borchardt and Schmalzried, 1972, (2) Fe^{++} in Mg$_2$SiO$_4$, Naughton and Fujikawa, 1959, (3) Ni^{++} in Fo$_{93}$Fa$_7$ olivine, Clark and Long, 1971, (4) Ni^{++} in Red Sea Olivine, Clark and Long (1971), (5) Mg^{++} in Mg$_2$SiO$_4$ powder, Jander and Stamm, 1932, (6) Mg^{++} in Mg$_2$SiO$_4$ from electrical conductivity, Pluschkell and Engell, 1968.

$K_1 = \dfrac{e^2 z^2 c}{k} = 1.9 \times 10^8$ (constant); and T = absolute temperature (°K). Substituting Equation 3 into Equation 5

$$\sigma_m = \frac{K_1}{T} D_0 \exp(-\Delta H^*/RT) \\ \exp(-P\Delta V^*/RT) \qquad (6)$$

$$\ln \sigma_m = \ln\left(\frac{K_1}{T}\right) + \ln D_0 - \frac{1}{RT}(\Delta H^* + P\Delta V^*) \qquad (7)$$

Using Equation 7 with the conductivity data of McDonald (1957) and Eckhardt et al. (1963) and a starting temperature versus depth profile of Clark and Ringwood (1964) and Ringwood (1966) an

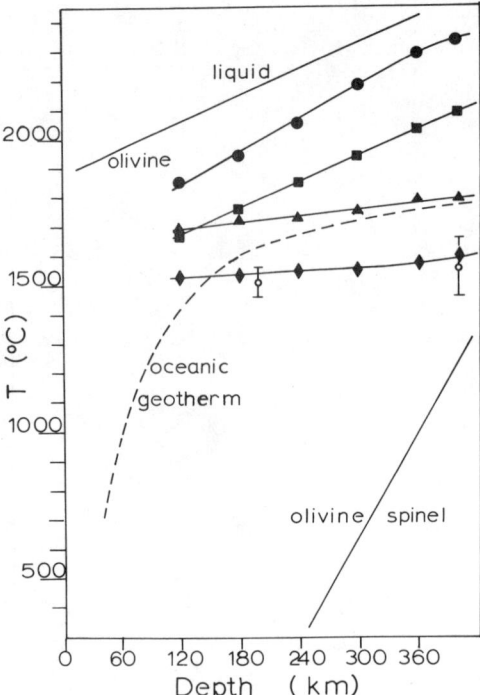

Fig. 14. Temperature versus depth in the mantle. Oceanic geotherm, Ringwood and Clark, (dashed line). Forsterite stability field, Akimoto and Fujisawa (1968), Davis and England (1964), (solid line). Experimentally determined values, Duba (1973), (open circles). Calculated values using experimentally determined diffusion coefficients, this study; Fo_{91} olivine, $\Delta V^* = 0.0$, McDonald's (1957) conductivity estimate (diamonds), Fo_{91} olivine, $\Delta V^* = 0.0$ Eckhardt et al. (1963) conductivity estimate (triangles), Fo_{91} olivine, $\Delta V^* = 5.50$ cm³/mole, McDonald's (1957) conductivity estimate (squares), Fo_{91} olivine, $\Delta V^* = 5.50$ cm³/mole, Eckhardt et al. (1963) conductivity estimate (circles). All calculations for diffusion parallel to c axis.

iterative procedure was used to calculate a new temperature versus depth profile. Iteration of temperature in Equation 7 was continued until

$$|\sigma_m \text{ (calculated)} - \sigma_m| \leqslant 0.05 |\sigma_m| \quad (8)$$

The temperature profiles calculated using Equation 7 are shown in Fig. 14 along with the olivine stability field and the oceanic geotherm of Ringwood (1966). The two points with the large error bars are recent estimates by Duba and Nichols (1973) based on electrical conductivity measurements on Red Sea olivine at 1 atmosphere and 8 kbar assuming the conductivity profile of McDonald (1957). The two profiles calculated for $Fo_{91}Fa_9$ olivine, the lower one using McDonald's estimate of σ_m and the upper one using the Eckhardt et al. estimate, bracket the geotherm of Ringwood (1966) below 150 km. The two calculations assuming a ΔV^* of 5.50 cm³/mole yield temperatures approximately 450°C higher at a depth of 300 km.

Uncertainties surrounding the importance of the pressure effect on the conductivity (Duba, 1972; Hughes, 1955), and uncertainties in the calculated σ_m profiles from geomagnetic and magnetotelluric measurements preclude definite conclusions as to the validity of the pressure-included versus pressure-excluded profiles; however, preliminary comments may be made. The ΔV^* used in these calculations was for single crystals, and for a polycrystalline aggregate pressure effects have been observed to be less (Schult and Schober, 1969; Hamilton, 1965; Bradley et al., 1962). The polycrystalline pressure effects may be a closer approximation to the mantle and thus the profiles calculated in this study using $\Delta V^* = 5.50$ cm³/mole may be too high. The magnetotelluric studies of the electromagnetic field may define σ_m versus depth in the upper 400 km of the mantle more accurately because shorter period vibrations are being recorded, and, therefore, better resolution is obtained at shallow depths. For these reasons, the profile marked with triangles in Fig. 14 was assumed to represent the best fit calculation based on the data from this study.

CONCLUSIONS

1. The interdiffusion coefficient for Fe and Mg in olivine decreases with increasing Mg cation mole fraction. The tem-

perature and composition dependence of $\tilde{D}$ parallel to $|001|$ is given by

$$\tilde{D} = (1.53\pm0.25 - 1.12N_2) \times 10^{-2} \times \exp\left(\frac{49.83\pm4.5+9.05N_2}{RT}\right)$$

where $900 \leq T \leq 1100$; $0.10 \leq N_2 \leq 0.60$; and $N_2 =$ cation mole fraction Mg.

2. The interdiffusion coefficient is a function of the crystallographic orientation. In the temperature range $900 \leq T \leq 1100$, $\tilde{D}$ [001] $> \tilde{D}$ [010]. In the temperature range $1200 < T < 1400$, $\tilde{D}$ [001] $> \tilde{D}$ [000] $> \tilde{D}$ [010].

3. The interdiffusion coefficient decreases with increasing pressure. A hydrostatic pressure of 35 kilobars decreases the interdiffusion coefficient by approximately a factor of ten.

4. Ionic conductivities calculated from the results of the diffusion experiments between 1200° and 1400°C agree with the experimental determinations of electrical conductivity in olivine. Below a depth of 100 kilometers in the mantle, ionic conduction is probably the dominant mechanism of electrical conduction.

Acknowledgments

I wish to thank H. J. Greenwood of the University of British Columbia and H. S. Yoder, Jr., F. R. Boyd, L. W. Finger, and P. M. Bell of the Geophysical Laboratory, Carnegie Institution of Washington, for their interest in the investigation and their helpful discussion. I am indebted to T. J. Shankland, D. R. Wones, D. Virgo, and T. Richards for supplying the olivine crystals used in this study.

The work was supported by the Carnegie Institution of Washington and Grant A-4222 of the National Research Council of Canada to H. J. Greenwood.

References Cited

Akimoto, S., and H. Fujisawa, Olivine-spinel solid solution equilibria in the system Mg_2SiO_4-Fe_2SiO_4, *J. Geophys. Res., 73,* 1467–1479, 1968.

Akimoto, S., E. Komado, and I. Kushiro, Effect of pressure on the melting of olivine and spinel polymorph of Fe_2SiO_4, *J. Geophys. Res., 72,* 679–686, 1967.

Albee, A. L., and L. Ray, Correction factors for electron probe microanalysis of silicates, oxides, phosphates, carbonates, and sulfates, *Anal. Chem., 42,* 1408–1414, 1970.

Bence, A. E., and A. L. Albee, Empirical correction factors for the electron-microanalysis of silicates and oxides, *J. Geol., 76,* 382–403, 1968.

Borchardt, V. G., and H. Schmalzried, Diffusion in Orthosilicaten, *Ber. Deut. Keram. Ges., 49,* 5–9, 1972.

Boyd, F. R., and J. L. England, Apparatus for phase-equilibrium measurements at pressures up to 50 kb. and temperatures up to 1750°C, *J. Geophys. Res., 65,* 741–748, 1960.

Bradley, R. S., A. K. Jamil, and D. C. Munro, The electrical conductivity of fayalite and spinel, *Nature, 193,* 965–966, 1962.

Buening, D. K., and P. R. Buseck, P_{O_2} dependence of vacancy formation in olivine (abstract), *Eos Trans. Amer. Geophys. Union, 54,* 488, 1973.

Clark, A. M., and J. V. P. Long, Anisotropic diffusion of Ni^{++} in olivine, in *Thomas Graham Memorial Symposium on Diffusion Processes,* Gordon and Breach, London, pp. 511–521, 1971.

Clark, S. P. Jr., and A. E. Ringwood, Density distribution and constitution of the mantle, *Rev. Geophys., 2,* 35–88, 1964.

Davis, B. T. C., and J. L. England, The melting of forsterite up to 50 kilobars, *J. Geophys. Res., 69,* 1113–1116, 1964.

Duba, A., Electrical conductivity of olivine, *J. Geophys. Res., 77,* 2483–2495, 1972.

Duba, A., and I. A. Nichols, The influence of oxidation state on the electrical conductivity of olivine, *Earth Planet. Sci. Lett., 18,* 59–64, 1973.

Duba, A., H. C. Heard, and R. N. Schock, The effect of temperature, oxygen fugacity and pressure on the electrical conductivity of olivine (abstract), *Eos Trans. Amer. Geophys. Union, 54,* 506, 1973.

Eckhardt, D., K. Larmer, and T. Madden, Long period magnetic fluctuations and mantle conductivity estimates, *J. Geophys. Res., 68,* 6279–6286, 1968.

Finger, L. W., and C. G. Hadidiacos, Electron-microprobe automation, *Carnegie Inst. Washington Yearb., 71,* 598–600, 1972.

Fisher, G. W., and L. G. Medaris Jr., Cell dimensions and x-ray determinative curve for synthetic Mg-Fe olivines, *Amer. Mineral., 54,* 741–753, 1969.

Greskovich, C., and V. S. Stubican, Interdiffusion studies in the system MgO-Cr_2O_3. *J. Phys. Chem. Solids, 30*, 909–917, 1969.

Hadidiacos, C., Temperature controller for high pressure apparatus, *Carnegie Inst. Washington Yearb. 71*, 620–622, 1972.

Hamilton, R. M., Temperature variation at constant pressure of the electrical conductivity of periclase and olivine, *J. Geophys. Res., 70*, 5619–5692, 1965.

Hughes, H., The pressure effect on the electrical conductivity of olivine, *J. Geophys. Res., 60*, 187–191, 1955.

Jander, W., and W. Stamm, Der innere Aufbau fester anorganischer Verbindungen bei höheren Temperaturen, *Z. Anorg. Allg. Chem., 207*, 289–307, 1932.

Kobayashi, Y., and H. Maruyamo, Electrical conductivity of olivine single crystals at high temperature. *Earth Planet. Sci. Lett., 11*, 415–419, 1971.

Lazarus, D., and N. H. Nachtrieb, in *Solids under Pressure*, W. Paul and D. M. Warschauer, eds., McGraw-Hill, New York, p. 43, 1963.

Matano, C., On the relation between the diffusion coefficients and concentrations of solid metals (the nickel-copper system), *Jap. J. Phys., 8*, 109–113, 1933.

McDonald, K. L., Penetration of the geomagnetic secular field through the mantle with variable conductivity, *J. Geophys. Res., 62*, 117–141, 1957.

Misener, D. J., Interdiffusion studies on the system Fe_2SiO_4-Mg_2SiO_4, *Carnegie Inst. Washington Yearb., 71*, 516–520, 1972.

Mizutani, H., and H. Kanamori, Electrical conductivities of rock forming minerals at high temperatures, *J. Phys. Earth, 15*, 25–31, 1967.

Naughton, J. J., and Y. Fujikawa, Measurement of intergranular diffusion in a silicate system, *Nature, 184*, 54–56, 1959.

O'Keefe, M., and T. J. Ribble, Interdiffusion and the solubility limits of Cr_2O_3 in the rutil phase TiO_2, *J. Solid State Chem., 4*, 351–356, 1972.

Pluschkell, W., and H. J. Engell, Ionen- und Electronenleitung im Magnesium-Orthosilikat, *Ber. Deut. Keram. Ges., 45*, 388–394, 1968.

Reed, S. J. B., and J. V. P. Long, Electron-probe measurements near phase boundaries, in *X-ray Optics and X-ray Microanalysis*, H. H. Pattee, V. E. Coslett, and Arne Engstrom, eds., Academic Press, New York, p. 317, 1963.

Ringwood, A. E., Mineralogy of the mantle, in *Advances in Earth Science*, P. M. Hurley, ed., M.I.T. Press., Cambridge, Mass., pp. 357–399, 1966.

Schult, A., and M. Schober, Measurements of electrical conductivity of natural olivine at temperatures up to 950°C and pressures up to 40 kilobars, *Z. Geophys., 35*, 105–112, 1969.

Shankland, T. J., Transport properties of olivine, in *The Application of Modern Physics to the Earth and Planetary Interiors*, S. K. Runcorn, ed., Interscience, New York, p. 175, 1969.

Wagner, C., The evaluation of data obtained with diffusion couples of binary single phase and multiphase systems, *Acta. Met., 17*, 99–107, 1969.

Whitney, W. P., and V. S. Stubican, Interdiffusion in the system MgO-$MgAl_2O_4$, *J. Amer. Ceram. Soc., 54*, 349–352, 1971.

Yoder, H. S. Jr., and Th. G. Sahama, Olivine, X-ray determinative curve, *Amer. Mineral., 42*, 475–491, 1957.

APPENDIX

List of the one-atmosphere diffusion experiments (900°–1100°C).

Experiment Number	Time (sec)	Temperature (°C)	Crystal Orientation
4	1.192×10^6	900 ± 5	c axis
14	8.46×10^5	950 ± 5	c axis
15	8.46×10^5	950 ± 5	b axis
1	3.56×10^5	1000 ± 5	c axis
6	1.14×10^6	1000 ± 5	c axis
59	3.64×10^5	1000 ± 5	b axis
40	8.95×10^5	1100 ± 5	c axis

List of the 1 atm experiments using forsterite–MgO and synthetic forsterite–Fo_{93} olivine diffusion couples.

Experiment Number	Time (sec)	Temperature (°C) ± 5	Crystal Orientation	Diffusion Couple
68	7.25×10^5	1200	a axis	Fo 1-MgO
69	7.25×10^5	1200	c axis	Fo 1-MgO
29	1.470×10^6	1250	a axis	Fo 1-MgO
30	1.470×10^5	1250	b axis	Fo 1-MgO
31	1.470×10^6	1250	c axis	Fo 1-MgO
50	8.80×10^5	1300	b axis	Fo 1-MgO
51	8.80×10^5	1300	c axis	Fo 1-MgO
52	1.210×10^6	1300	c axis	synthetic Fo 2
26	9.46×10^5	1350	a axis	Fo 2-MgO
27	9.46×10^5	1350	c axis	Fo 2-MgO
72	3.41×10^5	1400	a axis	Fo 2-MgO
73	3.41×10^5	1400	b axis	Fo 2-MgO
74	3.41×10^5	1400	c axis	Fo 2-MgO
71	3.41×10^5	1400	c axis	Fo 2-MgO

List of the high pressure diffusion experiments using forsterite-fayalite-MgO powder diffusion couples.

Experiment Number	Time (sec)	Temperature (°C) $\pm 10°$	Pressure (kbar) ± 2	Crystal Orientation
19-P	5.11×10^5	900	10	c axis
16-P	6.95×10^5	900	25	c axis
3-P	3.24×10^5	1000	10	c axis
1-P	3.42×10^5	1100	10	c axis
6-P	3.14×10^5	1100	10	c axis
21-P	4.84×10^5	1100	15	b axis
12-P (MgO-Fa)	1.548×10^6	1100	25	c axis
14-P	1.186×10^6	1100	25	c axis
18-P	4.07×10^5	1100	35	c axis

DIFFUSION OF TRITIATED WATER IN β-QUARTZ

E. W. Shaffer,[1] J. Shi-Lan Sang, A. R. Cooper, and A. H. Heuer
Department of Metallurgy and Materials Science
Case Western Reserve University
Cleveland, Ohio 44106

ABSTRACT

The diffusion of "water" in β-quartz was studied at temperatures from 720° to 850°C. Quartz single crystals were heated in tritiated water vapor at pressures of 225–595 mm Hg; the resulting tritium penetration profiles were determined by combining serial sectioning with liquid scintillation counting. The diffusion was composition dependent, *decreasing* with increasing "water" concentration. At a single temperature and water vapor pressure, diffusion coefficients varied by more than a factor of three. At constant tritium composition, the diffusion coefficients follow an Arrhenius behavior; activation energies of from 22 to 26 kcal/mole were found and they increased with increasing "water" concentration. The solubility of "water" in quartz was also determined and varied approximately linearly with water vapor pressure.

These results differ from those obtained by Roberts and co-workers on silica glass, who found that the diffusion coefficient of "water" *increased* with increasing concentration, and that the solubility varied as the square root of the water vapor pressure.

INTRODUCTION

The presence of dissolved "water"[2] in quartz and other silicates has been shown by Griggs (1967) and others (e.g., Baëta and Ashbee, 1970) to lead to a pronounced hydrolytic weakening. For mantle minerals, this is expected to lead to a strain-rate dependent strength of the mantle. While quartz is not a major mantle mineral, it is an important mineral in the continental crust and its solubility for "water" and the mechanism of the incorporation of "water" into the lattice, as well as knowledge of its diffusion kinetics, are of interest. In particular, the Frank-Griggs model of hydrolytic weakening in silicates—easy slip occurring only when Si-O-Si bridges adjacent to a dislocation are hydrolyzed by the migration of water—depends markedly on the diffusion kinetics. The purpose of this paper is to present preliminary results of such a "water" diffusion study in β-quartz.

Many properties of quartz depend sensitively on its water content. For example, although pure quartz is characterized by a very broad region of transparency at optical wavelengths, beginning with the ultraviolet and continuing to the near infrared region, almost all quartz crystals show infrared absorption bands around 3400 cm^{-1}. Wood (1960), Brunner and co-workers (1961) and Kats (1962) have investigated the impurities and lattice defects in α-quartz and proved that all bands observed around 3400 cm^{-1} were caused by OH vibrations. Similarly it has been suggested that there is a correlation between OH (i.e. "water") content and anelasticity in α-quartz (Dodd and Fraser, 1965). Sawyer (1972) has shown a direct proportionality between the OH content and the amplitude of the acoustic

---

[1] Now with Brockway Glass, Research and Development Laboratory, Brockway, Pa.

[2] Quotation marks are used here and throughout this paper since the exact form of the dissolved water is not known at this time.

loss peak (Q^{-1}). Therefore, knowledge of the "water" content of quartz is important when considering quartz crystals for use as piezoelectric oscillators.

The diffusion of hydrogen in quartz of constant "water" content has been studied by Kats et al. (1962). Their work indicated that the activation energy for diffusion in β-quartz (42.3 kcal/mole) is larger than that in α-quartz (19.3 kcal/mole). In contrast, Frischat (1970) found that for sodium diffusion, the transformation from α to β quartz at 573°C resulted in a decrease of activation energy from 20.2 to 11.5 kcal/mole and an increase of the diffusivity by a factor of 1.8. At the same time, Frischat also studied sodium diffusion along two directions, parallel and perpendicular to the c axis, and found a large diffusion anisotropy. Diffusion coefficients parallel to the c axis were $D = 7 \times 10^{-3}$ exp $(-11500/RT)$ cm^2/sec while those perpendicular to the c axis were $D = 4 \times 10^{-2}$ exp $(-27000/RT)$ cm^2/sec.

Haul and Dumbgen (1962) found that oxygen diffusivities in quartz crystals were considerably smaller than in SiO_2 glass. In addition, a strong anisotropy in the diffusion of H_2O^{18} in natural quartz was found by Choudhury et al. (1965).

No previous studies appear to have been concerned with determining the nature of the diffusing species and the diffusion mechanism(s) when crystalline quartz is heated in water vapor (some information of this type on "water" diffusion in silica glass is available from the work of Roberts and co-workers (Moulson and Roberts, 1960, 1961; Drury and Roberts, 1963; Roberts and Roberts, 1966). In order to increase our understanding of "water" diffusion in quartz, the solubility and diffusivity of tritiated water in β-quartz was studied using a liquid scintillation technique, which proved to be a suitable and sensitive method of determining tritium penetration profiles.

Experimental Methods

The specimens used in this work were synthetic, electronic-grade, quartz single crystals (obtained from Sawyer Research Products, Inc., Cleveland, Ohio). The predominant metallic impurities were Al, Fe, Na, and Li. With the exception of Al, which was present at the 60 ppm level, the concentration of each impurity was less than 10 ppm. The OH content of the samples was estimated from the intensity of the infra-red absorption peak at 2.7 μm, using the correlation of Dodd and Fraser (1965). All crystals had <9 ppm OH.

Quartz blanks 0.5 × 0.5 × 1.5 in., with the long axis parallel to the c axis, were obtained from the supplier and were used to cut basal specimens 0.5 × 0.5 × ~0.1 in. The orientations of all diffusion specimens were determined by Laue back-reflection techniques. Diffusion samples were glued to a metallic block and ground on a "Logitech" jig, which insured that the basal surfaces were ground and polished flat and parallel. Initial grinding was carried out with 400-, 1200-, and 3200-mesh emery on an aluminum lap, and final polishing with 0.3 and 0.06 μm alumina. The resulting surfaces were free from obvious defects when examined in reflected light at high magnification.

The diffusion apparatus consisted of the annealing chamber, a tritiated water reservoir, and a conventional vacuum system and furnace. A full description can be found elsewhere (Shaffer, 1973); only the salient details of the system will be mentioned here. The annealing chamber was a closed-ended fused silica tube, 18 inches long and 2 inches I.D. The upper portion of the tube was joined via a graded glass seal to the vacuum system. The samples were placed in a fused silica sample holder that was connected with nichrome wire to an iron slug. This permitted the samples to be raised or lowered into the hot zone of the apparatus by an external magnet without affecting the vacuum condition. Temperatures within the diffusion chamber were measured with a Pt/Pt-13% Rh thermocouple, which was placed within the chamber close to the diffusion specimens. Temperatures were maintained to ±2°C during the course of a diffusion anneal. During a diffusion run, all parts of the apparatus were wrapped with heating tape and kept at 150°C, to prevent condensation of the tritiated water (see below).

The tritiated water reservoir was attached to the diffusion apparatus by means of an O-ring joint, and was immersed in a temperature-controlled oil bath. The water vapor pressure in the system was determined by the temperature of the oil bath and was known to ±5 mm Hg.

A diffusion run consisted of heating the sample in the evacuated system until the diffusion temperature was reached. After the system reached thermal equilibrium, the system was isolated from the vacuum pumps and the tritiated water admitted (time zero). At the completion of the diffusion anneal, the sample was raised out of the hot zone of the furnace, using the external magnet, to a position where the temperature was ~585°C. The furnace power was then turned off, so that the sample cooled slowly (~15 minutes) through the α-β transformation. When the sample had cooled to ~570°C, it was withdrawn quickly to the coolest part of the apparatus (150°C), the tritiated water was recondensed in the reservoir and the power to the heating tape was shut off. This procedure, while not as desirable as an immediate quench, was necessitated by the tendency of quartz to crack at the α-β transformation if cooled too rapidly, and was successful in preventing visible cracking in any of the diffusion specimens.

The penetration of the tritiated water into the specimen was determined by measuring the activity of each new surface produced by grinding and polishing away a small thickness of material parallel to the basal plane, using the same procedure as for the initial specimen preparation; typically, ~3 μm was removed at a time. The quantity removed during sectioning was determined by thickness measurements using a precision micrometer and could be determined to ~1 μm. The counting equipment was a Packard Tri-Carb Liquid Scintillation Spectrometer model 3380. The scintillation solution used during these experiments was a standard Beckman Ready-Solv Solution VI. (Tritium is a β emitter (maximum energy of 18.6 keV) with a 12.26 year half life.) Counting was continued automatically until the count rate above background was known at the 0.1% confidence level.

Results

All experiments reported here were for diffusion parallel to the c axis; they were carried out in the temperature range of 720°–850°C, i.e., in the β-quartz stability field. In this temperature range, an appropriate amount of "water" penetration could be obtained in reasonable diffusion times. Figure 1 shows an exam-

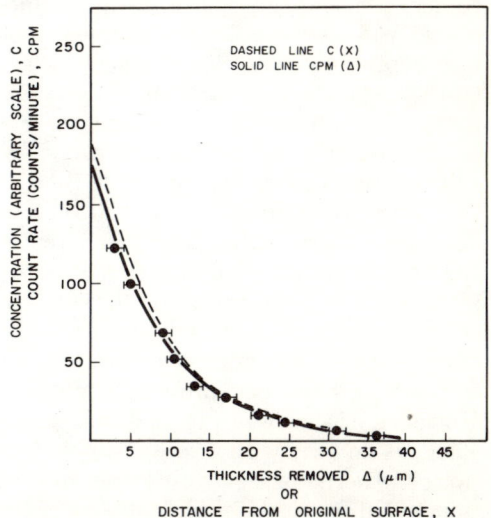

Fig. 1. Penetration profile (count rate vs thickness removed) for quartz single crystal annealed at 746°C and 460 mm Hg for 48 hours. The horizontal error bars are the uncertainties in the thickness of section removed; the uncertainty in the count rate is smaller than the symbol used for each datum.

ple of a diffusion profile determined for a specimen that had been heated in a tritiated water vapor atmosphere at 460 mm Hg pressure at 746°C for 48 hours.

As discussed above, the initial tracer concentration is zero throughout the specimen, and the surface is kept at a constant concentration, since the constant temperature bath surrounding the tritiated water reservoir supplied a constant vapor pressure. Therefore, departures from error function behavior, as shown in Fig. 2, must have arisen from a variation in diffusion coefficient with "water" concentration during the course of the diffusion anneal, and it was necessary to use the Matano-Boltzmann analysis to obtain the variation of the diffusion coefficient with "water" composition. Before presenting these results, however, it is necessary to discuss a correction that was applied to the "raw" penetration profiles.

To determine a tritium concentration from a given surface count rate, one has to know the range of tritium betas (electrons), because triton decay anywhere within this range will be included in the count rate, and the concentration determined will actually be an average over this depth interval. The range of electrons in quartz was calculated from the range of electrons in silicon and oxygen and also after an approximation by Libby (1947) and was found to be 3.3 μm. Since the tritium concentration can decrease substantially over this depth, the concentration determined will actually be an underestimate of the surface concentration. For determination of a composition-independent diffusion coefficient, this can generally be neglected. However, as shown in Fig. 2, diffusion coefficients do vary with "water" concentration; furthermore, neglecting this correction could cause a significant error in determining "water" solubilities. Thus, the following procedure was used to correct the raw profiles.

The count rates, CPM (Δ), (where Δ is thickness removed) depend on the concentration distribution, $C(x)$ (where x

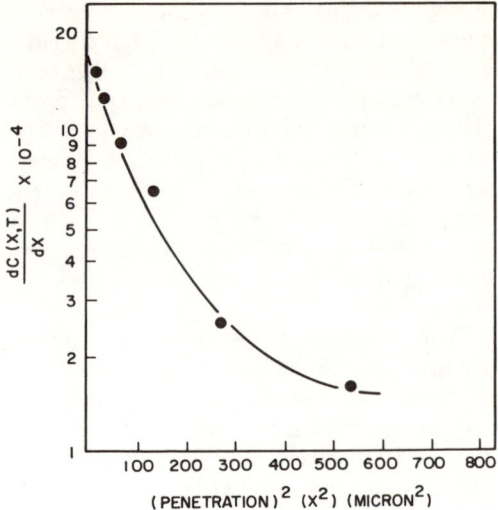

Fig. 2. Plot of the slope of the penetration profile of the curve of Fig. 1 vs (penetration)^2. On this plot, a linear relationship would indicate a composition independent diffusion coefficient.

is distance beneath the original interface), the efficiency of counting, the tritium decay rate, and the probability of emergence of a β created by a decay beneath the surface. Most simply it can be written that

$$\text{CPM}(\Delta) = KC(\Delta + \epsilon)$$

where K is a constant depending on the already mentioned factors and ϵ is an "effective" range. Since the probability of emergence decreases with increasing depth and since $C(x)$ always shows upward curvature (see Fig. 1), ϵ must be positive and less than half the actual range. It is reasonable and convenient to choose $\epsilon = 1$ μm, and to "displace" by 1 μm the CPM(Δ) (solid curve) to yield $C(x)$ (the dashed curve) on Fig. 1. Thus, such $C(x)$ curves were analyzed by the Matano-Boltzmann procedure to give D vs C curves at a given temperature, as shown in Fig. 3. (Note that the abscissa is given in both counts per minute and OH/SiO$_2$ $\times$ 10^5). The diffusion coefficients at a single temperature (746°C in Fig. 3) are seen to vary by more than a factor of 3 (from ~3 to ~10 $\times$ 10^{-12}

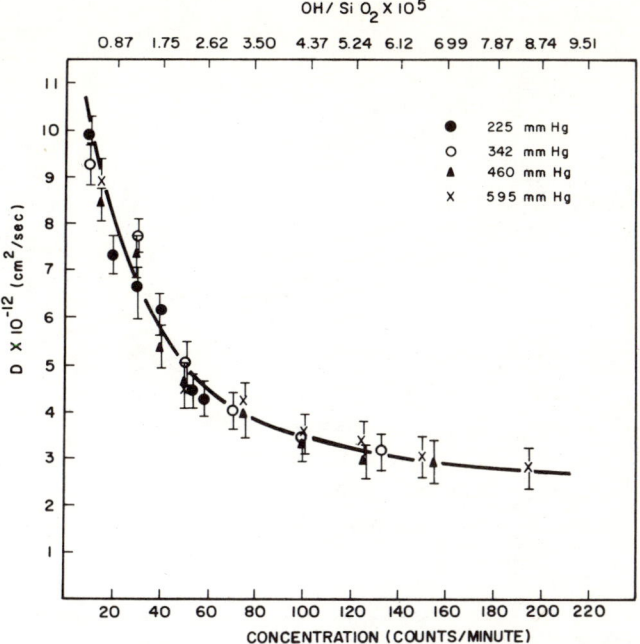

Fig. 3. Diffusion coefficient vs "water" concentration at 746°C.

cm²/sec). Diffusion anneals for other times at 746°C and 460 mm Hg (not presented here) fitted the D vs C curve of Fig. 3 within experimental error. Also, data for several water vapor pressures, a variable which in the range of pressure employed is expected to change the solubility (but not the diffusivity) of "water" in quartz, all fit a single D vs C curve (Fig. 3), indicating that the composition dependence of D is a real phenomenon and not a result of an inappropriate data analysis.

Diffusion anneals at a single pressure (460 mm Hg) were performed at a variety of temperatures from 725° to 850°C. In all cases, D vs C curves similar to Fig. 3 were obtained. At a constant composition, the diffusion coefficients followed an Arrhenius behavior

$$D = D_0 \exp -\frac{\Delta E}{RT},$$

as shown in Fig. 4. The slopes and intercepts obtained using a least squares analysis gave the activation energies, ΔE, and the pre-exponential values, D_0, shown in the figure, and suggest that ΔE also increases with increasing "water" concentration, at least at the one standard deviation confidence level.

The surface count rate of the (dotted) $C(x)$ curve of Fig. 1 should also yield a good estimate of the solubility of "water" in quartz for the temperature and pressure of the diffusion anneal, assuming that the uptake of water by the surface of the crystal occurs at a rate faster than the diffusion kinetics, as is likely. However, the surface count rate (177 cpm) of the CPM (X) curve shown in Fig. 1, from which the surface count rate of 192 cpm of the $C(X)$ curve was derived, is actually an extrapolated value and was much less than the measured surface count rate for this sample (430 cpm). This excess surface activity is due to absorbed water and was observed for all specimens; a similar phenomenon was found by Roberts and Roberts (1966) in their work on "water" diffusion in SiO_2 glass. While the surface count rate could be reduced

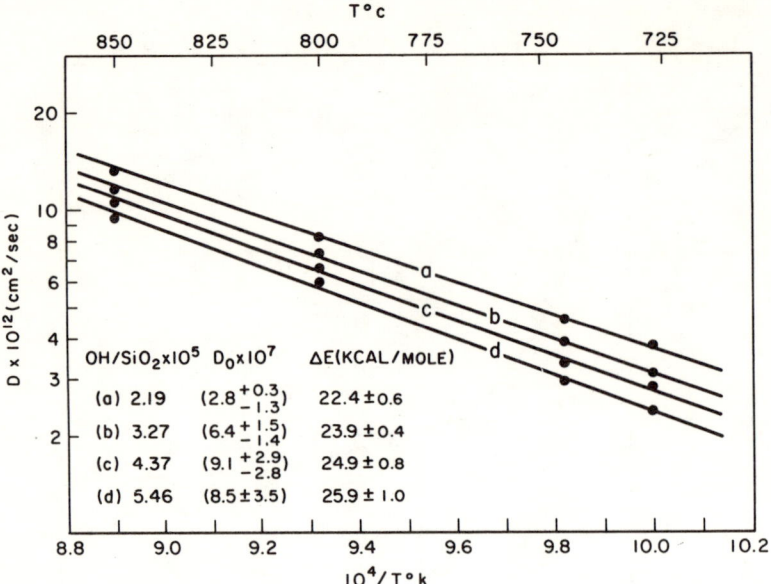

Fig. 4. Arrhenius plot of diffusion data at constant "water" concentration. All data taken at water vapor pressure of 460 mm Hg.

by extended pumping in vacuo at low temperatures (~200°C), it could never be reduced to as low a value as the extrapolated count rate. Thus, the solubility data reported in Fig. 5 depended on the extrapolation used for the $C(X)$ curve, which will now be discussed.

First, it can be seen that the D vs C curve shown in Fig. 3 tends to flatten out at the higher concentrations. It was therefore thought appropriate to use the count rate and diffusion coefficient obtained for the first section, typically at 2–3 μm below the original surface, to estimate the surface count rate, under the assumption that over this small depth interval, the diffusion was approximately "erfcian." Although this cannot be strictly true, this extrapolation is felt to be preferable to a simple visual extrapolation. With this proviso, the data shown in Fig. 5 suggest a linear relationship between solubility and water vapor pressure. However, the deviations from a linear relationship for the two lowest pressures are thought to be too great to ascribe to experimental error and it is possible that the solubility is a more complex function of water vapor pressure.

Discussion

The use of tritiated water as a tracer for studying "water" diffusion in quartz, by detection of the triton decay using liquid scintillation counting, has proved to be a sensitive and reproducible technique for obtaining the penetration profiles.

The diffusion coefficients were found to increase with *decreasing* concentration but tended to level off at higher concentrations. Note, however, that the composition region where the largest change in the D vs C curve occurs is where the apparent "water" composition (as measured by the tritium activity) is less than the OH composition of the starting material (~3×10^{-5} OH/SiO$_2$); this aspect of the composition dependence is not understood at present. In addition, this behavior is exactly the opposite of that found in SiO$_2$ glass, where D increases with *increasing* concentration. While it

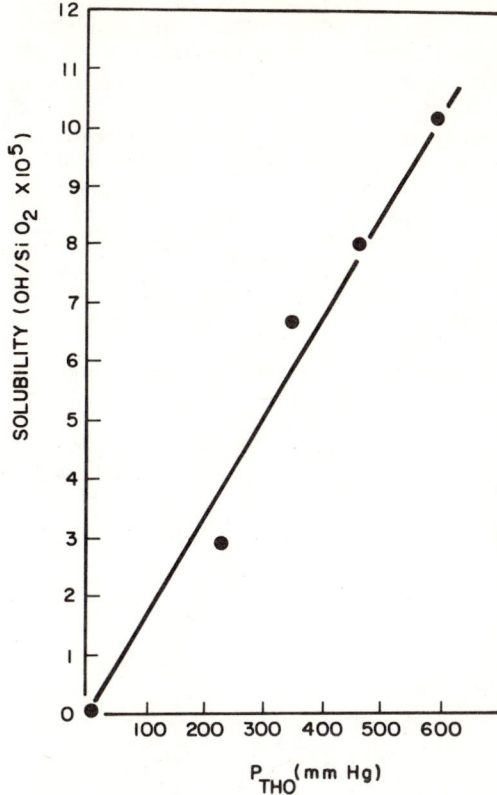

Fig. 5. Solubility of "water" in quartz versus water vapor pressure. All diffusion anneals were for 48 hours at 746°C.

would be tempting to speculate on possible mechanisms that could explain the difference in the functional dependence of D on C between β-quartz and silica glass, and which could also explain the particularly large D's at low "water" concentration, in terms of "channels" in the quartz crystal structure, "reactive" sites at vacancies or interstitials, etc., our present information is too limited to justify such speculations.

The linear relationship between solubility and water vapor pressure implies that equilibrium is dominated by the concentration of *molecular* water in the lattice and that reactions of the type

$$H_2O + {-}Si{-}O{-}Si{-} \rightarrow 2({-}Si{-}OH)$$

are less important in determining the solubility. However, it has already been suggested that the pressure dependence may be too complex to be treated by such a simple model. Further work, particularly using water tagged with O^{18}, is in progress and may shed light on this aspect of the problem.

Conclusions

Based on the present preliminary results, the following conclusions may be drawn: (1) The diffusion of water parallel to the c axis in quartz is composition dependent, decreasing with increasing "water" concentration. (2) Activation energies for diffusion varied between 22 and 26 kcal/mole, and appeared to increase with increasing "water" concentration. (3) The solubility of "water" in β-quartz varied approximately linearly with water vapor pressure. These results differ from those obtained for silica glass principally in that the diffusion coefficient for "water" in silica glass increases with increasing concentration and the solubility of "water" varies as the square root of the water vapor pressure.

Acknowledgments

This work was supported by the National Science Foundation under Grant GA-16549.

References Cited

Baëta, R. D., and K. H. G. Ashbee, Mechanical deformation of quartz. I. Constant strain-rate compression experiments, *Phil. Mag., 22*, 8th Ser., 601–623, 1970.

Brunner, G., H. Wondratschek, and F. Laves, Infrared investigations of the building-in of H into natural quartz, *Z. Elektrochem., 65*, 735, 1961.

Choudhury, A., D. Palmer, G. Ansel, H. Curier, and P. Baruch, Study of oxygen diffusion in quartz by using the nuclear reaction $O^{18}(p, \alpha) N^{15}$, *Solid State Commun., 3*, 119, 1965.

Dodd, D., and D. Fraser, The 300-3900 cm^{-1} absorption band and anelasticity in crystalline α-quartz, *J. Phys. Chem. Solids, 26*, 673, 1965.

Drury, T., and J. P. Roberts, Diffusion in silica glass following reaction with tritiated water vapor, *Phys. Chem. Glasses, 4,* 79, 1963.

Frischat, G. H., Sodium diffusion in natural quartz crystals, *J. Amer. Ceram. Soc., 53,* 357, 1970.

Griggs, D., Hydrolytic weakening of quartz and other silicates, *Geophys. J. Roy. Astron. Soc., 14,* 19, 1967.

Haul, R., and G. Dumbgen, Investigation of oxygen diffusion in TiO_2, quartz and quartz glass by isotope exchange, *Z. Elektrochem., 66,* 636, 1962.

Kats, A., Hydrogen in alpha-quartz, *Philips Res. Rep., 17,* 133, 1962.

Kats, A., H. Haven, and J. M. Stevels, Hydroxyl groups in α-quartz, *Phys. Chem. Glasses, 3,* 69, 1962.

Libby, W., Measurement of radioactive tracers, particularly C^{14}, S^{35}, T and other longer-lived low energy activities, *Anal. Chem., 19,* 2, 1947.

Moulson, A., and J. P. Roberts, Water in silica glass, *Trans. Brit. Ceram. Soc., 59,* 388, 1960.

Moulson, A., and J. P. Roberts, Water in silica glass, *Trans. Faraday Soc., 57,* 1208, 1961.

Roberts, G., and J. P. Roberts, An oxygen tracer investigation of the diffusion of "water" in silica glass, *Phys. Chem. Glasses, 7,* 82, 1966.

Sawyer, B., Q capability indications from infrared absorption measurements for Na_2CO_3 process cultured quartz, *IEEE Trans. Sonics Ultrason., SU19,* 41, 1972.

Shaffer, W. S., Diffusion of tritiated water in β-quartz, M.S. thesis, Case Western Reserve University, 1973.

Wood, D. L., Infrared absorption of defects in quartz, *J. Phys. Chem. Solids, 13,* 326, 1960.

DIFFUSION OF H_2O IN GRANITIC LIQUIDS: PART I. EXPERIMENTAL DATA; PART II. MASS TRANSFER IN MAGMA CHAMBERS

Herbert R. Shaw
U.S. Geological Survey
National Center
Reston, Virginia 22092

ABSTRACT

Part 1: Diffusivities of H_2O in granitic liquids have been determined from experimental hydration rates of obsidian at temperatures of 750°–850°C and concentrations of from 1% to 6% H_2O by weight; $100 \leq P_t (H_2O) \leq 2000$ bars. The diffusivities are shown to be strongly dependent on concentration of H_2O, ranging linearly from the order of 10^{-9} cm^2 sec^{-1} at vanishing H_2O concentration to the order of 10^{-7} cm^2 sec^{-1} at 6% H_2O in this temperature range. The temperature dependence in the above range is small, and activation energies are difficult to state precisely; the estimated range is 15 ± 5 kcal mole^{-1}. Both the temperature and the concentration dependence for diffusion of H_2O in obsidian are similar to published experimental data for diffusion of H_2O in silica glass. This similarity is used to suggest that experimental diffusivities of H_2 in silica glass indicate the orders of magnitude to be expected for transfer of this component in granitic liquids; diffusivities of H_2 are probably in the range 10^{-7} to 10^{-5} cm^2 sec^{-1} at magmatic temperatures. The relationship of viscosity and diffusivity as functions of H_2O concentration are shown to be incompatible with the Stokes-Einstein formula. The Ree-Eyring theory is used to illustrate structural distinctions between rate constants for diffusion and viscous flow.

Part II: Geological applications of the data in Part I are given relative to simplified boundary conditions that illustrate possible magnitudes of H_2O transport under magmatic conditions in both static and convecting magma chambers. Convection during the early stages of cooling of magma chambers is confined to relatively narrow boundary layers. Estimates are made of flow velocities and of the corresponding diffusive fluxes of heat and mass into or out of these boundary layers for magma viscosities ranging from basaltic to granitic. The duration of the boundary layer regime under typical plutonic conditions is estimated to be roughly one fifth of the theoretical solidification time computed from the theory of perfect conduction. Diffusive exchange of mass between the magma chamber and country rock during this time could significantly change the hydration state of the magma by gains or losses of H_2O or H_2 if there are strong initial contrasts between the chemical potentials of these components across the contact zone. Possible magnitudes of diffusive assimilation of other constituents are also estimated. Large changes are not predicted for the major element composition of the total magma volume, but in some cases diffusion probably plays a decisive role in chemical zoning and in distributions of isotopic and minor element composition. However, no specific conclusions are drawn relative to any actual magmatic province on the relative dominance of diffusive exchange versus internal mechanisms of differentiation involving crystal-liquid or liquid-vapor fractionation or effects of internal diffusive transport.

Introduction

The data of geochemistry are singularly deficient in the knowledge of physical, thermochemical, and transport properties of magmatic systems. A brief perusal of the *Handbook of Physical Constants* (Clark, 1966) illustrates the paucity of data on fundamental properties such as heat capacities, heat of fusion, thermal conductivities, densities, viscosities and chemical diffusivities representative of magmatic conditions. Since publication of that handbook some significant advances have been made, particularly in the knowledge of densities (Burnham and Davis, 1971; Bottinga and Weill, 1970) and viscosities (Bottinga and Weill, 1972; Shaw, 1972). Murase and McBirney (1973) report additional data for these and other physical properties of melts of rock compositions. However, data on chemical diffusivities are exceedingly scarce. Medford (1973) has completed a study on diffusion of calcium in a melt of basic composition, Carron (1969) gives data for diffusion of sodium in obsidian, and Shaw (1965) reported preliminary results of studies on diffusion of H_2O in a melt of granitic composition. Other data on silicate glasses have been summarized by Winchell (1969). The present paper describes extensions of my earlier work which, though still incomplete, may add to the growing interest in transport properties of magmatic constituents and the interaction of these constituents with the magmatic environment. Prior to these studies, the experimental data of Bowen (1921) were virtually the only widely known data on diffusion rates for melts resembling magmatic compositions. The results of those experiments were used by him and others to minimize the significance of diffusive processes in magmatic differentiation. It will be shown in Part II of this paper that, on the contrary, chemical diffusion might be a process of considerable significance in the chemical zoning of some magmatic systems.

Part I. Experimental Data

Description of Experiments

The simplest method for estimating rates of diffusion of a soluble component in a solvent is to measure the rate of solution, or sorption, of the component under specified boundary conditions. Rigorous determination of relative diffusion fluxes of all components in multicomponent mixtures is exceedingly difficult, but the "sorption method" avoids these difficulties by considering the flux of one component whose concentration gradient is large compared to all others. The method used here is parallel to those used in many other studies. Detailed discussion of the method and its application to other systems is given by Crank (1957).

As starting material for the present experiments, cylindrical core samples were drilled from homogeneous blocks of obsidian collected from the Valles Caldera, New Mexico. These obsidian blocks were the same specimens used earlier for viscosity measurements, and details of location and chemical analyses are given by Shaw (1963); the specimens used contained only rare crystallites and no bubbles. The cylinders were ground to a diameter of 0.25 cm on a centerless grinder and cut to various lengths; most of the runs used cylinders with lengths of about 0.3 cm. Description of the obsidian composition as "granitic" is based on its normative composition (C.I.P.W.): Q 34.0, or 27.1, ab 35.1, an 1.3, hy 0.5, mt 0.9, il 0.2, c 0.4.

The obsidian cylinders were sealed inside gold capsules with sufficient water to permit saturation at equilibrium but with little excess. This was done so that the glass cylinders would retain approximately the same dimensions during the runs, except for effects of thermal expansion, compressibility and the volume change of hydration. These dimensional factors introduce an error of no more than 25% in the calculation of diffusivity even if the original sample dimensions are used; this error is small compared with other uncertainties discussed later.

Sorption runs were made by placing the sample capsules inside cold seal pressure vessels, oriented horizontally, at a fixed total water pressure, P_t (H_2O), and heating to a predetermined temperature as quickly as possible, which took about half an hour. The capsules were held at constant temperature and pressure for a preselected time and then quenched. A set of such runs at increasing time intervals established a sorption isotherm at constant pressure and constant chemical potential of H_2O at the surface of the glass cylinder.[1] Run times ranged from about half an hour to several days. Because most of the runs on a given isotherm represent times greatly exceeding the time required to reach the run temperature, uncertainties in the "starting time" are almost negligible.[2] During the runs, temperature variations did not exceed about ±5°C in the longest runs; this is also the approximate uncertainty in temperature measurement. Pressure fluctuations were negligible, and the absolute uncertainty in pressure measurement was about 5% of the nominal setting (pressure gauges were checked periodically against a precision Heise bourdon gauge).

The amount of H_2O absorbed by the sample in each run was determined by the weight increase during the run and by ignition loss at 1000°C in air after the run; the former value was corrected for the initial H_2O content of the obsidian as determined in separate dehydration runs. A number of runs were made on rates of weight loss to establish that dehydration was almost complete on heating for two hours or more at 1000°C in air; residual H_2O in the glass after ignition is probably much less than 0.05% by weight judging from differences in results at 1000°C and 1200°C. The hydrated glasses were optically homogeneous and were free of bubbles and crystallites.

Table 1 lists the runs used to define limits of solubility as determined by apparent achievement of constant concentrations; the curve of solubility at 850°C is shown in Fig. 1. About one hundred additional runs were made to establish the forms of the sorption rate curves under different conditions. Most of the runs

[1] There was, of course, a tendency for the area of contact between the lower portions of the glass-melt cylinders and the container walls to increase as run durations increased, because of the decrease in viscosity with hydration (Shaw, 1963). The ends of the cylinders also eventually developed meniscus shapes that were convex toward the gas phase.

The slumping effect was partially countered by orienting the cylinders horizontally. However, molecular diffusion in the boundary regions of contact between the melt and metal container was expected to be fast relative to diffusion in the melt, particularly in view of the observation that the hydrous melts did not wet the gold metal (as was shown by meniscus shapes). Therefore, the assumption of a surface saturation condition was considered to be the most representative boundary condition in the absence of quantitative criteria for evaluating possible surface resistances. Evidence given later (Fig. 5 and its discussion in the text) indicates that the outer surfaces of the cylinders did in fact reach saturation very quickly. The effect of end curvature on surface area was not explicitly included as a correction. The maximum effect for conversion of a cylinder with a length twice the radius to a spherical shape at constant volume would decrease the surface area by only about 7%. The decrease of density (and increase of volume) during hydration (see Table 2) partly offsets the effect of shape on surface area, but the correction is smaller. The maximum volume expansion is about 3%, so the maximum increases in surface area and effective radius are about 2% and 1%, respectively.

[2] Diffusion rates during the first few hundred degrees of heating are undoubtedly many orders of magnitude smaller than the rates at run temperatures used in this study (750° to 850°C). Therefore, the effective starting time is ambiguous and is influenced mainly by the higher temperatures. However, initial heating rates to temperatures above about 600°C were made very fast by placing the pressure vessels into preheated furnaces. Consequently, it is thought that the uncertainties in effective starting time are only a few minutes, rather than the full 20 to 30 minutes required to reach steady run temperatures.

TABLE 1. Experimental data on solubility of H_2O in liquids at 850°C and initial H_2O contents of obsidian starting materials

Run No.	Batch	Time (hours)	ΣH_2O* (wt percent)
Initial Values:			
V-3a	DC-1	...	0.48
V-3b	DC-1	...	0.52
V-5	DC-1	...	0.45
			avg. 0.48
LD-1	DC-2	...	0.24
LD-2	DC-2	...	0.17
V-30	DC-2	...	0.21
			avg. 0.21
$P_t(H_2O) = 100$ bars:			
D-115	DC-2	196	1.3
			1.3
			avg. 1.3
$P_t(H_2O) = 340$ bars:			
D-54a	DC-1	97	2.1
			2.5
D-66a	DC-1	94	...
			2.5
D-66b	DC-1	94	...
			2.5
			avg. 2.4
$P_t(H_2O) = 500$ bars:			
D-117	DC-2	113	2.9
			2.9
D-118	DC-2	168	3.1
			3.0
D-125	DC-2	336	3.0
			avg. 3.0
$P_t(H_2O) = 660$ bars:			
D-29	DC-2	70	3.4
D-41a	DC-1	52	...
			3.7
D-41b	DC-1	52	...
			3.6
D-42	DC-1	49	...
			3.7
D-43	DC-1	72	3.2
D-55a	DC-1	72	...
			3.6
D-55b	DC-1	72	...
			3.4
D-56b	DC-1	72	...
			3.3
			avg. 3.5
$P_t(H_2O) = 1000$ bars:			
V-34	DC-2	150	4.1
			4.5
D-62	DC-1	24	4.3
			4.3
D-63	DC-1	51	4.2
D-64	DC-1	74	4.3
			4.5
			avg. 4.3

Run No.	Batch	Time (hours)	ΣH_2O* (wt percent)
$P_t(H_2O) = 2000$ bars:			
D-2	DC-2	30	5.9
			5.8
D-17	DC-2	21	5.9
			6.0
D-18	DC-2	43	5.9
			5.8
D-27	DC-2	50	...
			5.9
D-50	DC-1	24	5.8
D-51	DC-1	24	6.0
D-52	DC-1	24	6.1
D-72	DC-1	24	6.3
			avg. 6.0

*Italicized values were determined from estimates of weight change corrected for the assumed initial concentration of H_2O given by averages of the tabulated initial values. All other values were directly determined by dehydration of the hydrated glasses at 1000°C for about two hours. The averages are thought to be accurate to better than 5% at the highest concentrations and about 10% at the lowest concentrations.

were made at 850°C and "surface saturation conditions" set by vapor pressures ranging from 100 to 2000 bars; the sorption runs are illustrated in Fig. 2. A tabulation of runs is not given because of length, but the form of the data is the same as in Table 1. The scatter of data points in Fig. 2 is related to other uncertainties discussed later in the text. For these reasons no systematic regression analysis of the data was performed, and the dashed curves are based on a subjective evaluation of best values. However, the form of the data is clear except for some ambiguity in relative values of initial slopes of isotherms at different values of $P_t(H_2O)$. The meaning of the steps in the curves at low pressures is discussed in the next section.

Changes in Glass Volume during Hydration

Density measurements on glasses saturated with H_2O under different conditions are given in Table 2. The densities at room temperature range from about

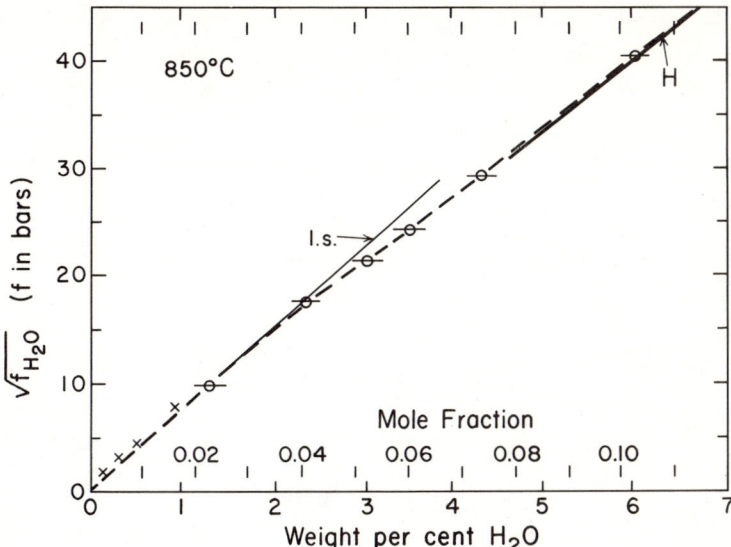

Fig. 1. Solubility of H_2O in obsidian versus square root of fugacity (f) of pure H_2O gas at 850°C, based on data of Table 1 and fugacities from Burnham and others (1969). The horizontal bars through the open circles give the estimated uncertainty in H_2O content; the dashed line through the points is fitted by eye to show the trends. Crosses are values estimated by interpolation from curves of Friedman and others (1963) at 850°C. The heavy line labeled H is an estimated transposition of H_2O solubility in a granitic composition from Hamilton and others (1964; curve 4, Fig. 4). The light line labeled $l.s.$ is the estimated limiting slope for low H_2O concentrations. The difference in slopes H and $l.s.$ and the inflection in the dashed curve are thought to be caused by changes in the structural states of the liquids at concentrations above about 2% H_2O by weight. However, it is not definitely proven that the states at low H_2O contents represent the most stable structural configurations (see discussion of volume relaxation in text). Concentrations expressed as mole fractions are based on a mean molecular weight containing 1 gram atom of oxygen (32.1 g $mole^{-1}$).

2.354 g cm^{-3} for obsidian starting materials containing 0.2% to 0.5% H_2O by weight to about 2.282 g cm^{-3} for hydrated obsidian containing 6.4% H_2O by weight. The variation, however, apparently is not linear; the density decrease per unit increase in H_2O content appears to become smaller at higher concentrations. This relationship is not shown graphically or analytically because the conditions and thermal histories of the initial and experimental glasses were not exactly the same. Systematic studies of relationships between H_2O concentration, run conditions and quenching history are required to give a rigorous definition of density variations as functions of equilibrium solubility. The data in Table 2 indicate the trend but do not necessarily identify the functional form of the variation accurately.

Two sets of runs were made as functions of time at 850°C and P_t (H_2O) values of 300 and 700 bars, respectively, to determine the rates of density change. This was done because the sorption curves of Fig. 2a appear to show solubility steps at the lower values of P_t (H_2O), and it was thought that the steps might represent metastable structural states. The values 300 and 700 bars were chosen to bracket the conditions for which this effect was most conspicuous relative to the uncertainties of H_2O determinations. The runs were made in a manner exactly the same as the sorption runs except that the glass products were saved for density determinations. Figure 3 shows the den-

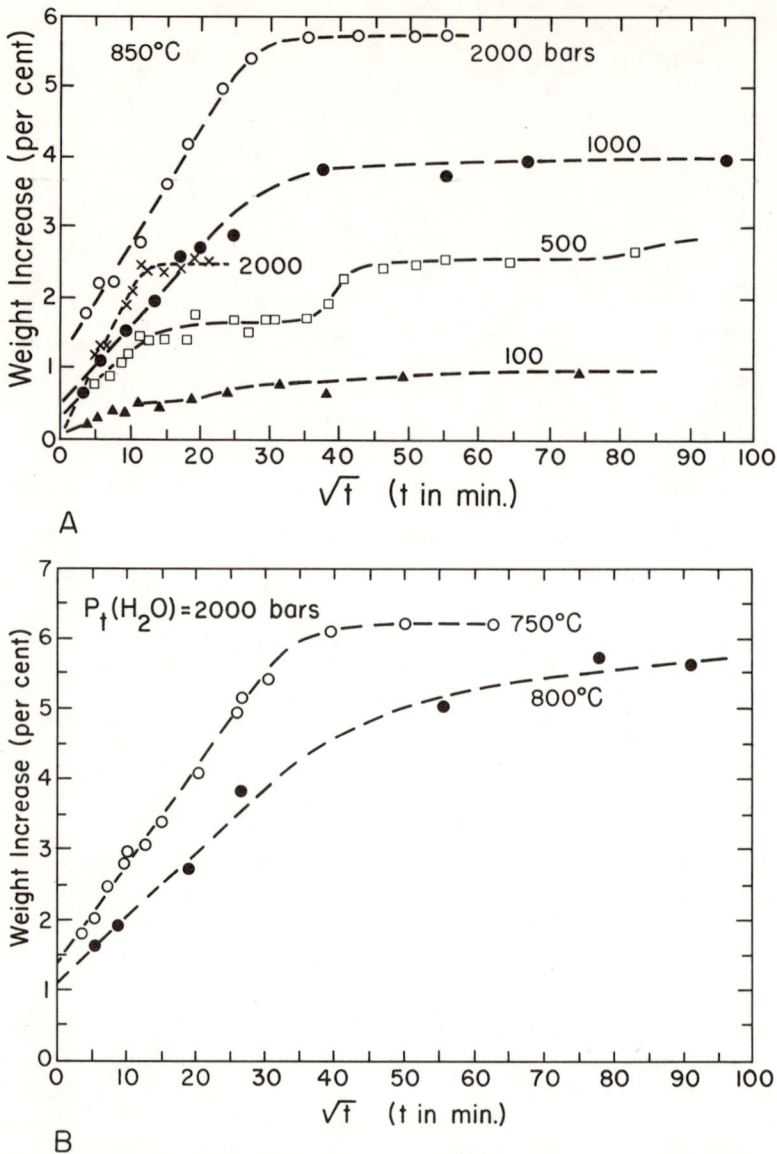

Fig. 2. (a) Data for rates of sorption of H_2O in obsidian cylinders at 850°C and values of total water pressure P_t (H_2O) labeled on the curves. The sizes of the data symbols correspond roughly to the estimated uncertainty in H_2O content. The dashed lines are drawn on the basis of subjective judgments of best values to aid comparisons of the relative slopes. The locations of points at $\sqrt{t} < 10$ depend on interpretations of effective starting times. Points were plotted on the basis of the time the sample was at temperatures greater than about 600°C. Correction for any greater lag time would shift these points farther to the left. The crosses and dotted line are for hydration of specimens previously hydrated to a uniform initial concentration of about 3.5% H_2O. Steps in the curves at low pressures are apparently caused by time-dependent volume relaxations of the glass structure (see discussion in text). The obsidian cylinders used for these runs had the original dimensions 0.25 cm diameter, 0.33 cm length. (b) Sorption data at P_t (H_2O) = 2000 bars for temperatures of 800° and 750°C. The apparent sorption rate is smaller at 800°C because the obsidian cylinders used in these runs were of greater length. The original dimensions were 0.25 cm diameter, 0.95 cm length for runs at 800°C; 0.25 cm diameter, 0.33 cm length for runs at 750°C.

TABLE 2. Densities of quenched hydrous glasses at 25°C, 1 atm

Run No.	Batch*	T (°C)	$P_t(H_2O)$ (bars)	Time (hours)	ΣH_2O† (wt percent)	Density‡ (g cm^{-3})
D-150	DC-2	850	300	72	2.2	*2.320*
D-151	DC-2	850	300	150	2.2	*2.319*
D-128	DC-2	850	700	98	3.7	*2.308*
D-129	DC-2	850	700	124	3.7	*2.307*
D-131	DC-2	850	700	146	3.7	*2.308*
V-34	DC-2	850	1000	150	4.3	2.305
V-16	DC-1	800	2000	136	6.2	2.290
V-31	DC-2	800	2000	100	6.2	2.284
V-28-3	DC-2	750	2000	40	6.4	2.282

* Two batches of starting material were used, each batch consisting of cylindrical specimens drilled from one of two different large blocks of obsidian from the Jemez Mountains, New Mexico (Shaw, 1963). Table 1 shows that the two batches differ in initial H_2O concentration: batch DC-1 contains about 0.5%, and batch DC-2 contains about 0.2% H_2O by weight. Densities measured on 4 specimens of DC-1 ranged from 2.350 to 2.357 with a mean value of 2.354 g cm^{-3}; densities measured on 8 specimens of DC-2 ranged from 2.351 to 2.357, also with a mean value of 2.354. The standard deviation of these measurements was about 0.003 g cm^{-3}, and the densities of the two batches do not differ significantly within this precision (see footnote 3). It was expected that batch DC-2 would have a higher density than batch DC-1, but the effect of higher H_2O concentration in DC-1 may be offset by a slightly larger proportion of incipient devitrification (even though the proportion of visible crystallites in both cases was below 1%).

† Concentrations of H_2O were not determined directly on specimens retained for density measurements. The indicated concentrations are apparent saturation values determined from Fig. 1 based on the data of Table 1. The runs used for density measurements were either of about the same duration or were of greater duration than the solubility runs. Solubility at temperatures below 850°C was corrected by an increase of 0.2% H_2O for each 50°C decrease of temperature; this is an approximate factor based on only a few solubility runs at these temperatures.

‡ Densities were determined using the Berman balance except for those in italic type, which were determined using a liquid column of calibrated density gradient as described in the text. Test pieces ranged from about 20 to 50 mg with most in the range 35 to 40 mg. The standard deviation of a single measurement with the Berman balance on a 40 mg sample of density about 2.3 g cm^{-3} is about 0.002 g cm^{-3} (E. H. Roseboom, 1968).

sity variations versus the square root of time plotted in the same units as the sorption data (Fig. 2).

In view of the uncertainties of density measurements using the Berman balance (Table 2), the density variations were determined by suspending the glass beads in a liquid column with a calibrated density gradient. The liquid was a mixture of bromoform and N,N-dimethylformamide (used as the light fraction because of its low volatility). Initially, twenty equal volume increments of liquids ranging in density from about 2.28 to 2.36 g cm^{-3} were carefully added to a graduated glass cylinder in the order of decreasing density, giving a stepwise distribution at increments of about 0.004 g cm^{-3}. The gradient was smoothed by periodic rotational stirring over a period of about 2 days and was checked against positions of internal reference markers (glass standards or mineral fragments for which independent determinations of density were made by repeated measurements on larger specimens using the Berman balance). Reference markers were at densities (g cm^{-3}) of 2.285 (glass standard for calibration of heavy liquids), 2.295 (bikitaite), 2.317 (gypsum), 2.327 (glass standard), and 2.357 (glass standard). Densities of the reference markers are thought to be accurate to better than 0.002 g cm^{-3}, and their maximum deviations from values given by a linear gradient in the liquid column was 0.001 g cm^{-3}. The length of column between the highest and lowest markers was about 80 cm, giving an average gradient of about 0.001 g cm^{-3} per cm height.

Readings on the hydrated obsidian beads (diameter and length about 0.3

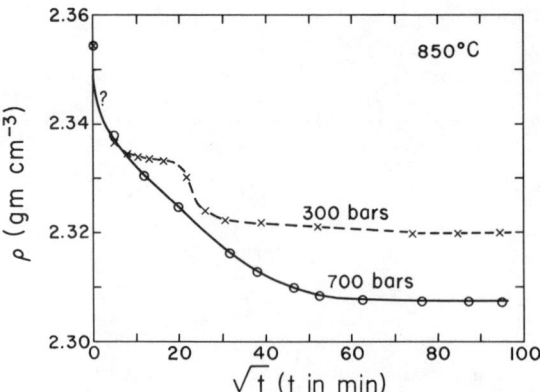

Fig. 3. Density (ρ) as a function of the square root of time at 850°C and P_t (H_2O) of 300 and 700 bars. Data points give the mean densities of quenched cylindrical beads held for the indicated times under the above conditions. Densities were determined by positions of the beads in a liquid column of graduated density, as described in the text.

cm) were made by introducing them into the column one at a time and recording their rest positions, which were again checked after several hours. Relative positions of the beads could be read to about 0.1 cm or better, so that the discrimination of density differences could be made to about 0.0001 g cm^{-3}.

Although the step in the 300 bar curve in Fig. 3 does not exactly duplicate the solubility step for data at 500 bars in Fig. 2, it is consistent with the interpretation that both the density and solubility steps are caused by rates of relaxation from higher to lower density structural states. Apparently these rates are greatly accelerated by saturation potentials exceeding a value somewhere between 500 and 700 bars P_t (H_2O).

Because the rates of structural changes in these glasses are complex functions of temperature, pressure, H_2O concentration, and any heterogeneities of the initial glassy states, a detailed description of the relaxation mechanisms will require a much more extensive program of study. Because methods of progressive sorption are too time consuming to be used as an effective technique for the study of structural properties, a more direct approach is needed.

According to the structural interpretation given above, the plateaus in the sorption curves of Fig. 2 represent transient saturation of intermediate structural states. Another interpretation that cannot be ruled out by the rate data is that the strain produced by the volume change of hydration introduces a transient impedance to diffusion at various depths of penetration which must become annealed before the next interval can be penetrated. The only obvious way to resolve these alternatives is to study details of concentration profiles by some other technique. In either case, above a certain level of hydration potential, given by P_t (H_2O) $\sim$700 bars at 850°C, the inhibiting structural effects apparently relax very fast so that hydration penetrates as a continuous diffusion wave. These conditions more nearly approximate diffusion in the liquid state, and the corresponding diffusivities are the principal concern of this paper relative to diffusive mass transport in magma chambers. More detailed consideration of the glassy states is reserved for future study.

Calculation of Diffusivity

The mathematical definition of diffusivity is obtained from an equation that describes the mass balance within a volume element of solution across which there is a concentration gradient of the diffusing constituent. Ignoring the complexities implied by effects of the diffusion on boundaries of the volume element and on other chemical gradients, the equation for a diffusion gradient in one dimension is usually written

$$\frac{\partial c}{\partial t} = D \frac{\partial^2 c}{\partial y^2} \qquad (1)$$

where c is any measure of concentration, t is time, y is distance and D is a constant of proportionality, called the diffusion constant or, as it is termed in this paper, the diffusivity; the units of D depend on the units of c, t and y. For cgs units and concentration expressed per unit volume, the units of D are $cm^2\ sec^{-1}$. Equation 1 is often called Fick's law of diffusion and is directly analogous to Fourier's law of heat conduction. However, in either case the equation is a rigorous description of mass or heat flow only if D is a constant and if the gradients of driving potentials (heat or mass) are proportional to the concentrations. The latter condition is satisfied for heat flow and is approximately satisfied in sorption experiments. The assumption that D is constant is often valid for heat flow and for many cases of chemical diffusion where the structure of the solution and the kinetic character of diffusing species is not strongly dependent on composition. However, such an assumption is not valid for diffusion of network modifying constituents in framework silicates. Here, the diffusivity of each constituent can be expected to depend on the level of concentrations defining the chemical gradient of the constituent of interest as well as on temperature, pressure, concentrations, and concentration gradients of other constituents (see Cooper, this volume).

In cases where D is a function of c, Equation 1 becomes "nonlinear" and may be impossible to solve analytically. Therefore, the usual expedient is to attempt to deal with conditions where variations in D are small or where they can be averaged so that, mathematically, D can be treated as a constant in Equation 1. However, if great numerical precision is required and D can be expressed explicitly in terms of c, analog computation or finite difference methods can be used to solve the mass balance equation (see Crank, 1957).

In the present paper values of diffusivity are expressed in two ways: first by assuming that D is constant and computing an apparent diffusivity from Equation 1 for a given set of conditions, and second by attempting to interpret the concentration dependence by intercomparison of results for different conditions and by comparison with results of other workers for diffusion of H_2O in silica glass. Values of diffusivity obtained directly from the experiments by using Equation 1 will be called, for brevity, "net diffusivities" and are symbolized by $\bar{D}$, where the bar indicates that the value represents an average apparent diffusivity as calculated from Equation 1 for a specified set of conditions. The D without the bar signifies a value explicitly defined by the concentration of H_2O expressed as though the solution were binary.

The best way of interpreting diffusivities and their concentration dependence is by analysis of concentration profiles as functions of time. However, unless H_2O is tagged isotopically, as was done in studies described later on H_2O solution at low vapor pressures in silica glass, no satisfactory method appears to exist to quantitatively map H_2O distributions in silicate glasses. Optical methods or determination by dehydration measurements on radial increments could not be used effectively because of the rather small specimen size. An electron microprobe method might be useful to indirectly demonstrate the form of chemical gradients, but the uncertainties in determining H_2O by difference appear to be large. Therefore, interpretation of the present data depends on expressing the diffusivity as a function of the rate of change of *total* absorption of H_2O as defined by the sorption curves of Fig. 2.

If the diffusivity is assumed to be constant, the analysis of the rate of sorption is directly parallel to that for the rate of heating of a cylinder. The mathematical theory of the former is treated in great detail by Crank (1957) and of the latter by Carslaw and Jaeger (1959); Darken and Gurry (1953) give a concise summary of various diffusion functions. The net diffusivity, $\overline{D}$, can be obtained directly from the slopes of the sorption curves by using Equation 1 expressed in the form (see Barrer and Brook, 1953)

$$\frac{d\left(\frac{c-c_0}{c_s-c_0}\right)}{dt^{1/2}} = \frac{2A}{V}\left(\frac{\overline{D}}{\pi}\right)^{1/2} \quad (2)$$

where c is the mean concentration at time t, c_0 is an initial uniform concentration, c_s is the limiting concentration at equilibrium (for a given constant value of H_2O potential at the surface of the specimen), A is surface area of the specimen and V is its volume. In Equation 2 the concentration can be expressed in any consistent units and the reduced concentration $(c-c_0)/(c_s-c_0)$ represents the fractional degree of saturation of the sample for given P, T, and $f_{H_2O}*$ (where $f_{H_2O}*$ is the equilibrium value of the fugacity of H_2O for the concentration c_s).

The sorption curves of Fig. 2 are replotted in terms of reduced concentration in Fig. 4. The initial slope is readily obtained by taking the difference in fractional concentration between $t=0$ and some value $t^{1/2}$, and then dividing by the latter value (after converting $\min^{1/2}$ to $\sec^{1/2}$). Numerical values of $\overline{D}$ determined in this way are given in Table 3.

The apparent intercepts of the sorption curves at $t=0$ depend on uncertainties in the starting time, or on the kinetic mechanisms by which the sample surface attains the assumed "surface saturation condition" c_s for a given value of $f_{H_2O}*$ defined by the experimental conditions. The effect is much too large to be explained by errors in effective starting time, and therefore the intercept is assumed to be a function of the initial surface properties of the sample. The experiments by Barrer and Brook (1953) on sorption of gases in zeolites showed similar effects which they also interpreted as being caused by rapid saturation of a finite surface layer. This conclusion in the present experiments is supported by two observations. First, curve 2 of Fig. 4 has an intercept near the origin. The runs made to define this curve used starting cylinders that were partially hydrated in previous runs, and therefore the original characteristics of sample surfaces had been annealed out (fire polishing of the original obsidian cylinders was not possible because of the intrinsic H_2O content and consequent vesiculation on heating at 1 atm above about 700°C). Second, the intercept values in Figs. 2 and 4 at 850°C increase with increasing values of

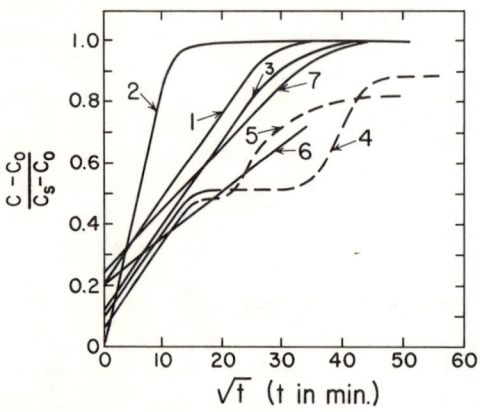

Fig. 4. Composite plot of sorption curves of Fig. 2 expressed in terms of reduced concentration, where c is the mean concentration (weight percent) at time t, c_0 is the initial concentration (assumed uniform), and c_s is the limiting concentration at equilibrium for given constant values of $P_t(H_2O)$ and T (from Table 1), as follows: (1) 850°C, 2000 bars, $c_0 = 0.2$; (2) 850°C, 2000 bars, $c_0 \cong 3.5$; (3) 850°C, 1000 bars, $c_0 = 0.5$; (4) 850°C, 500 bars, $c_0 = 0.2$; (5) 850°C, 100 bars, $c_0 = 0.2$; (6) 800°C, 2000 bars, $c_0 = 0.2$; (7) 750°C, 2000 bars, $c_0 = 0.2$. In all cases except curve 6 the dimensions of the glass cylinders were: $l = 0.33$ cm, $r = 0.13$ cm; for curve 6, $l = 0.95$ cm, $r = 0.13$ cm.

TABLE 3. Apparent diffusivities of H_2O in obsidian
from initial slopes of sorption curves in Fig. 4

T (°C)	$P_t(H_2O)$ (bars)	c_0 (wt percent)	c_s	$\overline{D}$ (cm² sec⁻¹)*
850	2000	0.2	6.0	2.4×10^{-8}
850	1000	0.5	4.3	2.4×10^{-8}
850	500	0.2	3.0	2.2×10^{-8}
850	100	0.2	1.3	2×10^{-8}
850	2000	~3.5	6.0	2×10^{-7}
800	2000	0.2	6.2	1×10^{-8}
750	2000	0.2	6.4	1.2×10^{-8}

* The number of significant figures is meant to indicate only the relative degree of confidence. The complexity of the diffusion process makes estimates of absolute accuracy difficult to assess without more complete data on the concentration dependence (see text). Correction for the maximum error caused by changes of shape of the ends of the sample cylinders due to surface tension effects would decrease the tabulated values by about 25%.

f_{H_2O}*. This observation is consistent with the assumption that a finite outer layer of the specimen volume is almost instantaneously saturated. This could only be due to the existence of numerous microcracks induced by sample preparation that are effectively healed by hydration and thermal annealing. Figure 5 shows the magnitude of the "instantaneous" change in concentration versus the value of saturation of the total sample, c_s. The relationship is roughly linear and gives a constant proportionality of surface layer to specimen volume of about 20%. The reproducibility of this ratio apparently reflects the fact that all of the specimen cylinders were initially ground as identically as possible so that they had very nearly the same distribution of microcracks. Much of the scatter in experimental points in Fig. 2 probably relates to variations in this factor. Once the surface layer became healed, hydration of the remaining glass proceeded by diffusion through the saturated layer. The effect gives almost the same result that would be obtained in a polished specimen without microcracks which was hydrated beginning at a much earlier time. The apparent shift of the time axis required to give the equivalent hydration state in an undamaged specimen is obtained by extrapolating the curves of Fig. 4 to zero concentration. These intercepts on the abscissa for some of the curves correspond to apparent time shifts greatly exceeding an hour. The fact that these relationships are systematic and involve time factors much larger than the uncertainties in experimental starting times virtually compels the above interpretation.

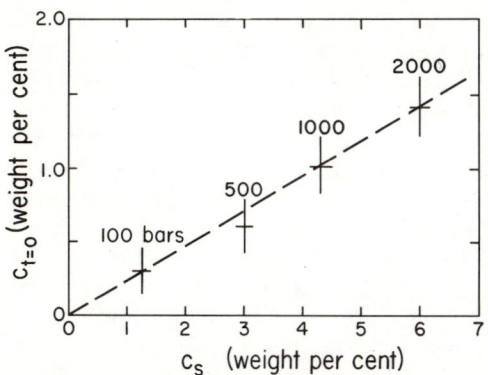

Fig. 5. Instantaneous sorption of H_2O in obsidian cylinders at 850°C ($c_{t=0}$) versus saturation concentrations (c_s) at 850°C for different vapor pressures $P_t(H_2O)$ labeled on the diagram. The values $c_{t=0}$ are taken from the ordinate intercepts of Fig. 2a corrected for the original H_2O content of the obsidian. The vertical bars are estimates of the uncertainty in intercept values and the horizontal bars give the approximate uncertainty in the saturation concentrations.

ESTIMATE OF CONCENTRATION DEPENDENCE

Crank (1957, Chap. 12) shows that if D increases with concentration, the linear portion of a sorption curve for finite specimens of the same shape will extend over a large fraction of the total concentration range. This linearity can be explained as follows: For diffusion in one dimension into a semi-infinite medium the amount absorbed is directly proportional to the square root of time (the so-called parabolic rate law) whatever the dependence of diffusion coefficients on concentration (Crank, 1957, p. 276), unless there is some change in properties of the medium with depth or time. In finite bodies the times for which this proportionality holds are governed by the shapes of the diffusion profiles. A perfectly sharp hydration front penetrating a finite sheet from both sides would obey the parabolic rate law until hydration is complete. Similarly, diffusion coefficients that increase strongly with concentration give concentration profiles that terminate more abruptly than do profiles for constant diffusivity (see later discussion of profiles in Fig. 8). Therefore, the parabolic rate law applies over greater depths of penetration and hence to higher levels of fractional saturation in such cases.

Sorption curves for cylinders and spheres with constant diffusivity begin to depart from the parabolic rate law when the total amount absorbed has reached about half the final value (for example, see Crank 1957, Figs. 5.4 and 6.2). By this criterion alone strong concentration dependence is indicated by the linearity of curves in Figs. 2 and 4. This conclusion is only qualitative, however, because more precise data are required to deduce the exact functional form of the concentration dependence on the basis of shapes of sorption curves. Verification of the increase of D with concentration was obtained by performing runs on specimens already partly hydrated (curve 2 of Fig. 4).

The runs at different values of $f_{H_2O}*$ starting from the same initial concentration did not resolve the question of concentration dependence because of the uncertainty of slopes for different isobars, as indicated by Fig. 2 and Table 3. The reason for the ambiguity between data for different isobars is that the average value $\overline{D}$ is heavily weighted by values of D at the higher concentrations if the variation in D with concentration is large; that is, the greatest changes in D take place at very small concentrations.

In other studies of concentration dependent diffusion, D can often be approximated by either a linear or an exponential function of c (see Crank, 1957). Studies made by Drury and Roberts (1963) on H_2O diffusion in silica glass demonstrate both linear and exponential variation of D with concentration (to about 0.1% by weight) at temperatures between about 700°C and 1300°C. A linear function applies above what appears to be a glass transition at a temperature in the neighborhood of 1100°C. These conclusions were based on studies of radiometric profiles in silica glass subjected to an atmosphere of tritiated H_2O. Similar forms of concentration dependence appear to be characteristic of polymeric systems (see Crank, 1957; Barrer and Brook, 1953). In the present paper, as in most other studies, the choice of a given function is a formal approximation used only to simplify the numerical discussion. It is assumed that the data on silica glass indicate a preference for a linear rather than exponential function for diffusion of H_2O into silicate liquids; other forms of concentration dependence probably would be required for diffusion into glasses at lower temperatures.

Assuming that the form of concentration dependence of H_2O diffusion in obsidian is a linear function similar to that in silica glass at high temperatures, diffusivities are expressed in the form

$$D = D_0 (1 + a\, c/c_s) \qquad (3)$$

where D_0 is the limiting diffusivity at $c = 0$, a is a constant of proportionality and c_s is the saturation limit at T for a value of $f_{H_2O}*$ defined by a vapor pressure equal to the total pressure. At 850°C the diffusivity of H_2O in silica glass approaches a limiting value at $c = 0$ of the order of 10^{-11} cm^2 sec^{-1}. The diffusivity in obsidian at 850°C for $c = 0$ must be much smaller than the average diffusivities, but probably is somewhat larger than that in silica glass at $c = 0$; that is, it is assumed that D_0 is smaller than 10^{-8} cm^2 sec^{-1} but larger than 10^{-11} cm^2 sec^{-1}. Therefore, the value of D_0 for diffusion of H_2O in obsidian at 850°C probably approaches the order of 10^{-9} or 10^{-10} cm^2 sec^{-1}. The value of $\overline{D}$ from curve 2 in Fig. 4 is of the order of 2×10^{-7} cm^2 sec^{-1}. For the sake of graphically illustrating the form of linear concentration dependence, this value is taken to represent the limiting value of D at $c = c_s$ (this assumption is probably within the limits of experimental uncertainty for D at concentrations near 6% H_2O by weight). With the above assumptions, one of several possible consistent sets of constants in Equation 3 would be $D_0 \simeq 10^{-9}$ cm^2 sec^{-1}, $c_s \simeq 6\%$ by weight and $a \simeq 200$. This somewhat arbitrary function is plotted in Fig. 6, showing

mainly that diffusivities would exceed 10^{-8} cm^2 sec^{-1} at concentrations greater than about 0.5% H_2O by weight. This would be true even if D_0 were one or more orders of magnitude smaller. Since the solubility of H_2O at 850°C and P_t (H_2O) = 100 bars already exceeds 1%, such a function would explain why values of $\overline{D}$ calculated for different isobars have roughly the same magnitudes. Unique values of D_0 and a cannot be derived unambiguously without more precise data for $\overline{D}$ over several different concentration ranges.

Although the above analysis of concentration dependence is crude, Fig. 6 is considered to give the best available representation of diffusivities of H_2O in granitic liquids. The uncertainty at the higher concentrations is probably half an order of magnitude, as shown by the shaded area; the anhydrous value, D_0, may be somewhat smaller than 10^{-9} cm^2 sec^{-1} but is probably much greater than 10^{-11} cm^2 sec^{-1}. The effect of changes in total pressure of a few thousand bars at constant H_2O concentration is probably negligible compared with the above uncertainties.

Temperature Dependence of Diffusivity

Diffusion in liquids usually obeys an activation function of Arrhenius form

$$D = A \exp(-E_D/RT) \qquad (4)$$

where A and E_D are experimental constants; the latter is usually called the activation energy. Values of these constants are not accurately defined by the experimental data. The magnitude of $\overline{D}$ decreases by a factor of about 2 between 850°C and 750°C at 2000 bars (Table 3), and this gives $E_{\overline{D}} \simeq 16$ kcal mole^{-1}. If it is assumed that the uncertainty in estimating initial slopes of sorption curves is as large as 20%, then the maximum range of uncertainties in $E_{\overline{D}}$ is about ±5 kcal mole^{-1}. Drury and Roberts (1963) give an average value of about 18.5 kcal

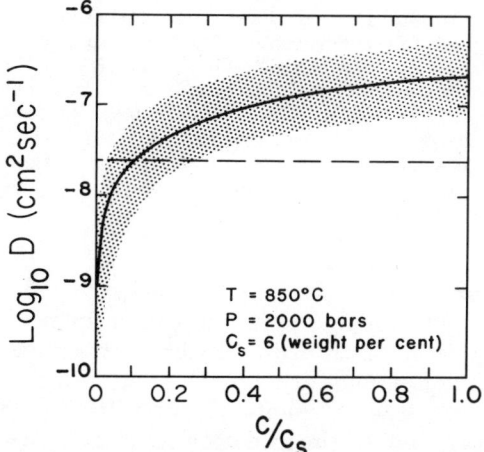

Fig. 6. Variation of diffusivity (D) with H_2O concentration at 850°C according to the linear function $D = D_0 (1 + a\, c/c_s)$ for $D_0 \simeq 10^{-9}$ cm^2 sec^{-1}, $a = 200$ and $c_s = 6\%$ by weight (see text for discussion). The dashed horizontal line is the value of net diffusivity at 850°C, 2000 bars, and an initial concentration $c_0 = 0.2\%$ H_2O from Table 3. The shaded area represents the estimated range of uncertainties in the diffusivity.

mole^{-1} for diffusion of H$_2$O in silica glass. Examination of their data shows that the temperature dependence abruptly changes slope at about 1100°C (presumably related to a glass transition); above this temperature the slope corresponds to $E_{\bar{D}} \simeq 13$ kcal mole^{-1} and below this temperature $E_{\bar{D}} \simeq 23$ kcal mole^{-1}. Since the data for obsidian are for conditions near those for melting of crystalline assemblages, and therefore presumably above the glass transition, the activation energy should be closer to the smaller of the above values. Use of this comparison suggests that the probable value for obsidian near liquidus temperatures is in the range $E_{\bar{D}} = 13 \pm 3$ kcal mole^{-1}.

Diffusivities of H$_2$O, H$_2$ and O$_2$ in Silica Glass

Figure 7 summarizes diffusivities of H$_2$O, H$_2$ and O$_2$ from data on permeation of silica glass, compared with the range of diffusivities for H$_2$O in obsidian obtained in the present study. The sources of data are given in the figure caption. The purpose of the comparison is to give some basis for estimating the orders of magnitude of diffusivity that may describe transfer of constituents that influence the oxidation state of magmas. Incidently, most available data on cation diffusivities in silicate melts also fall within the stippled range of Fig. 7 (see Medford, 1973; Koros and King, 1962; Williams and Heckman, 1964; Sucov and Gorman, 1965; Eitel, 1965; Winchell, 1969). Carron (1969) reports somewhat higher diffusivities for sodium in obsidians of granitic composition.

The data for hydrogen diffusion cannot be applied directly to granitic compositions, because reaction mechanisms in multicomponent systems may differ from those in a glass consisting predominantly of SiO$_2$. Bell et al. (1962) proposed that the rate of hydrogen permeation of silica glass is controlled by hydroxyl formation (with a balancing reduction of Si^{4+} to

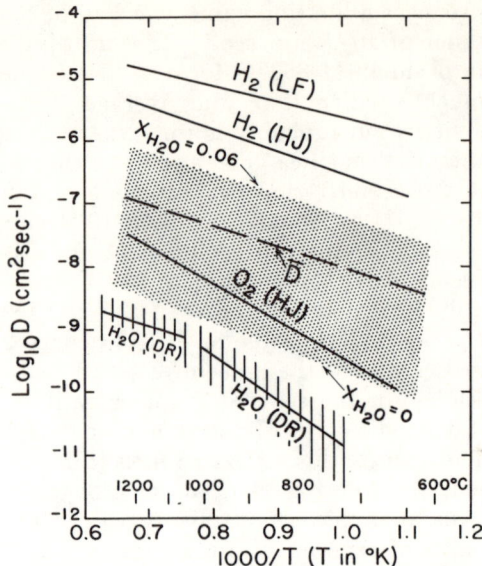

Fig. 7. Ranges of experimental diffusivities (D) of H$_2$, O$_2$ and H$_2$O in silica glass compared to the inferred diffusivity range for H$_2$O in obsidian with a concentration range of from 0 to 6% H$_2$O by weight (stippled area; X_{H_2O} represents mass fraction). The net diffusivities of H$_2$O for experimental hydration of obsidian in this study lie near the heavy dashed line (see Table 3). The hachured area represents the approximate range of diffusivities of H$_2$O in silica glass for concentrations ranging from 0 to 0.1% H$_2$O by weight. The symbols on the curves identify the following data sources: Lee and Fry (1966)—LF; Hetherington and Jack (1964)—HJ; Drury and Roberts (1963)—DR. Locations of the curves and temperature limits of reliability are only approximate; see original sources for exact ranges.

Si^{3+}). The data seem to show convincingly that hydrogen permeation in silica glass is controlled by hydroxyl reactions, but Lee and Fry (1966) point out that even trace amounts of impurities could take part in the oxidation-reduction reactions. Therefore, reaction kinetics could be different in the presence of major amounts of other cations.

In sorption of either H$_2$O or H$_2$ in silicate liquids of framework compositions, diffusion rates appear to be controlled both by the rates of transfer of the constituents in molecular form into a saturated layer and the rates of network modi-

TABLE 4. Comparison of viscosities* and diffusivities for obsidian-H_2O at 850°C

H_2O; Mass Fraction (X)	Liquid Viscosity (η, poise)	Diffusivity from Fig. 6 (D, cm²sec⁻¹)	Activation Energy; Flow (E_η, kcal mole⁻¹)	Activation Energy; Diff. (E_D, kcal mole⁻¹)	Activation† Ratio (E_η/E_D)	Stokes-Einstein Product; Obs. (ηD, dyne)	Stokes-Einstein‡ Radius; Calc. (r, cm)
0	3×10^{10}	1×10^{-9}	80	13 ± 3	8.0–5.0	30	3×10^{-16}
0.005	3×10^{9}	8×10^{-9}	76	13 ± 3	7.6–4.8	24	3×10^{-16}
0.04	2×10^{6}	7×10^{-8}	56	13 ± 3	5.6–3.5	0.14	6×10^{-14}
0.06	1×10^{5}	1×10^{-7}	48	13 ± 3	4.8–3.0	0.01	8×10^{-13}

* Viscosity data are from Shaw (1972).
† The range represents maximum and minimum values for E_D. If E_D decreases with H_2O concentration, the ratio could conceivably remain nearly constant at $E_\eta/E_D \sim 5$.
‡ The radius is calculated from the simple formula (see Equation 5 and discussions). The calculated values are physically impossible. However, the large increase with increasing X is consistent with transfer involving large domains at high concentrations.

fying reactions to form hydroxyl sites within this layer and at the hydration front. These mechanisms probably limit rates of hydrogen diffusion to values not greatly exceeding maximum rates for H_2O diffusion. Such factors must also control the rates of oxygen migration.

Although an upper limit cannot be placed on diffusion controlled by reaction rates, it seems unlikely that diffusion rates can exceed magnitudes characteristic of diffusion in the gaseous state. Diffusivities of H_2 in gases at pressures of 1 kbar or higher would be of the order 10^{-2} cm^2 sec^{-1} at magmatic temperatures according to correlations discussed in Bird et al. (1960, p. 505). Based on the above factors, rates of hydrogen diffusion under magmatic conditions are thought to be limited by a maximum diffusivity of the order of 10^{-5} cm^2 sec^{-1}, although this is no more than an educated guess. Direct measurements are needed for both granitic and basaltic compositions.

Relationships Between Diffusion and Viscous Flow

The well-known Stokes-Einstein formula relating the diffusivity of a chemical species in a solution to the viscosity of the solution is written (Ree and Eyring, 1958, p. 100)

$$D_i \eta = \frac{kT}{6\pi r} \quad (5)$$

where η is the viscosity, k is Boltzmann's constant, T is temperature (Kelvin) and r is the radius of the diffusing species. The formula was derived from hydrodynamic arguments for the balance between viscous drag on a sphere and the diffusion force defined by the gradient of chemical potentials in an ideal mixture. These assumptions are poor for structurally complex mixtures, particularly the assumption that there is a single molecular species of fixed size that governs the transfer rate.

Ree and Eyring (1958) generalized the Stokes-Einstein relation by allowing for nonideality of the mixture and by introducing rate constants for structural changes in the solution related to viscous flow. Their equation is written

$$D_i \eta = \frac{kT}{\lambda \xi} \frac{k'}{k''} \frac{d \ln a_i}{d \ln c_i} \quad (6)$$

where a_i is the activity, c_i the concentration, λ is the "jump" distance of the diffusing species, ξ is the coordination number of the species in a plane, k' is the rate constant for viscous flow in the solution and k'' is the corresponding rate constant in the pure solvent. For example, in dilute ideal solutions $k'/k'' \simeq 1$, and for diffusion of spherical molecules in a solution of similar size molecules the structural parameters are given by $\lambda \simeq (V/N)^{1/3}$ and $\xi = 6$ (closest packing of spheres), where V is the volume per mole of solution and N is Avogadro's number.

Use of Equation 6 for silicate solutions involves several unknown parameters, and no reliance can be placed on estimates of D_i from viscosity data. By way of illustration, values of the product $D\eta$ for different compositions are shown in Table 4. Clearly, there are probably very large changes in both structural and rate constants in this composition range. The significance of relationships such as Equation 6 stems mainly from the possibility of using the experimental data to test theoretical structure models. The only conclusion drawn in the present paper is that diffusion of constituents such as H_2O in silicate liquids is governed by local structural effects that bear no simple relation to those involved in wholesale viscous flow. Diffusion in these liquids may be governed by factors more like those involved in solid state diffusion in framework silicates, where the structure behaves partly like a rigid sieve (as in the case of absorption of gases in zeolites).

PART II. MASS TRANSFER IN MAGMA CHAMBERS

Diffusion in Static Magma Chambers

The following sections of the paper are addressed to one principal question: What is the magnitude of diffusional exchange that could conceivably take place between a magma chamber and the bounding media during the time of solidification? There is no unequivocal answer to this question, but some insight is given by simplified estimates of transfer rates defined by artificial boundary conditions.

The analogy mentioned earlier between diffusional transfer of heat and mass provides an intrinsic scaling factor for the time available for chemical transfer in magma. The thermal diffusivity governing conductive transfer of heat from magma into the country rock is of the order of 10^{-2} cm^2 sec^{-1} (see Jaeger, 1964; Shaw, 1965). Therefore the rate of a proportional change in temperature between two limiting values by diffusion of heat is orders of magnitude greater than that of a proportional change in composition by chemical diffusion, as was noted by Bowen (1928).

The idea of a "penetration distance" can be used to illustrate the above contrast. If the diffusivity is constant, the progress of the position in a semi-infinite medium at which the concentration (or heat content) has attained 50% of its final value for a constant boundary potential is given by

$$y_{1/2} \simeq \sqrt{Dt} \qquad (7)$$

where $y_{1/2}$ represents the distance of penetration of this concentration level; see graphs in Crank (1957) or in Carslaw and Jaeger (1959). If D increases greatly with concentration and a suitable average value $\overline{D}$ is used, the value $2y_{1/2}$ is a fair estimate of the distance to the diffusion "front," because in such cases the concentration profile is steeper and terminates more abruptly than it does for constant D, as is shown in Fig. 8 (see Crank,

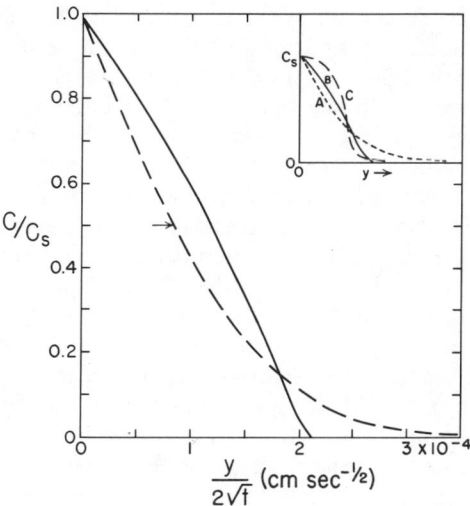

Fig. 8. The effect of concentration-dependent diffusivities on concentration profiles for sorption in a semi-infinite medium from a surface maintained at a fixed chemical potential of the diffusing substance. The solid curve is computed for the linear function $D = D_0 (1 + a\, c/c_s)$ given in Fig. 6, using methods discussed by Crank (1957; chapt. 12). The dashed curve gives the profile that would exist if the diffusivity were independent of concentration and had the value $D = 3 \times 10^{-8}$ cm^2 sec^{-1}. The arrow defines the distance $y_{1/2} \simeq \sqrt{Dt}$; for linear concentration dependence the diffusion front is at about $2y_{1/2}$ based on the equivalent net diffusivity, $\overline{D}$, treated as though it were constant. The inset schematically compares profile shapes after the same length of time for constant D (curve A), for D a linear function of c (curve B), and for D an exponential function of c in the form $D \doteq e^c$ (curve C).

1957; Haller, 1963). Although this relation provides a basis for estimating penetration of hydration fronts, it does not apply directly to dehydration. Dehydration rates differ because the concentration dependent factors that progressively increase diffusion rates in saturated layers near a boundary of constant high H$_2$O potential operate inversely to slow dehydration rates as the concentrations in these outer layers are decreased in response to a lowered H$_2$O potential. Such cases must be treated by the more complicated methods discussed in Chap. 12 of Crank (1957; see his figures 12.15 and

12.16); rates are not greatly different during initial stages of hydration or dehydration, but there would be major differences in the times for total hydration or dehydration of finite bodies between the same concentration limits.

Equation 7 shows that diffusion times increase as the square of the distance and inversely with the magnitude of D. For a value $\bar{D} \sim 10^{-8}$ cm^2 sec^{-1} the equation shows that a hydration front could move only about 1 cm in one year, or about 10 meters in one million years. However, in 1 m.y. heat losses are controlled by thermal penetration of country rock for distances exceeding 10 km, and even very large magma chambers would have become completely crystallized, unless they were subjected to a continuous addition of heat from a larger reservoir by convection or by a throughgoing flow of magma (see Jaeger, 1964). Therefore, transport distances for mass transfer by diffusion in magma chambers are limited to a few meters for most constituents (or possibly a few tens of meters in the case of hydrogen). Consequently, if there is no convection, compositional changes caused by diffusion are probably significant only at contact zones or relative to local transfers between internal heterogeneities of composition, or relative to exchanges between crystal, liquid and gas phases during crystallization and/or vesiculation.

The possibility of natural convection localized near the margins of a magma chamber could drastically change the above conclusions. Penetration distances are still limited by the magnitudes of chemical diffusivities and cooling times, but if convection currents are sharply defined, steep chemical gradients can be maintained by the action of these currents in sweeping away constituents gained from the walls and replenishing constituents lost to the walls. The theory of boundary layer convection is applied below to test whether or not diffusional transfer could have important effects on zoning of bulk compositions in magma chambers.

Diffusion in Convecting Boundary Layers

Most studies of convection in magma chambers start from the theory of free convection in a horizontal layer of fluid heated from below, primarily because the assumptions of analysis bear some resemblance to conditions in layered intrusions of sill-like geometry. By contrast, Shaw (1965) considered free convection in cylindrical or slot-like chambers with the long axis oriented vertically. Here, emphasis was placed on the "thermosyphon" concept in which a magma chamber is connected to a larger magma source at depth, as is appropriate to consideration of volcanic necks (McBirney, 1959) or to convective circulation in apophyses above large plutons. However, similar considerations apply even if there is no underlying reservoir. The main theoretical distinctions between the two approaches above relate to whether the convection is driven primarily by heat losses to horizontal or to vertical walls. The real situation is somewhere between these two limits, although the horizontal dimensions of most magma chambers, particularly large ones, exceed the vertical dimension (excluding possible feeder systems).

In the examples discussed below, attention is focused on processes that may operate near a vertical wall, and the discussion does not analyze the convective circulation in the entire chamber. This is done to simplify and shorten the discussion and to emphasize effects important to chemical exchange. The magnitude of this exchange relative to the bulk composition of the magma subsequently can be compared with solidification times and convective regimes determined by shape factors and the ratio of the total volume of the chamber to the contact area subject to the greatest chemical exchange. It also will be shown qualitatively that similar processes should operate over the entire area of magma contact with the

DIFFUSION

containing rocks during the early stages of solidification.

Thermal Boundary Layers

The transport equations given below contain the following main assumptions; the effects of more realistic conditions will be assessed later:

1. Density (buoyancy) gradients in the magma are governed by the temperature according to the magnitude of a thermal expansion coefficient, α_T; that is, density changes due to pressure and changes of phase are neglected. However, crystallization as a function only of temperature tends to augment such density gradients. Density changes related to chemical composition or vesiculation introduce other complications.

2. The vertical boundary across which heat is lost from the magma to surrounding rock remains approximately at a constant contact temperature, T_c. It will be shown later that this assumption is valid for part of the cooling history.

3. The magma temperature far from the cooling boundary remains at a constant temperature, T_∞; the subscript is a reminder that the model only describes rates of heat transfer from an infinite reservoir across an isothermal boundary. By definition this limits the model to the initial stages of cooling of finite bodies. The time interval over which this assumption approximately holds will be estimated.

4. The magma viscosity is constant. The accuracy of this assumption is governed by the assumptions concerning temperature.

With the above assumptions it can be shown that the average heat transfer rate across the vertical boundary is a function of two dimensionless variables, the Prandtl number, Pr, and the Grashof number, Gr, defined by the quotients (see Bird et al., 1960; Rohsenow and Choi, 1961)

$$\text{Pr} = \nu/\kappa \qquad (8)$$

$$\text{Gr} = \frac{g\alpha_T \Delta T L^3}{\nu^2} \qquad (9)$$

where ν is the kinematic viscosity ($\nu = \eta/\rho$, with units cm^2 sec^{-1}), κ is the thermal diffusivity (cm^2 sec^{-1}), g is the acceleration of gravity, α_T is the thermal expansion coefficient, ΔT is the temperature difference between the isothermal boundary and the fluid reservoir ($\Delta T = T_c - T_\infty$) and L is the height of cooling boundary. The product PrGr is called the Rayleigh number

$$\text{Ra} = \frac{g\alpha_T \Delta T L^3}{\nu\kappa} \qquad (10)$$

Physically, Pr is the rate at which momentum diffuses from the boundary layer into less disturbed fluid divided by the rate at which heat diffuses through the boundary layer to the isothermal wall (maintaining the temperature and density gradients). The Rayleigh number is the ratio of internal energy released by the buoyancy force to the kinetic energy dissipated by viscosity (see Chandrasekhar, 1961, p. 32). Balances among these ratios for given boundary conditions govern the velocity structures of convection, hence the boundary layer dimensions and heat transfer rates. This generalization is true whatever the shape and orientation of confining walls, although exact solutions of the balances are known only for simplified cases. The equations for development of a laminar boundary layer adjacent to a vertical isothermal wall are summarized below, followed by analogous expressions for turbulent flow.

The heat flux across the isothermal wall is defined by

$$J_H = -k_T(\partial T/\partial y)_{y=0} \qquad (11)$$

where k_T is the thermal conductivity and $\partial T/\partial y$ is the temperature gradient evaluated at the contact (positive y measured into the magma). If the temperature in the boundary layer is a quadratic function of y, then the gradient at the contact becomes

$$\left(\frac{\partial T}{\partial y}\right)_{y=0} \simeq \frac{2(T_\infty - T_c)}{\delta} = \frac{-2\Delta T}{\delta} \quad (12)$$

where δ is the boundary layer thickness as shown in Fig. 9 (see Rohsenow and Choi, 1961, p. 162). Therefore the heat flux is given by

$$J_H \simeq 2k_T \frac{\Delta T}{\delta} \quad (11a)$$

The heat flux can be expressed in a reduced form by normalizing relative to the vertical position, z, and temperature difference, defining the dimensionless group called the Nusselt number

$$\text{Nu} = \frac{J_H z}{k_T \Delta T} \quad (13)$$

or, using Equation 11a,

$$\text{Nu} = \frac{2z}{\delta} \quad (13a)$$

where $z = 0$ at the top of the isothermal boundary.

Solutions of the transport equations in terms of the variables Pr and Gr define the boundary layer thickness in the form (Rohsenow and Choi, 1961, p. 162)

$$\frac{\delta}{z} = 3.93 \, \text{Pr}^{-1/2}(0.952 + \text{Pr})^{1/4}\text{Gr}^{-1/4} \quad (14)$$

However, in viscous systems $\text{Pr} \gg 1$ and Equation 14 can be simplified to

$$\frac{\delta}{z} \simeq 4(\text{PrGr})^{-1/4} = 4 \, \text{Ra}^{-1/4} \text{ (laminar)} \quad (14a)$$

Heat transfer is therefore given by the function

$$\text{Nu} \simeq \frac{1}{2} \text{Ra}^{1/4} \text{ (laminar)} \quad (15)$$

The expression for a turbulent boundary layer is not as simple but is approximated in a similar manner for $\text{Pr} \gg 1$ (Rohsenow and Choi, 1961, p. 204)

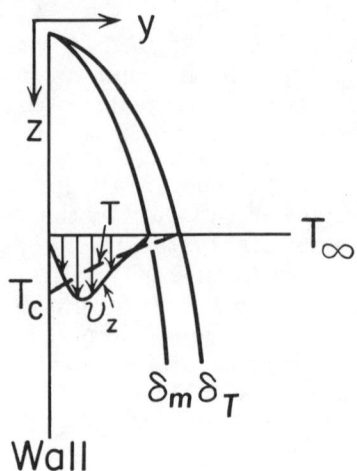

Fig. 9. Schematic illustration of meanings of boundary layer thickness, δ_m and δ_T, of momentum and temperature, respectively, for natural thermal convection on a vertical plate (modified from Rohsenow and Choi, 1961, Fig. 7.13). The heavy solid lines show the variation of boundary layer thicknesses in the horizontal direction, y, adjacent to a vertical cooling surface of arbitrary height (cooling adjacent to surfaces of closed form, as in magma chambers, will have nonvanishing boundary layer thicknesses at all points on the surface). The light horizontal line is a reference line for profiles of temperature, T, and vertical velocity, v_z, at a fixed vertical horizon. At distances, y, to the right of the intersection with curve δ_T along this profile, the temperature is very nearly equal to T_∞ (temperature at infinite distance from the cooling wall into the fluid; for magmatic convection, T_∞ is assumed to be the initial magma temperature). The dashed line shows the variation of temperature along the profile from T_∞ to the temperature of the cooling surface, T_c, within the thermal boundary layer (as though temperature were plotted along a third orthogonal coordinate and then rotated into the plane of the drawing). Similarly the downward velocities, v_z, along the same profile range effectively from zero outside the intersection with curve δ_m to a maximum within the boundary layer, and again to zero at the cooling wall. Solutions of the transport equations discussed in the text are based on the assumption that $\delta_m = \delta_T$; the validity of this approximation is substantiated by experimental data (see empirical equations in text and Figure 8.18 in Rohsenow and Choi, 1961).

$$\frac{\delta}{z} \simeq \frac{1}{2}\,\mathrm{Gr}^{-1/10}\,\mathrm{Pr}^{-1/2} \qquad (14b)$$

$$= \frac{1}{2}\,\mathrm{Ra}^{-1/10}\,\mathrm{Pr}^{-2/5} \text{ (turbulent)}$$

Experimental measurements of convective heat transfer have resulted in the empirical expressions (Rohsenow and Choi, 1961, p. 205, Fig. 8.18)

$$\mathrm{Nu} = 0.56\,\mathrm{Ra}_L{}^{1/4}$$
$$\text{(laminar; Ra} < \sim 10^9) \qquad (16)$$

$$\mathrm{Nu} = 0.13\,\mathrm{Ra}_L{}^{1/3}$$
$$\text{(turbulent; Ra} > \sim 10^9) \qquad (17)$$

where the subscript L indicates that the expressions represent averages over the plate height L. For values of Ra above 10^{11}, the correlation functions seem to vary somewhat depending on the value of Pr. Although few experimental data exist to test functions at large values of both Ra and Pr, Equations 16 and 17 give fair estimates of heat transfer rates, although the true rates could be somewhat higher. These correlations also approximately describe transfer rates for other geometries and orientations of cooled and heated boundaries if Ra is evaluated according to an appropriate length scale. Similar results were obtained by Kraichnan (1962) in a discussion of convective regimes for heat transfer between horizontal plates, where it is also shown that $\mathrm{Nu} = f\,(\mathrm{Pr}, \mathrm{Ra})$ although the additional dependence on Pr is not strong. Hess (1973) applied the theory of boundary layer convection between horizontal plates to convection during crystallization of the Stillwater Complex, and our general conclusions concerning the conditions of boundary layer convection appear to be consistent even though flow in the present discussion is referred to a vertical cooling surface.

The profiles of vertical velocities (downward in the case of cooling) across the boundary layers have the forms (Rohsenow and Choi, 1961, p. 161, p. 203)

$$v_z = C_l\,\frac{x}{\delta}\left(1 - \frac{x}{\delta}\right)^2 \text{ (laminar)} \qquad (18)$$

$$v_z = C_t\,\left(\frac{x}{\delta}\right)^{1/7}\left(1 - \frac{x}{\delta}\right)^4 \text{ (turbulent)} \qquad (19)$$

where C_l and C_t are functions of the vertical position, z, for a given set of fluid properties $(\nu \gg \kappa)$:

$$C_l \simeq 5\nu \mathrm{Pr}^{-1/2}\left(\frac{g\alpha_T \Delta T z}{\nu^2}\right)^{1/2} \qquad (20)$$

$$C_t \simeq 1.4\nu\,\mathrm{Pr}^{-1/3}\left(\frac{g\alpha_T \Delta T z}{\nu^2}\right)^{1/2} \qquad (21)$$

The x position of maximum vertical velocity in the laminar boundary layer is found from Equation 18 to be at $x = \delta/3$ from the contact. However, the velocity structure of the turbulent boundary layer is ambiguous because of vorticity and mixing across the layer. Therefore velocities must be viewed according to some statistical distribution such as that derived by Kraichnan (1962) for "mixing lengths" in horizontal boundary layers. Equation 19 shows that for present purposes we can only conclude that the highest average velocities are much closer to the contact than they are in the laminar case ($x < \delta/10$). Values of δ and v_z (max) are given in Table 5 for three values of viscosity characteristic of the magmatic range at liquidus temperatures (basalt, andesite and "wet granite," "dry granite"); other properties and boundary conditions represent values that may apply to some magma chambers during at least part of the convective lifetime.

With regard to chemical exchange the main conclusions from Table 5 are that boundary layer thicknesses are very small compared with the dimensions of many igneous intrusions, and the flow velocities in these layers can be, geologically, very fast in the early stages of cooling. In order to estimate how long such conditions can be maintained, the temperature distribution has to be related to balances defined by the heat flux and the total heat content of the magma. A more complete evaluation depends on analysis of the balance between the con-

TABLE 5. Estimated values of boundary layer thickness (δ) and vertical velocity (v_z) for free convection adjacent to a vertical contact for viscosities in the range for magmatic liquids (values in parentheses are calculated from equations for turbulent boundary layers; others assume laminar flow)

Kinematic Viscosity (ν, cm^{2}sec^{-1})	Prandtl Number (Pr)	Rayleigh Number (Ra)	Nusselt Number (Nu)	Heat Flux (J_H, cal cm^{-2}sec^{-1})	Boundary Layer (max.) (δ, cm)	Velocity (max.) (v_z, km yr^{-1})	Magma Type ($T \geq$ liquidus)
10^2	10^4	5×10^{14}	2500 (10,000)	1×10^{-3} (5×10^{-3})	80 (40)	500 (2100)	basalt (Hawaiian, tholeiite)
10^5	10^7	5×10^{11}	400 (1000)	2×10^{-4} (5×10^{-4})	500 (10)	15 (210)	andesite (dry); granite with H$_2$O > 5% by weight
10^9	10^{11}	5×10^7	40	2×10^{-5}	5000	0.15	anhydrous granite (ternary minimum composition)

Parameters assumed constant:
(Kinematic viscosities are based on absolute viscosity data from Shaw, 1972)

$g = 980$ cm sec^{-2}
$\alpha_t = 5 \times 10^{-5}$ deg^{-1}
$\kappa = 10^{-2}$ cm^2 sec^{-1}
$k = 5 \times 10^{-3}$ cal cm^{-1}sec^{-1}deg^{-1}
$\Delta T = 10°$C
$L = 10^5$ cm (1 km)

vective heat flux in the magma chamber and the heat flux controlled by conduction in the surrounding rocks.

The magnitude of the conductive heat flux into the surrounding rock (assumed to be of infinite extent and initial temperature $T = 0$) is given by the equation (Carslaw and Jaeger, 1959, p. 61)

$$J'_H = \frac{k\,T_c}{(\pi\kappa t)^{1/2}} \quad (22)$$

where T_c is the contact temperature assumed to be constant. In the absence of convection in the magma the contact temperature remains at $T_c \sim 0.6\,T_\infty$ for some time after emplacement (Jaeger, 1957). Note that if it were assumed that $T_c \simeq 0.8\,T_\infty$, then Equation 22 would give a value of initial heat flux that deviates only by about 25% from the value for either perfect conduction ($T_c \simeq 0.6\,T_\infty$) or perfect convection ($T_c \simeq T_\infty$). Assuming that convection is established (i.e. the chamber is so large that the assumption of a constant value of T_∞ is approximately satisfied), effective values of T_c as a function of time can be estimated by setting the convective and conductive fluxes equal, $J'_H = -J_H$ (the minus sign is used because the convective flux was originally defined as a loss to the walls). Accordingly, Equations 13 and 22 define the balances

$$\frac{kT_c}{(\pi\kappa t)^{1/2}} = \frac{-k\Delta T\,\mathrm{Nu}}{L} \quad (23)$$

and for the laminar case (Nu $\simeq \tfrac{1}{2}\,\mathrm{Ra}^{1/4}$)

$$\frac{T_c}{(\pi\kappa t)^{1/2}} = \frac{-\Delta T\,\mathrm{Ra}^{1/4}}{2L} \quad (23a)$$

After some rearrangement, the time dependence of contact temperature is defined approximately by

$$\left(\frac{T_\infty}{T_c} - 1\right)^{5/4} \simeq 2L\left(\frac{g\alpha_T L^3}{\nu\kappa}\right)^{-1/4}$$
$$\times (\pi\kappa t)^{-1/2} T_c^{-1/4} \quad (24)$$

The boundary layer thickness (laminar case) as a function of time can be calculated by combining Equations 13a and 23 using values of ΔT calculated from Equation 24; the resulting equation becomes

$$\delta = \frac{-2\Delta T(\pi\kappa t)^{1/2}}{T_c} \quad (25)$$

Values of ΔT and δ vs time are shown in Fig. 10 for the extremes of viscosities used in Table 5. The principal conclusion is that contact temperatures exceed solidus temperatures very soon after emplacement. The effect of initial quenching at the contact on viscosity will not greatly change the curves in Fig. 9 because viscosity enters as $\nu^{1/4}$ in Equation 24; the initial zone of quenched material adjacent to the contact is soon reheated by convection farther inside where T exceeds the solidus temperature. Boundary layer thicknesses are similar to, but are smaller than, those in Table 5 where an arbitrary value $\Delta T = 10°C$ was assumed. Note that in Equation 25 the length L cancels out, indicating that the calculated values of δ are averaged over all cooling surfaces. Velocities presumably will initially be higher than those in Table 5 until ΔT falls below $10°C$.

Unless buoyancy forces are cancelled by vesiculation or changes of chemical composition, convection lasts until crystallization causes viscosities to greatly exceed 10^{10} poise. For any type of magma this probably does not happen until the entire chamber is more than half crystallized (see Shaw, 1965, 1969). Therefore the time available for chemical exchange with boundary layers can be estimated from solidification times calculated from Equation 22 taking into account the shape and volume of the chamber, the specific heat, and the latent heat of crystallization of the magma. These calculations will not be given in detail, but it is noted that if boundary layer convection persists for about half the total solidification time, then the available time for chemical exchange is more than one-fifth

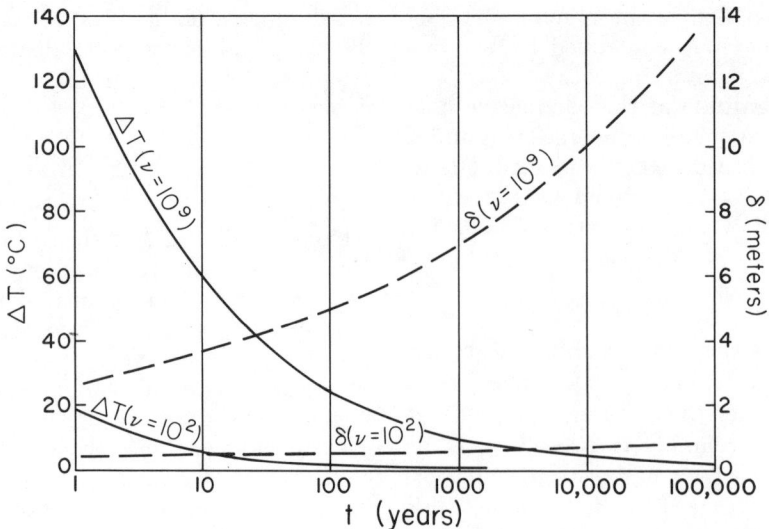

Fig. 10. Calculated variations of differences between interior magma temperatures and contact temperatures (ΔT) and boundary layer thicknesses (δ) with time during the initial stages of cooling of a magma chamber (see Equation 25 in text). Solutions are given for viscosities appropriate to basaltic ($\nu = 10^2$) and granitic ($\nu = 10^9$) compositions. As contact temperatures approach the initial magma temperature, $\Delta T \rightarrow 0$. If the initial magma temperature is as much as 100°C above the solidus temperature, contact temperatures become elevated above the solidus temperature in only a few years. Without convection, the initial contact temperature computed from heat conduction theory (Jaeger, 1957) is about 0.6 T_∞, where T_∞ is the temperature of magma far from the contact; in this discussion T_∞ is also assumed to be the initial magma temperature.

the total solidification time computed from the theory of perfect conduction (compare Jaeger, 1964, p. 455). As one example, which is also cited later, a deep-seated spherical chamber with a diameter of about 20 km, if cooled statically, would reach the solidus temperature at the center in a time of the order of 10^6 years (estimated from Fig. 1 of Jaeger, 1968, allowing for a latent heat of 100 cal gm^{-1}). Convection during the *entire* solidification-temperature interval would decrease this time to a minimum of around 400,000 yrs. So convection during half this solidification interval would last about 200,000 yrs and the subsequent cooling, mainly by conduction, to the solidus temperature would take an additional 600,000 yrs or so. Thus, convective cooling until the chamber is half crystallized lasts about one-fourth of the total solidification time, which is only about 20% less than the total time for cooling statically to the solidus temperature.[3]

Chemical Boundary Layers

Analysis of chemical exchange within convecting boundary layers is difficult because there is more than one concept of layer thickness as defined by profiles of temperature, velocity, and concentration normal to the chamber wall. Analysis of thermal boundary layers was simplified without much loss of accuracy by assuming that a single thickness, δ, described both the velocity and temperature distribution. The same assumption does

---

[3] This result also illustrates that approximate agreement of cooling curves for Hawaiian lava lakes with conduction models (Jaeger, 1968, p. 519) does not permit the additional conclusion that there has been no convective transport during solidification, as is also shown by estimates of convective cooling by Jaeger (1968, p. 533 ff.).

not necessarily hold for concentration gradients, because for small values of chemical diffusivity, D, the chemical boundary layer, δ_c, as defined by a chemical gradient analogous to that of temperature in Fig. 9, is much narrower than δ_T. The transport equations are analogous to those for heat transfer, but details of derivation are omitted and only two simplified cases will be considered. Case I assumes that the velocity distributions are given by the previous examples for thermal convection (that is, buoyancy forces remain unchanged in sign and are negligibly changed in magnitude by changes in chemical composition near the walls). Case II assumes that buoyancy forces are entirely governed by density changes due to composition, as though the fluid were originally static and isothermal. This assumption is unrealistic but approximates a situation in which flow velocities are locally dominated by the buoyancy effects of composition rather than thermal expansion. Accordingly, the analysis is simplified because convective instability is directly analogous to that for heat transfer, if one substitutes concentration for temperature, and if it is also assumed that the concentration at the contact is fixed (as in the hydration experiments of this paper).

Case I. Analogous examples are derived by Bird *et al.* (1960, p. 617) for sublimation of a substance into a laminar boundary layer flowing over a flat plate. This situation applies approximately for material dissolving from walls of the magma chamber if diffusion rates within the magma control the mass transfer rate into the chambers. The mass flux of a constituent into the chamber (assuming the solution is pseudobinary as in the hydration experiments) is given by

$$J_m = \frac{\rho v_\infty (X_c - X_\infty)}{3 \, \mathrm{Re}^{1/2} \mathrm{Sc}^{2/3}} \quad (26)$$

where X_c is the mass fraction of the diffusing constituent in the magma at the contact, X_∞ is the original mass fraction in the magma, v_∞ is the maximum velocity of flow over the plate outside the chemical boundary layer, Re is the Reynolds number and Sc is the Schmidt number defined as follows:

$$\mathrm{Re} = \frac{v_\infty L}{\nu} \quad (27)$$

$$\mathrm{Sc} = \frac{\nu}{D} \quad (28)$$

Estimates of mass flux are given in Table 6 for laminar convection velocities defined in Table 5. The assumption of laminar flow implies that material diffusing into or out of the boundary layer is either carried away or replenished (as in a through-flow system) or is added to or removed from an infinite reservoir of fluid so that X_∞ remains constant. However, in a finite chamber the material added or subtracted is relative to the fluid composition in the central regions, because the flow due to thermal boundary layers at the vertical walls is balanced by upward flow in the interior of the chamber. Therefore, the flux rate applies only as long as ΔX remains approximately constant. Turbulent flow increases the mass flux by steepening the velocity gradients near the wall and by diffusion of eddies normal to the boundary layer so that mixing of the exchanged mass with the central body of fluid is made more efficient.

The total mass exchange depends on the contact area, time, and the variation of ΔX caused by exchange. The implication cited earlier that thermal boundary layers exist over all contact surfaces (see discussion of Equation 25) suggests that chemical exchange exists over the entire contact area except for any area of the floor that becomes rapidly mantled by crystal accumulation (see Hess, 1973). In a pluton of the size previously cited ($r = 10$ km), convection was estimated to last 2×10^5 years (6×10^{12} sec) and the contact area would be of the order of 10^{13} cm². Therefore the total exchange for the three ex-

TABLE 6. Estimate of mass flux of a constituent diffusing into a laminar boundary layer flowing over a vertical contact, assuming that the concentration change does not significantly change thermal buoyancy forces. Flow velocities are from Table 5.

Chemical* Diffusivity ($\bar{D}$, cm²sec⁻¹)	Kinematic Viscosity (ν, cm²sec⁻¹)	Flow Velocity (ν, cm sec⁻¹)	Reynolds Number (Re)	Schmidt Number (Sc)	Mass Flux (J_m, g cm⁻²sec⁻¹)	Magma Type ($T \geq$ liquidus)
10^{-7}	10^2	1.6	1.6×10^3	10^{10}	3×10^{-11}	basalt
10^{-8}	10^5	5×10^{-2}	5×10^{-2}	10^{13}	2×10^{-12}	andesite; hydrous granite
10^{-8}	10^9	5×10^{-4}	5×10^{-8}	10^{17}	2×10^{-13}	anhydrous granite

Parameters assumed constant: $(X_c - X_\infty) = 0.01$
$\rho = 2.5$ g cm⁻³
$L = 10^5$ cm

The value of ρ is an average for magmatic liquids (the typical range is about 2.3 to 2.8 g cm⁻³). The difference in mass fractions represents a decrease of 1% by weight in concentration between liquid at the margin and liquid outside the boundary layer. Such a change in one constituent will have little effect on density except in the case of light volatile constituents like H_2O. The length L assumes that exchange takes place only over 1 km of the vertical contact. The same rates apply to addition or subtraction of material depending on the sign of ΔX if $\bar{D}$ remains nearly constant over ΔX.

* The values of $\bar{D}$ are estimated magnitudes applicable for H_2O. $\bar{D}$ for basalt is larger mainly because of the high liquidus temperature (for example, see Fig. 7 at 1200°C; data of Medford (1973) for diffusion of calcium in mugearite are also close to 10^{-7} cm²sec⁻¹).

amples of Table 6 for constant ΔX would be, respectively, of the orders of 10^{15}, 10^{14} and 10^{13} grams. The total mass of magma is of the order of 10^{19} grams, and if X_∞ is of the order of 0.01 (1% by weight) the original mass of this constituent is 10^{17} grams. Therefore the percentage change in average bulk composition is small. However, the amount of exchange resulting from convective transfer could be large if combined with other mechanisms that might concentrate a given constituent, such as crystal accumulation, vesiculation, or buoyant stratification of liquid compositions in the upper part of the chamber.

It is also of some interest to check conditions that tend to maximize the amount of assimilation because of a very high chemical potential at the contact (for example, solution of calcium from limestone, or solution of quartzofeldspathic constituents in basalt).[4] As an illustration, consider the possible change in SiO_2 concentration in a basaltic chamber of the above dimensions that is dissolving only quartz at the contact. Here the concentration difference is $\Delta X > 0.1$ and the flux rate approaches the order of 10^{-9} gm cm^{-2} sec^{-1} for a SiO_2 diffusivity of 10^{-7} cm^2 sec^{-1}, giving a total mass of the order of 5×10^{16} grams. The original mass (for an initial composition of 50 percent SiO_2) was 5×10^{18} grams, so the final bulk composition due to assimilation would be 50.5% SiO_2 by weight (the change in heat content of the magma caused by the heat of solution is negligible). An assumption of turbulent transfer could possibly increase this to a change of a few percent in SiO_2. Again, this magnitude of exchange could be more significant relative to local bulk compositions if other concentration mechanisms are also operative. It might also be significant relative to oxygen isotopes or to trace elements not originally present.

Case II. Here the treatment is directly parallel to that for heat transfer, if the buoyancy effects of composition are suitably expressed. The analogous expansion coefficient can be written in terms of a normalized partial molar volume as follows (binary case):

$$\alpha_c = \frac{1}{\overline{V}} \left(\frac{\partial \overline{V}}{\partial c} \right)_{T,P} \quad (29)$$

where $\overline{V}$ is the specific volume and c is concentration. Estimates based on density data for the hydrated glasses in this study indicate that $\alpha_c \sim 1$ for granitic compositions, for c expressed as mass fraction, X (larger values are indicated by the data of Burnham and Davis, 1971, depending on conditions). The analogous dimensionless variables in the transport equations become

$$\text{Nu}' = \frac{J_m L}{D \rho \Delta X} \quad (30)$$

$$\text{Pr}' = \nu/D \ (\equiv Sc) \quad (31)$$

$$\text{Ra}' = \frac{g \alpha_c \rho \Delta X L^3}{D \nu} \quad (32)$$

The boundary layer thicknesses and velocities for laminar flow are given by Equations 14a, 18, and 20 using the above definitions of Pr' and Ra', noting that the direction of transfer is upward for hydration layers because α_c and ΔX are positive. Values of boundary layer thickness, vertical velocity, and mass flux are given in Table 7 for hydration of an anhydrous granitic magma chamber with an arbitrary constant H_2O fugacity at the vertical contact. Two values of viscosity are used to show the effect of viscosity variation caused by the hydration. Although the initial kinematic viscosity is about 10^9 cm^2 sec^{-1}, the viscosity in the hydrated layer is much lower. Since the maximum velocity in the boundary layer is near the contact, the lower viscosity probably gives the more realistic transport rates. The magnitudes of total transfer would deliver on the order of 10^{16} g H_2O to the top of a chamber the size previously considered ($r = 10$ km). This corresponds to about 3×10^{17} grams of hydrated

---

[4] Composition changes caused by stoping are not included in these estimates.

TABLE 7. Laminar convection layer during hydration of an initially anhydrous granitic magma at a vertical contact with constant H_2O potential

Kinematic Viscosity (ν, cm^{2}sec^{-1})	Prandtl* Number (Pr')	Rayleigh* Number (Ra')	Nusselt* Number (Nu')	Mass Flux (J_m, g cm^{-2}sec^{-1})	Boundary Layer (δ, cm)	Maximum Velocity (v_z, km yr^{-1})
10^5	10^{13}	10^{20}	5×10^4	5×10^{-10}	4	0.3
10^9	10^{17}	10^{16}	5×10^3	5×10^{-11}	40	0.03

Parameters assumed constant:
$\alpha_c = 1$
$\rho = 2.5$ g cm^{-3}
$X_\infty = 0$
$X_c = 0.04$ (4 % H_2O by weight)
$f^*_{H_2O} = 700$ bars
$\bar{D} = 10^{-8}$ cm^{2}sec^{-1}
$L = 10^5$ cm

* The dimensionless groups are given the same names as in thermal convection problems to emphasize their analogous forms. The quotient Pr' is also called the Schmidt number (Equation 28).

magma (with 4% H_2O) or a volume of about 10^{17} cm³ (100 km³). The area in horizontal section is 3×10^{12} cm² so the cap of hydrated magma is 3×10^4 cm (0.3 km) thick. For lower values of $f_{H_2O}{}^*$ in Table 7 the conclusions are similar except for lower concentrations of H_2O in the hydrated layer.

Comparison of the upward velocities in Table 7 with downward velocities in Table 5 might suggest that the buoyancy effects would tend to cancel when thermal convection is taken into account. However, this would not be the case, because the hydration layer is much closer to the contact where velocities in the thermal boundary layer are much smaller than the maximum values. Also, Fig. 10 shows that within a time that is short compared to the duration of convection the temperature difference near the contact becomes small and hydration would control the density distribution. Therefore, the resultant velocity profile would show a form of countercurrent flow as illustrated schematically in Fig. 11. An additional effect of the thermal circulation is to supply continuously dry magma near the hydration front so that the concentration gradient is maintained. It is also noted that the composition gradient between the two currents will induce transport of other constituents into or out of the hydration layer depending on variations of the chemical activities with composition.

A more serious problem concerning longevity of the hydration source is whether the assumption of constant f_{H_2O} at the contact is realistic. If the country rock contains hydrous minerals, fugacities of a few hundred bars might exist until the available reactants become dehydrated. Estimates of temperatures in the contact zone suggest that a zone of relatively high vapor pressure could exist to a distance of a kilometer or so from the contact. If the rock contains roughly 1% H_2O, then 1 km of rock would supply the amount absorbed by the magma. Of course, there is also a tend-

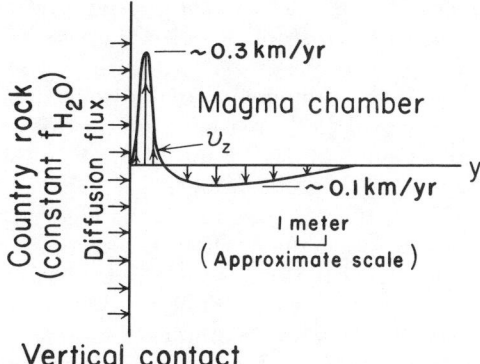

Fig. 11. Schematic illustration of countercurrent chemical and thermal convection. Velocities (v_z) are based on estimates in Tables 5 and 7. The chemical boundary layer is narrower than the thermal boundary layer, so chemical gradients leading to upward buoyancy forces near the wall may overbalance the small downward velocities of thermal convection that would exist in the absence of the density changes induced by the chemical gradient. The width and maximum downward velocity of the thermal boundary layer illustrated is for anhydrous granite magma (Table 5); the width and maximum upward velocity of the chemical boundary layer is based on absorption of 4% H_2O in the boundary layer from the country rock (Table 7).

ency for H_2O to migrate away from the contact zone toward cooler rock, but under deep-seated conditions the total transfer rate away from the contact is limited by the great distances involved, even for an effective diffusivity larger than that in the magma. For example, the data given by Brace et al. (1968) suggest that permeabilities under deep-seated conditions correspond to effective diffusivities of 10^{-4} or 10^{-5} cm² sec⁻¹. For these values the effective penetration distance during the time convection lasts in the magma is less than 1 km, suggesting that the controlling path for loss of H_2O from the contact zone is absorption by the magma.

Following solidification of the magma chamber, the gradient of f_{H_2O} is reversed during cooling because of the previous depletion in the contact zone. Retrograde reactions there and in the cooling pluton could then reestablish assemblages similar to those that existed initially, although the proportions of hydrous minerals would be decreased in proportion to the amount of H_2O concentrated at the top of the magma chamber.

If the initial value of f_{H_2O} is higher in the magma chamber than in surrounding rocks, the situation is reversed. Similar transport equations would apply, but the convection velocity gradients would be similar to those for thermal boundary layers. Two effects would dominate: The viscosity would increase in the chemical boundary layer with decreasing H_2O concentration, and the solidus temperature would be elevated relative to that of the hydrous magma. The combination of these two effects would cause rapid solidification of magma close to the contact. Continued supply of hydrous magma to the contact by thermal convection combined with loss of H_2O to a dry and permeable country rock would greatly accelerate the solidification rate, possibly bringing the solidification time close to the minimum value for perfect convective cooling.

Summary of Part II

It is shown that convective heat transfer in a cooling magma chamber can greatly increase the effectiveness of chemical diffusion as a mechanism by which the solidifying magma could become chemically zoned. Diffusion of H_2O in basalt and granite magmas and diffusion of SiO_2 in basalt magma are treated as specific examples. The behavior of other constituents depends on several factors that influence gradients of chemical potentials and permeabilities of wall rock composed of mineral assemblages that may act as sources or sinks of either actual or possible components of the magma. The changes in major element chemistry are likely to be small, but convection-induced changes could be important to interpretations of chemical variations in large magma bodies in which residual liquids are maintained in convective circulation for long times. In such cases, the local changes in H_2O content, minor element content, or isotopic composition might be profound, and any interpretation of the chemistry must take into account the possibility of convection-driven chemical diffusion both internally and between magma and country rocks.

The extent to which hydration-dehydration reactions occur during contact metamorphism is greatly increased by convective models of heat and mass transfer. Wall rocks dehydrated during contact metamorphism should show evidence of retrograde reactions, at least adjacent to areas of high water concentration in the magma, due to the reversal of the H_2O potential gradient across the contact during final cooling of the magma; near the upper contacts of shallow intrusives, retrograde reactions will also be influenced by encroachment of groundwater, which will eventually affect the contact and subsolidus mineral assemblages and compositions.

Acknowledgments

I thank R. L. Smith for supplying the obsidian used in this study and E. C. Robertson for careful preparation of the glass cylinders. I also thank D. B. Stewart and D. R. Wones for their generous cooperation in the use of experimental apparatus. Neither the diffusion study nor the earlier work on viscosity of obsidian could have been accomplished without the support of these people. The manuscript was read by E-an Zen, D. R. Wones, and T. L. Wright (all of the U.S. Geological Survey), C. Goetze (Massachusetts Institute of Technology), D. E. Anderson (University of Illinois), and A. Hofmann (Department of Terrestrial Magnetism, Carnegie Institution of Washington). I am grateful for their constructive suggestions.

References Cited

Barrer, R. M., and D. W. Brook, Molecular diffusion in chabazite, mordenite and levynite, *Trans. Faraday Soc.*, *49*, 1049–1059, 1953.

Bell, T., G. Hetherington, and K. H. Jack, Water in vitreous silica. Part 2: Some aspects of hydrogen-water-silica equilibria, *Phys. Chem. Glasses*, *3*, 141–146, 1962.

Bird, R. B., W. E. Stewart, and E. N. Lightfoot, *Transport Phenomena*, Wiley, New York, 780 pp., 1960.

Bottinga, Yan, and D. F. Weill, Densities of liquid silicate systems calculated from partial molar volumes of oxide components, *Amer. J. Sci.*, *269*, 169–182, 1970.

Bottinga, Yan, and D. F. Weill, The viscosity of magmatic silicate liquids: a model for calculation, *Amer. J. Sci.*, *272*, 438–475, 1972.

Bowen, N. L., Diffusion in silicate melts, *J. Geol.*, *29*, 295–317, 1921.

Bowen, N. L., *The Evolution of the Igneous Rocks*, Dover, New York, 334 pp., 1928.

Brace, W. F., J. B. Walsh, and W. T. Frangos, Permeability of granite under high pressure, *J. Geophys. Res.*, *73*, 2225–2236, 1968.

Burnham, C. W., and N. F. Davis, The role of H_2O in silicate melts I. P-V-T relations in the system $NaAlSi_3O_8$-H_2O to 10 kilobars and 1000°C, *Amer. J. Sci.*, *270*, 54–79, 1971.

Burnham, C. W., J. R. Holloway, and N. F. Davis, Thermodynamic properties of water to 1000°C and 10,000 bars, *Geol. Soc. Amer. Spec. Pap. 132*, 96 pp., 1969.

Carron, J., Recherches sur la viscosité et les phenomenes de transport des ions alcalins dans les obsidiennes granitiques, *Paris, Ecole Normale Superieure, Trav. Lab. Geol.*, no. *3*, p. 112, 1969.

Carslaw, H. S., and J. C. Jaeger, *Conduction of Heat in Solids*, 2nd ed., Oxford University Press, London, 510 pp., 1959.

Chandrasekhar, S., *Hydrodynamic and Hydromagnetic Stability*. Oxford University Press, London, 654 pp., 1961.

Clark, S. P., Jr., *Handbook of Physical Constants*, Geol. Soc. Amer., *Memoir 97*, 1966.

Cooper, A. R., Jr., Vector space treatment of multicomponent diffusion, this volume.

Crank, J., *The Mathematics of Diffusion*, Oxford University Press, London, 347 pp., 1957.

Darken, L. S., and R. W. Gurry, *Physical Chemistry of Metals*, McGraw-Hill, New York, 535 pp., 1953.

Drury, T., and J. P. Roberts, Diffusion in silica glass following reaction with tritiated water vapour, *Phys. Chem. Glasses*, *4*, 79–90, 1963.

Eitel, W., *Silicate Science*, Vol. 2, Academic Press, New York, 707 pp., 1965.

Friedman, Irving, W. Long, and R. L. Smith, Viscosity and water content of rhyolite glass, *J. Geophys. Res.*, *68*, 6523–6535, 1963.

Haller, W., Concentration-dependent diffusion coefficient of water in glass, *Phys. Chem. Glasses*, *4*, 217–220, 1963.

Hamilton, D. L., C. W. Burnham, and E. F. Osborn, The solubility of water and effects of oxygen fugacity and water content on crystallization in mafic magmas, *J. Petrology*, *5*, 21–39, 1964.

Hess, G. B., Heat and mass transport during crystallization of the Stillwater Igneous Complex, *Geol. Soc. Amer. Mem. 132*, 503–520, 1973.

Hetherington, G., and K. H. Jack, The oxidation of vitreous silica, *Phys. Chem. Glasses*, *5*, 147–149, 1964.

Jaeger, J. C., The temperature in the neighborhood of a cooling intrusive sheet, *Amer. J. Sci.*, *255*, 306–318, 1957.

Jaeger, J. C., Thermal effects of intrusions, *Rev. Geophys.*, *2*, 443–466, 1964.

Jaeger, J. C., Cooling and solidification of igneous rocks, in *The Poldervaart Treatise on Rocks of Basaltic Composition*, Vol. 2, H. H. Hess and A. Poldervaart, eds., Interscience, New York, pp. 503–536, 1968.

Koros, P. J., and T. B. King, The self-diffusion of oxygen in a lime-silica-alumina slag, *Trans. Met. Soc. AIME*, *224*, 299–305, 1962.

Kraichnan, R. H., Turbulent thermal convection at arbitrary Prandtl numbers, *Phys. Fluids*, *5*, 1374–1389, 1962.

Lee, R. W., and D. L. Fry, A comparative study of the diffusion of hydrogen in glass, *Phys. Chem. Glasses*, *7*, 19–28, 1966.

McBirney, A. R., Factors governing emplacement of volcanic necks, *Amer. J. Sci.*, *257*, 431–448, 1959.

Medford, G. A., Calcium diffusion in a mugearite melt. *Can. J. Earth Sci.*, *10*, 394–402, 1973.

Murase, T., and A. R. McBirney, Properties of some common igneous rocks and their melts at high temperature, *Geol. Soc. Amer., Bull.*, *84*, 3563–3592, 1973.

Ree, T., and H. Eyring, The relaxation theory of transport phenomena, in *Rheology*, Vol. 2, F. R. Eirich, ed., Academic Press, New York, pp. 83–144, 1958.

Rohsenow, W. M., and H. Y. Choi, *Heat, Mass, and Momentum Transfer*, Prentice-Hall, New Jersey, 537 pp., 1961.

Roseboom, E. H., oral communication, 1968.

Shaw, H. R., Obsidian-H_2O viscosities at 1000 and 2000 bars in the temperature range 700° to 900°C, *J. Geophys. Res.*, *68*, 6337–6343, 1963.

Shaw, H. R., Comments on viscosity, crystal setting and convection in granitic magmas, *Amer. J. Sci., 263,* 120–152, 1965.

Shaw, H. R., Rheology of basalt in the melting range, *J. Petrology, 10,* 510–535, 1969.

Shaw, H. R., Viscosities of magmatic silicate liquids: an empirical method of prediction, *Amer. J. Sci., 272,* 870–893, 1972.

Sucov, E. W., and R. R. Gorman, Interdiffusion of calcium in soda-lime-silica glass at 800° to 1308°C, *J. Amer. Ceram. Soc., 48,* 426–429, 1965.

Williams, E. L., and R. W. Heckman, Sodium diffusion in soda-lime-aluminosilicate glasses, *Phys. Chem. Glasses, 5,* 166–171, 1964.

Winchell, P., The compensation law for diffusion in silicates, *High Temp. Sci., 1,* 200–215, 1969.

PART II. REACTION KINETICS

Many reactions involving minerals are characterized by the nucleation of new phases or grains as well as their growth. The rate at which nuclei are formed, their rate of growth by diffusion, or both may limit the overall kinetics of a reaction. The dependence of nucleation and of growth rates on temperature is generally different and may even have opposite sign. Nucleation and growth rates should therefore be independently evaluated before kinetic data, for many reactions can be extrapolated to lower temperature by the Arrhenius relation.

Growth of nuclei or original grains is dependent on diffusion in the solid phases or through a surrounding fluid phase. Hence diffusion data such as discussed in Part I are applicable for calculating growth rates if the growth mechanism and certain geometrical factors are known. The papers in this section by Chai and by Chai and Anderson identify the mechanism of calcite recrystallization in aqueous solutions and model the change in grain size in order to calculate the amount of mass transfer during recrystallization. Nucleation of new grains does not contribute significantly to the process under the conditions of their experiments.

When a reaction involves nucleation as well as growth, kinetic data alone are usually insufficient to model the reaction. In addition, the present state of knowledge is insufficient for nucleation rates to be calculated for most mineral reactions, and would presuppose a detailed knowledge of how the nucleation occurs. One approach is to study the dependence of the nucleation rate on other parameters by performing isothermal experiments in which it is reasonable to expect that the growth rate is constant. Alternatively, if nucleation is restricted to the early part of a reaction, subsequent heat treatment of the material at different temperatures may be used to determine the activation energy for growth. These approaches are employed in the paper by McCallister.

Some solid state reactions may not involve a nucleation event even though a new phase is formed. Site ordering or disordering in a crystal may proceed by a homogeneous transformation without the formation of nuclei. Although the rate data for the transformation may be consistent with such a mechanism, it is difficult to distinguish this positively from a nucleation mechanism if the separation of the nuclei is only a few unit cells. Kinetic data for Al/Si disordering in alkali feldspars are treated in the paper by Sipling and Yund.

An alternative mechanism to classical nucleation for some exsolution reactions is the so-called spinodal decomposition discussed in numerous articles in the recent literature (see references in the paper by Yund). However, regardless of the exsolution mechanism, elastic strain plays an important role in limiting the composition of the exsolved phases if this strain is not relieved along the interface between the phases. The application of this concept to the alkali feldspars is treated in the paper by Yund.

Many factors may affect the distribution of nuclei and their rate of formation. For example, elastic strain in a mineral as a result of deformation may control the nature of the nucleation. This and many other factors have not been evaluated for most mineral reactions.

Numerous other techniques are important for studying mineral reactions. For example, transmission electron microscopy is especially valuable for providing information on the size, shape, orientation, and distribution of the nuclei. The omission of papers dealing with these and other techniques is due entirely to the need to limit the scope of this conference and should in no way detract from their importance. A separate conference devoted entirely to evaluating the mechanisms and kinetics of mineral reactions would be necessary to emphasize properly the variety of ways in which this general problem is being treated in current research.

R. A. Yund

COHERENT EXSOLUTION IN THE ALKALI FELDSPARS

R. A. Yund
Department of Geological Sciences
Brown University
Providence, Rhode Island 02912

ABSTRACT

Exsolution has been studied in single crystals of intermediate compositions of the microcline–low-albite series which were produced by alkali exchange and annealed between 500° and 750°C in the dry state at atmospheric pressure. The character and behavior of the exsolution process were determined from x-ray precession photographs of quenched samples at room temperature. The exsolved phases are coherent along a plane parallel to or near ($\bar{6}01$). The compositions of the exsolved phases as a function of temperature define a coherent solvus which lies within the classical or chemical solvus because of the elastic strain due to the coherency. The maximum of the coherent solvus is near Or_{33} and approximately 710°C. Exsolution is not observed outside the region bounded by the coherent solvus. Similar behavior is observed in crystals of the sanidine-analbite series, with exsolution restricted to below 600°C.

REVIEW OF COHERENT PHASE RELATIONS

This study is concerned with determining the nature of and the conditions required for exsolution in the alkali feldspars. The concepts used to explain the observed relations emphasize the importance of the coherent interface between the exsolved phases. Cahn (1962) discussed this situation for isotropic solids and the extension to anisotropic solids is not difficult on a qualitative level. Although coherency is not restricted to a spinodal decomposition mechanism, it is useful to consider coherency in relation to this mechanism since elastic strain plays an important role in both of these concepts. A more detailed discussion of spinodal decomposition can be found in Cahn (1968) or in the brief review by Yund and McCallister (1970).

The strain-free or chemical spinodal defines a curve on a temperature-composition phase diagram below which compositional fluctuations are unstable. These fluctuations would grow spontaneously if it were not for the strain associated with the mismatch in their lattices due to their compositional difference. When this strain term is included, the temperature-composition region in which spontaneous growth of these compositional fluctuations can occur is reduced; the curve bounding this region is called the coherent spinodal. The relative positions of a classical or chemical solvus (a), a strain-free or chemical spinodal (b), and a coherent spinodal (c) for a hypothetical system are shown in Fig. 1. Although there will be additional curves similar to c for different orientations of the compositional waves, we will consider the case where there is only one coherent spinodal in the temperature interval of interest.

Spinodal decomposition requires coherency of the phases, but classical, homogeneous nucleation may also produce coherent phases. Regardless of the mechanism by which exsolution occurs, the nature of the interface between the exsolved phases determines the composi-

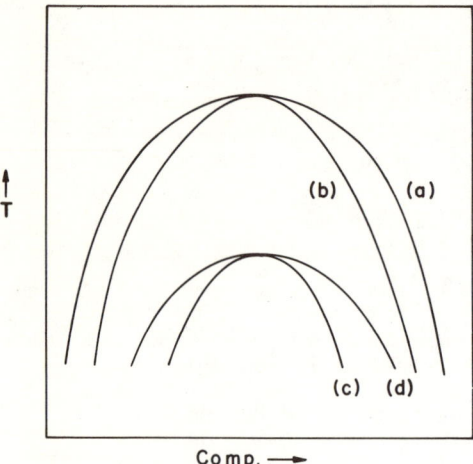

Fig. 1. A chemical (classical) solvus (*a*), strain-free or chemical spinodal (*b*), coherent spinodal (*c*), and coherent solvus (*d*) for a hypothetical system. After Cahn (1962).

tional relations of the phases. A coherent boundary between any two lamellae of the exsolved phases (which differ only in composition) is shown schematically in Fig. 2b. The unconstrained lattices are shown in Fig. 2a. In Fig. 2b the lattices are forced to match exactly across the interface, and the elastic strain within each phase is assumed to be homogeneous, that is, it is not restricted to a narrow region of the boundary. The lattice direction normal to this interface is continuous and straight. However, the spacing along this direction adjusts in each phase to minimize the total elastic strain energy. Neither the lattice spacings within nor those normal to the interface will be equal to their values in the unconstrained lattice. Any departure from total coherency, such as the introduction of dislocations, produces local strain and distortions along the interface. The boundary is then better described as being semicoherent. Complete loss of coherency is common in many exsolution processes.

The importance of coherency can be described with the aid of Fig. 1. If the boundaries remain coherent as the compositional difference of the phases in-creases, the coherency strain will increase until it balances the change in the Gibbs energy (chemical energy) of the system. This will result in compositions which are closer together than the strain-free compositions. The locus of these compositions defines a coherent solvus as shown by curve *d* in Fig. 1. This is in contrast to the chemical solvus (*a*) which is determined in classical phase equilibrium studies. A change in the temperature of the coherent phases requires an adjustment in their compositions along curve *d*. If the temperature for any bulk composition is raised above this curve, the

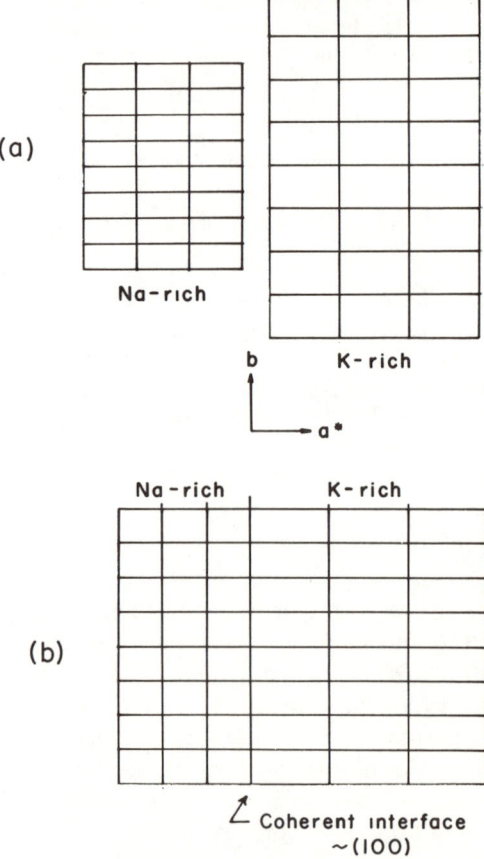

Fig. 2. Schematic representation of lattices for unconstrained Na-rich and K-rich feldspar (*a*) and of a coherent boundary between these phases (*b*). Note that the strain in *b* is homogeneous within each phase and all lattice dimensions are different from those in *a*.

distinction between the two phases is lost (reversion) even though the temperature of the sample is still below curve a. Thus the coherent solvus is analogous to the chemical or classical solvus. Both define the compositions of coexisting phases in a critical region. The difference is due to whether the phases are coherent or not. If the phases are initially coherent but later become incoherent, the compositional relations change from the coherent solvus to the chemical solvus.

The situation is actually somewhat more complicated than the above discussion indicates. Conjugate compositions for the coherent solvus depend on the bulk composition if the elastic constants vary with composition (Cahn, 1962). However, for the purpose of the present discussion we will ignore this effect.

Experimental Methods

Natural alkali feldspars were selected and their alkali ratios modified by exchange in molten chlorides. The microcline–low-albite series of experiments was done using a portion of the Amelia microcline which Lin and Yund (1972) used for alkali diffusion measurements. Preparation of end-member compositions and subsequent annealing of mixtures of these (Orville, 1967) did not result in homogeneous compositions even after four days at 950°C. The lack of chemical homogeneity in these preparations is probably due to the relatively large grain size (88–105 μm) and poor grain to grain contact during annealing. In order to achieve homogeneous alkali feldspars of intermediate compositions and of a size suitable for single crystal x-ray diffraction examination, it was necessary to exchange the feldspar in a molten mixture of KCl and NaCl at 1000°C for five days. The partition of alkalies between feldspar and melt was determined empirically in order to prepare the desired intermediate feldspar compositions.

A disordered sample of adularia from Kristallina, Switzerland (American Museum of Natural History no. 26545) was used for the sanidine-analbite series of experiments. It was first disordered at 1120°C for 15 days before exchange in a mixture of molten alkali chlorides.

The cell parameters for the exchanged and disordered samples were determined by a least-squares refinement of the x-ray powder diffraction data using the computer program of Evans, et al. (1963) and the method outlined by Wright and Stewart (1968).

The isothermal annealing experiments were done in unsealed silica-glass or platinum tubes with the samples exposed to the atmosphere. The total temperature variation and uncertainty is less than ±5° of the reported temperature.

Results and Interpretation

Nature of the Observations

Preliminary experiments indicated that exsolution of intermediate compositions could be achieved under certain conditions, but there was some variability in the exsolution rates for different grains from the same sample. This variability, combined with the poorer resolution and lack of information concerning coherency from the x-ray powder diffraction, made the use of single-crystal x-ray techniques necessary.

After annealing for a few hours below 600°C, there is an elongation of the reflections parallel to a^* on an hk0 level precession photograph. This elongation indicates that the crystal is no longer compositionally homogeneous and a two-phase structure has started to develop. With further annealing each elongated reflection separates and is replaced by two distinct reflections which correspond to the sodium- and potassium-rich phases. However, streaking between these reflections indicates that the phases are not yet homogeneous. With further annealing the separation between the reflections increases and they become sharper as the compositional gradients in each phase are eliminated. The distance between the two reflections in a pair increases outward from the center of the photograph, which indicates that this separation is due to composition and is not the same as the side bands observed by Owen and McConnell (1974). The time required for virtually complete separation of the reflections depends on the annealing temperature because it is determined by the diffusivities of the alkali ions. (See Fig. 4 and later discussion of this figure.)

Along b^* and approximately c^* there is no elongation or separation of the reflections. They remain sharp throughout the annealing. Thus the lattices of the two phases are constrained to be the same in these directions—the plane of coherency. Even though b^* and c^* are less sensitive to composition than a^*, some separation or broadening of the reflections along these directions is observed when the phases are incoherent. However, there was no indication of loss of coherency in any of the crystals annealed in this study.

Orientation of the Coherent Interface

The approximate orientation of the interface can be determined from single-crystal photographs by using the method outlined by Laves (1952). Measurements of precession photographs for both the microcline–low-albite series and the sanidine-analbite series indicate that the plane of coherency is parallel to $(\bar{6}01)$ within experimental error of 2 to 4 degrees. Because a^* is within approximately 9° of the normal to this plane and this parameter is the most sensitive to composition, hk0 level precession photographs were used for most of the observations in this study.

The orientation of a coherent interface is controlled by the strain associated with the lattice mismatch and by the elastic anisotropy. If a nonisometric crystal has two perpendicular directions for which the change in length per unit composition is small compared to that for the direction normal to this plane, development of the interface parallel to the directions of similar spacing will result in the smallest elastic strain. This is the basis for the optimal phase boundary argument of Bollmann and Nissen (1968). The elastic anisotropy of the phases may require additional adjustment in the orientation of the interface to minimize the strain energy. However, if the strain due to the variation in the lattice parameters with composition is not clearly minimal for one plane, then even the approximate orientation of the lamellar structure cannot be predicted without consideration of the elastic anisotropy. Of course, the above discussion is independent of whether the lamellae form by spinodal decomposition or classical nucleation.

For alkali feldspar, the lattice strain along b and c is only about 1.3% and 0.9%, respectively, whereas along a it is 5.4% for a compositional interval of 100 mole percent. Clearly the elastic strain will be minimized for an interface nearly parallel to the bc plane. The exact orientation cannot be predicted by this simple argument which does not properly consider the low symmetry of feldspar and which ignores the elastic anisotropy. However, the elastic anisotropy of feldspar is not large enough to cause a major change in the orientation of the interface that corresponds to the minimal elastic strain. The observed orientation of $(\bar{6}01)$ is consistent with this general prediction.

Composition of the Phases

Measurement of reciprocal lattice parameters such as a^* can be used to determine only approximate or apparent compositions of the phases because the coherency distorts the lattice spacings (Fig. 2b). Smith (1961) recognized this problem and proposed a method for correcting the apparent compositions determined from cell parameters. This method assumes that the cell volumes are unchanged even though the lattice spacings

TABLE 1. Cell parameters of exchanged microcline (Mi_{45} and Mi_{33}) and a disordered adularia (Sa_{100})

Sample	a	b	c	α	β	γ
Mi_{45}	8.392 (6)	12.928 (4)	7.210 (4)	91° 19′ (3)	115° 44′ (4)	87° 42′ (3)
Mi_{33}	8.297 (8)	12.889 (7)	7.188 (4)	92° 31′ (6)	116° 23′ (7)	87° 38′ (5)
Sa_{100}	8.588 (2)	13.030 (3)	7.180 (2)	...	116° 03′ (1)	...

One standard deviation of the last digit is shown in parentheses.

of the coherent phases are different from those of incoherent phases of the same compositions. This assumption is incorrect, and J. Tullis (in press) used crystal elasticity to calculate the complete strain tensor in order to determine the changes in the cell parameters due to coherency. Her method is strictly valid only for monoclinic phases and can be applied to the microcline–low-albite series only as an approximation. Although we will continue to report apparent compositions determined from a^* and Orville's (1967) data for the variation of a^* with composition, the best estimates of the true compositions from Tullis's calculations are indicated in parentheses for the first two entries in Table 3.

The Microcline–Low-Albite Series

The cell parameters for two of the intermediate compositions produced from this material by alkali exchange are listed in Table 1. On a b-c plot (Wright and Stewart, 1968), there is no indication within the experimental errors of a change in the Al-Si order associated with the alkali exchange. Thus the chemical solvus for microcline–low-albite determined by Bachinski and Müller (1971) is applicable to the present samples. The solvus based on Bachinski's alkali exchange experiments is shown as curve a on Fig. 3. (See also discussion in last section of this paper.)

The experimental results for different crystals annealed at 600°C for 4 to 1608 hours are given in Table 2. Four of the individual crystals were annealed, a precession photograph taken, and then the sequence repeated at a different temperature or for a longer time. These results are given in Table 3, and portions of the precession photographs for five of these samples are shown in Fig. 4. Albite twinning causes the doubling of the reflections (right and left) along a^* (vertical axis). Most of the pericline twinning is lost during alkali exchange but some remains and causes the vertical streaking along b^* (horizontal axis). Exsolution gives rise to a doubling of reflections which is best seen as a vertical separation of the reflections along a^*. The apparent separation of the reflections along b^* is due to resolution of copper $K\alpha_1$ and $K\alpha_2$.

Exsolution can be observed in these crystals after only a few hours above

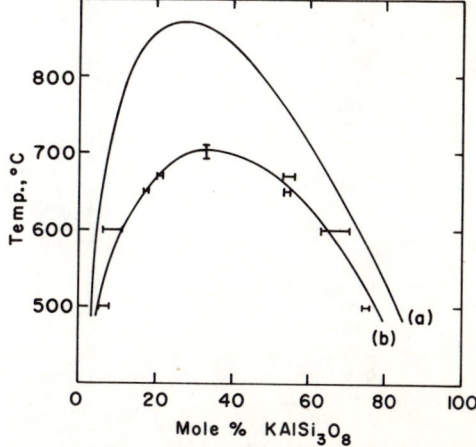

Fig. 3. Approximate location of the coherent solvus (b) for the microcline–low-albite series. The vertical and horizontal bars define this curve. The compositions are uncorrected for coherency strain (see text). The chemical solvus (a) is from Bachinski and Müller (1971).

TABLE 2. Results of isothermal annealing of single crystals of the microcline–low-albite series at 600°C

Initial Composition (mole % $KAlSi_3O_8$)	Time (hours)	Apparent Compositions (mole % $KAlSi_3O_8$)	
		Na-rich	K-rich
45	4	No exsolution in 3 crystals	
		18	56 4th crystal
45	192	17	63
45	480	6	65
		11	71
		11	64
		11	63
		7	68
45	1608	11	64
		9	65
33	4	No exsolution in 3 crystals	
		Start of exsolution in 4th	
33	48	17	63
33	168	15	67
		11	64
33	672	12	71
12	192	No exsolution	

TABLE 3. Results for the sequential heating and examination of single crystals of the microcline–low-albite series

Temperature (°C)	Time (hours)	Apparent Compositions (mole % $KAlSi_3O_8$)	
		Na-rich	K-rich
	Crystal No. 1, 33 mole %		
670	48	20 (22)*	54 (51) (a)†
600	48	11 (13)	77 (71) (b)
670	48	22	53 (c)
700	48	nearly homogeneous	
500	48	some exsolution	
500	96	5	76‡ (d)
670	96	21	56
710	63	homogeneous	(e)
	Crystal No. 2, 33 mole %		
750	24	no exsolution	
700	24	start of exsolution	
650	24	18	55
	Crystal No. 3, 33 mole %		
700	6	no exsolution	
690	6	no exsolution	
680	6	start of exsolution	
670	6	reflections not separated	
660	6	19	43
660	12	21	48
650	6	17	53
	Crystal No. 4, 55 mole %		
500	1008	8	74
650	59	homogeneous	
600	168	12	63

*Numbers in parentheses are corrected compositions (Tullis, in press).
†Letters correspond to the precession photographs in Fig. 4.
‡Not complete exsolution; see Fig. 4 (d).

650°C, and even at 600° some crystals have partially exsolved in four hours. Exsolution is significantly slower at 500°C. The time required to reach steady state compositions is not more than about 24 hours at the above 600°C. A crystal of 33 mole percent $KAlSi_3O_8$ annealed at 500°C for four days still shows streaking between the reflections parallel to a^* (Fig. 4e). This indicates that major

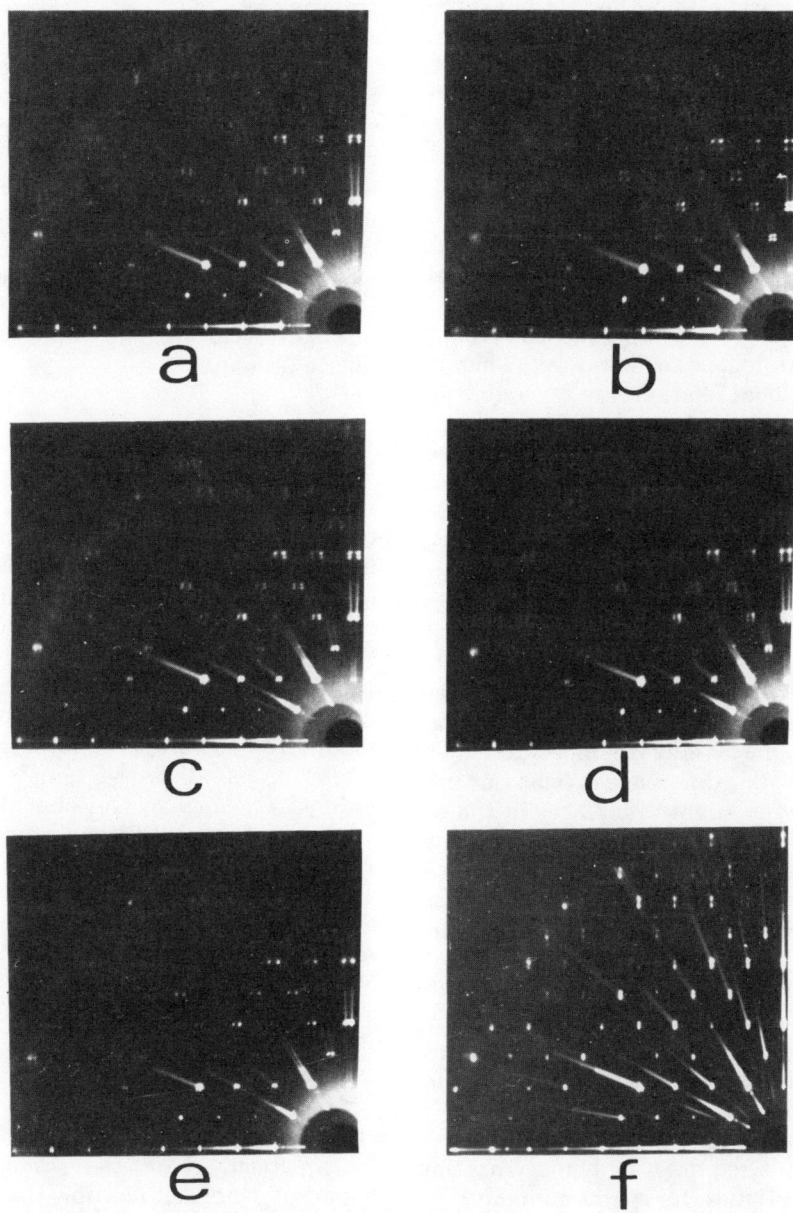

Fig. 4. Portions of precession photographs of annealed samples. a^* is vertical, b^* is horizontal. Photographs a through e are from the microcline–low-albite series and correspond to the samples identified in Table 3. Photograph f is a sample from the sanidine–analbite series (Or_{41}) which was annealed at 500°C for 2000 hours.

compositional gradients still exist within the phases.

Except for very short annealing times, the x-ray reflections are characteristically sharp for samples annealed above 600°C. The sharpness and the absence of streaking argue that the two phases are essentially homogeneous and that the interface is compositionally sharp. The absence of any noticeable asterism of the reflections also argues for homogeneous strain rather than local strain along the interface.

It is clear from the data in Table 3 that the compositional relations of the exsolved phases vary in a regular manner and are reproducible regardless of whether they are approached from a lower or a higher temperature. Thus the compositional relations are reversible and define a coherent solvus. The approximate location of this boundary is shown by curve b on Fig. 3. These are the apparent compositions determined from a^* and are uncorrected for coherency strain.

Although the uncertainty in determining the apparent compositions of the phases is ± 2–3 mole percent, the data in Table 2 illustrate that there is a greater variation than this in the compositions of the exsolved phase for different crystals of the same average composition. The reason for this difference in behavior may be related to slight differences in the degree of Al-Si order in the crystals. Since the Gibbs energy varies sharply with the Al-Si ordering whereas the strain energy associated with the coherency does not, the conjugate compositions of the coherent phases change markedly with the degree of Al-Si order in the crystal. This is clearly demonstrated by the temperature depression of the coherent solvus for the sanidine-analbite series which is considered in the next section.

The time required for homogenization, the diffusivities for potassium and sodium in feldspar (Bailey, 1971; Lin and Yund, 1972; Petrović, 1972; and Foland, this volume), and a numerical solution for the diffusion equation with moving planar interfaces (Tanzilli and Heckel, 1968) can be used to estimate the lamellae thickness. By this method the thickness is estimated to be on the order of a few hundred angstroms. Confirmation and detailed study of the lamellae by transmission electron microscopy are needed.

No exsolution was observed above about 710°C for Or_{33} even after four weeks. Judging from the shape of the chemical solvus, this composition is probably close to the maximum temperature of the coherent solvus as shown on Fig. 3. Although the supersaturation is still large, incoherent nucleation is extremely slow and does not provide a competing mechanism in the dry state at atmospheric pressure.

Sanidine-Analbite Series

Some preliminary results are available for the sanidine-analbite series using intermediate compositions produced from a disordered adularia. The cell parameters of this disordered material (Table 1) indicate that it belongs to the high-sanidine structural state (Wright and Stewart, 1968). The study of this series is more difficult because of the lower temperature required for exsolution and because some grains prepared at the same time (disordered, exchanged, and annealed) show evidence of exsolution and others do not. For a bulk composition of Or_{41}, no grains appear to have exsolved at 600° or 640°C after 4800 and 7200 hours, respectively. After about 2000 hours at 500° and 550°C, between 10% and 30% of the grains have exsolved (see Fig. 4f). Powder x-ray diffraction was used to estimate these percentages.

The apparent compositions for one crystal exsolved at 550°C were 25 and 54 mole percent $KAlSi_3O_8$. At 500°C the apparent compositions for two exsolved crystals were 17–67 and 21–62 mole percent $KAlSi_3O_8$. One of these two crystals rehomogenized when heated at 550°C for 144 hours; the other did not rehomogenize at 550° but it did rehomogenize at 570°C. One crystal was reannealed at 500°C for 168 hours and showed the beginning of exsolution—streaking parallel to a^*. After 336 hours separate reflections were visible although there was still streaking between them.

Until the reason for the difference in the behavior of these crystals is understood, it is impossible to outline the compositional-temperature limits for exsolution as was possible for the microcline–low-albite series. These data appear to indicate however, a similar exsolution behavior for the high-sanidine–analbite series.

Conclusions and Discussion

Bachinski and Müller's (1971) solvus for the microcline–low-albite series was determined using various experimental techniques; however, none of their experiments involved exsolution from a single homogeneous phase. (Müller's "unmixing" experiments used two-phase mechanical mixtures.) Thus, they have determined the chemical or classical solvus. Their independent determinations of this solvus yield somewhat different results which are probably due in part to a different degree of Al-Si order in their starting materials. The lower of their two curves, based on Bachinski's alkali exchange experiments, is shown on Fig. 3. The difference between this chemical solvus and the coherent solvus determined in this study is clearly demonstrated. Comparison with their other curve based on Müller's experiments increases the difference. Furthermore, the difference is actually greater than shown on Fig. 3 because the true compositions of conjugate pairs on the coherent solvus are actually closer together when corrected for the coherency strain. Although Tullis's (in press) method can be used to approximately correct these compositions (see Table 3), it seems advisable to retain the apparent compositions until an exact correction can be applied.

Most experimental studies of the sanidine-analbite series have not included exsolution experiments and hence have determined the chemical solvus. Smith and MacKenzie (1958) did exsolution experiments with a cryptoperthite (Spencer P) and obtained a solvus which was very different from those determined previously. It is clear now that their solvus (their Fig. 3) is actually the coherent solvus. The preliminary data given in this paper are in fair agreement with their values for the sodium-rich limb of this solvus. Their compositions were determined using the $(\overline{2}01)$ reflection and are approximately 10 mole percent more potassium rich than the values determined here. It is clear that the distinction between the chemical and coherent solvus curves is real, but additional experimental study is needed to establish the coherent solvus more accurately. The coherent solvus for the sanidine-analbite series has greater mineralogical significance because exsolution probably occurs before Al-Si ordering in most instances. The occurrence of coherent submicroscopic perthites (Smith, 1961; Laves, 1952; Brown et al., 1972; Brown and Willaime, 1972; and others) demonstrates that under certain geological conditions the coherency between the phases can persist indefinitely.

The type of exsolution behavior observed in the alkali feldspars is different from that observed in many solid solutions. In other materials the interface between the phases often becomes incoherent before the compositions of the phases reach the coherent solvus. In addition, incoherent nucleation and growth is often a competing mechanism and thus exsolution is observed anywhere below the chemical solvus.

For the alkali feldspars in this study, the coherent nature of the interface is indicated by the x-ray single-crystal photographs which show no separation of the reflections along b^* and approximately c^* (which are directions in or near the interface). In addition, the compositions of the exsolved phases do not reach the chemical solvus. However, the best measure of how coherent the interfaces remain is provided by the reversibility of the compositional relations with successive heating at different temperatures. If the boundaries became incoherent, or even semicoherent, reversibility would be lost and higher temperatures would be required to achieve rehomogenization following exsolution and annealing at a lower temperature.

The fact that the phases are coherent is not in itself an argument for spinodal decomposition. Classical nucleation can also produce coherent phases. Nearly regular, parallel lamellae on any scale are not diagnostic of the exsolution mechanism either. Homogeneous, copious nucleation as well as spinodal decomposition would produce this texture.

The principal observation in support of spinodal decomposition in this study is the gradual change from a homogeneous phase with single sharp h00 reflections, followed by elongation of the reflections approximately parallel to a^*, and the final development of two distinct reflections. However, these observations could be explained by classical nucleation with an initial small difference in composition of the phases (low-amplitude square waves) as well as the gradual growth in amplitude of sinusoidal waves as required by spinodal decomposition theory. Although it is tempting to suggest that spinodal decomposition may be operative in the initial exsolution process of the alkali feldspars, the evidence provided by this study is not conclusive.

Owen and McConnell (1971 and 1974) have studied hydrothermally annealed samples by transmission electron microscopy techniques. Our studies complement each other in many ways, but there are also some significant differences. They observe satellite reflections ("side bands") in their electron diffraction patterns. Satellite reflections were not observed in this study. The reason for these differences in the diffraction patterns must be related to the nature of the lamellar structure, but clarification of this point is needed. Another difference is the degree of Al-Si ordering. Their samples were probably slightly ordered but closer to the sanidine-analbite series.

ACKNOWLEDGMENTS

This study was supported by National Science Foundation Grant GA-21145. I especially wish to thank Dr. J. Tullis for many helpful discussions during the course of this study.

REFERENCES CITED

Bachinski, S. W., and G. Müller, Experimental determinations of the microcline–low-albite solvus, *J. Petrology, 12,* 329–356, 1971.

Bailey, A., Comparison of low-temperature with high-temperature diffusion of sodium in albite, *Geochim. Cosmochim. Acta, 35,* 1073–1081, 1971.

Bollmann, W., and H.-U. Nissen, A study of optimal phase boundaries: the case of exsolved alkali feldspars, *Acta Crystallogr., A24,* 546–557, 1968.

Brown, W. L., C. Willaime, and C. Guillemin, Exsolution selon l'association diagonale dans une cryptoperthite: étude par microscopie électronique et diffraction des rayons X, *Bull. Soc. Fr. Minéral. Cristallogr., 95,* 429–436, 1972.

Brown, W. L., and C. Willaime, An explanation of exsolution orientations and residual strain in cryptoperthites, *Proc. NATO Special Conf. on Feldspars,* Manchester, 1972, in press.

Cahn, J. W., Coherent fluctations and nucleation in isotropic solids, *Acta Met., 10,* 907–913, 1962.

Cahn, J. W., Spinodal decomposition, *Trans. Met. Soc. AIME, 242,* 166–180, 1968.

Evans, H. T., Jr., D. E. Appleman, and D. S. Handwerker, The least squares refinement of crystal unit cells with powder diffraction data by an automatic computer indexing method (abstr.), Amer. Crystallogr. Ass. Meeting, Cambridge, Mass., March 28, 1963, 42–43.

Foland, K. A., Alkali diffusion in orthoclase, this volume.

Laves, F., Phase relations of the alkali feldspars, II. The stable and pseudostable phase relations in the alkali feldspar system, *J. Geol.*, *60*, 549–574, 1952.

Lin, T.-H., and R. A. Yund, Potassium and sodium self-diffusion in alkali feldspar, *Contrib. Mineral. Petrol.*, *34*, 177–184, 1972.

Orville, P., Unit cell parameters of the microcline–low-albite and the sanidine–high-albite solid-solution series, *Amer. Mineral.*, *52*, 55–86, 1967.

Owen, D. C., and J. D. C. McConnell, Spinodal behaviour in an alkali feldspar, *Nature Phys. Sci.*, *230*, 118–119, 1971.

Owen, D. C., and J. D. C. McConnell, Spinodal unmixing in an alkali feldspar. *The Feldspars, Proc. NATO Special Conf. on Feldspars*, 1972, Manchester University Press, 424–439, 1974.

Petrović, R., Alkali ion diffusion in alkali feldspar, Ph.D. thesis, Yale University, 1972.

Smith, J. V., Explanation of strain and orientation effects in perthites, *Amer. Mineral.*, *46*, 1489–1492, 1961.

Smith, J. V., and W. S. MacKenzie, The alkali feldspars: IV. The cooling history of high-temperature sodium-rich feldspars, *Amer. Mineral.*, *43*, 872–889, 1958.

Tanzilli, R. A., and R. W. Heckel, Numerical solutions to the finite, diffusion-controlled, two-phase, moving-interface problem (with planar, cylindrical, and spherical interfaces), *Trans. Met. Soc. AIME*, *242*, 2313–2321, 1968.

Tullis, J., Elastic strain effects in coherent perthitic feldspars, *Contributions to Mineralogy and Petrology*, in press.

Wright, T. L., and D. B. Stewart, X-ray and optical study of alkali feldspar: I. Determination of composition and structural state from refined unit-cell parameters and 2V, *Amer. Mineral.*, *53*, 38–87, 1968.

Yund, R. A., and R. H. McCallister, Kinetics and mechanisms of exsolution, *Chem. Geol.*, *6*, 5–30, 1970.

KINETICS OF Al/Si DISORDERING IN ALKALI FELDSPARS

Philip J. Sipling and Richard A. Yund
Department of Geological Sciences
Brown University
Providence, Rhode Island 02912

ABSTRACT

The rate of the microcline–sanidine transformation was studied for the Hugo microcline, its K-exchanged equivalent, and K-exchanged Amelia albite. The reaction was followed by variation in $2\theta_{131} - 2\theta_{1\bar{3}1}$ with time under isothermal conditions in the temperature range 970°–1120°C at atmospheric pressure. The disordering rate increased with a decrease in grain size and with an increase in the Na/K ratio. The average activation energy of 99 ± 10 kcal/mole for these potassium-rich feldspars is significantly higher than the 74 ± 1 kcal/mole for the Amelia albite (McKie and McConnell, 1963).

Analysis of these data and those of Müller (1970) confirms that the Al/Si disordering proceeds as a homogeneous transformation involving simultaneous transfer of Al from $T_1(0)$ into the three other tetrahedral sites.

Under hydrothermal conditions the rate of the homogeneous transformation is not measurably altered. Thus it remains to be demonstrated that water alters the rate of microcline disordering except by dissolution and reprecipitation.

Introduction

Although there has been previous experimental work on the microcline-sanidine transformation, it was the purpose of this investigation to study quantitatively the kinetics of this disordering reaction. Previous studies of this transformation in alkali feldspars include dry isothermal heating experiments reported by Goldsmith and Laves (1954), Baskin (1956), and Müller (1970). Similar experiments for the low-albite–high-albite transformation were described by Baskin (1956), Schneider (1957), and McKie and McConnell (1963). All of these studies report the formation of metastable intermediate structural states as a result of disordering of Al and Si atoms in the tetrahedral sites. Previous quantitative kinetic studies on alkali feldspars have been limited to the sodium end-member; rates for dry disordering are given by McKie and McConnell (1963) and rates for ordering under hydrothermal conditions are given by McConnell and McKie (1960). In this study we started with potassium-rich compositions and investigated the kinetics of the microcline-sanidine transformation under dry conditions as a function of temperature, grain size, and composition.

Hydrothermal conversion of potassium-rich feldspars from low to high structural states was reported by Goldsmith and Laves (1954) and Tomisaka (1962). Under similar conditions of temperature and pressure, Goldsmith and Laves (1954) observed the dissolution of microcline and the reprecipitation of sanidine, whereas Tomisaka (1962) reported the presence of intermediate structural states. We have performed several hydrothermal experiments for comparison with these conflicting results.

Procedure

The starting materials were microcline from the Hugo pegmatite, Black Hills, South Dakota, and albite from Amelia County, Virginia. After homogenization of the slightly perthitic microcline, the composition was Or_{89} according to the microcline–low-albite d-spacing curve of Orville (1967). Potassium end-members were prepared for both the microcline and albite by exchanging the alkalies with molten KCl in a platinum crucible at 900°C for 48 hours. The albite was recovered after the first exchange, rinsed well, and exchanged again.

The samples were heated at temperatures between 970° and 1120°C at a pressure of one atmosphere in unsealed Pt capsules. Temperature control was ±5°C and quenching took less than one minute. A portion of the sample was removed at regular intervals for a diffractometer smear mount and the specimen returned to the furnace.

The transformation was monitored by measuring $\Delta 2\theta$ ($\equiv 2\theta_{131} - 2\theta_{1\bar{3}1}$) using CuK_α radiation. The recorded $\Delta 2\theta$ value is the mean of four oscillations at $\frac{1}{4}°2\theta$/minute. With an increase in heating time these reflections move toward each other, decrease in intensity, and broaden relative to those of the starting material. Additional information was obtained using single crystal precession techniques. These were employed to determine the variation in the rate of transformation among crystals run in the same charge.

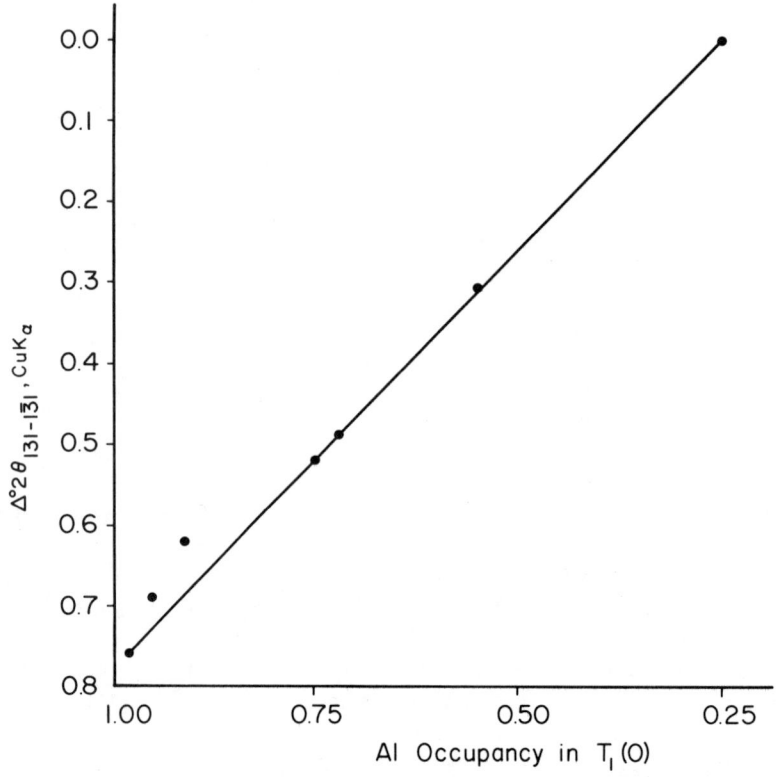

Fig. 1. Comparison of the disordering parameter $\Delta 2\theta_{131-1\bar{3}1}$ with the Al occupancy in the $T_1(0)$ tetrahedral site determined by using the method of Stewart and Ribbe (1969). The point for maximum disorder is the theoretical point for sanidine. All other points are experimental points.

REACTION KINETICS

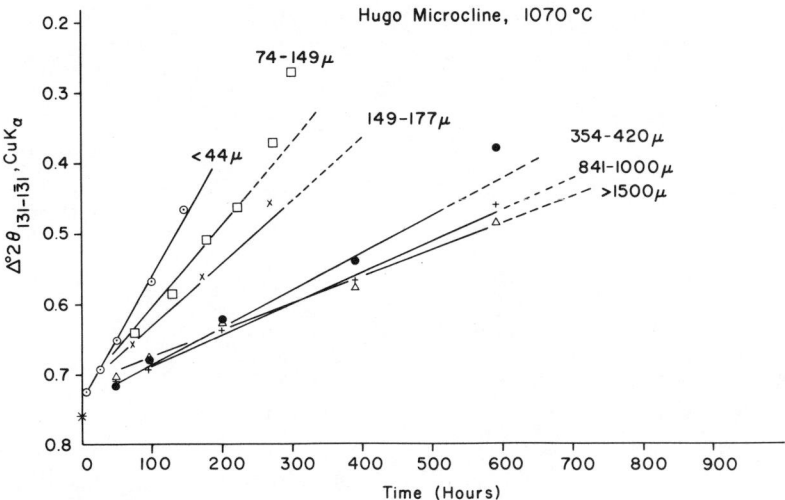

Fig. 2. Rate of disordering as a function of grain size. All data are for the Hugo microcline annealed at 1070°C. The asterisk represents the original Δ2θ value (0.76).

Results and Discussion

Anhydrous Experiments

Disordering Parameter. The use of the Δ2θ parameter as a measure of the Al/Si distribution was tested by comparing it to the calculated $T_1(0)$ site occupancy. Al occupancy in $T_1(0)$ was calculated by the method of Stewart and Ribbe (1969) using cell refinements for Hugo microcline annealed at 1045°C for various times. These results are shown in Fig. 1 with a straight line connecting the point for the starting material and the theoretical point for maximum disorder. The fit of the other points to this line supports the use of Δ2θ as a direct quantitative measure of the disordering reaction.

Effect of Grain Size. In order to determine the effect of grain size on the rate of disordering, several size fractions were sieved from a powdered portion of the Hugo microcline and isothermally heated to 1070°C. The data for six different grain sizes are shown on Fig. 2. Several observations of the results shown on Fig. 2 can be made. First, none of the curves extrapolate back to the initial Δ2θ value. There appears to be a small but real initial period of rapid disordering. The cause of this "initial jump" is not known. Secondly, for α greater than approximately 0.4 (see later section) there is an apparent increase in disordering rate. With broadening as well as merging of the reflections, it is difficult to measure the true peak position after the transformation has proceeded to this extent. More important, the rate of transformation should decrease with time for either a homogeneous or a nucleation and growth type of transformation (Christian, 1965, p. 16). Therefore, extrapolation of the linear portions of the curves should provide a minimum time for total transformation. Only the data points defining the linear portions of the curves were used to calculate the disordering rates which are given in Table 1.

As shown on Fig. 2, the disordering rate is clearly dependent on the grain size of the sample. The disordering rate for the $<44~\mu$ fraction was four to five times that for the $>1500~\mu$ fraction, with intermediate rates for intermediate grain sizes.

Although the effect of grain size is clear, the cause is not. This effect would be more compatible with a nucleation-

TABLE 1. Disordering rates

Sample	Temperature (°C)	Rate (° Δ 2θ/hour) (<44 μ)	(>1500 μ)
Hugo microcline	1070	0.00177 (11)	0.00038 (3)
Hugo microcline	1045	0.00127 (3)	0.000263 (15)
Hugo microcline	1020	0.00058 (5)	0.00011 (2)
Hugo microcline	995	0.000241 (10)	0.000062 (10)
Hugo microcline	970	0.000061 (6)	0.000020 (7)
K-Hugo microcline	1120	0.00159 (7)	
K-Hugo microcline	1070	0.00037 (3)	
K-Hugo microcline	1045	0.00022 (1)	
K-Hugo microcline	1020	0.00013 (1)	
K-Hugo microcline	995	0.000061 (5)	
K-Hugo microcline	970	0.000028 (8)	
K-Amelia albite	1120	0.00146 (13)	
K-Amelia albite	1070	0.00056 (5)	
K-Amelia albite	1020	0.00012 (1)	

Sample	Temperature (°C)	Size Fraction (μ)	Rate (° Δ 2θ/hour)
Hugo microcline	1070	74–149	0.00118 (7)
Hugo microcline	1070	149–177	0.00088 (5)
Hugo microcline	1070	354–420	0.00053 (4)
Hugo microcline	1070	841–1000	0.00045 (2)

and-growth process than with a homogeneous transformation. The larger proportion of surface area in the smaller grains would provide more nucleation sites, but would not affect a homogeneous transformation unless it makes it easier for the Al and Si atoms to interchange. However, the data from this study and that of Müller (1970) indicate that the transformation is probably homogeneous (see later section). A deformed or disturbed outer rim of the crystal might account for the observed increase in rate with decreasing grain size, if this rim were thick enough. Whatever the explanation, the results clearly indicate the need to consider grain size in solid state kinetic studies.

Effect of Composition. In order to investigate the effect of the Na/K ratio on the rate of disordering, both a Hugo microcline and an Amelia low albite were exchanged in molten KCl to essentially Or_{100}. The rate data for these materials are shown on Fig. 3, using the parameter α instead of $\Delta 2\theta$; α is defined as the volume fraction transformed and is calculated from the initial and final values for $\Delta 2\theta$. Use of this parameter makes it possible to compare the present data with those for Amelia low albite (McKie and McConnell, 1963) for which $\Delta 2\theta$ increases as the sample becomes progressively disordered.

As shown on Fig. 3, the disordering rate for the Hugo microcline is about four times that for the K-exchanged Hugo microcline of the same grain size and heating temperature. It is significant that the rates for the K-exchanged Hugo microcline and the K-exchanged Amelia low albite are approximately equal. The rate for unexchanged Amelia albite given by McKie and McConnell (1963) is plotted for comparison with that for Hugo microcline of a similar grain size and heating temperature. Although these grain sizes and heating temperatures are not identical, the results clearly indicate that the Amelia low albite disorders more rapidly than the potassium-rich microclines.

The data in Fig. 3 show a clear relation between the Na/K ratio and the disordering rate of the alkali feldspar. One possible explanation for this compositional effect concerns the location and coordination of the alkali ion in the feld-

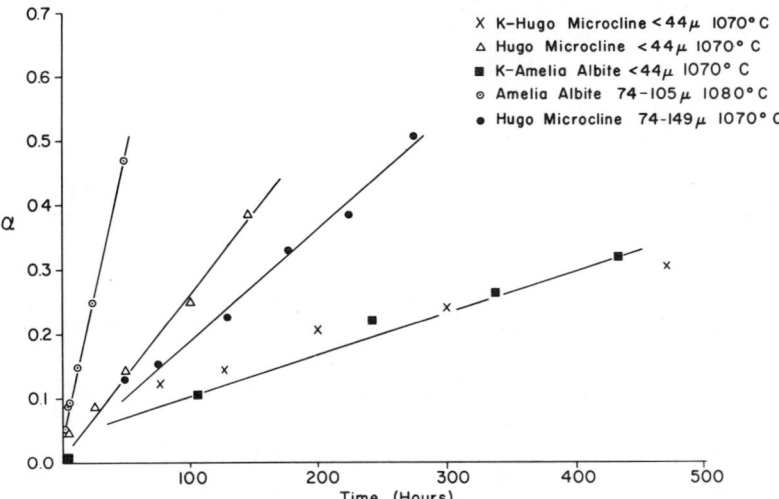

Fig. 3. Rate of disordering as a function of composition. The curve for the K-Hugo microcline has been omitted for clarity.

spar structure (Ribbe et al., 1969). The tetrahedral-oxygen (T-O) bond lengths for those oxygens also bonded to Na in low albite are generally longer than the equivalent bonds in maximum microcline (Jones and Taylor, 1968). These longer T-O bonds in albite could be responsible for the faster Al/Si interchange as well as a lower activation energy for disordering.

The displacement of the Na atom toward one end of its ellipsoidal site (Ribbe et al., 1969) could further weaken those T-O bonds nearest to the Na atom. This displacement could further enhance those effects just described.

Although the critical disordering temperature for low albite and microcline are not known, the temperature for albite (Smith, 1972) is probably somewhat higher than that for microcline (Goldsmith and Laves, 1954; Steiger and Hart, 1967). Thus the difference in their critical temperatures is not the explanation for the faster disordering rate of albite.

The disordering rate may also be a function of parameters other than grain size and composition. Müller (1970) reports that different regions of the same crystal transform at different rates. In the present study, the 149–177 μ size fraction annealed at 1070°C for 341 hours appeared to be only approximately half transformed from the $\Delta 2\theta$ measurement, whereas a single crystal c-axis precession photograph of a selected crystal indicated complete transformation. The reason for the variation in rate of different crystals is not known.

Activation Energy. Activation energies for the transformation were calculated from the rate data; they are listed in Table 2 and are shown on an Arrhenius diagram in Fig. 4. The activation energy for the K-Amelia albite was calculated on the basis of only three temperatures; these data are not shown on Fig. 4. The 970° and 1070°C data for the <44 μ Hugo microcline do not fit a straight line as well as the data for the other samples. Except for this sample, the activation energies agree within the uncertainties. Because of the relatively large uncertainties, it is not clear whether this difference is real. On the basis of the present data we prefer to use the average of all four values, 99 kcal/mole. Rather than using the standard method to estimate the uncertainty of this average, we prefer a more generous estimate of ±10 kcal/

TABLE 2. Calculated activation energies

Sample	Grain Size	Activation Energy (kcal/mole)
Hugo microcline	<44 μ	115±13
Hugo microcline	841–1000 μ	100± 9
K-Hugo microcline	<44 μ	91± 3
K-Amelia albite	<44 μ	90±10

mole. It is clear that this average of 99 ± 10 kcal/mole is significantly higher than the 74.3 ± 1.4 kcal/mole (McKie and McConnell, 1963) for the Amelia albite.

The activation energy is a measure of the energy required for the Al/Si interchange under anhydrous conditions. If the atomic mechanism for the disordering of Al and Si is the same as that for the long range migration of Al and Si atoms, which seems likely, then the activation energies should be similar. Therefore, the value reported here is a useful indication of the activation energy for aluminum-silicon interdiffusion in potassium-rich feldspars until diffusion experiments have been performed.

Using the disordering rates determined in this study, it is interesting to extrapolate the data to temperatures of geological interest. This assumes that the same mechanism is operative at the lower temperatures. It is also important to remember that a linear extrapolation of the disordering rates provides a minimum estimate for the time of total transformation.

The calculations were made using the data for the 841–1000 μ fraction of the Hugo microcline because this is probably the best choice geologically, especially in terms of the grain size. For total conversion at a temperature of 500°C, times of 167 and 18 billion years are needed for activation energies of 100 and 91 kcal/mole, respectively. At 600°C, the times required are 100 and 20 million years for the above activation energies.

Based on the data of Goldsmith and Laves (1954) and Steiger and Hart

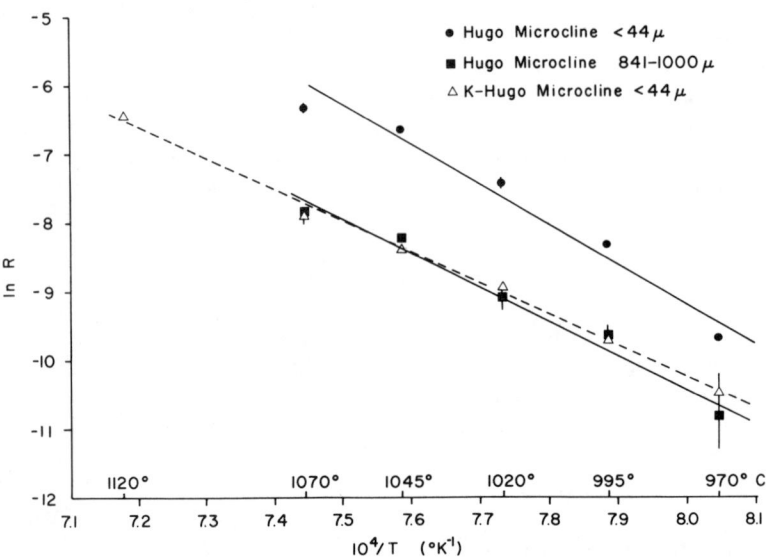

Fig. 4. Arrhenius diagram showing data for three different samples. The dashed line is for the K-Hugo microcline.

(1967), the actual transformation temperature for microcline-sanidine is 500°C or slightly lower. Based on the above calculations, the microcline-sanidine reaction at metamorphic temperatures near the transformation temperature would require a different mechanism for total conversion in reasonable geologic time.

Nature of the Transformation. Since the work of Goldsmith and Laves (1954), the microcline to sanidine transformation has been known to represent a diffusive transfer of Al and Si atoms which interchange among the tetrahedral sites. In their dry heating experiments, the presence of sanidine was not observed until completion of the transformation; monoclinic and triclinic reflections were not observed simultaneously during the progress of the reaction. Thus a homogeneous transformation rather than a nucleation-and-growth mechanism is indicated by these observations.

We have performed a rate analysis employing the method of Sharp et al. (1966) on the data of Müller (1970) as well as representative data from the present study. These data are shown in Fig. 5 in which α is plotted versus $t/t_{0.5}$ where $t_{0.5}$ corresponds to time for half transformation ($\alpha = 0.5$). Two theoretical curves from Sharp et al. (1966) are shown on Fig. 5. Curve 1 represents a solid state reaction which follows first order kinetics (homogeneous transformation) and curve 2 represents the Avrami equation for a nucleation-and-growth mechanism which considers impingement resulting from growth of nuclei. The rate data are clearly consistent with a homogeneous reaction.

Because the microcline-sanidine transformation appears to be a homogeneous type reaction, it is still necessary to consider how the transformation proceeds. On the basis of annealing experiments on one sample of K-Itatiaia albite on which cell dimensions could be obtained accurately, Müller (1970) concluded that the transformation proceeds by transfer of Al atoms from $T_1(0)$ sites simultaneously into the other three tetrahedral sites with the concurrent exchange of Si atoms. This was substantiated by the linearity of b^*-γ^* and c^*-γ^* curves.

In order to substantiate this behavior for the Hugo sample, we calculated tetrahedral site occupancies by the method of Stewart and Ribbe (1969) as stated in a previous section. Accurate cell refinements could be obtained up to approximately 25% transformation during which time the reflections were quite sharp and could be measured accurately. After approximately 25% transformation, there is a slight variation in the transition rate in different crystals of a polycrystalline sample. Therefore, the cell refinement data are not as accurate beyond this extent of transformation. Within the 25% range, however, the site occupancies of $T_1(m)$, $T_2(0)$, and $T_2(m)$ are equal within the uncertainties.

Therefore, the data for the microcline-sanidine transformation in the Hugo sample are consistent with a disordering process following a homogeneous type reaction in which the Al atoms are simultaneously transferred from the $T_1(0)$ site into the other three tetrahedral sites.

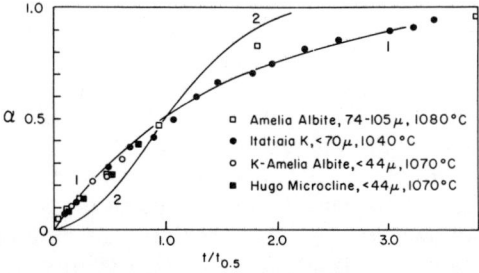

Fig. 5. Extent of reaction versus the reduced time scale, $t/t_{0.5}$. Curve 1 is the theoretical curve for a homogeneous reaction, and curve 2 is the curve for the Avrami equation for a nucleation and growth reaction. The Amelia albite data are those of McKie and McConnell (1963) and the Itatiaia K data those of Müller (1970).

Hydrothermal Experiments

Several hydrothermal annealing experiments were performed using the Hugo microcline. The experimental conditions were 700° or 800°C and 1 or 2 kbar. Two grain sizes were used, $<44~\mu$ and $354–420~\mu$. The solid/H_2O ratio was varied from 1.0 to 10.0 by weight.

The purpose of these experiments was to compare the results with the conflicting observations of Tomisaka (1962) and Goldsmith and Laves (1954). Tomisaka (1962) reported that intermediate structural states were attained in his hydrothermal experiments. Goldsmith and Laves (1954) observed that hydrothermal annealing of microcline at sufficiently high temperatures always produced sanidine by a process of dissolution and reprecipitation. Our experimental results agree with those of Goldsmith and Laves (1954). Optical and x-ray diffraction analyses always indicated the presence of both triclinic and monoclinic phases but no intermediate structural states.

A possible alternative explanation for Tomisaka's (1962) results is that the change in triclinicity was the result of homogenization of originally perthitic microcline, perhaps perthitic on a submicroscopic scale. The composition of his feldspar (Or:Ab:An = 68.5:29.9:1.6) and the fact that it was obtained as phenocrysts in porphyritic granite argue that the sample was perthitic. The available Na was sufficient to lower the triclinicity to the values observed. For example, an Or_{70} Hugo microcline prepared by Na-K exchange gave a Δ value as defined by Goldsmith and Laves (1954) of approximately 0.59 compared to its initial value of 0.94 (Orville, 1967). This change is greater than that reported by Tomisaka (1962). Tomisaka's initial Δ value, 0.8556, increased to a maximum of 0.9900, and then decreased to a minimum of 0.7630.

Therefore, we conclude that the shift in the 131 and $1\bar{3}1$ reflections observed by Tomisaka was probably due to homogenization of the alkalies rather than to any change in the Al-Si distribution. To our knowledge, it has never been unambiguously demonstrated that water alters the rate of microcline disordering except by dissolution and reprecipitation, which is significant in laboratory experiments only above 600°C. The data on oxygen exchange between potassium feldspar and water reported by Yund and Anderson (this volume) also support this conclusion.

Conclusions

The rate of the microcline-sanidine transformation under anhydrous conditions is dependent on grain size and composition as well as temperature. The rate of disordering increases with a decrease in grain size and an increase in the Na/K ratio. We realize that different samples may disorder at different rates (see Goldsmith and Laves, 1961). However, if the mechanism of disordering is the same for all samples, the activation energy should be constant. This is why we believe that a comparison of activation energies is more meaningful than a comparison of the absolute disordering rates.

The activation energy for the Al/Si disordering in potassium-rich feldspars is approximately 25 kcal/mole higher than in sodium-rich feldspars. The average activation energy of 99 kcal/mole determined in this study is a useful guide for the activation energy for the Al/Si interdiffusion in alkali feldspars until diffusion data are available. Extrapolation of the disordering rates to temperature of geologic interest suggests that at metamorphic temperatures near the transformation temperature ($\sim$500°C), microcline would not disorder by this mechanism. Because this calculation assumes a knowledge of the absolute disordering rate as well as the activation energy, the conclusion must be considered only as an estimate for other samples. For this reason and because of the possible difference between ordering and disordering kinetics, it is not possible to comment on the more interesting question of how much time is required for ordering a sanidine.

Rate analysis of these data for the Hugo microcline and that of Müller (1970) for the specimen labeled Itatiaia K indicate that the transformation proceeds as a homogeneous reaction. Cell refinements of the partially transformed Hugo confirm Müller's (1970) conclusion that the reaction involves simultaneous transfer of Al from $T_1(0)$ into the three other tetrahedral sites. Stewart and Ribbe (1969) reached the same conclusion for ordering in albite, using MacKenzie's (1957) data.

Finally, hydrothermal experimental work supports the observations of Goldsmith and Laves (1954) and not of Tomisaka (1962). It remains to be demonstrated that water alters the rate of microcline disordering except by dissolution and reprecipitation.

Acknowledgments

This study was supported by the Advanced Research Projects Agency and National Science Foundation grant GA-21145. We are especially grateful to Dr. P. M. Orville (Yale) for supplying the Hugo microcline sample, to Dr. J. Tullis for her criticism of the manuscript, and to Dr. B. J. Giletti for several helpful discussions.

References Cited

Baskin, Y., Observations on heat-treated authigenic microcline and albite crystals, *J. Geol.*, *64*, 219–224, 1956.

Christian, J. W., *The Theory of Transformation in Metals*, Pergamon Press, London, 1965.

Goldsmith, J. R., and F. Laves, The microcline-sanidine stability relations, *Geochim. Cosmochim. Acta*, *5*, 1–19, 1954.

Goldsmith, J. R., and F. Laves, The sodium content of microclines and the microcline-albite series, *Cursillos y Conferencias Inst. Lucas Mallada, C.S.I.C.*, *8*, 81–96, 1961.

Jones, J. B., and W. H. Taylor, Bond lengths in alkali feldspars, *Acta Crystallogr., Sect. B.*, *24*, 1387–1392, 1968.

MacKenzie, W. S., The crystalline modifications of $NaAlSi_3O_8$, *Amer. J. Sci.*, *225*, 481–516, 1957.

McConnell, J. D. C., and D. McKie, The kinetics of the ordering process in triclinic $NaAlSi_3O_8$, *Mineral. Mag.*, *32*, 436–454, 1960.

McKie, D., and J. D. C. McConnell, The kinetics of the low–high transformation in albite, I, Amelia albite under dry conditions, *Mineral. Mag.*, *33*, 581–588, 1963.

Müller, G., Der Ordnungs-Unordnungs-Übergang in getemperten Mikroklinen und Albiten, *Z. Kristallogr.*, *132*, 212–227, 1970.

Orville, P. M., Unit-cell parameters of the microcline–low albite and the sanidine–high albite solid solution series, *Amer. Mineral.*, *52*, 56–86, 1967.

Ribbe, P. H., H. D. Megaw, and W. H. Taylor, The albite structures, *Acta Crystallogr., Sect. B.*, *25*, 1503–1518, 1969.

Schneider, T. T., Röntgenographische und optische Untersuchung der Umwandlung Albit-Analbit-Monalbit, *Z. Kristallogr.*, *109*, 245–271, 1957.

Sharp, J. H., G. W. Brindley, and B. N. Narahari Achar, Numerical data for some commonly used solid state reaction equations, *J. Amer. Ceram. Soc.*, *49*, 379–382, 1966.

Smith, J. V., Critical review of synthesis and occurrence of plagioclase feldspars and a possible phase diagram, *J. Geol.*, *80*, 505–525, 1972.

Steiger, R. H., and S. R. Hart, The microcline-orthoclase transition within a contact aureole, *Amer. Mineral.*, *52*, 87–116, 1967.

Stewart, D. B., and P. H. Ribbe, Structural explanation for variations in cell parameters of alkali feldspar with Al/Si ordering, *Amer. J. Sci., Schairer Vol. 267A*, 444–462, 1969.

Tomisaka, T., On order-disorder transformation and stability range of microcline under high water vapour pressure, *Mineral. J.* (Japan), *3*, 261–281, 1962.

Yund, R. A., and T. F. Anderson, Oxygen isotope exchange between potassium feldspar and KCl solutions, this volume.

KINETICS OF ENSTATITE EXSOLUTION FROM SUPERSATURATED DIOPSIDES

Robert H. McCallister
Department of Geosciences
Purdue University
West Lafayette, Indiana 47907

ABSTRACT

The precipitation of enstatite from supersaturated diopsides was studied at 1125°C and one atmosphere pressure. For these experiments two methods were used in preparing the starting materials, and depending on the method of preparation the results differed significantly. This difference is believed to be related to subtle variations in the homogeneity of the starting materials. In the less homogeneous material, exsolution occurred by nucleation of enstatite and growth of these nuclei by volume diffusion. Consistent with this mechanism, the number of nuclei formed increased with increasing supersaturation and the activation energy for growth was determined to be 28 kcal/mole. The more homogeneous material did not exsolve even after 450 hours at 1125°C, as determined by x-ray powder diffraction.

INTRODUCTION

For many years geoscientists have recorded and discussed the presence of complex exsolution textures in pyroxenes (Hess, 1941; Poldervaart and Hess, 1951; Bown and Gay, 1960; Boyd and Brown, 1969). Also, with the return of the rocks from the Apollo missions and the utilization of electron microscopy, submicron as well as coarser exsolution textures were observed in many of the lunar pyroxenes (Ross et al., 1970; Christie et al., 1971; Smith et al., 1972; Lally et al., 1972; Brown et al., 1973).

These observations led to various suggestions with regard to the mechanism of exsolution and relative cooling histories of these pyroxenes (Christie et al., 1971; Champness and Lorimer, 1971; Papike et al., 1971; Ghose et al., 1972; Nord et al., 1973; Takeda et al., 1973).

In spite of the numerous studies concerned with the phase equilibria of various pyroxene systems, kinetic studies of their exsolution rates had not been made. In order to understand better the mechanism(s) of exsolution in pyroxenes and to supply a foundation for other studies in this area, an experimental investigation was initiated to examine the exsolution kinetics in the diopside-enstatite system. This system was selected for two reasons:

1. The phase equilibria were originally determined by Boyd and Schairer (1964). Recent work, Kushiro (1972), Yang and Foster (1972), and Yang (1973), has demonstrated the existence of a field of pigeonite stability.

2. This is an iron-free system; thus, oxygen fugacity was one less variable to consider in the initial study.

Since most natural pyroxenes contain some iron, the kinetics of exsolution determined in this study cannot be extrapolated directly to exsolution in natural systems. However, the mechanisms of precipitation as determined should have application to pyroxene systems in general.

Preparation of Starting Materials

Diopside solid solutions were synthesized by two methods.

Method I: Gels of the desired compositions were prepared using the technique developed by Luth and Ingamells (1965). The procedure involved using nitrate solutions of Ca and Mg and tetraethyl orthosilicate as a source of silica. Due to rapid formation and persistence of metastable phases, the gels could not be crystallized directly. Instead they were initially hydrothermally crystallized at 800°–850°C and 0.5–1.0 kbar in 5 mm diameter Au or Pt capsules for approximately one week. The resultant diopside-enstatite mixture was then annealed dry for four weeks at 1350°C to obtain a single phase as determined by x-ray powder diffraction.

Method II: A dried gel was melted at 1450°C for four hours, quenched and ground. This procedure was followed several times to insure homogeneity of glass. As a final step the glass was annealed at 1350°C for five days to produce a single phase pyroxene as determined by x-ray powder diffraction.

The significance of the two methods of synthesis described above will be pointed out in the text where appropriate. The compositions which were utilized throughout the study correspond to 52.0, 58.2, and 65.0 mole percent diopside (Fig. 1).

Isothermal Rate Study

Kinetic data were obtained by exsolving Method I compositions at 1125°C. Crimped 2.5 mm diameter Pt capsules were used to contain the charge and the individual runs were heated in Kanthal-

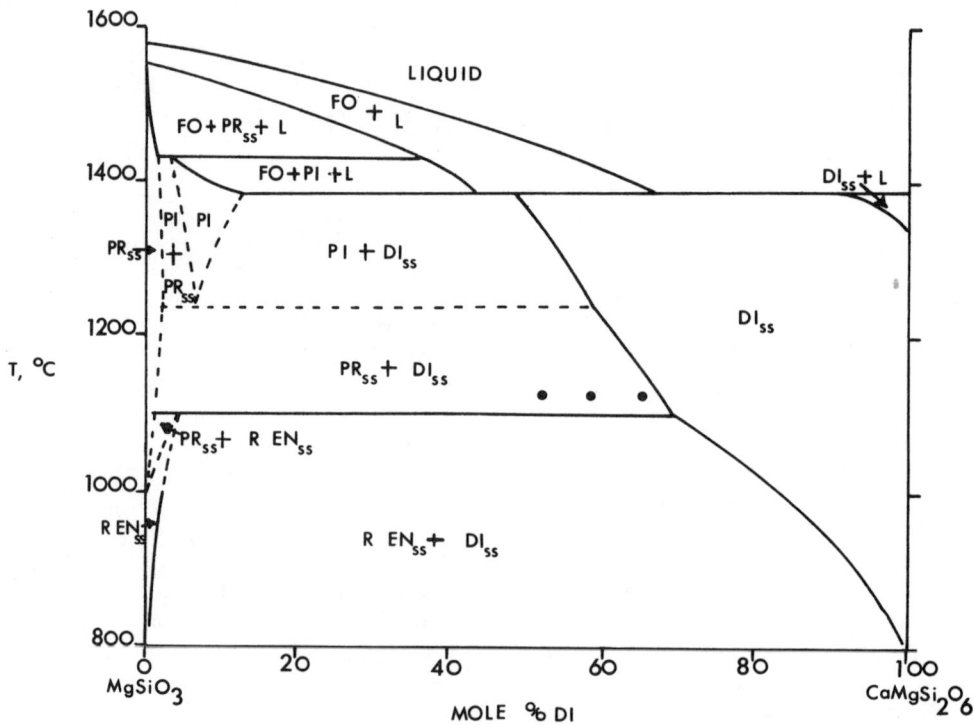

Fig. 1. The phase diagram of the $MgSiO_3$-$CaMgSi_2O_6$ system at one atmosphere (after Kushiro, 1972). Abbreviations: FO, forsterite solid solution; PR_{ss}, protoenstatite solid solution; PI, pigeonite solid solution; DI_{ss}, diopside solid solution; R EN_{ss}, rhombic enstatite solid solution; L, liquid. Dots correspond to 52.0, 58.2 and 65.0 mole percent diopside. Note that Kushiro's diagram is given in terms of weight percent diopside.

Al resistance furnaces. The fractional approach to equilibrium (ζ) was determined from the following equation:

$$\zeta = \frac{C_t - C_0}{C_e - C_0}$$

where C_t = the composition of the diopside at time t; C_0 = the initial composition of the diopside; and C_e = the equilibrium composition for the diopside-rich side of the solvus. The value of C_t was determined by measuring the (311) x-ray reflection of the diopside and comparing it to a spacing curve of $°2\theta$ (311) vs composition. The (111) reflection of LiF was used as an internal standard, and its value was determined to be 38.739° 2θ CuKα when compared to the (102) reflection of Lake Toxaway quartz ($a = 4.9131$ Å, $c = 5.4046$ Å). The best fit for the spacing curve data was obtained through use of the "Least Squares Cubic" program of York (1966). The equation has the following form: $°2\theta$ (311) = $(0.0114 \pm 0.0002)X + 40.196 \pm 0.029$, where X is mole percent diopside on the MgSiO$_3$-CaMgSi$_2$O$_6$ join. The value for C_e at 1125°C was found to be 67.5 mole percent diopside, and this agrees to within 0.5 mole percent diopside with the value obtained by Boyd and Schairer (1964).

The data are given in Table 1 and are plotted in Fig. 2A–D. The uncertainty in ζ was calculated from the general propagation of errors given by Pugh and Winslow (1966). In Fig. 2 all the curves have been drawn by eye, as only their gross shape is significant in terms of the observations listed below:

1. The general shape of the curves for the 52.0 and 58.2 compositions is characteristic of the nucleation and growth process. The lower rates of transformation at shorter times correspond to nuclei formation. The intercept of the relatively linear portion of the transformation curve with the time axis is called the incubation period and is a measure of the ease or difficulty with which nucleation takes place, longer incubation times corresponding, in part, to higher nucleation barriers. The reduction in growth rate at longer times is due to the mutual overlap of regions transforming from separate nuclei (soft impingement) (Christian, 1965, pp. 16–22).

2. The 52.0 composition seems to have a shorter incubation period than the 58.2; however, because of the lack of data for the 52.0 composition at shorter times, this is only a suggestion based on the general trend. Possibly more significant is the fact that the 52.0 and 58.2 compositions reach equilibrium after approximately the same annealing time even though the 52.0 mole percent diopside is initially further from equilibrium. Both of these observations are also consistent with a nucleation and growth mechanism. The increasing exsolution rate at higher supersaturations is due to a higher nucleation rate (Christian, 1965, p. 389). The nucleation rate increases as a result of two factors (Fine, 1964, pp. 9–11): (a) The energy barrier for nucleation decreases with increasing supersaturation. (b) The size of the critical nucleus needed for growth decreases with the increasing supersaturation.

Thus, at higher supersaturations a critical nucleus has a greater probability of formation than at lower supersaturation. Consequently, the overall nucleation and transformation rates increase with increasing supersaturation.

3. The 65.0 composition, which is only 2.5 mole percent supersaturated, did not reach equilibrium after 2987.5 hours. Both the large scatter in the data and the overall decreased rate of transformation are related to the low supersaturation. At such a low supersaturation the critical nucleus needed for growth has a very low probability of formation.

Mechanism

There are three commonly recognized mechanisms of exsolution: continuous, discontinuous and spinodal decomposition (Christian, 1965, pp. 606–669). Both

TABLE 1. Results of Isothermal Rate Study at 1125°C

Run No.	Final Mole % Diopside	ζ	Time (hrs.)
	$C_0 = 52.0 \pm 0.5$ mole % diopside		
Pyx-119	55.3±0.3	0.21±0.03	91.5
Pyx-71	58.1±0.3	0.40±0.03	191
Pyx-72	56.7±0.3	0.31±0.03	359.5
Pyx-73	63.1±0.3	0.72±0.03	696.5
Pyx-74	64.8±0.2	0.82±0.03	1025.5
Pyx-84	65.9±0.4	0.88±0.04	1556.5
Pyx-85	67.4±0.4	0.99±0.04	2008.5
	$C_0 = 58.2 \pm 0.5$ mole % diopside		
Pyx-151	58.0±0.1	0.00±0.06	10
Pyx-133	58.0±0.4	0.00±0.07	51.5
Pyx-134	58.7±0.3	0.06±0.06	166.5
Pyx-75	59.5±0.2	0.14±0.05	191
Pyx-76	60.1±0.2	0.21±0.05	359.5
Pyx-135	60.5±0.5	0.25±0.07	505
Pyx-77	62.7±0.3	0.48±0.05	696.5
Pyx-152	61.0±0.2	0.30±0.05	698
Pyx-78	65.0±0.2	0.73±0.05	1025.5
Pyx-153	64.8±0.3	0.71±0.05	1488
Pyx-87	66.0±0.2	0.84±0.05	1556.5
Pyx-88	67.8±0.2	1.00±0.06	2008.5
Pyx-154	66.3±0.2	0.87±0.05	2987.5
	$C_0 = 65.0 \pm 0.5$ mole % diopside		
Pyx-137	65.6±0.3	0.24±0.19	166.5
Pyx-79	65.0±0.2	0.00±0.21	191
Pyx-80	65.6±0.3	0.26±0.19	359.5
Pyx-138	65.3±0.4	0.13±0.23	505
Pyx-156	65.4±0.3	0.07±0.21	698
Pyx-82	65.6±0.3	0.28±0.18	1025.5
Pyx-157	66.2±0.4	0.50±0.20	1488
Pyx-90	66.0±0.4	0.39±0.20	1556.5
Pyx-91	65.6±0.3	0.24±0.19	2008.5
Pyx-158	66.3±0.3	0.52±0.19	2987.5
	$C_0 = 52.0 \pm 0.5$ mole % diopside		
Pyx-238*	51.8	0.00	92
Pyx-243*	51.8	0.00	259.5
Pyx-242*	51.3	0.00	464.5

*These runs were made with material which was initially synthesized using Method II. All other data in Table 1 were obtained from material which was initially synthesized according to Method I.

continuous and discontinuous precipitation involve a distinct nucleation event. Spinodal decomposition, however, is a pre-precipitation phenomenon which occurs as a result of the growth of initial compositional inhomogeneities by volume diffusion. Nucleation is not involved in this latter mechanism.

When spinodal decomposition is operative, x-ray side bands are developed around first-order Bragg reflections (Douglass, 1969) of the host phase. Also, because this is a spontaneous process, there is no incubation period and the transformation is essentially initiated at time = 0. Side bands were not visible on x-ray powder patterns of the most supersaturated diopside which had been annealed for short periods. In addition, the incubation period for the exsolution reaction was on the order of 30 hours. Both of these observations suggest that spinodal decomposition is not operative for the compositions studied.

Discontinuous precipitation involves the nucleation of colonies of duplex cells, consisting of $\alpha_{saturated}$ (the equilibrium composition of α) and β (the new phase)

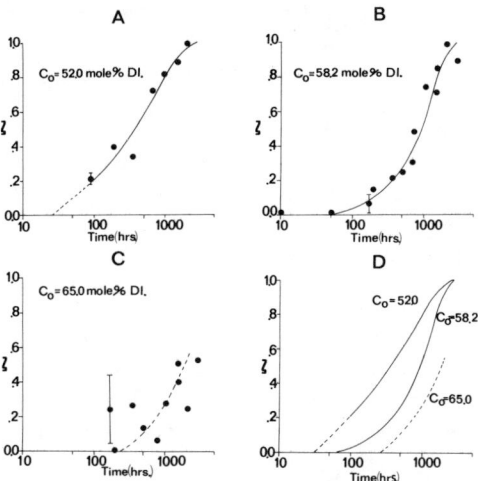

Fig. 2(A–C). Fractional approach to equilibrium (ζ) as a function of time for $C_0 = 52.0$, 58.2, and 65.0 mole percent diopside, respectively. Bars represent one standard deviation. (D) Composite of Figures 2A, 2B, and 2C with the data excluded.

phases, on grain boundaries and other imperfections within the host phase ($\alpha_{\text{supersaturated}}$). Growth of these $\alpha_{\text{sat.}}$-β cells occurs at the incoherent interface between the cell and host and follows a linear growth law. In continuous precipitation β nuclei are formed and grow by volume diffusion. As the reaction proceeds the host is continuously depleted in β component until the equilibrium α phase is achieved. These mechanisms and their possible mineralogical significance have been described previously by Yund and McCallister (1970).

In the case of enstatite exsolution from supersaturated diopsides we suggest that the operative mechanism is nucleation and growth by the continuous mode. Because the exsolved phase was not observable optically, evidence for the above conclusion is provided by the x-ray diffraction data. If discontinuous precipitation were involved, the diffractometer trace would show peaks corresponding to the $\alpha_{\text{sat.}}$ and β phases in addition to the $\alpha_{\text{supersat.}}$ peaks. As the transformation continued, the $\alpha_{\text{sat.}}$ and β peaks would grow at the expense of the host's reflections. What was observed in the diopside-enstatite system was a systematic shift of the (311) reflection of diopside to a value corresponding to the equilibrium composition with no appreciable broadening of the peak. This same behavior has been noted by Yund and Hall (1970) for the exsolution of pyrite from pyrrhotite and Yund et al. (1972) for the exsolution of kalsilite from nepheline. In both cases the mechanism was identified as continuous precipitation.

Theoretical analysis of the rate data can be made using a relation proposed by Avrami (1941):

$$\zeta = 1 - \exp(-kt^n)$$

where k is theoretically a constant only if the growth rate is linear. This is important because if volume diffusion is involved, the relation should be valid only for the early stages of the transformation. In some cases, however, it is found to adequately describe diffusion controlled growth for zeta as large as 0.9 (Christian, 1965, p. 481). The value of n is dependent upon the growth rate, nucleation rate, and the type and distribution of nuclei. For the case in which all nuclei are present at $t = 0$, zero nucleation rate, the value of n is indicative of the growth rate. For linear growth (discontinuous precipitation), the theoretical value of n is 3 as opposed to 3/2 for diffusion controlled growth (Christian, 1965, p. 489). All values of t used to obtain n for the 52.0 and 58.2 data have incubation periods of 25 and 170 hours, respectively, subtracted from them. This is to ensure that n will depend on the growth rate and not on a complicated nucleation and prenucleation history. Values of n equal to 0.9 and 0.7 for the 58.2 and 52.0 compositions, respectively, suggest diffusion controlled growth. The fact that they are less than 3/2 probably indicates heterogeneous nucleation on grain margins and crystal imperfections. As a point of comparison, the values obtained for n in the nepheline-kalsilite system varied between 0.5 and 0.8 depending upon initial supersaturation (Yund et al., 1972), and n was approximately equal to one for pyrite exsolution from pyrrhotite (Yund and Hall, 1970).

ACTIVATION ENERGY FOR GROWTH

Assuming the operative mechanism under dry conditions for the initial compositions studied is continuous precipitation, we determined the activation energy for growth in the following manner. A portion of the 52.0 Method I preparation was annealed at 1125°C until it had reached a value for zeta of approximately 0.21 ($C_t = 55.3$ mole percent diopside). Portions of this material were annealed at 1175°C and 1025°C (the latter temperature is within the orthorhombic enstatite$_{ss}$-diopside$_{ss}$ field as opposed to the protoenstatite$_{ss}$-diopside$_{ss}$ field; however, no discontinuity in the activation energy was observed. The runs at 1175°C and 1025°C were made for times such that zeta did not exceed 0.7. For greater values of zeta, soft impingement could be a problem, and this would cause a decrease in the rate of transformation. The value at 1125°C was calculated from the smoothed data in Fig. 2A between $\zeta = 0.21$ and 0.60.

The data are plotted in Fig. 3. ΔC is defined as the difference in mole percent diopside between the initial and second annealing. Thus, ΔC divided by the second annealing time defines a growth rate, provided that additional nucleation does not occur during this period. The slope of the Arrhenius plot in Fig. 3 corresponds to $Q/2.303R$ where R is the gas constant and Q is the activation energy for growth. The value obtained for Q is 28 kcal/mole. The linear relationship suggests that additional nucleation is not taking place.

Assuming that the growth rate is not limited by mobility of the enstatite interface within the diopside host, the activation energy corresponds to that for interdiffusion. The only available data which can be used for comparison are from Linder (1955). He gave a value of 78 kcal/mole for the activation energy of

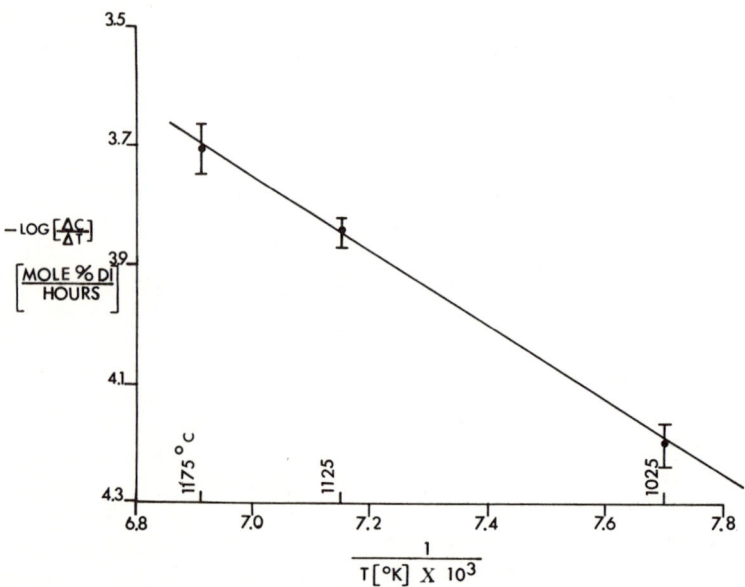

Fig. 3. Arrhenius plot. The slope of the line represents the activation energy for growth of enstatite nuclei under dry, one atmosphere conditions and corresponds to 28 kcal/mole. Error bars represent one standard deviation.

Ca self-diffusion in wollastonite ($CaSiO_3$). The fact that this value is considerably higher than that obtained for Ca-Mg exchange in diopside may not be inconsistent in light of the difference between self-diffusion and interdiffusion, in addition to the structural differences between wollastonite and diopside.

An interesting observation can be made by comparing the data of Virgo and Hafner (1969) on the ordering and disordering of Fe^{2+} and Mg in orthopyroxene and these data. Virgo and Hafner (1969) obtained activation energies for disordering and ordering of 20 and 15–16 kcal/mole, respectively, whereas the author's value for Ca-Mg exchange was found to be 28 kcal/mole. Admittedly these two processes are not identical. For the case of order-disorder the exchange takes place between adjacent coordination polyhedra, but in exsolution interdiffusion occurs over appreciable atomic distances. Assuming that the type of process and not the scale is important, the difference in the activation energies may be a reflection of the cations involved in the exchange and/or the structural difference between clinopyroxene and orthopyroxene.

Homogeneity of Starting Materials

A portion of the 52.0 composition was synthesized according to Method II. Even after 450 hours this preparation had not exsolved any of its excess enstatite, while the Method I starting material in the same time had reached a zeta value of approximately 0.5. Both starting materials were of the same average grain size, chemical homogeneity as determined by the electron microprobe (which is capable of resolution on the micron scale), and value for the (311) x-ray reflection. A possible explanation for the marked difference in the exsolution rate of the two preparations is the presence of enstatite subnuclei in the Method I material. These subnuclei would simply be very localized, high amplitude fluctuations in composition which effectively increase the nucleation rate. Such fluctuations in a highly supersaturated composition would represent a greater percentage of the solute atoms needed to form a critical nucleus than at lower supersaturation. Thus, the effect of a decreasing rate of exsolution with a lowering of supersaturation would not be lost.

It is interesting to note that Boyd and Schairer (1964) who crystallized glasses which were prepared from oxide mixes did not observe an effect of supersaturation. In fact, they show that bulk compositions closer to the solvus reach equilibrium faster. This suggests the possibility that there may have been inhomogeneities present which were approximately the size of the critical nucleus of the least supersaturated composition. In this case the effect of supersaturation on nucleation would be lost and compositions closer to the solvus would reach equilibrium faster.

The results for the different starting materials do not modify the conclusions concerning the mechanisms for this reaction or the activation energy for growth. However, they do point out that the nucleation rate and hence the overall exsolution rate can be strongly affected by the synthesis procedure used in the preparation of the starting materials.

Summary

The data presented for the isothermal rate and activation energy studies suggest that (1) the mechanism of exsolution, at least for the compositions studied, is the continuous mode of precipitation and (2) the activation energy for growth is approximately 28 kcal/mole. Additional work with compositions of higher supersaturation may supply evidence that the mechanism under these conditions is spinodal decomposition. From the phase-equilibria data of Davis and Boyd (1966) and Kushiro (1969) it appears possible to synthesize diopside solid solutions at high pressure (30 and 20 kbars, respectively) which will have considerably higher supersaturation at one atmosphere and 1125°C than it was possible to achieve with the equipment utilized in this study. It will also be necessary to determine interdiffusion coefficients in diopside and other pyroxenes before any details of cooling histories can be supplied for terrestrially and extraterrestrially exsolved pyroxenes.

The marked variation in exsolution rate for the 52.0 mole percent diopside prepared by Methods I and II suggests a possible variation in the homogeneity of synthetic starting materials. The fact that there is a difference between the two preparations is extremely important if the results of rate studies done in the laboratory are to be used as a basis for extrapolation to natural conditions. As a consequence, the characterization and control of any inhomogeneities must be investigated before direct comparisons can be made between synthetic and natural pyroxene exsolution.

One final point worth mentioning is that although it was not possible to observe the exsolved phase using phase contrast microscopy, alternatives currently under investigation are electron microscopy and selected area electron diffraction. These techniques could aid in determining the size and shape of the exsolved phase as well as any relationship between the orientation of the host and guest phases. The electron microscopy is also being used to characterize the various starting materials discussed in this paper.

Acknowledgments

The author wishes to thank Drs. R. A. Yund, H. O. A. Meyer, M. Ross, I. Kushiro, and H. S. Yoder, Jr., and Mr. N. Z. Boctor for critically reviewing this manuscript. Financial assistance for this study was provided by National Science Foundation grants GA-1268 and GA-21145 to R. A. Yund while the author was a graduate student at Brown University, Providence, R.I.

References Cited

Avrami, M., Kinetics of phase change. 3. Granulation, phase change, microstructure, *J. Chem. Phys.*, *9*, 177–184, 1941.

Bown, M. G., and P. Gay, An x-ray study of exsolution phenomena in the Skaergaard pyroxenes, *Mineral. Mag.*, *32*, 379–388, 1960.

Boyd, F. R., and G. M. Brown, Pyroxene exsolution, *Mineral. Soc. Amer. Spec. Pap. 2*, 211–216, 1969.

Boyd, F. R., and J. F. Schairer, The system $MgSiO_3$-$CaMgSi_2O_6$, *J. Petrology*, *5*, Part 2, 275–309, 1964.

Brown, G. M., A. Pecket, C. H. Emleus, J. G. Holland, and R. Phillips, Petrologic observations on Apollo 16 and 15 samples (abstract), in *Lunar Science IV*, J. W. Chamberlain and C. Watkins, eds., Lunar Science Institute, Houston, pp. 94–96, 1973.

Champness, P. E., and G. W. Lorimer, An electron microscopic study of a lunar pyroxene, *Contrib. Mineral. Petrol.*, *33*, 171–183, 1971.

Christian, J. W., *The Theory of Transformations in Metals and Alloys*, Pergamon Press, Oxford, 1965.

Christie, J. M., J. S. Lally, A. H. Heuer, R. M. Fisher, D. T. Griggs, and S. V. Radcliffe, Comparative electron petrography of Apollo 11, Apollo 12 and terrestrial rocks, *Proc. Second Lunar Sci. Conf.*, *Geochim. Cosmochim. Acta*, Suppl. 2, Vol. 1, pp. 69–90, 1971.

Davis, B. T. C., and F. R. Boyd, The join $Mg_2Si_2O_6$-$CaMgSi_2O_6$ at 30 kilobars pressure and its application to pyroxenes from kimberlites, *J. Geophys. Res.*, *71*, no. 14, 3567–3576, 1966.

Douglass, D. L., Spinodal decomposition in Al/Zn alloys. Part 2. X-ray diffraction studies, *J. Mater. Sci.*, *4*, 130–137, 1969.

Fine, M. E., *Introduction to Phase Transformations in Condensed Systems*, Macmillan, New York, N.Y., 1964.

Ghose, Subrata, George Ng, and L. S. Walter, Clinopyroxenes from Apollo 12 and 14: exsolution, domain structure and cation order, *Proc. Third Lunar Sci. Conf.*, *Geochim. Cosmochim. Acta* Suppl. 3, Vol. 1, pp. 507–532, 1972.

Hess, H. H., Pyroxenes of common mafic magmas, *Amer. Mineral.*, *26*, 513–535 and 573–594, 1941.

Kushiro, I., The system forsterite-diopside-silica with and without water at high pressures, *Amer. J. Sci.*, *267 A*, 269–294, 1969.

Kushiro, I., Determination of liquidus relations in synthetic silicate systems with electron probe analysis: The system forsterite-diopside-silica at 1 atmosphere, *Amer. J. Sci., 57,* nos. 7-8, 1260–1271, 1972.

Lally, J. S., R. M. Fisher, J. M. Christie, D. T. Griggs, A. H. Heuer, G. L. Nord, and S. V. Radcliffe, Electron petrography of Apollo 14 and 15 rocks, *Proc. Third Lunar Sci. Conf., Geochim. Cosmochim. Acta* Suppl. 3, Vol. 1, pp. 401–422, 1972.

Linder, R., Studies on solid state reactions with radiotracers: *J. Chem. Phys., 27,* 410, 1955.

Luth, W. C., and C. O. Ingamells, Gel preparation of starting materials for hydrothermal experimentation, *Amer. Mineral., 50,* 255–258, 1965.

Nord, G. L., J. S. Lally, J. M. Christie, A. H. Heuer, R. M. Fisher, D. T. Griggs, and S. V. Radcliffe, High voltage electron microscopy of igneous rocks from Apollo 15 and 16 (abstract), in *Lunar Science IV,* J. W. Chamberlain and C. Watkins, eds., pp. 564–566, 1973.

Papike, J. J., A. E. Bence, G. E. Brown, C. T. Prewitt, and C. H. Wu, Apollo 12 clinopyroxenes: exsolution and epitaxy, *Earth Planet. Sci. Lett., 10,* 307–315, 1971.

Poldervaart, A., and H. H. Hess, Pyroxenes in the crystallization of basaltic magmas, *J. Geol., 59,* 472–489, 1951.

Pugh, E. M., and G. H. Winslow, *The Analysis of Physical Measurements,* Addison-Wesley, Reading, Mass., 1966.

Ross, M., A. E. Bence, E. J. Dwornik, J. R. Clark, and J. J. Papike, Lunar clinopyroxenes: chemical composition, structural state and texture, *Science, 167,* 628–631, 1970.

Smith, D. K., P. A. Thrower, and W. P. Hoffman, Petrography of Apollo sample 14310, *Proc. Third Lunar Sci. Conf., Geochim. Cosmochim. Acta* Suppl. 3, Vol. 1, pp. 207–230, 1972.

Takeda, Hiroshi, I. Ridley, A. M. Reid, and Robin Brett, Inverted pigeonite from a clast of rock 15459 (abstract), in *Lunar Science IV,* J. W. Chamberlain and C. Watkins, eds., pp. 701–703, 1973.

Virgo, D., and S. S. Hafner, Order-disorder in orthopyroxenes, *Mineral. Soc. Amer. Spec. Pap. 2,* 67–81, 1969.

Yang, Houng-Yi, Crystallization of iron-free pigeonite in the system anorthite-diopside-enstatite-silicate at atmospheric pressure, *Amer. J. Sci., 273,* No. 6, 488–497, 1973.

Yang, Houng-Yi, and W. R. Foster, Stability of iron-free pigeonite at atmospheric pressure, *Amer. Mineral., 57,* Nos. 7–8, 1232–1241, 1972.

York, D., Least squares fitting of a straight line, *Can. J. Phys., 44,* 1079–1086, 1966.

Yund, R. A., and H. T. Hall, Kinetics and mechanism of pyrite exsolution from pyrrhotite, *J. Petrology, 11,* No. 2, 381–404, 1970.

Yund, R. A., and R. H. McCallister, Kinetics and mechanisms of exsolution, *Chem. Geol., 6,* 5–30, 1970.

Yund, R. A., R. H. McCallister, and S. M. Savin, An experimental study of nepheline-kalsilite exsolution, *J. Petrology, 13,* No. 2, 255–272, 1972.

MASS TRANSFER OF CALCITE DURING HYDROTHERMAL RECRYSTALLIZATION

Bruce H. T. Chai
Department of Geology and Geophysics
Yale University
New Haven, Connecticut 06520

ABSTRACT

An increase in grain size during progressive metamorphism is a general feature of recrystallized monomineralic rocks such as limestone. In a chemically closed system, material is transferred from one grain to another such that the average grain size increases and the total number of grains decreases. Hydrothermal recrystallization of calcite has been found to follow Ostwald ripening kinetics. The normalized size distribution curve is independent of time and physicochemical variables within the system. The fractional volume of the original material transferred through the fluid medium is calculated for a given increment of the average grain size, using an empirical time-independent size distribution function. This calculation provides a minimum value for the amount of fractional isotope or trace element exchange between the solid crystalline phase and the fluid medium during diagenesis and metamorphism.

INTRODUCTION

Two distinct processes are commonly observed during prograde metamorphism —the formation of new minerals and an increase in grain size. The metamorphic facies concept based on the growth of new minerals and of equilibrium mineral assemblages is well developed, but the kinetics of metamorphic recrystallization, with or without the formation of new mineral phases, has not received adequate attention. Harker (1939) suggested that grain size could be used as an indication of metamorphic grade, but quantitative studies of grain size distributions in metamorphic rocks have only been conducted in the last ten years, by Galwey and Jones (1963, 1966), Jones and Galwey (1964, 1966), Grigorév (1965), Kretz (1966), Spray (1969), Jones, Morgan, and Galwey (1972), and others. Recrystallization kinetics in multiphase assemblages is a complicated problem, as it involves both nucleation and crystal growth phenomena. Development of models for kinetic processes in metamorphic rocks has therefore lagged behind that of equilibrium models. In this and related studies (Chai, 1972; Anderson and Chai, this volume) an attempt is made to focus on the problem of secondary recrystallization, as defined by Beck (1954) and by Barrett and Massalski (1966), of monomineralic rocks such as limestone and chert.

Secondary recrystallization of monomineralic crystal aggregates does not involve the formation of a new mineral phase, and therefore depends on grain growth rather than on nucleation processes. The average grain size of an assemblage increases through the gradual disappearance of smaller and less stable crystals. This paper presents a mathematical model by which the total amount of material transferred from less stable crystals to more stable ones during hydrothermal recrystallization can be evaluated. The model will have important applications to studies of material exchange kinetics in many geologic situations. For example, during the transport of materials through a fluid medium isotopic and trace element equilibrium between solid and fluid phases can easily be reestablished. Knowing the total amount of material transported, and assuming equilibrium between the species in the fluid phase and the surface of crystals, one can determine a lower limit for the amount of exchange of isotopes and/or trace elements between these phases. In a separate paper (Anderson and Chai, this volume) kinetic data of oxygen isotope exchange between calcite and water are used to substantiate this model.

RECRYSTALLIZATION BY THE OSTWALD RIPENING PROCESS

The necessary condition for a stable interface between two phases is that the surface tension be positive. The surface free energy will decrease when the contact surface diminishes at constant temperature and pressure. Therefore, the decrease of contact area is a natural process in the attainment of thermodynamic equilibrium. The decrease in surface free energy as a driving force for grain enlargement in metamorphic rocks has been discussed by Leith (1905), Becke (1913), Eskola (1939), Harker (1939), and others.

In a chemically closed system, the increase in size of some grains requires a decrease in size leading to the disappearance of others. The process is thus one of continuous dissolution and growth of preexisting grains. The result of this process is that the total number of grains in the system decreases and the average grain size increases. The phenomenon was first observed by Ostwald (1900) and others (Hulett, 1901, 1904; Kendric, 1912; Bigelow and Trimble, 1927); and it has come to be called "Ostwald ripening." A theoretical model for the kinetics of Ostwald ripening was first developed by Lifshitz and Slyozov (1961) and by Wagner (1961). A brief summary is given below.

OSTWALD RIPENING MODEL

In a system of dispersed particles of widely different sizes, dissolution and reprecipitation arise from the variation of solubility with particle size. Assuming that all particles are spherical, the solubility relationship can be described by the Thomson-Freundlich equation for two particles:

$$\ln \frac{S_1}{S_2} = \frac{2\sigma v}{KT} \left(\frac{1}{r_1} - \frac{1}{r_2} \right) \quad (1)$$

where σ = the interfacial tension between the phases; v = the volume of one molecule in the solid phase; K = the Boltzman constant; T = temperature in °K; r_1, r_2 = radius of two different sizes of particles; and S_1, S_2 = solubility (or solubility product) of the two particles.

The larger particle is always less soluble and is thus more stable than the smaller one. With time, therefore, the smaller crystal will dissolve and disappear and the larger one will grow. For a multisized aggregate in a solution of given concentration there is only one crystal size which is in true equilibrium with the solution. The radius of this crystal, r^*, is the critical radius and is defined by the Gibbs-Thomson equation to be:

$$r^* = \frac{2\sigma v}{KT \ln \frac{(\text{IAP})}{(\text{KSP})}} \quad (2)$$

where IAP = ion activity product of the species in solution, and KSP = solubility product of a crystal of infinite size.

All grains with radii greater than r^* will be supersaturated with respect to solution and therefore will grow, while all grains with radii smaller than r^* will be undersaturated with respect to solution and will dissolve. In a closed system, as the process proceeds, the average grain size will increase, the IAP of the solution will therefore decrease, the total number of crystals within the system will decrease, and the critical radius will increase.

If an initial system contains a large number of grains with a narrow size distribution, the Ostwald ripening theory predicts that the system will asymptotically approach a steady state. Two distinct features characterize the steady state. First, the size distribution approaches a shape which is time-independent but depends on the particular growth kinetics and the boundary conditions of the system. Second, the rate at which the critical radius increases can be expressed by the general relationship:

$$r^{*n} - r^{*}_0{}^n = K^* (t - t_0) \qquad (3)$$

where r^*_0 = critical radius at t_0; r^* = critical radius at t; K^* = rate constant for the critical radius; n = a positive integer; and t_0 = time required to reach steady state.

The relationship between the critical radius and the average radius is not simple and depends on the particular growth mechanism of the system. For example, $r^* = \bar{r}$ is valid only if the diffusion through the fluid medium is the rate-determining step (Wagner, 1961). However, because of the character of the time-independent size distribution, the increase of average grain size follows the same rate law,

$$\bar{r}^n - \bar{r}^n{}_0 = K_{av.} (t - t_0) \qquad (4)$$

where $\bar{r}$ = average radius at t; $\bar{r}_0$ = average radius at t_0; and K_{av} = rate constant for the average radius.

Because the critical radius cannot be measured directly, the average radius is generally used to specify or define the growth rate. In both Equations 3 and 4, n and K^* (or K_{av}) depend on the particular growth mechanism and on the boundary conditions of the system. For example, n equals 2 if a first order reaction on the crystal surface is the rate limiting step; n equals 3 if the limiting step is either volume diffusion in the fluid phase or growth by screw dislocations in the solid phase; and n equals 4 when surface diffusion is the rate limiting step. (Lifshitz and Slyozov, 1961; Wagner, 1961; Hanitzsch and Kahlweit, 1968, 1969; Kahlweit, 1970).

EXPERIMENTAL RESULTS

Hydrothermal recrystallization of calcite at 650°C and 2 kbar in distilled water, in NaCl, CaCl$_2$, and MgCl$_2$ solutions of various concentrations, and in $2N$ NaCl solution containing 10% CO$_2$ is summarized in Fig. 1. The starting material is the primary standard calcite powder of Mallinckrodt Chemical Works. The average grain diameter $\bar{d}_0$ is about 2.5μ. For most of the runs, the final average grain diameters $\bar{d}$ are within the range of 10 to 100μ, depending on the physicochemical conditions and time. It is clear that the growth rate follows the cube root of time ($n = 3$). From Figs. 2 and 3 it is apparent that the size distribution (normalized to the average grain size) approaches a time-independent steady state distribution which also

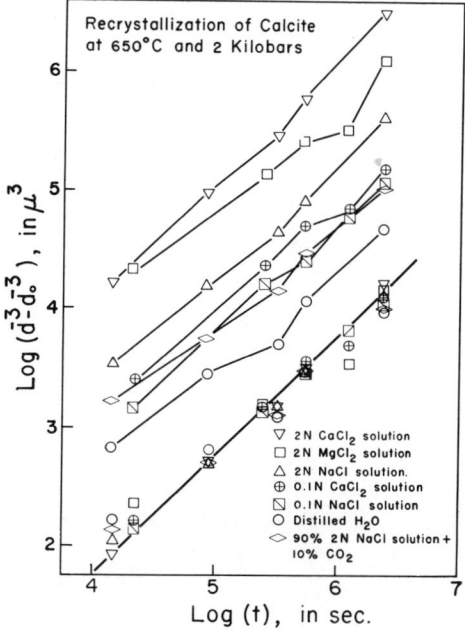

Fig. 1. Hydrothermal recrystallization of calcite at 650°C and 2 kbar in different solutions. The best fit to the unit slope line is also illustrated.

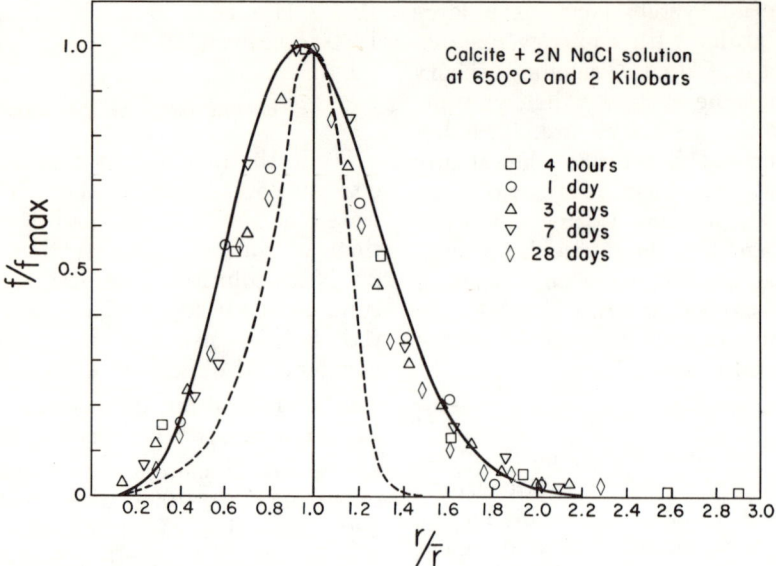

Fig. 2. Size distribution of hydrothermally recrystallized calcite at 650°C, 2 kbar in 2N NaCl solution. The solid curve is a semiempirical approximation line (Equation 5). The dotted curve is the curve predicted by Lifshitz-Slyozov-Wagner theory only when bulk diffusion in the fluid phase is the rate limiting step. The size has been normalized by the maximum frequency.

is independent of the physical and chemical conditions of the system. The time-independent normalized size distribution strongly suggests that, as predicted by the theory, steady state ripening applies to all systems studied.

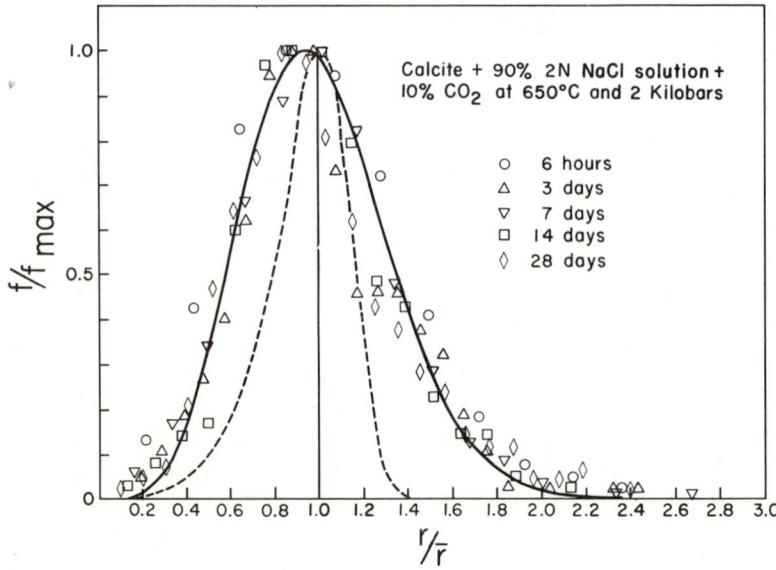

Fig. 3. Size distribution of hydrothermally recrystallized calcite at 650°C, 2 kbar in 90% 2N NaCl solution + 10% CO_2. The solid and dotted curves are the same curves as in Fig. 2. The coordinates are also normalized as in Fig. 2.

A problem is that the time-independent size distribution pattern on Figs. 2 and 3 is somewhat broader than that predicted by the Lifshitz-Slyozov-Wagner theory for the particular case where volume diffusion in the fluid phase is the rate determining step. Similarly broader distributions have also been observed in other systems (Ardell and Nicholson, 1966; Smith, 1967; Krop, Mizia and Korbel, 1967; James and Fern, 1969; Rastogi and Ardell, 1971; Marchant and Gordon, 1972; Sauthoff, 1971, 1973).

The important point is that the broadening does not negate the validity of the mass transfer model presented in the following section. The only constraint of the mass transfer model is that the size distribution function be convergent and time-independent under the given physicochemical conditions. For the calcite system that was studied, the following semiempirical equation, which is plotted in Figs. 2 and 3, approximates the observed size distribution (after normalizing both coordinates):

$$f\left(\frac{r}{\bar{r}}\right) = \left(\frac{32}{\pi}\right)^2 \left(\frac{r}{\bar{r}}\right)^4 \exp\left[2 - \frac{64}{9\pi}\left(\frac{r}{\bar{r}}\right)^2\right] = \frac{n}{n_{\max}} \quad (5)$$

where n = frequency density at a given r, and $n_{\max}$ = maximum frequency density.

In the subsequent discussions, Equation 5 will be used to evaluate the mass transfer of calcite during hydrothermal recrystallization. By employing the kinetic relation of Equation 4, the transfer of mass can be directly related to time.

Mass Transfer Model

During the hydrothermal recrystallization of calcite by the Ostwald ripening process, material is continuously transferred from smaller to larger grains. If the Thomson-Freundlich Equation 1 is valid throughout the ripening process, the mass transfer will be unidirectional—there will be no "net" mass transfer of material from the larger grain to the smaller one. The original relative order of particle sizes will therefore not be changed, even though both the absolute and the relative sizes of individual crystals change with time. Maintenance of the relative size order during Ostwald ripening has been rigorously demonstrated by Markworth (1970).

Growth of a calcite grain is assumed to be uniform over its active surface, and this assumption has been supported by the microscopic study of recrystallized grains. Therefore, at time t_1, when the critical radius is r_1^*, grains larger than r_1^* will be overgrown by the materials dissolved from the grains with radii smaller than r_1^*. At time t_2^*, when the critical radius increases to r_2^*, grains with radii greater than r_2^* will continue to grow. However, grains with radii between r_1^* and r_2^* will start to dissolve even though they were growing at time t_1. During the dissolution process, material which was previously added to the crystals will, of course, dissolve first.

The recrystallization history for individual grains is illustrated in Fig. 4. The

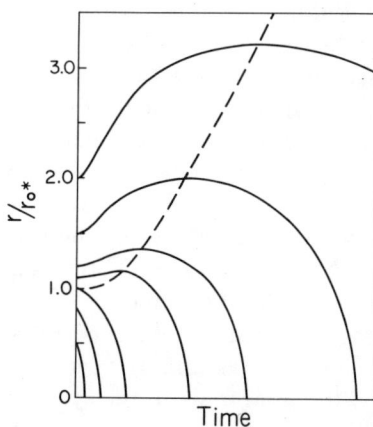

Fig. 4. Variation of particle radius with time (in arbitrary units). The dotted line indicates the increase of critical size with time. The solid lines depict the recrystallization history of individual grains. Notice that the line of critical size cuts the maxima of the growth curves of individual crystals.

initial critical size is r_0^*, so that grains smaller than r_0^* will never be able to grow. The dissolution rate depends on the size of the grain: the smaller the size the faster the dissolution. Grains with radii equal to r_0^* will neither grow nor dissolve at time zero, but as the critical radius increases, they will eventually dissolve. Only grains with radii larger than r_0^* will have a chance to grow. The growth rate of an individual grain therefore depends on its size relative to the critical size at that moment. There is therefore a continuous competition among all the grains remaining in the system at any moment. In principle, only the largest single grain at the beginning of the process will survive and it will gain all of the material dissolved from the other grains as an overgrowth. This grain will still contain a core of starting composition which has never been involved in the dissolution-reprecipitation process. At any moment during the ripening process, all grains remaining in the system will either have a core of original material which is covered by newly grown material or will consist of only the original material. There can be no grains consisting only of newly grown material.

In the present calculations, we are not concerned with the total amount of material transported through the fluid medium, because the total includes material recycled many times. We are concerned only with the amount of "core" or original material which has not been involved in the dissolution and reprecipitation process. The net amount of material transported can be obtained by subtraction of the remaining "core" material from the total mass in the system.

Consider any function $f(r/\bar{r})$ which satisfies the constraints that it is a continuous, time-independent, dimensionless, and convergent function. We have

$$\int_0^\infty f(r/\bar{r}) \, d(r/\bar{r}) = \text{const.} \quad (6)$$

The total number of particles N in the system at a particular time is

$$N = \int_0^\infty n(r) \, d(r) = \bar{r} \, n_{\max} \int_0^\infty f(r/\bar{r}) \, d(r/\bar{r}) \quad (7)$$

The total volume V of the solid phase in the system is

$$V = \frac{4\pi\xi}{3} \int_0^\infty r^3 n(r) \, d(r)$$

$$= \frac{4\pi\xi}{3} \bar{r}^4 n_{\max} \int_0^\infty (r/\bar{r})^3 f(r/\bar{r}) \, d(r/\bar{r})$$

$$= \frac{4\pi\xi}{3} \bar{r}^3 \frac{\int_0^\infty (r/\bar{r})^3 f(r/\bar{r}) \, d(r/\bar{r})}{\int_0^\infty f(r/\bar{r}) \, d(r/\bar{r})}$$

$$[\bar{r} \, n_{\max} \int_0^\infty f(r/\bar{r}) \, d(r/\bar{r})]$$

$$= \frac{4\pi\xi}{3} \bar{r}^3 \left\langle \frac{r^3}{\bar{r}^3} \right\rangle N$$

$$= \frac{4\pi\xi}{3} \langle r^3 \rangle N \quad (8)$$

where ξ = a geometric factor, and $\langle r^3 \rangle$ = average of the cube of the radius. Because

$$\frac{\langle r^3 \rangle}{(\bar{r})^3} = \frac{\int_0^\infty (r/\bar{r})^3 f(r/\bar{r}) \, d(r/\bar{r})}{\int_0^\infty f(r/\bar{r}) \, d(r/\bar{r})}$$

$$= \text{Constant} \, (C) \quad (9)$$

the function on the right-hand side of Equation 9 is a dimensionless, convergent function which will be constant as long as $\bar{r}$ has a positive definite value. Substituting 9 into 8, we obtain

$$V = \frac{4\pi\xi}{3} \cdot C \cdot (\bar{r})^3 \cdot N \quad (10)$$

At time $t = t_0$,

$$V = \frac{4\pi\xi}{3} \cdot C \cdot (\bar{r}_0)^3 \cdot N_0 \quad (11)$$

Since V is constant in a closed system

$$N_0(\bar{r}_0)^3 = N(\bar{r})^3. \quad (12)$$

Equation 12 gives the direct relationship between the average grain size and the total number of grains in the system. The relationship is independent of the actual size distribution, as long as the distribution satisfies the constraints previously stated.

To evaluate the actual mass transfer, we need to know the exact form of the size distribution function and the integral. Using the semiempirical relationship, Equation 5, we obtain the normalized cumulative size distribution function of dimensionless size ρ,

where

$$\rho = \frac{r}{\bar{r}}$$

and

$$N_{cumu} = \frac{\int_0^z f(\rho) \, d(\rho)}{\int_0^\infty f(\rho) \, d(\rho)}$$

$$= \frac{2}{\sqrt{\pi}} \int_0^{\frac{8z}{3\sqrt{\pi}}} e^{-\mu^2} \, d\mu -$$

$$\left[\frac{8 \times 256}{81\pi^2} z^3 + \frac{16}{3\pi} z \right] e^{-\frac{64}{9\pi} z^2} \quad (13)$$

where N_{cumu} = fraction of number of particles within the size range from 0 to z; z = any special value of ρ with $0 \leq z \leq \infty$; and μ = integration parameter of the error function.

The normalized cumulative volume distribution function is

$$V_{cumu} = \frac{\int_0^z \rho^3 f(\rho) \, d(\rho)}{\int_0^\infty \rho^3 f(\rho) \, d(\rho)}$$

$$= 1 - \left[\frac{1}{6} \left(\frac{64}{9\pi} \right)^3 z^6 + \frac{1}{2} \left(\frac{64}{9\pi} \right)^2 z^4 \right.$$

$$\left. + \left(\frac{64}{9\pi} \right) z^2 + 1 \right] e^{-\frac{64}{9\pi} z^2} \quad (14)$$

where V_{cumu} = the volume fraction of particles within the size range from 0 to z.

The two cumulative functions N_{cumu} and V_{cumu} are plotted in Figs. 5 and 6.

Knowing the character of ordering and the time-independent dimensionless size

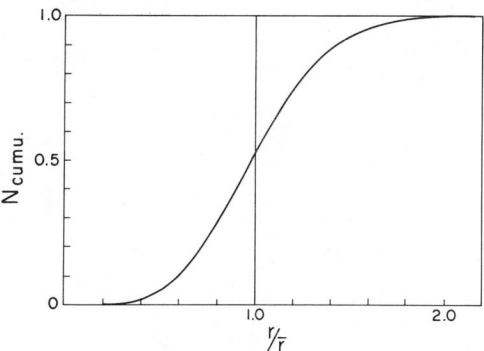

Fig. 5. Plot of normalized cumulative grain size distribution with the cumulative numbers (N_{cumu}) versus dimensionless grain size ($r/\bar{r}$).

distribution during the ripening process, we can evaluate the mass transfer in the following way: The size distribution can be subdivided into several sections (Fig. 7), so that the number of particles in each section has the following relationship:

$$\frac{A}{A+B+C+\cdots} = \frac{B}{B+C+D+\cdots}$$

$$= \cdots = \frac{N_n}{\sum\limits_n^\infty N_n} = \text{constant}, \quad (15)$$

where N_n is the number of particles in section n.

Dissolution and disappearance of particles commences with the smallest. For example, after a certain period t_A, particles in section A have been completely

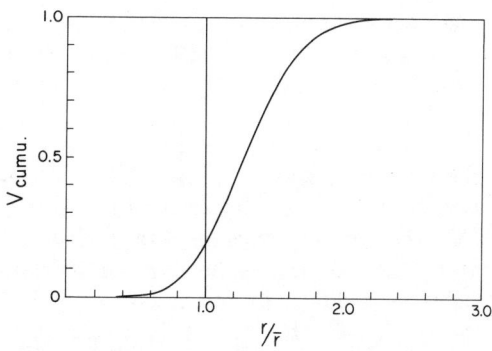

Fig. 6. Plot of normalized cumulative volume distribution (V_{cumu}) versus dimensionless grain size ($r/\bar{r}$).

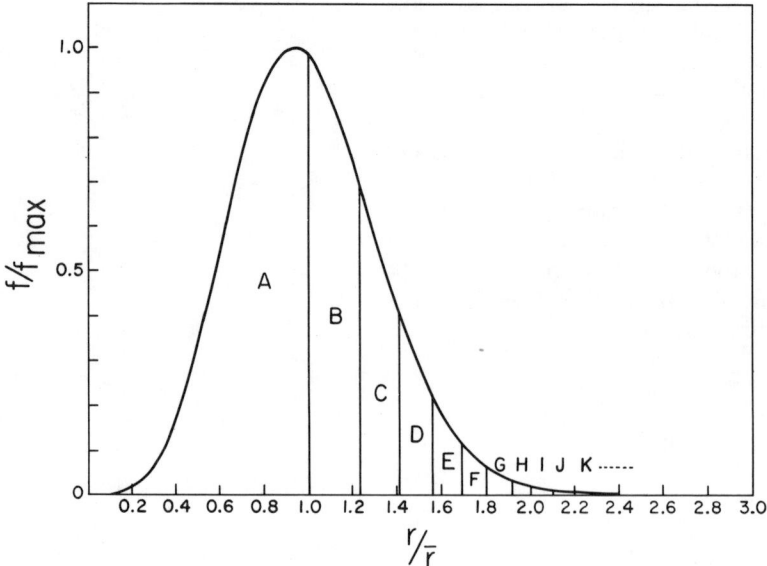

Fig. 7. Plot of normalized empirical grain size distribution function curve (Equation 5) which has been cut into sections having the relationship

$$\frac{N_n}{\sum\limits_{n}^{\infty} N_n} = 0.52373.$$

dissolved. Because the remaining particles retain the same size distribution (after having been normalized at t_A), the shape of the distribution curve of the original section B is changed and is now (at t_A) exactly the same shape as section A at t_0. Section C will now have the same shape as section B at t_0, and so on. After this change, more than 52% ($N_{\text{cumu}}|_0^1 = 0.524$) of the total number of particles has disappeared and material in section A ($V_{\text{cumu}}|_0^1 = 0.193$ or 19.3%) has been transferred to the remaining sections of the system.

Initially, section A contains 19.3% of the total material, section B, 26.5%, section C, 20.4%, etc. After time t_A, section B_A (section B occupying the position A at t_A) retains the same number of particles as it did at t_0, but it no longer has the same volume fraction as it had at t_0. This means that at least $26.5 - 19.3 = 7.2\%$ of the material in the original section B was dissolved and transferred.

Material is gained for section C when it shifts to C_B at t_A. Therefore, the total mass transfer after this stage is

$$\Delta V_A \simeq V_A + (V_B - V_{B_A})$$
$$= 19.3 + (26.5 - 19.3) = 26.5\%.$$

This is an approximation of minimum transfer, because we assume that all of the material remaining in section B_A was original material. Since the process is continuous rather than stepwise, as we have here assumed, there is a certain fraction of mass in section B_A being recycled. One step after t_B, all particles in section B_A have completely disappeared, and the total mass transfer will be:

$$\Delta V_B \simeq (V_A + V_B) + \frac{7.2}{26.5} V_C$$
$$= 51.3\%$$

Again, an approximation is made to evaluate the amount of original material of section C transferred. In the present case, the general formula for the estima-

TABLE 1. Volume fraction of the solid phase transferred by Ostwald ripening process and the corresponding increment of average grain size

Stages	Volume Fraction of Each Section	Corresponding Cutting Point (z)	Fraction of Total No. of Grains Remaining in the System	Cumulative Volume Transfer ($\Delta V/V$)	Relative Average Grain Size Increment	
					$\bar{r}/\bar{r}_0$	$\bar{r}_0/\bar{r}$
Initial	...	0	1.00000	0	1.0000	1.0000
A	0.19328	1.0000	0.47627	0.26467	1.2805	0.7809
B	0.26467	1.2390	0.22683	0.51291	1.6485	0.6076
C	0.20377	1.4141	0.10803	0.69949	2.1056	0.4749
D	0.14005	1.5638	5.15×10^{-2}	0.82424	2.7094	0.3691
E	0.08333	1.6896	2.45×10^{-2}	0.89914	3.4579	0.2892
F	0.05207	1.8035	1.17×10^{-2}	0.94456	4.4141	0.2265
G	0.02743	1.9119	5.56×10^{-3}	0.96889	5.6745	0.1762
H	0.01595	2.0099	2.65×10^{-3}	0.98296	7.2652	0.1376
I	0.00895	2.1000	1.26×10^{-3}	0.99093	9.2586	0.1080
J	0.00524	2.1916	6.01×10^{-4}	0.99459	11.9529	0.0837

tion of total mass transfer of original material at each stage is

$$\Delta V_n = \sum_{1}^{n} V_i + \frac{7.2}{26.5} V_{n+1} \quad (16)$$

where n = number of stages; ΔV_n = total fraction of mass transfer of original material at stage n; and V_i = volume fraction of section i.

The cumulative volume transfer at each stage is listed in Table 1. The corresponding average radius calculated from Equation 12 at each stage is also listed in Table 1. Figure 8 shows the fractional volume transfer of original material versus the increase of average grain size. The fraction of material which was not involved in the dissolution-repreciptation process equals 1.0 minus the fractional volume transfer.

The error in the approximation at each stage probably does not exceed more than 2% to 3% and arises from the estimation of the mass transfer of original material in section $(n + 1)$ at stage n. The approximation becomes better at each additional stage because, as can be seen in Table 1, the volume fraction, V_n, decreases rapidly with increasing n.

Discussion

The mass transfer calculation outlined above could be applied to many geological situations. Recrystallization in diagenetic and metamorphic environments, where material transferred to the surroundings is limited and where closed systems can be approximated, probably follows the Ostwald ripening mechanism. The process should be important, for example, in the recrystallization of carbonate and chert formations. Even though the dissolution-and-repreciptation process is generally accepted for closed system recrystallization, the relevant quantitative data have never been reported.

One merit of the present mass transfer calculation is that it is independent of time, and of all physicochemical variables so far studied in the calcite system. The only parameters needed are the ini-

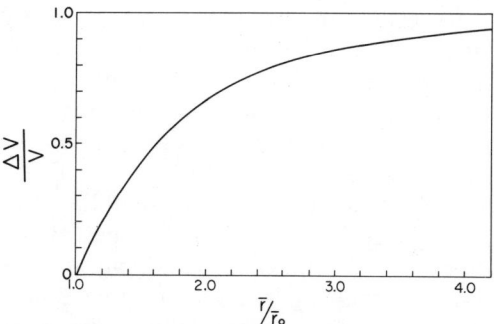

Fig. 8. The total fractional volume transfers of original material as a function of increasing average grain size ($\bar{r}/\bar{r}_0$).

tial and final average grain sizes and the dimensionless time-independent size distribution functions. An empirical size distribution function which fits the calcite recrystallization data was here applied to illustrate the mass transfer calculation. Another merit of this mass transfer calculation is that it applies to a system in which some physical and chemical parameters are changing during recrystallization. The system will deviate from the steady state in proportion to the amount of disturbance. Yet it will again asymptotically reach a new steady state and the steady-state particle-size distribution will be preserved. According to the present result, two-thirds of the total mass will be transferred by Ostwald ripening if the average radius is doubled. This is true for both constant and changing physicochemical conditions. Recrystallization under the latter condition is probably very common in contact metamorphic aureoles, regional metamorphism, and hydrothermal recrystallization.

The mass transfer calculation may be applied to estimate the amount of heterogeneous exchange of isotopes or of trace elements between solid and fluid phases. During the heterogeneous exchange process, materials that are in solution or are exposed on the surface will exchange easily with the solution. Assuming that the exchange can take place only when material is directly in contact with solution and that surface equilibrium is readily reached, the fraction of isotopes or of trace elements exchanged will be equal to the fractional mass transfer by Ostwald ripening. In practice, this calculation only sets a lower limit to the fractional exchange, because other processes such as solid diffusion could contribute additional fractional exchange, and would proceed together with recrystallization. However, recrystallization will decrease the efficiency of solid diffusion exchange because in a dissolving crystal the exchanged part is continuously removed by dissolution, while on the growing crystals the central core is increasingly isolated from solution by the overgrowth. In summary, if there is a measurable grain size change, it seems logical to attribute the exchange to recrystallization.

In a rock thin section, the average grain size and the size distribution can be obtained by converting the two-dimensional measurements to three-dimensional values (Fullman, 1953; Underwood, 1970). We need to know the initial average grain size in order to calculate the fractional volume transfer. This can be done by directly converting the isotope or trace element fractional exchange to fractional volume transfer, assuming that the exchange is completely due to recrystallization. Then $\bar{r}_{0(calc)}$ can be directly evaluated from Fig. 8. But $\bar{r}_{0(calc)}$ may not be identical with the true initial average grain size $\bar{r}_0$. Because recrystallization is a continuous process, it proceeds regardless of whether there is any change in the isotopic or trace element composition of the fluid phase. A fluid phase with a different isotopic composition need not appear at the end, but could first appear at the beginning or middle stages of metamorphism. Fractional exchange with the host rock depends on the quantity and composition of the fluid phase and on the grain size of the crystals. The crystal size is very critical in controlling the rate of exchange by recrystallization. Computations based on the kinetic data of calcite recrystallization show that at 500°C and 2 kbar, in $2N$ NaCl solution, recrystallization from an average grain size of 1 mm to 2 mm requires about 0.5 million years at about 5% porosity. However, recrystallization from an average grain size of 10 μ to 100 μ takes only about 74 years at the same porosity.[1] From Fig. 8, it can be seen that doubling the average grain size only exchanges about 66.5% of the total material, but if the average grain size increases ten times, about 99% of the total solid material

---

[1] Recrystallization rate also depends upon porosity.

will be exchanged. Therefore the exchange process can be easily accomplished in a very short time if the initial crystal size is small and if the fluid phase is present in sufficient quantity.

Incomplete fractional exchange of an isotope or trace element could be caused either by a slow rate of recrystallization or by a limited amount of fluid phase passing through the host rock, or both. A slow recrystallization rate could be due to an initial coarse grain size, to a change in the composition and concentration of the fluid phase, to a lower temperature of recrystallization, or to inhibition of recrystallization by some chemical species, such as organic carbon (Robinson, 1971) or phosphate (Berner and Morse, 1973). The solution is in equilibrium with the surface layer but not with the total crystal. Figure 4 makes it clear that large crystals will show large exchange effects because the new material added is in equilibrium with the transporting fluid, while small crystals in the same system will retain their original chemical and isotopic composition. Examples have been discussed by Engel, Clayton, and Epstein (1958) in studying the oxygen isotopes in Leadville limestone (Mississippian, Colorado), and by Pinckney and Rye (1972) in studying the oxygen isotopes in the Hill Mine of the Cave-in-Rock district. In both cases, the evidence indicates that the hydrothermal solution was present in large quantities and the extent of isotopic exchange correlates well with the extent of recrystallization. Incomplete fractional exchange due to insufficient amount of fluid phase may arise either because there was insufficient fluid at the source, as suggested for the contact metamorphism of Yule marble (Engel, Clayton, and Epstein, 1958), or because of low porosity and permeability in the host rock, such as limestone at Providencia, Mexico (Pinckney and Rye, 1972, cited from the unpublished data of Rye). In both cases, recrystallization still proceeds in a more or less closed system and most of the isotopic composition of the host rock is preserved.

Kinetic recrystallization data in combination with the mass transfer calculation presented here can probably provide some insight into the very controversial problem of retrograde (or cooling) isotopic reequilibration (Taylor, 1967). As discussed above, the rate of recrystallization is critical to exchange kinetics. If physical and chemical conditions remain unchanged, the recrystallization rate depends on the size of the crystals in the system. Starting with a fine-grained host rock, the rate of recrystallization will increase with increasing temperature during metamorphism. At the same time, the rate will slow down because of the increase in grain size. The size of crystals at the highest temperature will therefore depend on the heating rate, the temperature achieved in the system, and the total length of heating. During the cooling part of the metamorphic episode, the temperature decrease and the increase in grain size both will reduce the recrystallization rate. Retrograde effects depend, therefore, on the cooling rate and on the average grain size reached during the maximum temperature of metamorphism. Other conditions being equal, it will take more than eight times longer for a rock with an average grain size of 2 mm to show the same retrograde effects as a rock with an average grain size of 1 mm. Retrograde effects can then be evaluated from the mass transfer calculation and the thermal history of metamorphism.

The calculations and discussions presented above have been restricted to monomineralic assemblages. The possibility of extending the present study to multiphase systems, which are more common geologically, depends on the effects of impurities on the recrystallization mechanism. Solid phase impurities have been found by Kaleda (1955) and Robinson (1971) to inhibit the recrystallization process in limestones. The mutual interference of growth rate in a system with two solid phases and a liquid phase has been studied by White (1968) and his co-workers for a range of ceramic compositions. In all the systems they studied, it was found that the growth rate of one phase decreased as the concentration of the other phase increased. They conclude that this is because interfacial energies between grains of different phases are smaller than the grain boundary energies between grains of the same phase. Therefore, the minor phase prevented the major phase from achieving minimum curvature and inhibited its growth. Nevertheless, if the solid phase impurities in the system do not react or exchange material with the solid phase of interest, and if the size distribution of the phase of interest, in the presence of impurities, remains a time-independent function (which may not be the same function as determined here for the pure phase), the method discussed in this paper will still be valid.

Conclusion

A mathematical mode is here suggested for evaluating the fractional mass transfer of a crystalline phase during recrystallization. The calculation is based on the facts that the kinetic recrystallization of calcite under hydrothermal condition follows the Ostwald ripening mechanism, and that the normalized particle size distribution function is independent of time and the physicochemical conditions studied. The same method can be applied to other particle size distribution curves as long as they are also time-independent functions. The size ordering of the ripening mechanism ("the big get bigger and the small get smaller") is assumed to be strictly followed in the experimental system as predicted mathematically by the model. The model does not apply to systems in which the particle size distribution changes with time. However, if the size distribution of each successive stage is known and the preservation of size ordering is assumed, it is still possible to evaluate the fractional mass transfer. The calculation sets a minimum for the fractional exchange of isotopes or trace elements during recrystallization, provided that the fluid phase is in excess.

The present study is an attempt to understand the heterogeneous reaction kinetics in a simple geological system. There are still many problems to be pursued in order to understand the recrystallization in monomineralic rock assemblages and to apply the present model to general geological problems. Further studies will be necessary.

Acknowledgments

The writer expresses special thanks to Professor P. M. Orville for the guidance in this research. This paper was much improved by the critical reviews by Drs. T. F. Anderson, R. A. Berner, D. Burt, R. McCallister, B. J. Skinner, and R. A. Yund. This work was supported by the National Science Foundation under Grant GA-36410X.

References Cited

Anderson, T. F., and B. H. T. Chai, Oxygen isotope exchange between calcite and water under hydrothermal conditions, this volume.

Ardell, A. J., and R. B. Nicholson, The coarsening of γ' in Ni-Al alloys, *J. Phys. Chem. Solids, 27,* 1793–1804, 1966.

Barrett, C. S., and T. B. Massalski, *Structure of Metals,* McGraw-Hill, New York, 3rd ed., 654 pp., 1966.

Beck, P. A., Annealing of cold worked metals, *Advan. Phys., 3,* 245–324, 1954.

Becke, F., Über Mineralbestand und Struktur der kristallinischen Schiefer, *Denkschr. Akad. Wiss., 75,* 1–53, 1913.

Berner, R. A., and J. W. Morse, Dissolution kinetics of calcium carbonate in sea water, IV. Theory of calcite dissolution, *Am. J. Sci., 274,* 108–134, 1974.

Bigelow, S. L., and H. M. Trimble, The relation of vapor pressure to particle size, *J. Phys. Chem,. 31,* 1798–1816, 1927.

Chai, B. H. T., Experimental study of hydrothermal recrystallization of calcite, *Geol. Soc. Amer. Abstr. with Programs, 4,* 467–468, 1972.

Engel, A. E. J., R. N. Clayton, and S. Epstein, Variation in isotopic composition of oxygen and carbon in Leadville limestone (Mississippian, Colorado) and in its hydrothermal and metamorphic phases, *J. Geol., 66,* 374–393, 1958.

Eskola, P., in *Die Entstehung der Gesteine,* T. F. W. Barth, C. W. Correns, and P. Eskola, eds., Springer-Verlag, Berlin, pp. 263–411, 1939.

Fullman, R. L., Measurement of particle size in opaque bodies, *Trans. AIME, 197*(3), 447–452, 1953.

Galwey, A. K., and K. A. Jones, An attempt to determine the mechanism of a natural mineral-forming reaction from examination of the products, *J. Chem. Soc., London,* 5681–5686, 1963.

Galwey, A. K., and K. A. Jones, Crystal size frequency distribution of garnets in some analysed metamorphic rocks from Mallaig, Inverness, Scotland, *Geol. Mag., 103,* 143–152, 1966.

Grigorév, D. P., *Ontogeny of Minerals,* Daniel Davey & Co., Inc., New York, 250 pp., 1965.

Hanitzsch, E., and M. Kahlweit, Zur Umlösung aufgedampfter Metallkristalle, I., *Z. Phys. Chem. (Frankfurt am Main), 57,* 145–155, 1968.

Hanitzsch, E., and M. Kahlweit, Zur Umlösung aufgedampfter Metallkristalle, II., *Z. Phys. Chem. (Frankfurt am Main), 65,* 290–305, 1969.

Harker, A., *Metamorphism,* Methuen, London, 1939.

Hulett, G. A., Beziehungen zwischen Oberflächenspannung und Löslichkeit, *Z. Phys. Chem. Stoechiom., Verwandschaftslehre, 37,* 385–406, 1901.

Hulett, G. A., Löslichkeit und Korngrösse, Erwiderung an Herrn Prof. F. Kohlrausch, *Z. Phys. Chem., 47,* 357–367, 1904.

James, P. F., and F. H. Fern, Kinetics of precipitation in alloys of uranium-aluminum and uranium-iron-aluminum, *J. Nucl. Mater., 29,* 203–216, 1969.

Jones, K. A., and A. K. Galwey, A study of possible factors concerning garnet formation in rocks from Ardara, Co. Donegal, Ireland, *Geol. Mag., 101,* 76–92, 1964.

Jones, K. A., and A. K. Galwey, Size distribution, composition and growth kinetics of garnet crystals in some metamorphic rocks from West of Ireland, *Quart. J. Geol. Soc. London, 122,* 29–44, 1966.

Jones, K. A., G. J. Morgan, and A. K. Galwey, The significance of the size distribution function of crystals formed in metamorphic reactions, *Chem. Geol., 9,* 137–143, 1972.

Kahlweit, M., Precipitation and aging, in *Physical Chemistry, vol. 10, Solid State Chemistry,* H. Eyring, D. Henderson, and W. Jost, eds., Academic Press, New York, pp. 719–760, 1970.

Kaleda, G. A., K voprosu o perekristallizatsii karbonatnykh porod (The recrystallization of carbonate rocks), in *Voprosy mineralogii osadochnykh obrazovanii (The mineralogy of sedimentary formations), Book 2,* Izdanie L'vovskogo Universiteta, Lvov, Ukrainian S.S.R., pp. 209–214, 1955.

Kendrick, F. B., Some lecture experiments on surface tension, *J. Phys. Chem., 16,* 513–518, 1912.

Kretz, R., Grain-size distribution for certain metamorphic minerals in relation to nucleation and growth, *J. Geol., 74,* 147–173, 1966.

Krop, K., J. Mizia, and A. Korbel, Size distribution function of precipitate Co particles in Cu-1%Co solid solution, *Acta Met., 15,* 463–468, 1967.

Leith, C. K., *Rock cleavage, U.S. Geol. Surv. Bull. no. 239,* 216 pp., 1905.

Lifshitz, I. M., and V. V. Slyozov, The kinetics of precipitation from supersaturated solid solutions, *J. Phys. Chem. Solids, 19,* 35–50, 1961.

Marchant, D. O., and R. S. Gordon, Grain size distributions and grain growth in MgO and MgO-Fe_2O_3 solid solutions, *J. Amer. Ceram. Soc., 55,* 19–24, 1972.

Markworth, A. J., The kinetic behavior of precipitate particles under Ostwald ripening conditions, *Metallography, 3,* 197–208, 1970.

Ostwald, W., Über die vermeintliche Isomeric des roten und gelben Quecksilberoxyds und die Oberflächenspannung fester Körper, *Z. Phys. Chem. Stoechiom. Verwandschaftslehre, 34,* 495–503, 1900.

Pinckney, D. M., and R. O. Rye, Variation of O^{18}/O^{16}, C^{13}/C^{12}, texture, and mineralogy in altered limestone in the Hill mine, Cave-in-Rock district, Illinois, *Econ. Geol., 67,* 1–18, 1972.

Rastogi, P. K., and A. J. Ardell, The coarsening behavior of the γ' precipitate in nickel-silicon alloys, *Acta Met., 19,* 321–330, 1971.

Robinson, D., The inhibiting effect of organic carbon on contact metamorphic recrystallization of limestones, *Contrib. Mineral. Petrol., 32,* 245–250, 1971.

Sauthoff, G., The internal reduction of Ag^+ centers in NaCl crystals, *Acta Met., 19,* 665, 1971.

Sauthoff, G., Growth kinetics and size distribution of ordered domains in Cu_3Au, *Acta Met., 21,* 273–279, 1973.

Smith, A. F., The isothermal growth of manganese precipitates in a binary magnesium alloy, *Acta Met.*, 15, 1867–1873, 1967.

Spray, A., *Metamorphic Textures,* Pergamon Press, Oxford, 350 pp., 1969.

Taylor, H. P., Jr., Oxygen isotope studies on hydrothermal mineral deposits, in *Geochemistry of Hydrothermal Ore Deposits,* H. L. Barnes, ed., Holt, Rinehart and Winston, New York, pp., 109–142, 1967.

Underwood, E. E., *Quantitative Stereology,* Addison-Wesley Pub. Co., Reading, Mass., 274 pp., 1970.

Wagner, C., Theorie der Alterung von Neiderschlägen durch Umlösen, (Ostwald-Reifung), *Z. Elektrochem.,* 65, 581–591, 1961.

White, J., Phase distribution in ceramics, in *Ceramic Microstructures: Their Analysis, Significance and Production,* R. M. Fulrath and J. A. Pask, eds., Wiley, New York, Chap. 35, pp. 728–762, 1968.

OXYGEN ISOTOPE EXCHANGE BETWEEN CALCITE AND WATER UNDER HYDROTHERMAL CONDITIONS

Thomas F. Anderson
Department of Geology
University of Illinois
Urbana, Illinois 61801

and

Bruce H. T. Chai
Department of Geology and Geophysics
Yale University
New Haven, Connecticut 06520

ABSTRACT

Oxygen isotope exchange between calcite and water under hydrothermal conditions was studied from 305° to 700° at 2 kilobars. Based on oxygen isotope exchange kinetics and detailed microscopic examination and grain dimension analysis of the initial starting material and the reaction products, it is concluded that dissolution and reprecipitation is the mechanism by which isotope exchange occurs. The experimental isotope exchange data at 500° and 600° are consistent with predictions of Chai's (this volume) model of volume transport during steady-state Ostwald ripening. However, there is a systematic positive deviation of the experimental results from the predicted values. This may be due to uncertainties in the grain size distribution in the early stages of recrystallization, to initial rapid surface readjustment, or to the presence of fluid inclusions in the reaction products. The steady state is not reached at 305° and 400°C, and at 700°C recrystallization is not under steady state. Nonetheless, the extent of fractional isotope exchange correlates with the observed extent of recrystallization even when recrystallization does not follow steady-state ripening.

INTRODUCTION

Although oxygen isotope exchange between carbonate minerals and aqueous solutions has been studied extensively in the past two decades, the mechanism of exchange is not quantitatively well understood. Hamza (1972) observed that his results on isotope exchange between calcite crystals and pure water at 400° and 600°C fit diffusion-controlled kinetics. He points out, however, that the calculated diffusion coefficients are several orders of magnitude greater than the self-diffusivities of carbon and oxygen in calcite that were reported by Haul and Stein (1955) and by Anderson (1969) for $CaCO_3$-CO_2 isotope exchange. On the other hand, numerous studies on equilibrium fractionation in the $CaCO_3$-H_2O system (for example, see Clayton, 1961; Northrup and Clayton, 1966; O'Neil, Clayton, and Mayeda, 1969) indicate that dissolution-reprecipitation controls the rate of exchange between finely dispersed calcium carbonate and aqueous solutions.

This investigation was undertaken to provide more detailed experimental results on the rates and mechanisms of oxygen isotope exchange between calcite and water under hydrothermal conditions. The principal objectives were to determine the kinetics of oxygen isotope exchange over a wide temperature range and to evaluate the importance of dissolution-reprecipitation processes involved in the oxygen exchange reactions. A kinetic model of hydrothermal recrystallization proposed by Chai (1972, and this volume) will be used in this analysis.

METHODS

Conventional hydrothermal techniques were employed in the exchange reactions. The isotope compositions of the calcite starting material and reaction products were determined by mass-spectrometric analysis of CO_2 liberated from the samples by phosphoric acid treatment (McCrea, 1950). The isotope composition of the starting water was determined using the technique of Epstein and Mayeda (1953). The calcite starting material was the 25–44 micron size fraction of crushed single crystals of Iceland spar. The initial water was distilled water having an O^{18}/O^{16} ratio approximately 75 permil greater than that of the starting calcite. Typical charges consisted of 35 mg calcite and 200 mg water sealed in platinum or gold tubes. Runs were made at 305°–700°C for 11 to 2000 hours at 2 kbar pressure. All reaction products were examined microscopically.

RESULTS

The isotope exchange results are shown in Table 1. The fractional approach to

TABLE 1. Results of the isotope exchange experiments

T (°C)	Run	Time (hr)	Fractional Approach to Isotope Equilibrium	Average Grain Size (μ)
				30.2 (starting material)
305	31	44	.057	...
	32	169	.100	...
	33	400	.122	...
	34	710	.124	...
	35	1111	.208	...
400	41	44	.133	...
	42	100	.115	...
	43	285	.182	...
	44	545	.213	...
	45	911	.217	...
500	51	45	.174	26.6
	52	100	.213	27.6
	52.5	145	.264	28.3
	53	190	.307	28.6
	54	433	.359	29.6
	55	677	.143	28.9
600	61	44	.298	26.7
	62	97	.367	27.2
	63	178	.420	28.7
	64	278	.409	28.7
	65	590	.569	32.4
700	71	11	.349	25.8
	72	45	.793	41.7
	72.5	72	.698	30.5
	73-1	101	.314	31.9
	73-2	100	.701	...
	74-1	180	.878	45.4
	74-2	176	.769	...
	75-1	278	.702	79.9
	75-2	281	.830	...

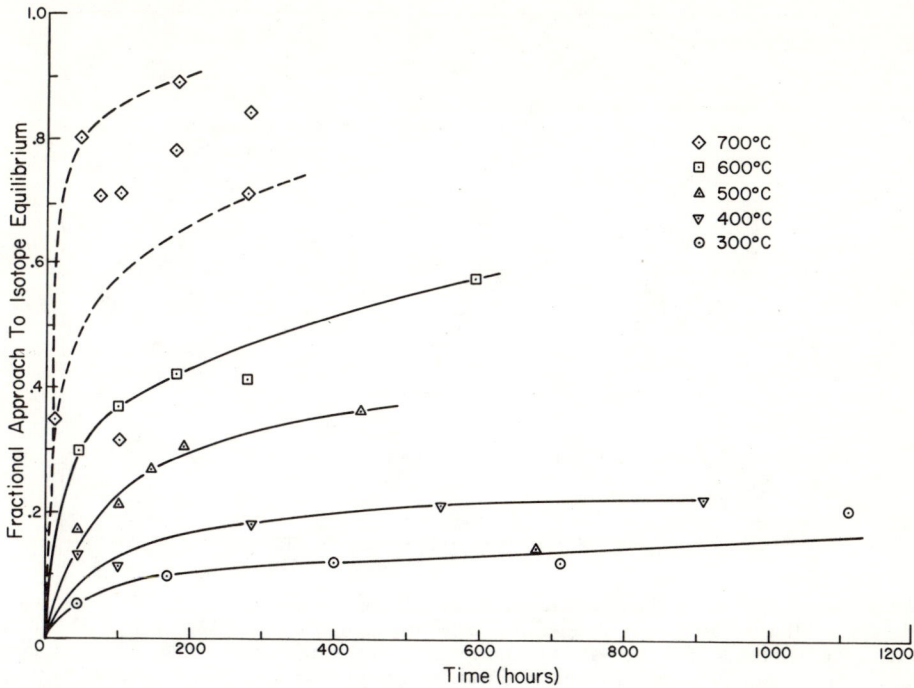

Fig. 1. Experimental data plotted as fractional approach to isotope equilibrium (fractional isotope exchange) versus reaction time for different temperatures.

isotope equilibrium is the ratio of the change in isotope composition of the calcite to the change calculated for isotope equilibrium in the calcite-water system at the temperature of the run. The temperature dependence of the equilibrium oxygen isotope fractionation factor is taken from O'Neil, Clayton, and Mayeda (1969). Figure 1 shows that the approach to isotope equilibrium with reaction time in the runs at 305°, 400°, 500°, and 600°C follows a fairly regular pattern. At 700°C equilibrium is approached at a faster rate, but the time dependence of the results is not consistent.

Chai (1972, and this volume) has shown that the recrystallization of dispersed calcite crystals under hydrothermal conditions is described by an Ostwald ripening mechanism (Ostwald, 1900) in which the large crystals grow and smaller crystals dissolve. The average grain size, therefore, will increase with the degree of recrystallization. Chai further observed that by this mechanism the grain size distribution tends toward a steady-state distribution which is independent of time and experimental conditions, provided that the total number of crystal grains is large enough to yield a nearly continuous size distribution. The normalized distributions of grain sizes for the starting materials and for the recrystallized calcite crystals at 500°, 600°, and 700° are shown in Figs. 2–5. The size distribution of each sample was obtained from the direct measurement of 500 to 1000 calcite grains from photomicrographs. The solid curve in these figures represents the steady-state distribution.

Photomicrographs of the starting material and of representative reaction products are presented in Fig. 6. A brief description of changes in grain size and morphology in the reaction products is

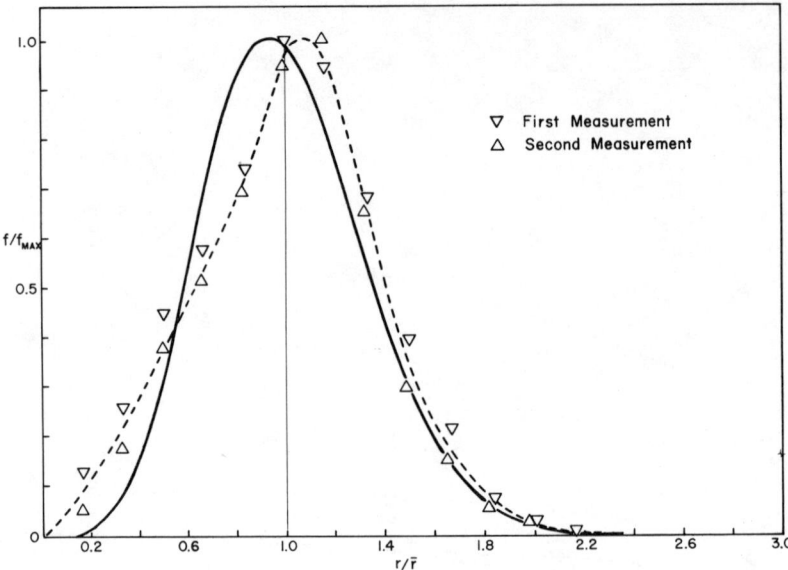

Fig. 2. Normalized grain size distribution of the calcite starting material. All measured grains (500–1000 total) were arbitrarily divided into separate fractions, i.e., 2.5–7.5 μ, 7.5–12.5 μ, etc. f/f_{max} = the frequency of the size fraction normalized to the frequency of the most abundant fraction. $r/\bar{r}$ = the grain size radius of the size fraction normalized to the average grain radius. Two separate measurements of grain size distribution were made; the dotted curve is fitted to these measurements. The solid curve is the calculated steady-state size distribution from Chai (this volume).

given and the correlations of these changes with the isotope exchange results are summarized below.

305° and 400°C

The general texture, average grain size, and grain size distribution of these reac-

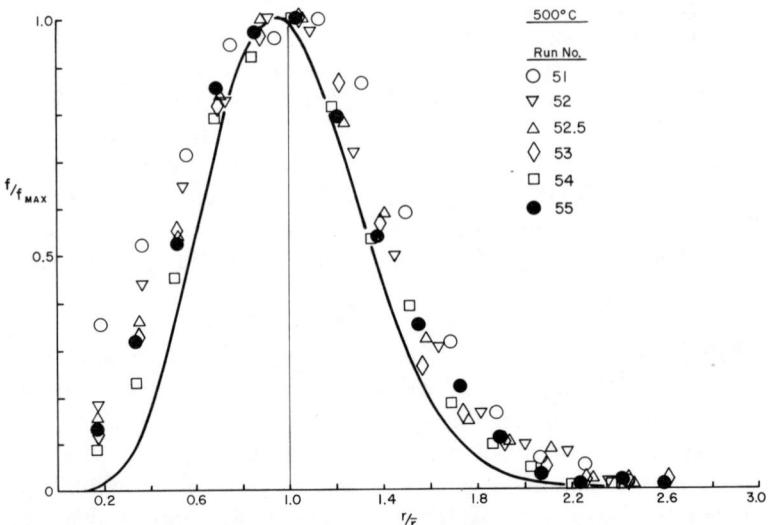

Fig. 3. The normalized grain size distribution for the reaction products at 500°C. The experimental conditions for each run can be obtained from Table 1.

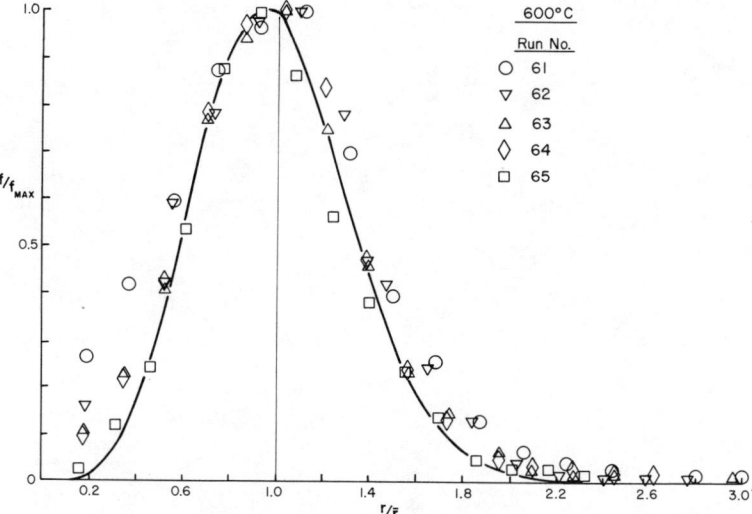

Fig. 4. The normalized grain size distribution for 600°C.

tion products are very similar to those of the starting material. Nonetheless, limited recrystallization is evident in the smoothing of rough grain surfaces and in the presence of fluid inclusions along surface fractures. This type of surficial recrystallization is apparently capable of mobilizing significant quantities of calcite, since fractional isotope exchange values of up to 0.20 were obtained in this temperature range.

500° and 600°C

Substantial textural changes, including grain rounding and smoothing of surface features, were observed in these reaction products. Fluid inclusions are more numerous, and the size of individual inclusions increases with temperature (from 305° to 700°C). Although the average grain size of the reaction products increases with reaction time (Table 1, Fig.

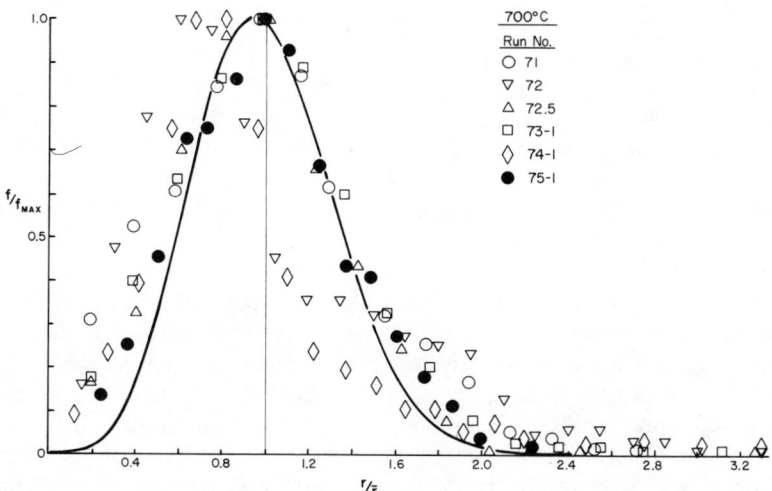

Fig. 5. The normalized grain size distribution for 700°C.

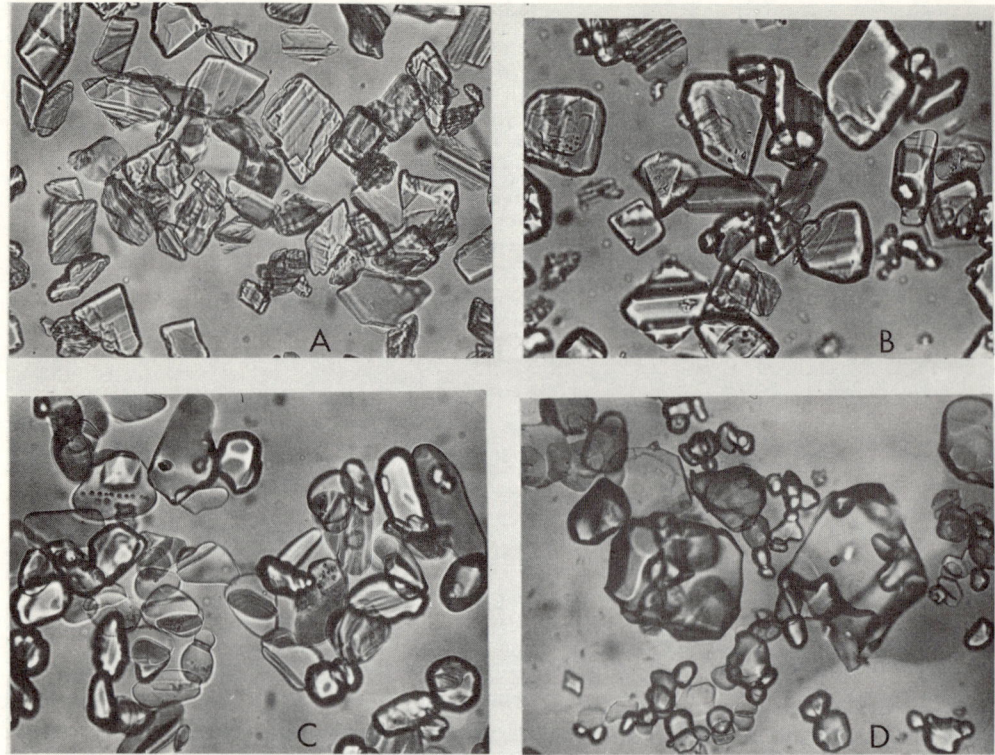

Fig. 6. Photomicrographs of the starting material and representative reaction products. (a) Starting material (magnification = 380 ×); (b) Run 44. 400°C, 545 hours (magnification = 380 ×); (c) Run 61. 600°C, 44 hours (magnification = 380 ×); (d) Run 74-1. 700°C, 180 hours (magnification = 170 ×).

7), these values are less than the average grain size of the starting material. We attribute this effect to initial disaggregation and perhaps dissolution along fractures in the mechanically crushed starting material, which would produce comparatively small grains early in the run. Quench products consisting of very fine (approximately 1 μm) flakes of calcite were observed in the 600°C runs. However, they were easily distinguished from the reaction products, and their abundance was much too small to account for the low average grain size.

The size distributions of the reaction products at 500° and 600°C (Figs. 3 and 4) change from a somewhat broadened pattern for short reaction times to an approximate steady-state distribution for the longer runs.

700°C

Recrystallization is very extensive at this temperature. The reaction products generally contain large crystals up to 400 μm in diameter, some of which have rhombic crystal faces. Rapid recrystallization is apparently due to the much higher solubility of calcite at 700°C compared to lower temperatures (Table 2). The occurrence of greater quantities of quench products in the 700°C run must also be related to the higher solubility of calcite. Except for runs 71 and 75-1, the size distribution of the reaction products does not approach the steady-state distribution. These nonsteady-state distributions probably result from abnormal grain growth and from the breakdown of the statistically continuous size distribu-

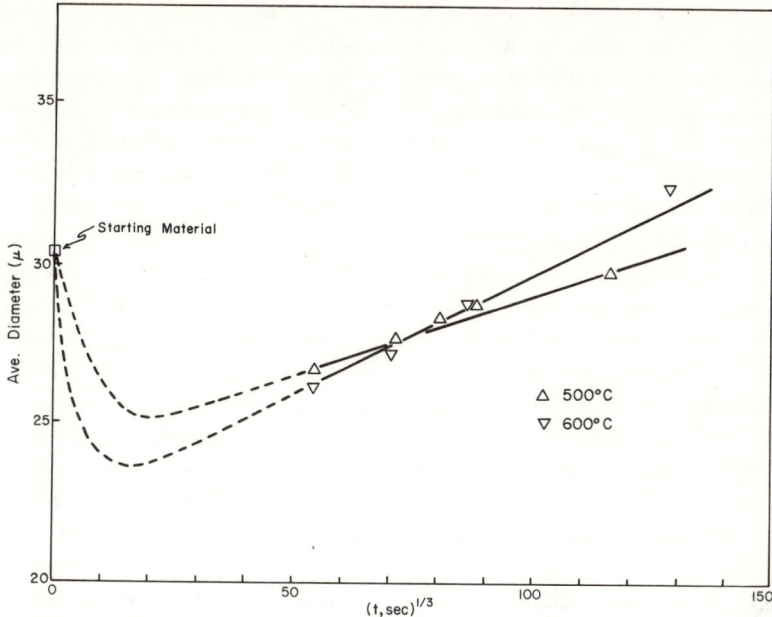

Fig. 7. Variation in average grain diameter ($\bar{d}$) of the reaction products at 500° and 600° with the cube root of the reaction time.

tion by the occurrence of appreciable volume fractions of calcite in a relatively few large grains.

Discussion and Conclusions

The results of microscopic examination of the reaction products indicate that dissolution and reprecipitation control oxygen isotope exchange between calcite and water under hydrothermal conditions. With few exceptions, fractional isotope exchange correlates with the extent of observable recrystallization, including changes in both grain morphology and average grain size. Thus, our results corroborate the conclusions of Clayton and co-workers (Clayton, 1961; Northrup and Clayton, 1966; O'Neil, Clayton, and Mayeda, 1969) on the mechanism of isotope equilibration between calcite and water.

Chai (this volume) has shown that the volume fraction of the original solid (excluding recycled material) that has been transported in the solution is a function of the increase in average grain size, provided that recrystallization is governed by steady-state ripening. The primary objective of this study was to determine whether this model of steady-state ripening adequately predicts the kinetics of oxygen isotope exchange between calcite and water under hydrothermal conditions.

TABLE 2. The solubility of calcite in water at 2 kbar*

Temperature (°C)	Solubility (g/liter)	Mass Fraction of Calcite in Solution in Our Experiments†
305	0.13	.001
400	0.25	.001
500	1.59	.009
600	5.62	.032
700	22.39	.128

*From the data of Ellis, 1963; Malinin, 1963; Malinin and Kanukov, 1971; and Sharp and Kennedy, 1965.
†The typical charge in our runs was 35 mg calcite and 200 mg water.

A steady-state size distribution is observed in most of the runs at 500° and 600°C (Figs. 3 and 4). In addition, after the initial disaggregation or dissolution of grains, the average grain size of the reaction products at these temperatures increases more or less linearly with the cube root of the reaction time (Fig. 7), in agreement with Chai's (this volume) calculation for steady-state ripening. In order to compare our experimental results with Chai's model we assume that the initial average grain size of the calcite at each temperature is approximated by the extrapolation of the linear trend in Fig. 7 to $t^{1/3} = 0$. These values of $\bar{r}_0$ are 24.0 μm at 500°C and 22.1 μm at 600°C.

Figure 8 compares the calculated volume fraction of calcite that has passed through dissolution and reprecipitation in steady-state ripening with the fractional isotope exchange data at 500° and 600°C. The experimental results follow the trend of the calculated curve. However, the experimental data are displaced positively by an amount corresponding to a volume fraction (or fractional isotope exchange) of about 0.10. Several factors may contribute to this displacement: (1) The initial average grain sizes ($\bar{r}_0$) of the reaction products may be lower than estimated. However, if the $\bar{r}_0$ values are lowered enough to produce a significantly better agreement between experiment and calculations, then a linear relation between $\bar{r}$ and $t^{1/3}$ is not obtained. (2) Quench products may make up a much greater fraction of the reaction products at 500° and 600°C than we have estimated. (3) The presence of fluid inclusions of O^{18} enriched water can result in anomalously high values of fractional isotope exchange. When this water is released by acidification of the reaction product, it may exchange isotopes with the evolved CO_2, thereby increasing its O^{18}/O^{16} ratio. The corresponding (positive) deviation in the calculated fractional isotope exchange from the true

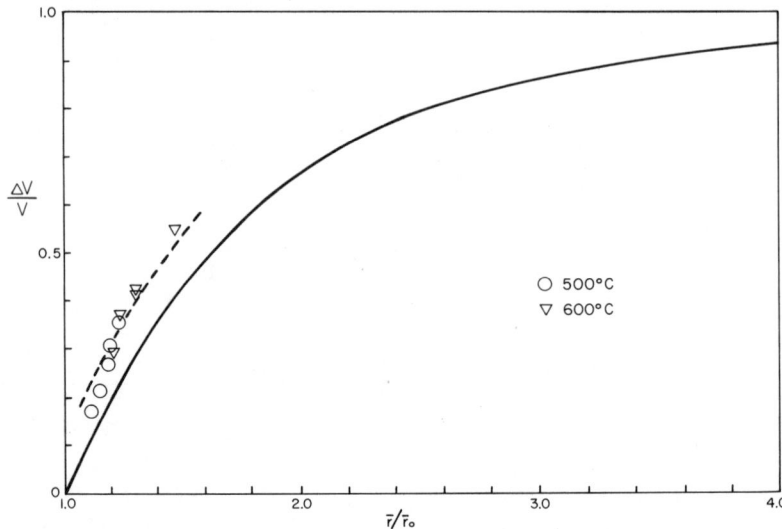

Fig. 8. Volume fraction of calcite that has passed through the dissolution-reprecipitation process versus the average grain radius normalized to the initial average grain radius. The solid line represents the model calculated by Chai (this volume) for steady-state ripening. The points are the experimental isotope exchange data at 500° and 600°C.

value will depend upon (a) the isotope composition of the fluid inclusions, (b) the volume of inclusions per volume of reaction product and (c) the extent of CO_2-H_2O isotope exchange. However, since the volume fraction of fluid inclusions in the reaction products is less than 0.01, we estimate that the fractional isotope exchange values given in Table 1 should be too high by no more than 0.02. (4) Rapid isotope exchange during surficial crystallization (similar to that at 305–400°C) of the disaggregated starting material may occur. However, this type of surficial readjustment should become relatively less important at higher temperatures (and correspondingly higher recrystallization rates), since the dissolved material may not necessarily reprecipitate on the original grain.

The general agreement between the model of steady-state ripening proposed by Chai and our experimental results is encouraging. In our opinion, the systematic discrepancy between experiment and calculation is probably related to the nature of the mechanically crushed starting material. Surficial dissolution-reprecipitation as well as disaggregation in the early stages of hydrothermal recrystallization and the incorporation of fluid inclusions introduce considerable uncertainty into the application of the model of steady-state ripening. In future experiments, we plan to use different calcite starting materials, including crystalline precipitates and annealed crystals, in order to avoid problems of surficial recrystallization and also to achieve steady-state ripening at lower temperatures.

Acknowledgments

This research was supported by National Science Foundation grants NSF GA 1680 (T. F. Anderson, University of Illinois) and NSF GA 36410X (P. M. Orville, Yale University).

References Cited

Anderson, T. F., Self-diffusion of carbon and oxygen in calcite by isotope exchange with carbon dioxide, *J. Geophys. Res.*, *74,* 3918–3932, 1969.

Chai, B. H. T., Experimental study of hydrothermal recrystallization of calcite, *Geol. Soc. Amer. 1972 Annual Meeting, Abstracts with Programs,* *4,* 467–468, 1972.

Chai, B. H. T., Mass transfer of calcite during hydrothermal recrystallization, this volume.

Clayton, R. N., Oxygen isotope fractionation between calcium carbonate and water, *J. Chem. Phys.,* *34,* 724–726, 1961.

Ellis, A. J., The solubility of calcite in sodium chloride solutions at high temperatures, *Amer. J. Sci.,* *261,* 259–267, 1963.

Epstein, S., and T. Mayeda, The variation in O^{18} content of waters from natural sources, *Geochim. Cosmochim. Acta,* *4,* 213–224, 1953.

Hamza, M. S. A., Isotopic fractionation studies between CO_2, water vapor and the outmost molecular layers of the carbonate minerals, calcite, dolomite and witherite and oxygen diffusion in calcite, unpublished Ph.D. thesis, Columbia University, 100 pp., 1972.

Haul, R. A. W., and L. H. Stein, Diffusion in calcite crystals on the basis of isotopic exchange with carbon dioxide, *Trans. Faraday Soc.,* *51,* 1280–1290, 1955.

McCrea, J. M., On the isotopic chemistry of carbonates and a paleotemperature scale, *J. Chem. Phys.,* *18,* 849–857, 1950.

Malinin, S. D., An experimental investigation of the solubility of calcite and witherite under hydrothermal conditions, *Geochemistry (USSR), 1963,* 650–667, 1963.

Malinin, S. D., and A. B. Kanukov, The solubility of calcite in homogeneous H_2O-NaCl-CO_2 system in the 200°–600°C temperature interval, *Geochem. Int. 1971,* 668–679, 1971.

Northrup, D. A., and R. N. Clayton, Oxygen isotope fractionations in systems containing dolomite, *J. Geol., 74,* 174–196, 1966.

O'Neil, J. R., R. N. Clayton, and T. K. Mayeda, Oxygen isotope fractionation in divalent metal carbonates, *J. Chem. Phys., 51,* 5547–5558, 1969.

Ostwald, W., Über die vermeintliche Isomerie des roten und gelben Quecksilberoxyds und die Oberflächenspannung fester Körper, *Z. Phys. Chem. Stoechiom. Verwandschaftslehre,* *34,* 495–503, 1900.

Sharp, W. E., and G. C. Kennedy, The system CaO-CO_2-H_2O in the two phase region calcite + aqueous solution, *J. Geol., 73,* 391–403, 1965.

PART III. TRANSPORT AND REACTION IN ROCKS

Metamorphism involves for the most part interaction of minerals within a preexisting assemblage or reaction of contiguous assemblages. In progressive metamorphism it is assumed that there is a sequence of reactions, each the result of an approach to equilibrium in response to changes in conditions. The approach to equilibrium is in part attained by transport of material. The process whereby the diffusing species migrate can be formalized with nonequilibrium thermodynamics, provided that local equilibrium is maintained. Nearest neighbor grains are usually assumed to be in both homogeneous and heterogeneous equilibrium. Fisher and Elliott have devised a method for determining whether or not the assumption of local equilibrium is valid and whether the process is diffusion controlled or reaction controlled. The deductions of the field geologist concerning the predominant role of diffusion and the observations of the electron probe expert on local equilibrium appear to be supported by their calculations of order of magnitude.

Metasomatic transport may occur by diffusion through solid crystals, by diffusion through a static intergranular pore fluid, or by moving the fluid in which the material is dissolved. Fletcher and Hofmann argue that the first of these mechanisms is inefficient compared with the other two and illustrate the effects of diffusion and infiltration on the resulting metasomatic profile for several hypothetical fluid-solid equilibria. Their evaluation of the available data on rock permeability, fluid viscosity, and diffusivity leads them to the conclusion that infiltration may be effective for large-scale transport but diffusion is not.

A specific case of the infiltration process is illustrated by Frantz and Weisbrod using the K_2O-Al_2O_3-SiO_2-H_2O-HCl system as an example. This first attempt at evaluating quantitatively the effects of transport in a fluid responding to a pressure gradient treats the problem generated by Lindgren's "law of constant volume." They take into account the important volume term by introducing a porosity parameter and show that the infiltration process will stop when the pores are filled with solid phases unless the rock expands.

The pressure gradient which drives the transport in the fluid is often generated by the deformation of the rock. The regularity of sheared and unsheared zones leads to a differentiation. Such metamorphic differentiation may also result from the compositional gradients of preexisting layering or from previously developed cleavage, for example. Vidale illustrates the metamorphic differentiation from observations in Dutchess Co., New York, and notes the strong similarity of composition of certain layers to fissure fillings. The relationship supports her view that mineral solubility differences are a rough measure of the tendency of material to diffuse under a pressure gradient.

The simple alteration of layers is one expression of differentiation, and more complex zoning may ocur. Burt deals with the striking sequence of zones often observed in Ca-Fe-Si exoskarns where mineral enrichment leads to ore deposits. He uses the diffusion models, made popular by Korzhinskii, based on chemical potentials. The μ-μ diagrams illustrate the possible zoning sequences and are useful in predicting where in a given sequence of zones a potential ore-bearing assemblage may occur. The diagrams imply that certain species in solution in some skarns could diffuse "uphill" against their concentration gradients. Such uphill diffusion is also discussed from a theoretical point of view in the paper by Cooper (Part I).

The interaction of minerals through the medium of an interstitial fluid and the mass transport through that fluid are not always easily recognized from observation of mineral compositions and textures and of element abundances, because it may not be possible to reconstruct the initial composition of the system. This ambiguity has kept the discussion about the origin of granites alive for many decades. It may be resolved in some cases when isotopic abundances are used as tracers of geological processes. Radiogenic isotopes bear information not only about the timing of geological events but also about the source material of a given rock and the amount of isotopic exchange with other rocks. For example, the difference in the isotopic composition of strontium between carbonatite and sedimentary limestones demonstrates that the limestone strontium cannot be the source of the carbonatite strontium.

Brandt introduces a new approach to obtaining information from radiogenic isotopes. He uses apparent K-Ar ages of minerals from a contact aureole to determine both the activation energy of argon diffusion in biotite and hornblende and the temperature at the igneous contact.

Isotopic differences between rocks may be generated by radioactive decay or by temperature-dependent isotopic fractionation. In either case, the resulting isotopic "fingerprint" may be used to trace mass transport in geologic processes. Taylor has studied the large-scale interaction of epizonal igneous intrusions with meteoric groundwaters in the San Juan volcanic field of mid-Tertiary age and compared this with rocks of Precambrian age. One mechanism he envisages involves grain-boundary migration of H_2O up a thermal gradient through the hot contact zone into the magma, convective circulation distributing H_2O throughout the magma, and considerable recycling of H_2O. He presents other models involving less direct exchange between meteoric water and magma. Of particular interest is his observation that plagioclase crystals may retain their compositional zoning while exchanging most or all of their oxygen atoms with an external reservoir. He concludes that the cation chemistry of a rock need not always be altered during the oxygen exchange process. The explanation for this behavior might lie in a relatively low activation energy for oxygen diffusion in feldspar (see Yund and Anderson, Part I).

Shieh has examined in detail the consequences of oxygen isotope exchange in contact aureoles and emphasized the limited nature of the effect when no massive outward movement of water from the intrusive takes place or when the country rock has already been dehydrated by a prior metamorphism. He shows further that rocks of increasing grades of regional metamorphism contain progressively lighter oxygen. In this case, the scale of the implied oxygen migration is enormous. Migmatites from the Grenville province resemble normal granites in their oxygen composition but are much lower in O^{18} than alleged sedimentary source rocks. Shieh's interpretation of these data, namely that extensive oxygen exchange occurred with the lower crust or upper mantle, will no doubt enliven the debate about the evolution of the crust.

H. S. Yoder, Jr.
A. W. Hofmann

CRITERIA FOR QUASI-STEADY DIFFUSION AND LOCAL EQUILIBRIUM IN METAMORPHISM

George W. Fisher and David Elliott
Department of Earth and Planetary Sciences
Johns Hopkins University
Baltimore, Maryland 21218

ABSTRACT

The methods of approximation theory are used to show that two interacting metamorphic assemblages α and β, will each be in local equilibrium whenever

$$\frac{2 L_{ij}{}^n}{X_s \nu_i^{(r)} L^{(r)} (x^\alpha - x^\beta)} \ll 1$$

and that diffusion between assemblages α and β will closely approximate steady-state diffusion whenever

$$\frac{9 X_s \Delta V_i (c_i{}^\beta - c_i{}^\alpha)}{4 (x^\alpha - x^\beta)} \ll 1.$$

In these dimensionless ratios, $L_{ij}{}^n$ and $L^{(r)}$ are the phenomenological coefficients for diffusion of component i and for reaction (r), respectively; $c_i{}^s$ is the concentration of i in the grain boundary network in assemblage s, $\nu_i^{(r)}$ is the stoichiometric coefficient of i in reaction (r); ΔV_i is the volume change in the zone separating assemblages β and α as a result of transferring one mole of i from assemblage β to α; X_s is the thickness of the reaction zone where assemblage β is actively growing at the expense of α; and $x^\alpha - x^\beta$ is the thickness of the zone separating assemblages α and β. Order of magnitude calculations suggest that these conditions will be met for the major components participating in most metamorphic reaction processes.

INTRODUCTION

Many metamorphic processes depend upon diffusion of material between two or more interacting assemblages, commonly through an intervening layer of rock compatible with the assemblages on either side; for example, the replacement of one mineral by another at an isograd (Carmichael, 1969), the growth of mineral segregations (Fisher, 1970), and the development of reaction zones between chemically incompatible beds (Vidale, 1969). Figure 1 shows a hypothetical process of this type, with a layer of mineral **C** growing in a rock initially composed of **A** and **B**, by a mechanism requiring the diffusion of components **1** and **2** through a monomineralic layer of **B**.

Fisher (1973) has proposed a kinetic treatment for such processes, utilizing the formalism of nonequilibrium thermodynamics. His approach is based on two assumptions: (1) that the diffusing species are everywhere in local equilibrium[1] with the adjacent minerals, so that the rate of the overall process de-

---

[1] "Local equilibrium" is used here in the petrological sense, meaning that a rock can be divided into regions several grain diameters across or larger, within which all phases are in both homogeneous and heterogeneous equilibrium (cf. Thompson, 1959). In the thermodynamic literature, the phrase "local equilibrium" is widely used in a less restrictive sense, meaning that at every point in a system all thermodynamic functions exist, and that the local entropy is given by the Gibbs equation, as in an equilibrium system (cf. de Groot and Mazur, 1962, p. 23). This less restrictive usage does not require that the system be in either heterogeneous or homogeneous equilibrium, even locally.

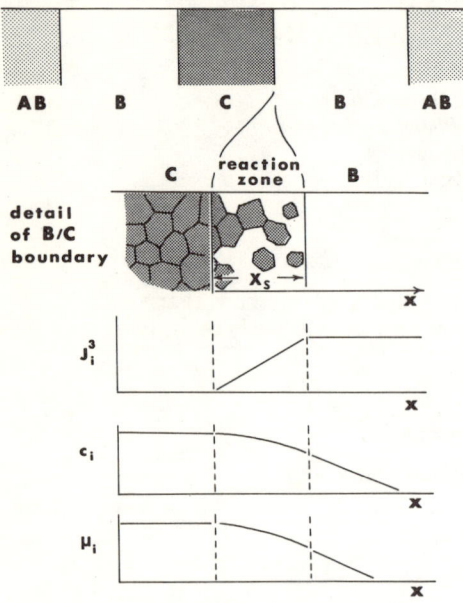

Fig. 1. A hypothetical planar metamorphic segregation, in which the assemblage **BC** is growing at the expense of the assemblage **AB** in a reaction zone X_s thick, by a reaction requiring outward diffusion of components **1** and **2**. The values of J_i^3, c_i, and μ_i are shown near the reaction zone. It is assumed that (1) grain boundary diffusion predominates over lattice diffusion; (2) minerals **A**, **B**, and **C** and all diffusing species can be represented by components **1**, **2**, and **3**, and **A**, **B**, and **C** have fixed compositions; (3) all ionic species present along grain boundaries are in local homogeneous equilibrium; (4) component **3** diffuses much more slowly than **1** and **2**, so that the reaction at the **B/BC** boundary is $\mathbf{B} \longrightarrow \mathbf{C} + \nu_1^{(A)} \mathbf{1} + \nu_2^{(A)} \mathbf{2}$, and that at the **AB/B** boundary is $\mathbf{A} + \nu_1^{(C)} \mathbf{1} + \nu_2^{(C)} \mathbf{2} \longrightarrow \mathbf{B}$.

pends upon the rate of diffusion; and (2) that the fluxes and potential gradients within the system will continually adjust to values closely approximating those of the steady state, a process conveniently termed *quasi-steady* diffusion. Provided that these assumptions hold, relative mass transfer can be predicted for a wide variety of metamorphic processes; and for some processes the relative mass transfer can be calculated directly from the stoichiometry of the pertinent reactions without explicit knowledge of the diffusion coefficients or other thermodynamic information (Fisher, 1973).

Accordingly, it is desirable to establish definite criteria which can be used to determine when the quasi-steady approximation and the assumption of local equilibrium are justifiable and when they are not. Criteria of this sort can be formulated by first normalizing the variables in the kinetic equations in such a way that the terms containing the variables are roughly equal to one; then the coefficients of these terms can be manipulated into a series of dimensionless ratios, whose magnitude can be used to determine which terms can be neglected and which cannot (Kline, 1965). This paper derives the dimensionless ratios needed to evaluate whether or not the quasi-steady and local equilibrium assumptions are realistic for metamorphic processes of the type illustrated in Fig. 1.

Throughout, we will use the formalism of nonequilibrium thermodynamics, which has a number of advantages over conventional diffusion equations for petrologic discussions: (1) Diffusion rates are expressed as a function of chemical potential gradients, which are much easier to determine in rocks than concentration gradients. (2) The matrix of phenomenological coefficients is symmetrical, so that fewer coefficients need be determined. (3) Potential gradients driving steady state diffusion may be easily determined using the principle that the rate of entropy production has a minimum value in the steady state (Fisher, 1973). All measurements will be given in SI units.

A Criterion for Quasi-Steady Processes

Consider the growth of a layer of mineral **C** at the expense of another assemblage **A** and **B** by a reaction requiring diffusion of material through a layer of pure **B** (Fig. 1). We assume that diffusion through the grain boundary network predominates over diffusion through crystal lattices; that minerals **A**, **B** and **C** and all of the species present in the grain boundary network can be represented by three neutral components, **1**, **2** and **3**; and that the minerals have fixed compositions. If component **3** diffuses much more slowly than **1** and **2**, **3** will be conserved at each point in the system, and the reaction at the **B/C** boundary may be written[2]

(A) $B \longrightarrow C + \nu_1^{(A)}1 + \nu_2^{(A)}2$;

the components released by (**A**) are assumed to form various ionic species in the grain boundary network. Similarly, the reaction at the **AB/B** boundary may be written

(C) $A + \nu_1^{(C)}1 + \nu_2^{(C)}2 \longrightarrow B$.

The available experimental data (Garrels, 1959; Helgeson, 1971) suggest that homogeneous reactions among species in the grain boundary network will be much faster than heterogeneous reactions like (**A**) and (**C**), so we will assume that the ionic species in the grain boundary network are everywhere in local homogeneous equilibrium. In most metamorphic processes of the type considered here, the ionic species in the grain boundary network of the matrix **AB** will have been in contact with minerals **A** and **B** long before the **C** layer began to grow, so we will assume that all species in the **AB** matrix are in heterogeneous equilibrium as well. Accordingly, the rate of growth of **C** will depend upon two steps: (1) the rate of diffusion of **1** and **2** through the grain-boundary network, and (2) the rate of production of **1** and **2** by reaction (**A**).

Provided that electroneutrality is maintained, and all diffusing species are in local homogeneous equilibrium, the rate of diffusion of all species can be represented by diffusion of the two neutral components (Fisher, 1973):

$$J_i{}^3 = -\sum_{j=1}^{2} L_{ij}{}^3 \frac{d\mu_j}{dx} \quad (i = 1, 2), \quad (1)$$

where $J_i{}^3$ represents the number of

[2] All reactions will be designated by the symbol for the phase not participating in the reaction (cf. Zen, 1966).

moles of **i** per second passing a unit cross section which is oriented perpendicular to the mean direction of flow, and which is moving with the mean local velocity of component **3**; $L_{ij}{}^3$ is a phenomenological coefficient analogous to a diffusion coefficient, relating $J_i{}^3$ to $d\mu_j/dx$; and $d\mu_j/dx$ is the local instantaneous value of the gradient in the chemical potential of component **i**. For simplicity, we assume that all diffusion is one-dimensional, with the x direction taken perpendicular to the **C** layer.

The rate of production of component **i** by reaction (**A**) is given by

$$J_i{}^{(A)} = \nu_i{}^{(A)} L^{(A)} A^{(A)}, \quad (2)$$

where $J_i{}^{(A)}$ is the number of moles of **i** released by reaction (**A**) per second, $L^{(A)}$ is a phenomenological coefficient analogous to a reaction rate constant for reaction (**A**), and $A^{(A)}$ is the affinity of reaction (**A**) (Katchalsky and Curran, 1967, pp. 92–94).

Viewed on a microscopic scale, the **B/C** boundary is a reaction zone of some finite thickness; this zone must be at least as thick as a grain boundary, and in most metamorphic structures like that in Fig. 1, it is one to five grain diameters thick (cf. Loberg, 1963). We will designate the thickness of this zone X_s, taking the outermost edge at the outermost grain of **C** and the innermost edge at the point where $J_i{}^3$ equals zero (Fig. 1). Consider a slab of this zone with unit cross section, containing a network of grain boundary channels having a total cross-section area $\hat{a}$ measured in a plane perpendicular to x; the volume of the grain boundary network in the reacting slab is then $\hat{v} = \hat{a} X_s$. The rate of accumulation of **i** in the grain boundary network within the reacting slab must equal the difference between the net amount of **i** produced in the reacting slab by reaction (**A**), and the net amount of **i** diffusing out of the slab:

$$\int_{\hat{v}} \frac{\partial c_i}{\partial t} d\hat{v} = \int_{\hat{v}} J_i{}^{(A)} d\hat{v} - J_i{}^3 \hat{a}, \quad (3)$$

where c_i is the concentration of **i** in the grain boundary network, and t is time measured from the beginning of the growth of the **C** layer. Inserting Equations 1 and 2 into 3,

$$\int_{\hat{v}} \frac{\partial c_i}{\partial t} d\hat{v} = \nu_i^{(A)} \int_{\hat{v}} L^{(A)} A^{(A)} d\hat{v} +$$

$$\hat{a} \sum_{j=1}^{2} L_{ij}^3 \frac{\partial \mu_j}{\partial x}. \quad (4)$$

By definition,

$$A^{(A)} = \mu_C + \nu_1^{(A)} \mu_1^{BC} + \nu_2^{(A)} \mu_2^{BC} - \mu_B, \quad (5)$$

where μ_i^{BC} is the *actual* value of μ_i in the reacting slab. At equilibrium, $A^{(A)} = 0$, so

$$\mu_B - \mu_C = \nu_1^{(A)} \mu_1^{BC*} + \nu_2^{(A)} \mu_2^{BC*} \quad (6)$$

where μ_i^{BC*} is the equilibrium value of μ_i in contact with **B** and **C**. Substituting (5) and (6) into (4), we obtain

$$\int_{\hat{v}} \frac{\partial c_i}{\partial t} d\hat{v} = \nu_i^{(A)} L^{(A)}$$

$$\int_{\hat{v}} \sum_{j=1}^{2} \nu_j^{(A)} (\mu_j^{BC} - \mu_j^{BC*}) d\hat{v}$$

$$+ \hat{a} \sum_{j=1}^{2} L_{ij}^3 \frac{\partial \mu_j}{\partial x} \quad (7)$$

In a true steady state, the rate of accumulation of **i** in the grain boundary network of the slab (given by the left-hand side of Equation 7) must be zero, and the net rate of production of **i** within the slab (the first term on the right) must exactly balance the net flux out of the slab (the second term on the right). We may define a quasi-steady state as one in which the accumulation term is negligibly small in comparison with both the diffusive term and the production term. The methods of approximation theory (Kline, 1965) will now be used to formulate the conditions under which the accumulation term can be neglected.

We begin by redefining the principal variables of Equation 7 in a dimensionless form, chosen in such a way that we can estimate the order of magnitude of the derivatives corresponding to those in (7) (*cf*. Kline, 1965, pp. 101-118):

$$\bar{c}_i \equiv \frac{c_i - c_i^{AB*}}{c_i^{BC} - c_i^{AB*}} \qquad \bar{x} \equiv \frac{x - x^{BC}}{x^{AB} - x^{BC}}$$

$$\bar{\mu}_i \equiv \frac{\mu_i - \mu_i^{BC}}{\mu_i^{AB*} - \mu_i^{BC}} \qquad \bar{t} \equiv \frac{t}{t^*},$$

where c_i^{AB*} and μ_i^{AB*} are the equilibrium values for the assemblage **AB**, c_i^{BC} and μ_i^{BC} are the actual values for the assemblage **BC**, $x^{AB} - x^{BC}$ is the thickness of the zone of **B**, and t^* is the time for the zone of **B** to grow to two-thirds of its final thickness measured after all growth has ceased.

At the **B/C** boundary, μ_i is by definition μ_i^{BC}, so $\bar{\mu}_i$ is zero; at the **AB/B** boundary, $\mu_i = \mu_i^{AB*}$ by our initial assumption, so $\bar{\mu}_i$ there will be one. Similarly, $\bar{x}$ will run from zero at the **B/C** boundary to one at the **AB/B** boundary. For most processes, therefore, $d\bar{\mu}_i/d\bar{x}$ will have an *order of magnitude* of approximately one, which we will indicate by the notation $d\bar{\mu}_i/d\bar{x} \approx 1$.

Similarly, c_i will initially have the value c_i^{AB*}, so that $\bar{c}_i$ will be zero at time $t = 0$, when the **C** layer is just starting to grow. As the layer of **C** grows, c_i at any point will by definition approach c_i^{BC}, so that $\bar{c}_i$ will approach one. The variable $\bar{t}$ will vary from near zero in the early stages of the process to values near one in the later stages; when all growth ceases, $\bar{t} = 3/2$ by definition of t^*. Because $\bar{c}_i$ has an upper bound and $\bar{t}$ does not, a plot of $\bar{c}_i$ against $\bar{t}$ will be convex upward, and $0 \leq d\bar{c}_i/d\bar{t} \leq 1$ during most of the growth process.

Substituting these dimensionless variables into the first and last terms of (7)

$$\frac{(c_i^{BC} - c_i^{AB*})}{t^*} \int_{\hat{v}} \left(\frac{\partial \bar{c}_i}{\partial \bar{t}}\right) d\hat{v} = \nu_i^{(A)} L^{(A)}$$

$$\int_{\hat{v}} \sum_{j=1}^{2} \nu_j^{(A)} (\mu_j^{BC} - \mu_j^{BC*}) d\hat{v} +$$

$$\hat{a} \left[L_{i1}^3 \frac{\mu_1^{AB*} - \mu_1^{BC}}{x^{AB} - x^{BC}} \left(\frac{\partial \bar{\mu}_1}{\partial \bar{x}}\right) \right.$$

$$\left. + L_{i2}^3 \frac{\mu_2^{AB*} - \mu_2^{BC}}{x^{AB} - x^{BC}} \left(\frac{\partial \bar{\mu}_2}{\partial \bar{x}}\right) \right]. \quad (8)$$

Next we make all of the terms in (8) dimensionless by dividing through by $\hat{a} L_{11}^3 (\mu_i^{AB^*} - \mu_i^{BC}) / (x^{AB} - x^{BC})$,

$$\frac{(c_i^{BC} - c_i^{AB^*})(x^{AB} - x^{BC})}{\hat{a} t^* L_{11}^3 (\mu_1^{AB^*} - \mu_1^{BC})}$$

$$\int_{\hat{v}} \left(\frac{\partial \bar{c}_i}{\partial \bar{t}}\right) d\hat{v} = \frac{\nu_i^{(A)} L^{(A)}(x^{AB} - x^{BC})}{\hat{a} L_{11}^3 (\mu_1^{AB^*} - \mu_i^{BC})}$$

$$\int_{\hat{v}} \sum_{j=1}^{2} \nu_j^{(A)} (\mu_j^{BC} - \mu_j^{BC^*}) d\hat{v}$$

$$+ \left[\left(\frac{\partial \bar{\mu}_1}{\partial \bar{x}}\right) + \frac{L_{12}^3 (\mu_2^{AB^*} - \mu_2^{BC})}{L_{11}^3 (\mu_1^{AB^*} - \mu_1^{BC})} \left(\frac{\partial \bar{\mu}_2}{\partial \bar{x}}\right) \right]. \quad (9)$$

Each term in Equation 9 has exactly the same significance and the same relative magnitude as in Equation 7. The left-hand side again represents the rate of accumulation of **i** in the grain boundary network of the slab; the first right-hand term, the rate of production of **i** within the slab; and the second right-hand term, the flux of **i** out of the slab. The advantage of rewriting (7) in the form of (9) is that the relative order of magnitude of each of the terms can be estimated simply by evaluating the coefficient of each of the derivative terms. For example, $\bar{\mu}_i$ and $\bar{x}$ were defined so that $\partial \bar{\mu}_i / \partial \bar{x} \approx 1$; therefore, the relative order of magnitude of the two diffusion terms enclosed in brackets depends entirely upon the dimensionless ratio $L_{12}^3 (\mu_2^{AB^*} - \mu_2^{BC}) / L_{11}^3 (\mu_1^{AB^*} - \mu_1^{BC})$, not upon the values of the derivatives. If

$$\frac{L_{12}^3 (\mu_2^{AB^*} - \mu_2^{BC})}{L_{11}^3 (\mu_1^{AB^*} - \mu_1^{BC})} << 1, \quad (10)$$

the second diffusion term will be negligible relative to the first, and the flux out of the slab will closely approximate $\partial \bar{\mu}_1 / \partial \bar{x}$, of order one. We shall assume that this is the case.[3]

Now $\bar{c}_i$ and $\bar{t}$ were defined so that $0 \leq (\partial \bar{c}_i / \partial \bar{t}) \leq 1$ for most of the time the **C** layer is growing. At any time during this interval, we will estimate the mean value of $\partial \bar{c}_i / \partial \bar{t}$ in the volume $\hat{v}$ to be approximately half of the value of $\partial \bar{c}_i / \partial \bar{t}$ at the right-hand face of the slab, $(\partial \bar{c}_i / \partial \bar{t})_{X_s}$. Accordingly, we may substitute $\frac{1}{2}(\partial \bar{c}_i / \partial \bar{t})_{X_s}$ for $\partial \bar{c}_i / \partial \bar{t}$ in Equation 9, and integrate the left-hand side. In addition, we will approximate the mean value of $(\mu_j^{BC} - \mu_j^{BC^*})$ by $\frac{1}{2}(\mu_j^{BC} - \mu_j^{BC^*})_{X_s}$, allowing us to integrate the production term in (9). Remembering that $\hat{v} = \hat{a} X_s$, we obtain

$$\frac{X_s (c_i^{BC} - c_i^{AB^*})(x^{AB} - x^{BC})}{2 t^* L_{11}^3 (\mu_1^{AB^*} - \mu_1^{BC})} \left(\frac{\partial \bar{c}_i}{\partial \bar{t}}\right)_{X_s}$$

$$= \left[\frac{\partial \bar{\mu}_1}{\partial \bar{x}}\right] + \frac{X_s \nu_i^{(A)} L^{(A)}(x^{AB^*} - x^{BC})}{2 L_{11}^3 (\mu_1^{AB^*} - \mu_1^{BC})}$$

$$\sum_{j=1}^{2} \nu_j^{(A)} (\mu_j^{BC} - \mu_j^{BC^*})_{X_s}. \quad (11)$$

The criterion for a quasi-steady state is simply that the left-hand side of (11) be small compared with either the diffusion term or the production term; because $\partial \bar{\mu}_i / \partial \bar{x} \approx 1$ and $0 \leq \partial \bar{c}_i / \partial \bar{t} \leq 1$, the left-hand will be small relative to the diffusion term whenever[4]

$$\frac{X_s (c_i^{BC} - c_i^{AB^*})(x^{AB} - x^{BC})}{2 t^* L_{11}^3 (\mu_1^{AB^*} - \mu_1^{BC})} << 1. \quad (12)$$

All the parameters in (12) except t^* can be measured in the rock itself or estimated thermodynamically. To estimate t^* we need to determine whether the rate of the overall process depends mainly upon the rate of diffusion or the rate of reaction (**A**).

---

[3] If (10) does not hold, two cases are possible: The left-hand side of (10) may be approximately one, or it may be greater than one. If it is approximately one, the diffusive flux out of the slab will be approximately two, still of *order* unity. If the left-hand side of (10) is greater than one, an equation analogous to (9) should be obtained by dividing (8) through by $\hat{a} L_{12}^3 (\mu_2^{AB^*} - \mu_2^{BC}) / (x^{AB} - x^{BC})$, instead of by $\hat{a} L_{11}^3 (\mu_1^{AB^*} - \mu_1^{BC}) / (x^{AB} - x^{BC})$ and the inverse of (10) may then be used to eliminate one of the diffusion terms.

[4] The inequality (12) is a condition sufficient to assure the applicability of the quasi-steady approximation to a particular metamorphic process of the type considered here, but it is not a necessary condition for a quasi-steady state, because $\partial \bar{c}_i / \partial \bar{t}$ may be small enough to make the left-hand side of (11) small even when (12) does not hold.

Criteria for Distinguishing between Diffusion-Controlled and Reaction-Controlled Processes

If reaction (**A**) is fast relative to the rate of diffusion through the zone of pure **B**, the reaction zone will be in local equilibrium and $\mu_i = \mu_i^{BC*}$. The rate of the overall process will then depend upon the rate of diffusion through zone **B**, given by Equation 1, and the process may be termed diffusion-controlled. On the other hand, if reaction (**A**) is slow relative to diffusion, the potentials in the reacting zone will be determined by those in the surrounding rock, and $\mu_i = \mu_i^{AB*}$. In this case, the rate of the overall process will be that of reaction (**A**), given by Equation 2, and the process will be reaction-controlled.

To determine whether the structure in Fig. 1 is diffusion-controlled or reaction-controlled, we assume that the process has reached a steady state; then the left-hand side of (11) will be zero, and

$$\left(\frac{\partial \bar{\mu}_i}{\partial x}\right) = \frac{X_s \nu_i^{(A)} L^{(A)} (x^{AB} - x^{BC})}{2 L_{ii}^3 (\mu_i^{BC} - \mu_i^{AB*})}$$

$$\sum_{j=1}^{2} \nu_j^{(A)} (\mu_j^{BC} - \mu_j^{BC*}) X_s$$

or

$$\left[\frac{2 L_{ii}^3}{X_s \nu_i^{(A)} L^{(A)} (x^{AB} - x^{BC})}\right] \left(\frac{\partial \bar{\mu}_i}{\partial x}\right)$$

$$= \frac{\nu_1^{(A)} (\mu_1^{BC} - \mu_1^{BC*}) X_s}{(\mu_i^{BC} - \mu_i^{AB*})}$$

$$+ \frac{\nu_2^{(A)} (\mu_2^{BC} - \mu_2^{BC*}) X_s}{(\mu_i^{BC} - \mu_i^{AB*})} \quad (13)$$

Since $\partial \bar{\mu}_i / \partial x \approx 1$, the value of the left-hand side of (13) depends upon the dimensionless ratio in brackets. If

$$\left| \frac{2 L_{ii}^3}{X_s \nu_i^{(A)} L^{(A)} (x^{AB} - x^{BC})} \right| \ll 1, \quad (14)$$

both sides of (13) must be much smaller than one. Potential differences between interacting assemblages in metamorphic rocks are commonly small (*cf.* Fisher, 1970), so the denominator ($\mu_i^{BC} - \mu_i^{AB*}$) should never be very large. Accordingly, the right-hand side of 13 can assume small values only if

$$\nu_1^{(A)} (\mu_1^{BC} - \mu_1^{BC*}) \cong - \nu_2^{(A)} (\mu_2^{BC} - \mu_2^{BC*}),$$

or

$$\frac{\mu_1^{BC} - \mu_1^{BC*}}{\mu_2^{BC} - \mu_2^{BC*}} \cong - \frac{\nu_2^{(A)}}{\nu_1^{(A)}}. \quad (15)$$

Equation 15 represents the slope of a line in a chemical potential plot connecting the actual potentials in the reacting slab ($\mu_1^{BC} \mu_2^{BC}$) with a pair of equilibrium potentials ($\mu_1^{BC*} \mu_2^{BC*}$). But μ_1^{BC*} and μ_2^{BC*} lie on the univariant equilibrium representing assemblage **BC**, which has the slope

$$\frac{\partial \mu_1}{\partial \mu_2} = - \frac{\nu_2^{(A)}}{\nu_1^{(A)}}. \quad (16)$$

Therefore, μ_1^{BC} and μ_2^{BC} also lie on the equilibrium **BC**, the system is in local equilibrium, and the process will be diffusion-controlled.

On the other hand, if

$$\frac{2 L_{ii}^3}{X_s \nu_i^{(A)} L^{(A)} (x^{AB} - x^{BC})} \gg 1, \quad (17)$$

the right-hand side of (13) must be much larger than one. This condition implies that $\mu_i^{BC} = \mu_i^{AB*}$, as will occur whenever the potentials in the reaction zone are imposed by diffusion out of the surrounding rock, and the process is reaction-controlled.

Thus evaluation of the dimensionless ratio in inequalities (14) and (17) provides a simple method for determining whether a given process is diffusion-controlled or reaction-controlled.[5] Values of the $L_{ij}{}^n$ can be estimated from the corresponding diffusion coefficients ($D_{ij}{}^n$) using the approximate relation (Katchalsky and Curran, 1967, p. 101)

$$L_{ij}{}^n = \frac{D_{ij}{}^n c_i}{RT}, \quad (18)$$

but unfortunately there appear to be virtually no reliable values of the $D_{ij}{}^n$ for grain boundary diffusion in metamorphic rocks. Figure 2 shows measured diffusion coefficients for cations in a variety of diffusion paths. Kovalev (1971) determined self-diffusion coefficients for Fe, Ca, and S in granodiorite. His measurements probably included some transport by infiltration as well as by diffusion, and his coefficients may be too high. Still, they are in the upper range of grain boundary diffusion coefficients in metals and do not appear unreasonably high. Much of the total mass transfer along grain boundaries in metamorphic rocks may occur in channels one or two atomic spaces thick (see review in Elliott, 1973). These channels are highly disordered, with a high concentration of vacancies, and may approach the structure of a silicate glass. If so, self-diffusion data for silicate glasses may be a rough guide to silicate grain boundary diffusion. Diffusion coefficients for many silicate glasses are in the lower range of grain boundary diffusion coefficients for metals, and, when extrapolated to lower temperatures, approximate the measured diffusion coefficients through surface reaction layers in silicates (Fig. 2). On the

[5] Similar dimensionless ratios are widely used in chemical engineering. For example, the Damköhler Group II, $UL/D_v c_i$ (where U is the reaction rate per unit volume, L is a characteristic length, D_v, the molecular diffusivity, and c_i, the molar concentration of i) gives the ratio of a chemical reaction to the rate of molecular diffusion (Boucher and Alves, 1959); and the group D/kd (where D is the diffusion coefficient, k, a surface reaction rate constant, and d, the thickness of the diffusion layer) gives the effect of the surface reaction rate on a reaction involving diffusion through a surface layer (Bird, Stewart, and Lightfoot, 1966, p. 531).

basis of this rather sketchy data, it seems probable that $D_{ij}{}^n$ for grain boundary diffusion in silicates is on the order of 10^{-9} to 10^{-13} m²/sec (10^{-5} to 10^{-9} cm²/sec) at 500°C. From Equation 18, $L_{ij}{}^n$ for diffusion through a grain boundary network one molar in i will then be on the order of 10^{-12} to 10^{-16} mol·sec/m³ (10^{-18} to 10^{-22} mol·sec/cm³).

Evaluation of $L^{(A)}$ can be approached in two ways. Many silicate dissolution reactions appear to depend on diffusion through a thin surface layer coating the reactant mineral (Helgeson, 1971; Luce and others, 1972). For reactions of this type, $J_i^{(A)}$ is equal to the flux of ions through the surface layer divided by the thickness of the grain boundary into which i is diffusing. Therefore, comparing (1) and (2),

$$L^{(A)} = \frac{\hat{L}_{ij}{}^n}{x_l x_g \nu_i^{(A)}}, \quad (19)$$

where $\hat{L}_{ij}{}^n$ is the coefficient for diffusion of i through the surface layer, x_l thick, and x_g is the thickness of the grain boundary channel. Substituting (19) into (14) and noting that $\hat{L}_{ij} \simeq L_{ij}{}^n$ (by extrapolation of results of Luce et al., 1972; see Fig. 2), the criterion for local equilibrium becomes

$$\frac{x_l \, x_g}{X_s (x^{AB} - x^{BC})} \ll 1. \quad (20)$$

We may determine whether a metamorphic process involving a reaction of this type is diffusion-controlled (meaning that the rate of the overall process is determined by the rate of diffusion through the layer of pure **B**) or reaction-controlled (meaning that the rate of the process is determined by the rate of transfer of material between mineral grains and adjacent grain boundaries) by evaluating the left-hand side of (20). In a typical metamorphic structure, ($x^{AB} - x^{BC}$) may be about 10 mm, and x_l one micron. The value of X_s can be no less than the thickness of a grain boundary, and more commonly it is on the order of 1 mm; here we use the latter value. The grain boundary thickness

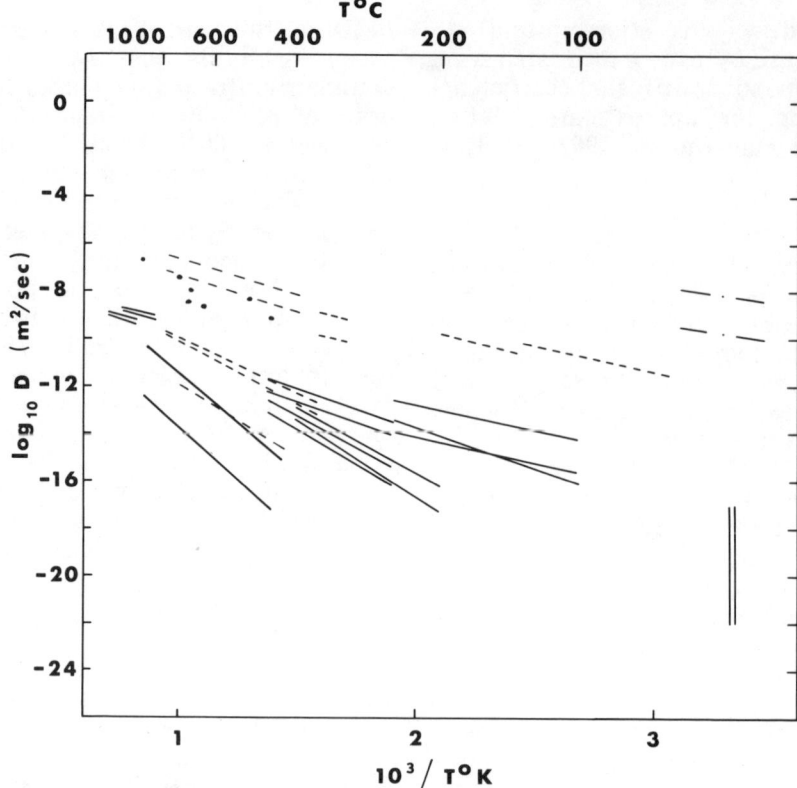

Fig. 2. Arrhenius plot of $\log_{10} D$ versus $1/T$ for grain boundary self-diffusion in metals (dashed lines, from Gibbs and Harris, 1970), cation self-diffusion in silicate glasses and melts (solid lines, data compiled from Diffusion Data, and from Varshneya and Cooper, 1972), in grain boundaries of granodiorite (dots, after Kovalev, 1971), in aqueous solutions (dot-dash lines represent typical range of coefficients from Robinson and Stokes, 1959) and in surface reaction layers on silicates (vertical bar, after Luce and others, 1972).

may range between two extremes (see review in Elliott, 1973): the effective grain boundary thickness may be that of a high diffusivity channel, about one nanometer (10 Å) thick, or it may be the thickness of a partially disordered zone rich in impurities flanking the high diffusivity channel, and about one micron thick. Accordingly, x_g should be no less than 10^{-6} mm and no more than 10^{-3} mm. Assuming these values, the left-hand side of (20) may vary from 10^{-10} to 10^{-7}. Such a process will therefore be diffusion-controlled. In order for a process to be reaction-controlled, the left-hand side of (20) must be much greater than one. For this condition to hold, the product of the reaction zone thickness and the thickness of the zone of pure **B** must be substantially *less* than the product of the grain boundary thickness and the thickness of the surface layer coating the reacting minerals; obviously, this is possible only in the earliest stages of growth, and most processes involving a dissolution reaction of this type will attain local equilibrium as soon as the structure is large enough to be seen microscopically. It is important to note that this conclusion is independent of any uncertainty about the value of L_{ij}^n, which does not appear in (20).

Other reaction mechanisms are possible, though; van Lier and others (1960) report that the rate of dissolution of quartz is controlled by a chemical reaction at the quartz-solution interface, rather than by diffusion through a surface layer. Extrapolating their surface rate constants for the dissolution of quartz to 500°C on an Arrhenius plot, and again assuming a grain boundary thickness between one micron and one nanometer, we obtain a volumetric rate constant on the order of 10^2 to 10^5 kg/m^3·sec (10^{-1} to 10^2 gm/cm^3·sec). If $k^{(A)}$ is between those values, calculations analogous to those of Katchalsky and Curran (1967, pp. 92–94) yield a value for $L^{(A)}$ between 10^{-2} and 10 mol·sec/m^5 (10^{-12} to 10^{-9} mol·sec/cm^5). If these values are correct, the left-hand side of (14) will range between 10^{-12} and 10^{-5}, again assuming L_{ij}^n on the order of 10^{-12} to 10^{-16} mol·sec/m^3 (10^{-18} to 10^{-22} mol·sec/cm^3), X_s, one millimeter, and ($x^{AB} - x^{BC}$), 10 mm. Once again, the process appears to be diffusion-controlled.

At higher temperatures, diffusion control becomes even more likely, because most solid-solid reactions have larger activation energies (around 50 kcal/mole) than most grain boundary diffusion (around 20 kcal/mole), so that the left-hand side of (18) will tend to decrease even further with increasing temperature. At lower temperatures, the difference in the activation energies will tend to favor reaction control, but the left-hand side of (14) will be substantially less than one under most metamorphic conditions; using the above numerical values for the parameters in (14), the left-hand side approaches one only at temperatures below 25°C.

These provisional evaluations of (14) and (20) suggest that metamorphic processes which approach a steady state will be diffusion-controlled, so that local equilibrium will prevail. Indeed, diffusion control is so strongly favored that it seems reasonable to extend this conclusion even to processes which are rather far from a steady state; accordingly, we will henceforth assume that $\mu_i^{BC} = \mu_i^{BC*}$ for all processes. It should be remembered, though, that the values of the phenomenological coefficients in (14) are not well known; data on diffusion and metamorphic reaction kinetics are badly needed! Lower values of $L^{(A)}$, higher values of L_{ij}^n, or inhibitions of reactions by foreign ions along the grain boundary (Terjesen and others, 1961) could increase the value of the left-hand side of (14), favoring reaction control.

Evaluation of t^*

Assuming that the overall process is diffusion-controlled, we may now estimate the value of t^* in condition (12). Using Equation 1,

$$\frac{\partial x^{AB}}{\partial t} = -\Delta V_i^{AB} \sum_{j=1}^{2} L_{ij}^3 \left(\frac{\partial \mu_j}{\partial x}\right)_{AB/B}$$

$$\frac{\partial x^{BC}}{\partial t} = -\Delta V_i^{BC} \sum_{j=1}^{2} L_{ij}^3 \left(\frac{\partial \mu_j}{\partial x}\right)_{C/B} \quad (21)$$

where ΔV_i^{AB} is the change in volume of the zone of pure **B** as a result of reaction of one mole of **i** with the assemblage **AB**, and ΔV_i^{BC} is the change in volume of the zone of **C** as a result of liberation of one mole of **i** by the assemblage **BC**. The values of ΔV_i^{AB} and ΔV_i^{BC} depend primarily upon the modal distribution of reactants for reactions (**C**) and (**A**); for example, if reactants for (**C**) are widely dispersed in the surrounding rock, consumption of one mole of **i** will convert a large volume of rock from the assemblage **AB** to **B**. For components diffusing outward from **BC** to **AB**, the shape of the potential profile in the zone of **B** requires that

$$\left|\frac{\partial \mu_j}{\partial x}\right|_{AB/B} \leq \left|\frac{\mu_j^{AB*} - \mu_j^{BC*}}{x^{AB} - x^{BC}}\right|;$$

$$\left|\frac{\partial \mu_j}{\partial x}\right|_{C/B} \geq \left|\frac{\mu_j^{AB*} - \mu_j^{BC*}}{x^{AB} - x^{BC}}\right|.$$

Hence, from Equation 21,

$$\frac{\partial x^{AB}}{\partial t} \leq \Delta V_i^{AB} \left(L_{i1}^3 \frac{\mu_1^{AB*} - \mu_1^{BC}}{x^{AB} - x^{BC}} + L_{i2}^3 \frac{\mu_2^{AB*} - \mu_2^{BC*}}{x^{AB} - x^{BC}}\right)$$

$$\frac{\partial x^{BC}}{\partial t} \geq \Delta V_i^{BC} \left(L_{i1}^3 \frac{\mu_1^{AB*} - \mu_1^{BC}}{x^{AB} - x^{BC}}\right.$$

$$+ L_{i2}{}^3 \frac{\mu_2{}^{AB*} - \mu_2{}^{BC*}}{x^{AB} - x^{BC}} \Bigg). \quad (22)$$

In order for the zone of **B** to grow,

$$\frac{\partial x^{AB}}{\partial t} \geq \frac{\partial x^{BC}}{\partial t}.$$

Therefore, dropping the $L_{i2}{}^3$ terms because of (10),

$$\left| \frac{\partial (x^{AB} - x^{BC})}{\partial t} \right| \leq$$

$$\left| (\Delta V_i{}^{AB} - \Delta V_i{}^{BC}) \right.$$
$$\left. L_{i1}{}^3 \frac{\mu_1{}^{AB*} - \mu_1{}^{BC*}}{x^{AB} - x^{BC}} \right|. \quad (23)$$

Rearranging (23) and integrating,

$$t^* \geq \left| \frac{(x^{AB} - x^{BC})^2{}_{t^*}}{2 L_{i1}{}^3 \Delta V_i (\mu_1{}^{AB*} - \mu_1{}^{BC*})} \right|, \quad (24)$$

where $\Delta V_i = \Delta V_i{}^{AB} - \Delta V_i{}^{BC}$, and $(x^{AB} - x^{BC})_{t^*}$ is the value of $x^{AB} - x^{BC}$ at time t^*. Substituting (24) into (12), and remembering that by definition of t^*, $(x^{AB} - x^{BC})_{t^*}$ equals two-thirds of the final value of $(x^{AB} - x^{BC})$, the condition for a quasi-steady state becomes

$$\left| \frac{9 X_s \Delta V_i (c_i{}^{BC} - c_i{}^{AB*})}{4 (x^{AB} - x^{BC})} \right| \ll 1. \quad (25)$$

Trial calculations suggest that this condition will be met by most metamorphic processes; for the Vastervik segregations (Loberg, 1963, Fig. 10; Fisher, 1970) the left-hand side is on the order of 10^{-6}. Condition (25) will be likely to fail only in the case of very small structures, structures formed in the presence of large concentration gradients, or in the case of components for which ΔV_i is very large. Accordingly, it appears safe to assume that in most metamorphic processes, the concentrations of species within the reacting slab will be approximately constant during growth of the new assemblage. Therefore, the affinities and the rate of reaction must be nearly constant, and diffusion out of the slab must be quasi-steady, and the kinetics of the process may be evaluated by the relatively simple steady-state approach (Fisher, 1973). Once again, it is important to note that this conclusion is independent of our uncertainty about the values of $L_{ij}{}^n$ and $L^{(A)}$, which do not appear in (25).

Conclusion

By reformulating the conservation equations governing growth of metamorphic minerals and structures in dimensionless form, we have derived the dimensionless ratios needed to evaluate whether a growth process will be diffusion-controlled or reaction-controlled, and whether or not growth can be approximated by steady-state diffusion.

If the ratio

$$\left| \frac{2 L_{ij}{}^n}{X_s \nu_i{}^{(r)} L^{(r)} (x^\alpha - x^\beta)} \right|$$

is substantially less than one (say 0.1 or less) for a process involving diffusion of any component **i** between assemblages α and β, μ_i will be fixed by equilibria with the local assemblages. For component **i**, diffusion between assemblages will be the rate determining step, and the process will be diffusion-controlled with respect to component **i**. On the other hand, if the ratio is much greater than one (say 10 or more) for another component, **j**, then μ_j will be nearly uniform throughout the entire structure, and its value will be that imposed by the original assemblage. For component **j**, the rate of reaction will be the rate determining step, and the process will be reaction-controlled with respect to component **j**.

Order-of-magnitude calculations suggest that most metamorphic growth processes will be diffusion-controlled, and hence that the common assumption of local equilibrium in metamorphic rocks (Thompson, 1959; Orville, 1962; Carmichael, 1969; and Fisher, 1970) is justified in most cases.

Assuming that local equilibrium prevails, growth of assemblage β at the expense of assemblage α will be approximated by steady-state diffusion provided that

$$\left| \frac{9 X_s \Delta V_i (C_i{}^\beta - C_i{}^\alpha)}{4 (x^\alpha - x^\beta)} \right| \ll 1.$$

Trial calculations suggest that most metamorphic processes will conform to this condition, and therefore that potential gradients in most metamorphic processes can be determined by minimizing the rate of entropy production (Fisher, 1973). The main exceptions will be components for which ΔV_i is very large, as it will be for components which are strongly partitioned in favor of a mineral growing in assemblage β, and for all components in reactions consuming a mineral that is widely dispersed in the assemblage being replaced.

Acknowledgments

We have benefited greatly from constructive reviews of an early draft of this manuscript by R. C. Fletcher and A. R. Cooper. Financial support for this research was provided by Earth Science Section National Science Foundation Grants GA-29685 (to G.W.F.), and GA-10279 (to D.E.).

References Cited

Bird, R. B., W. E. Stewart, and E. N. Lightfoot, *Transport Phenomena*, Wiley, New York, 780 pp., 1966.

Boucher, D. R., and G. E. Alves, Dimensionless numbers, *Chem. Eng. Progr.*, 55, no. 9, 55-64, 1959.

Carmichael, D. M., On the mechanism of prograde metamorphic reactions in quartz-bearing pelitic rocks, *Contrib. Mineral. Petrol.*, 20, 244-267, 1969.

de Groot, S. R., and P. Mazur, *Nonequilibrium Thermodynamics*, North-Holland Publishing Co., Amsterdam, 510 pp., 1962.

Elliott, David, Diffusion flow laws in metamorphic rocks, *Geol. Soc. Amer. Bull.*, 84, 2645-2664, 1973.

Fisher, G. W., The application of ionic equilibria to metamorphic differentiation: an example, *Contrib. Mineral. Petrol.*, 29, 91-103, 1970.

Fisher, G. W., Nonequilibrium thermodynamics as a model for diffusion-controlled metamorphic processes, *Amer. J. Sci.*, 273, 897-924, 1973.

Garrels, Robert M., Rates of geochemical reactions at low temperatures and pressures, in *Researches in Geochemistry*, Vol. 1, P. H. Abelson, ed., Wiley, New York, pp. 25-37, 1959.

Gibbs, G. B., and J. E. Harris, Diffusion at solid-solid interfaces, in *Interfaces*, R. C. Gifkins, ed., Australian Institute of Metals in Association with Butterworths, Melbourne, pp. 53-76, 1969.

Helgeson, H. C., Kinetics of mass transfer among silicates and aqueous solutions, *Geochim. Cosmochim. Acta*, 35, 421-469, 1971.

Katchalsky, A., and P. F. Curran, *Nonequilibrium Thermodynamics in Biophysics*, Harvard Univ. Press, Cambridge, 248 pp., 1967.

Kline, Stephen J., *Similitude and Approximation Theory*, McGraw-Hill, New York, 229 pp., 1965.

Kovalev, G. N., O diffuzii po granitsam zeren v gornoy v prisutstvii vody (On diffusion through the boundaries of grains in a rock in the presence of water), *Dokl. Akad. Nauk SSSR*, 197, no. 6, 1410-1412, 1971.

Loberg, B., The formation of a flecky gneiss and similar phenomena in relation to the migmatite and vein gneiss problem, *Geol. Fören. Stockholm Förh.*, 85, 3-109, 1963.

Luce, R. W., R. W. Bartlett, and G. A. Parks, Dissolution kinetics of magnesium silicates, *Geochim. Cosmochim. Acta*, 36, 35-50, 1972.

Orville, P. M., Alkali metasomatism and feldspars, *Norsk Geol. Tidsskrift*, 42 (Feldspar volume), 283-316, 1962.

Robinson, R. A., and R. M. Stokes, *Electrolyte Solutions*, 2nd ed., Butterworth's, London, 559 pp., 1959.

Terjesen, S. G., O. Erga, G. Thorsen, and P. Ve, Phase boundary processes as rate-determining steps in reactions between solids and liquids, *Chem. Eng. Sci.*, 14, 277-288, 1961.

Thompson, J. B., Jr., Local equilibrium in metasomatic processes, in *Researches in Geochemistry*, Vol. 1, P. H. Abelson, ed., Wiley, New York, 427-457, 1959.

van Lier, J. A., P. L. de Bruyn, and J. Th. G. Overbeek, The solubility of quartz, *J. Phys. Chem.*, 64, 1675-1682, 1960.

Varshneya A. K., and A. R. Cooper, Diffusion in the system K_2O-SrO-SiO_2: II cation self-diffusion coefficients, *J. Amer. Ceram. Soc.*, 55, 220-223, 1972.

Vidale, R., Metasomatism in a chemical gradient and the formation of calc-silicate bands, *Amer. J. Sci.*, 267, 857-874, 1969.

Zen, E'an, Construction of pressure-temperature diagrams for multicomponent systems after the method of Schreinemakers—a geometric approach, *U.S. Geol. Surv. Bull. 1225*, 56 pp., 1966.

SIMPLE MODELS OF DIFFUSION AND COMBINED DIFFUSION-INFILTRATION METASOMATISM

R. C. Fletcher and A. W. Hofmann
Department of Terrestrial Magnetism
Carnegie Institution of Washington
Washington, D.C. 20015

Introduction

Three mechanisms for chemical transport in masses of rock are (1) solid diffusion through the crystalline phases, (2) diffusion through an intergranular medium or pore fluid, and (3) infiltration, that is, convective transport due to the motion of the pore fluid relative to the solid framework. An important variant of intergranular transport is diffusion, or convection through fluid-filled larger channels such as fissures. Several authors (see for example Fyfe et al., 1958) have convincingly argued that solid diffusion can be effective only on the scale of centimeters because diffusion rates in silicates are slow. This is borne out by several of the experimental contributions presented at this conference. For example, Foland (this volume) found Na diffusion in feldspar to be comparatively rapid with a diffusion coefficient of about 2×10^{-11} cm^2 sec^{-1} at 800°C. A convenient rough measure of the characteristic transport distance in diffusion is the expression $(Dt)^{1/2}$ which yields 80 cm for a duration of 10^6 years. For potassium, with a diffusion coefficient of about 2×10^{-13} cm^2 sec^{-1} (Foland, this volume), this characteristic distance is only about 2.5 cm. At lower temperatures, these distances will be still smaller.

Because of the inefficiency of solid diffusion, attention has focused on the other two mechanisms, those involving an intergranular medium. Korzhinskii (1951, 1952, 1970) has proposed two simple models: diffusion in a stationary pore fluid and infiltration of the pore fluid. In both cases the fluid is assumed to be in local equilibrium with the solid phases. These models are useful idealizations of real metasomatic processes, even in cases where much of the transport occurs through veins. They would be applicable, for example, to the analysis of metasomatic replacements in the rocks adjacent to the veins.

It may be argued that these models are unrealistic in that the grain boundary phase may not be a fluid phase in the ordinary sense of the word, but this is contradicted by the ubiquity of fluid inclusions in metamorphic rocks and the necessity of moving large volumes of fluids (H_2O, CO_2) in regional metamorphism. The assumption of local equilibrium may also be questioned, but the mineral assemblages in zoned skarn deposits, for example, indicate that local equilibrium controlled the formation of these assemblages (Burt, this volume). A detailed discussion of the conditions necessary for local equilibrium is given by Fisher and Elliott (this volume).

Korzhinskii (1951, 1952, 1970) derived transport equations for these metasomatic models and used the equations to obtain general properties of the resulting metasomatic columns. Hofmann (1972) reviewed the infiltration case from the point of view of chromatographic theory, as developed by Wilson (1940), De Vault (1943) and Glueckauf (1947), and applied the transport equations to alkali metasomatism in feldspars.[1]

In this paper we consider the transport of a single component by pure diffusion and combined diffusion and infiltration under the condition of local equilibrium. Local equilibrium is specified by an isotherm relating the concentration of the component in the solid to that in the fluid at fixed temperature and pressure. We derive the equations for transport in one dimension and use these to compute concentration profiles in a porous rock adjoining a well-mixed fluid reservoir.

We shall use "component" here in the thermodynamic sense although the component may consist of more than one migrating species. For example, a component KCl might consist of the species K^+, Cl^-, and KCl, which are linked by electroneutrality and the dissociation equilibrium. A detailed discussion of the use of components and the application of the phase rule in metasomatic systems has been given by Thompson (1959). The important point, for our purposes, is that the amount of the migrating component, or species that make up the component, that is removed from, or released into, the transporting medium by the solid phase(s) must be known everywhere in the system. This requirement is most easily met if local equilibrium prevails and if the migrating component obeys a unique and single-valued function $s = s(c)$, (where s is the concentration of the component in the solid and c is the concentration in the fluid), which can be determined by equilibrium experiments. This approach is applicable to transport of certain trace elements that obey Henry's law, and to simple ion exchange systems such as the compositional system $KAlSi_3O_8$-$NaAlSi_3O_8$-$NaCl$-KCl-H_2O (see Orville, 1963) under the restriction that the total alkali chloride concentration is held constant. This latter restriction can be met rigorously in a pure infiltration process, but diffusion will in general cause the chloride ion concentration to change somewhat in response to other concentration gradients. The application of our results to ion exchange systems is thus valid only if the diffusion-induced variation in chloride-ion concentration can be neglected.

In multicomponent systems, the isotherm $s = s(c)$ may be thought of as the composition path, which depends on the multicomponent equilibrium, the boundary conditions, and the diffusion coefficients. Methods to find this path have been given by Glueckauf (1949) for the case of two-component infiltration and by Cooper (this volume) for multicomponent diffusion.

We give results for a variety of hypothetical fluid-solid isotherms. Some of the isotherms chosen are similar in shape to the experimentally determined ion exchange isotherms for the systems plagioclase–aqueous chloride solution and alkali feldspar–aqueous chloride solution (Orville, 1963, 1972). Both the shape of the isotherm and the metasomatic mechanism have a strong influence on the resulting concentration-distance profile, so that under favorable circumstances the shape of such a profile could be used to determine the metasomatic mechanism in a given geological setting.

The transport equations and the computed profiles are also useful in that they elucidate a number of metasomatic features of more general interest. Among these are the relationship between the parabolic rate law (see, for example, Helgeson, 1971) in diffusion processes and the linear rate law in infiltration

---

[1] Hofmann (1972) erroneously stated that the filtration coefficient in Korzhinskii's infiltration equations violates the mass balance requirement. A clarification of this point is given in Korzhinskii (1973) and Hofmann (1973). Briefly, in the usual chromatographic treatment the solute particles are assumed to move with the same velocity as the solvent fluid and all retardation of the solute relative to the solvent is caused by extraction of solute by the solid phases. Korzhinskii's treatment assumes, in addition, that the solute particles are retarded by an unspecified filtration effect so that, as a result, the velocity of the solvent is greater than that of the solute.

processes, the effect of this relationship on the efficiency of the overall rate of migration, and the close approach to steady-state concentration profiles between advancing diffusion fronts.

Finally, we use the rate laws and estimate transport parameters to set limits on the mechanisms of large-scale metasomatic processes such as granitization and the reconstitution of a residual upper mantle that has been depleted in certain elements by withdrawal of melt.

Analysis

Consider one-dimensional transport in a porous solid with quantities uniform on planes normal to the coordinate direction z. Within the region considered, the concentrations are smooth functions of z and time t.

The mass balance is derived for a lamina in the rock of fixed volume and with fixed geometric surfaces that are normal to the direction of flow. Consider the net mass flow of a single component across surfaces $z + dz/2$ and $z - dz/2$ into the lamina of thickness dz per unit time. Let $J(z)$ be the flux of the component (mass flow across a unit surface of the porous solid in unit time, gm/cm²sec, for example) across the surface at z, and $\partial J/\partial z$ its gradient at z. At the two surfaces the fluxes are $J + (\partial J/\partial z)(dz/2)$ and $J - (\partial J/\partial z)(dz/2)$, respectively. The net addition of the component to the lamina per area A per infinitesimal increment of time dt is then

$$\frac{\partial \rho}{\partial t} A\, dz dt = [J - (\partial J/\partial z)(dz/2)] A\, dt \\ - [J + (\partial J/\partial z)(dz/2)] A\, dt$$

where ρ is the mass of the component per unit volume of porous solid, or

$$\partial \rho/\partial t = -\partial J/\partial z \qquad (1)$$

Let the component be partitioned between the solid, in which its concentration (or density, gm/cm³, for example) is s, and the pore fluid, in which its concentration is c. Let the volume fraction of pore fluid, β, assumed equal to the porosity, be uniform and constant in the region of interest. Then we can write

$$\partial \rho/\partial t = (1 - \beta)\, \partial s/\partial t + \beta\, \partial c/\partial t$$

and if $\beta \ll 1$,

$$\partial \rho/\partial t \cong \partial s/\partial t \qquad (2)$$

The flux J will be supposed to result solely from transport in the fluid. In general it will consist of a flux due to diffusion, J_D, and a flux due to infiltration, J_I, so that

$$J = J_D + J_I \qquad (3)$$

Using Fick's first law

$$J_D = -D\,(\partial c/\partial z) \qquad (4)$$

where D is the effective diffusivity in the porous solid due to diffusion transport in the fluid alone. The effective diffusivity can be written

$$D = \tau\, \beta\, D_i \qquad (5)$$

where D_i is the intrinsic diffusivity in the fluid and τ is a parameter related to the tortuosity of the diffusion path. We assume D is uniform in the region of interest. If $\bar{v}$ is the average velocity of a typical fluid particle in the z direction,

$$J_I = \beta\, \bar{v}\, c \qquad (6)$$

Assuming the geometry of the pore space to be isotropic, β is also the fraction of cross-section area occupied by fluid, and $\beta \bar{v}$ is the volume flux (cm³/cm²sec, for example) or fluid per unit cross section of porous solid lying normal to the z direction.

Substitution of 2, 3, 4, and 6 into 1 yields

$$\partial s/\partial t = D\partial^2 c/\partial z^2 - \beta\, \bar{v}\, \partial c/\partial z \qquad (7)$$

The component will be supposed to transfer sufficiently rapidly between fluid and solid so that an equilibrium distribution is locally attained everywhere. The equilibrium partitioning between fluid and solid at a fixed temperature and pressure is expressed by the isotherm

$$s = s(c) \qquad (8)$$

If the derivative ds/dc is a continuous function of c, it is convenient to eliminate s from (7) using

$$\partial s/\partial t = (ds/dc)\, \partial c/\partial t \qquad (9)$$

Denote dimensionless quantities by capital letters and write

$$z = Z\,z',\ t = T\,t',\ c = C\,c'\ \text{and}\ s = S\,s',$$

where z' and t' are a characteristic length and time, respectively, and s' and c' are characteristic concentrations. Introducing (5) and (9), (7) may be written in the dimensionless form

$$\partial C/\partial T = (dC/dS)(N_1 \partial^2 C/\partial Z^2 - N_2\, \partial C/\partial Z) \qquad (10)$$

where

$$N_1 = \tau\, \beta\, D_i\, (c'/s')t'/(z')^2$$

and

$$N_2 = \beta\, \bar{v}\, (c'/s')t'/z'.$$

The dimensionless isotherm is

$$S = S(C) \qquad (11)$$

and we shall choose s' and c' in such a way that the dimensionless concentrations range from 0 to 1.

Transport Models

Initial and Boundary Conditions

Together with the field equations, (10) and (11) or (7) and (8), a transport model requires the specification of boundary and initial conditions on the fluid concentration C. (S is then specified from the isotherm 11.) For transport into a semi-infinite medium occupying the region $Z > 0$ from a well-mixed fluid reservoir in which the concentration C_R is constant, the boundary condition is

$$C(0, T) = C_R \qquad (12)$$

We suppose that the component is initially absent from the porous medium, so that

$$C(Z, 0) = 0,\ Z > 0 \qquad (13)$$

For this model, the reservoir concentration must be the largest value of C attained, so that in (12) we take $C_R = 1$. Equations 12 and 13 together with the field equations completely specify the class of transport models that we shall consider. An individual model will correspond to a choice of isotherm and of the two dimensionless parameters N_1 and N_2. The application of such a model to an actual transport process also will require the specification of the absolute values of the physical parameters entering N_1 and N_2.

Most of the solutions discussed are obtained from a numerical finite-difference procedure which is described in the Appendix. In some cases, analytic solutions have been used; these are specifically noted.

Diffusion

When the pore fluid is stationary relative to the solid phases, $N_2 = 0$ and Equation 10 assumes the form of Fick's second law modified by the term dC/dS, the slope of the isotherm. Figure 1 shows metasomatic diffusion profiles for three simple solid-liquid isotherms. The isotherms, shown in the inset, all exhibit continuous solid solution. The (dimensionless) fluid concentration C is given as a function of the (dimensionless) solid concentration S.

The linear isotherm (Fig. 1, isotherm B) is representative of trace element equilibria which follow Henry's law or of ideal solid solutions. An example of an isotherm of type E (concave upward) would be plagioclase in equilibrium with $2N$ aqueous (Na, Ca) Cl solution at 700°C and 2 kbar (Orville, 1972), where C represents Ca/(Ca+Na) in the fluid and S represents Ca/(Ca+Na) in the plagioclase. An example of a convex isotherm of type A would be the equivalent plagioclase equilibrium at low pressure (Orville, personal communication, 1966).

For the initial- and boundary-value problem described in the previous section, the solution in the case of diffusion alone

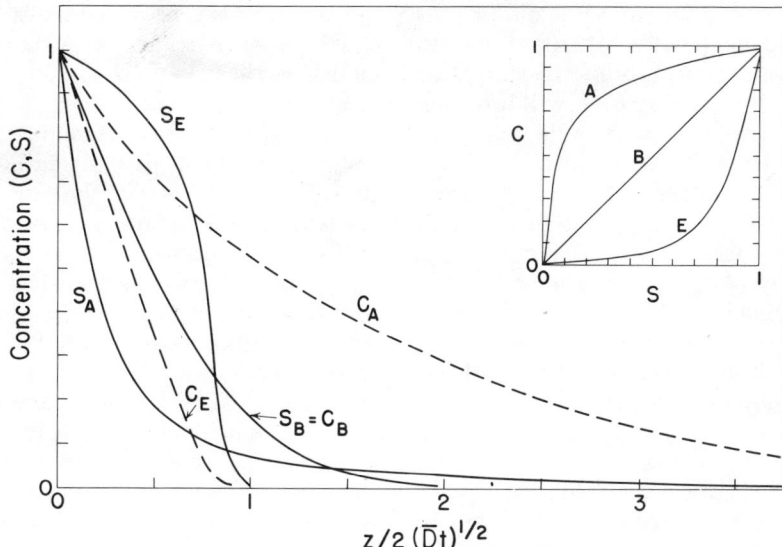

Fig. 1. Concentration profiles for diffusion into a semi-infinite medium. The dimensionless solid concentration S_A, S_B, and S_E (solid lines) and fluid concentrations C_A, C_B, and C_E (dashed lines) are plotted versus dimensionless depth of penetration $z/2(\overline{D}t)^{1/2}$. The profiles correspond to the three equilibrium isotherms A, B, and E (C versus S shown in the inset) having continuous solid solution.

for the linear isotherm of slope $dC/dS = 1$ is

$$C = S = \operatorname{erfc} \frac{z}{2\sqrt{\overline{D}\,t}}$$
$$= \operatorname{erfc} \frac{Z}{2\sqrt{N_1\,T}} \qquad (14)$$

where $\overline{D} = (c'/s')\,D$ (see Crank, 1956, Equation 3.13, and Table 2.1 for a tabulation of the error function). A single concentration profile plotted against the dimensionless parameter $z/2\sqrt{\overline{D}t}$ is valid for any $t > 0$. This representation is applicable to other isotherms as well. The solution (14) has the property that the profile is concave upward for all $z > 0$. This is true of the fluid concentration profiles for any isotherm (see Figs. 1 and 3) if transport occurs by diffusion only and if the diffusion coefficient is constant in the fluid. Notice, however, that the solid concentrations may be concave or convex depending on the shape of the isotherm.

The profiles for the curved isotherms A and E may be discussed conveniently in relation to the "standard" profile $S_B = C_B$. The fluid concentrations C_A move considerably faster than the standard profile. The reason for this may be seen intuitively by inspection of Equation 10 with N_2 set equal to zero. The concentration C increases at a rate proportional to the slope of the isotherm. Because the slope of isotherm A is greatest at the lowest concentrations, these concentration values move much more rapidly than equivalent concentrations C_B. This effect is also evident in the solid-concentration profile S_A. Values of $S_A < 0.05$ move faster than the standard profile, but the higher concentrations are retarded as the slope of the isotherm decreases.

The effects observed on the diffusion curves for isotherm E are complementary. The low slope of the isotherm at low concentrations retards the movement of these concentrations and causes the entire profile to "stub its toe." Diffusion at higher concentrations is faster, so that the overall profile resembles a steady-state diffusion gradient. Except for the short, concave portion at the low-concentration end, the fluid-concentration profile C_E is a nearly straight line, but the solid concentration profile is convex as required by the isotherm.

True steady-state diffusion is shown in Fig. 2 for the same isotherms. This case corresponds to diffusion in a slab bounded by planes where the concentrations are fixed at $C = S = 1$ and $C = S = 0$, respectively, when the steady state has been reached. We shall see further below that the steady-state profile is a good approximation to concentration gradients *between* moving boundaries (caused by solubility gaps). Therefore, steady-state solutions are useful well beyond geological settings where special boundary conditions might lead to a true steady state. This usefulness is further enhanced by the fact that such solutions are mathematically simple. The steady state requires that $dC/dT = 0$, and from Equation 10 it follows immediately that $d^2C/dZ^2 = 0$, so that $dC/dZ = $ const. throughout the diffusion zone. Thus the straight line in Fig. 2 represents the fluid concentrations for all three isotherms shown in Fig. 1. The solid concentrations S_A and S_E have curvatures opposite those of the corresponding isotherms. The curvature of solid-concentration steady-state profiles can be useful in distinguishing between metasomatic mechanisms. For example, steady-state diffusion through the solid (with no transport through the fluid) would produce a straight profile rather than a curved one if the diffusion coefficient is constant in the diffusion zone.

When the diffusion current gives rise to new mineral phases, a replacement front is formed where the solid concentrations S are discontinuous and the fluid concentration gradients dC/dZ are also discontinuous. These fronts correspond to *moving boundaries*. In order to illustrate the close approach to the steady state that is achieved by the concentration profiles between such boundaries, we have computed a diffusion profile for an isotherm containing two solubility gaps.[2] An example of such an isotherm might be the plagioclase series exsolved into three discrete phases at low temperatures. Figure 3 shows that the diffusion profiles between the origin and the first front and between the two fronts are very nearly straight lines, that is, they resemble steady-state profiles. Each front moves with a rate $dZ/dT = K/\sqrt{T}$ where K is a constant that depends on the effective diffusivity, the boundary concentration values, the width of the solubility gap, and the slopes of the isotherm. Physically, a solubility gap acts as a moving sink for the diffusing component. This sink moves sufficiently slowly so that the concentration profile has time to achieve a quasi-steady state.

For comparison with the diffusion fronts shown in Fig. 3, the solution for

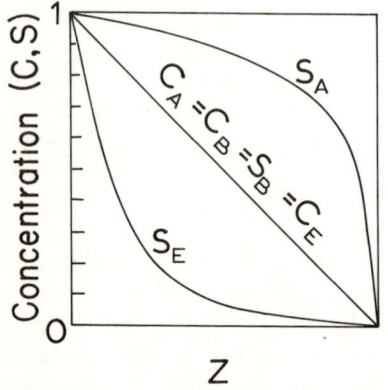

Fig. 2. Concentration profiles for steady-state diffusion through a slab of fixed thickness. The profiles correspond to the three isotherms A, B, and E shown in Fig. 1 (inset).

[2] The solution shown in Fig. 3 was obtained by Neumann's method (see Crank, 1956, p. 109). A separate solution is obtained for each segment between moving boundaries. This is necessary because the differential equation does not apply at the boundaries where the concentrations are not differentiable. A separate mass balance equation for each moving boundary is then used to make the individual solutions compatible with the overall initial and boundary conditions.

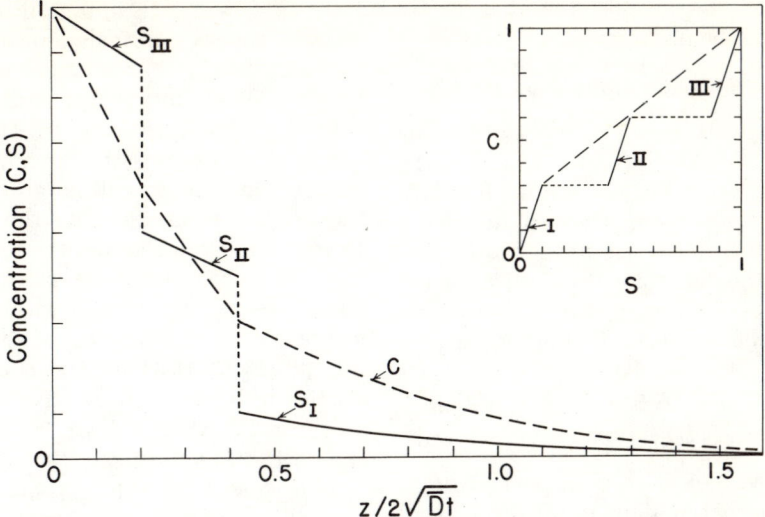

Fig. 3. Diffusion into semi-infinite medium for an isotherm (inset) with two solubility gaps. The three solid phases are labeled S_I, S_II, and S_III. The dotted lines mark the solubility gaps. The dashed line on the isotherm is an auxiliary line used in the construction of the infiltration profile shown in Fig. 4. For other nomenclature see Fig. 1.

pure infiltration is given in Fig. 4 for the same isotherm. Details of how this solution is obtained may be found in the paper by Hofmann (1972). The rate of movement of the boundary of phase III replacing phase I is given by the slope of the auxiliary line on the isotherm (dashed, diagonal line on the isotherm in Fig. 3). Because the auxiliary line "misses" phase II in the configuration of this particular isotherm, phase II is absent in the infiltration case. Clearly, the conspicuous differences between the concentration profiles of Figs. 3 and 4 are diagnostic of the transport if the shape of the isotherm is sufficiently well known.

Infiltration Superposed on Diffusion

With the superposition of a variable amount of infiltration on diffusion, it is no longer useful to represent concentration profiles for the semi-infinite medium in terms of the dimensionless depth $z/2(\bar{D}t)^{1/2}$. We shall use absolute values for the parameters in the present section, retaining the assignment $N_1 = 1$ which emphasizes a fixed contribution from diffusion. We use $\tau = 1$, $\beta = 10^{-3}$, $D_i = 10^{-4}$ cm^2/sec and $c'/s' = 0.1$; the choice of these is discussed later.

If $z' = 1$ cm, the choice $N_1 = 1$ implies $t' = 10^8$ sec $= 3.16$ yr. The relative strength of the contribution of infiltra-

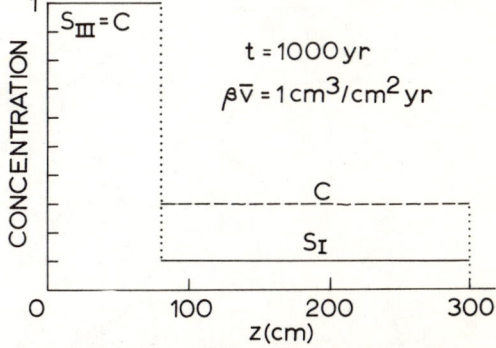

Fig. 4. Pure infiltration profile for the isotherm shown in Fig. 3 (two solubility gaps). The two solid phases present are S_I and S_III. S_II is absent because, as the slope of the auxiliary line on the isotherm (see Fig. 3, inset) indicates, the potential front $S_\text{II}|S_\text{I}$ is overtaken by the faster moving front $S_\text{III}|S_\text{I}$. The dotted lines mark the concentration discontinuities. $\beta\bar{v}$ is the volume flux of the fluid (see Equation 6).

tion to the net transport is indicated by the value of the parameter

$$N_2 = \beta \bar{v} \text{ (cm}^3/\text{cm}^2\text{sec)} \times 10^7,$$

from (10) and the above values. For fluxes of fluid, $\beta\bar{v}$, of 3.16, 1 and 0.316 cm^3/cm^{2}yr, N_2 takes the values 1, 0.316, and 0.1, respectively. One might expect an insignificant contribution from infiltration to the net transport if N_2, as computed above, is very small. This notion is misleading as we shall shortly see.

Infiltration and Diffusion with a Linear Isotherm

Again, we consider transport from a reservoir into the semi-infinite medium. To illustrate some of the basic features of combined infiltration and diffusion, we first consider the simple linear isotherm $S = C$. Figure 5 shows concentration profiles as a function of time for a case with "moderate" infiltration, $N_2 = 0.158$ ($\beta\bar{v} = 0.5$ cm^3/cm^{2}yr), for depths of penetration of from 10 cm to about 1 m. Initially, transport is dominated by diffusion as shown by the profile for $t = 100$ years, which is similar in shape to a strictly concave diffusion profile (curve $S_B = C_B$ in Fig. 1). This effect is caused by the initial steep concentration gradient. As time increases, the profile assumes a characteristic S shape with a nearly flat concentration gradient near the origin. Consequently, the transport from the reservoir across the plane $z = 0$ into the porous solid comes to be dominated by infiltration, with diffusion negligible. For comparison, the profiles for pure infiltration are shown. The sharp infiltration front moves with a velocity $dz_f/dt = (c'/s') \beta\bar{v}$, where z_f is the position of the front. When diffusion is superposed on infiltration, the sharp front is degraded to an S shape which flattens continuously with time. After a moderate time the mean depth of penetration of the component is determined almost exactly by the rate of infiltration.

The effect of increasing infiltration relative to a fixed amount of diffusion is shown in Fig. 6. Profiles at $t = 500$ yr are shown for fluid fluxes of 0, 0.1, 0.5, 1, 2 and 3.16 cm^3/cm^{2}yr. Because

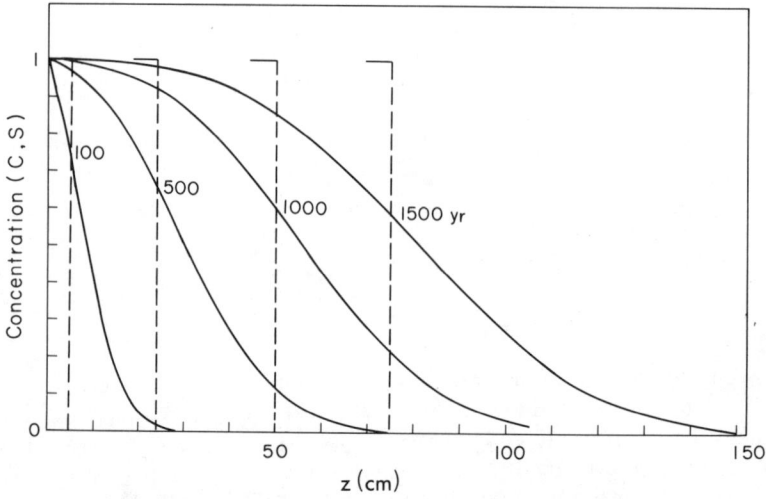

Fig. 5. Combined diffusion and infiltration for the linear isotherm $S = C$. Concentration profiles (solid lines) are shown as a function of distance z for different times $t = 100, 500, 1000,$ and 1500 years. The infiltration flux is $\beta\bar{v} = 0.5$ cm^3/cm^2 yr. The effective diffusivity is given by $\bar{D} = D(c'/s') = 10^{-8}$ cm^2/sec. Pure infiltration profiles ($\bar{D} = 0$) for the same values of t and $\beta\bar{v}$ are indicated by dashed lines.

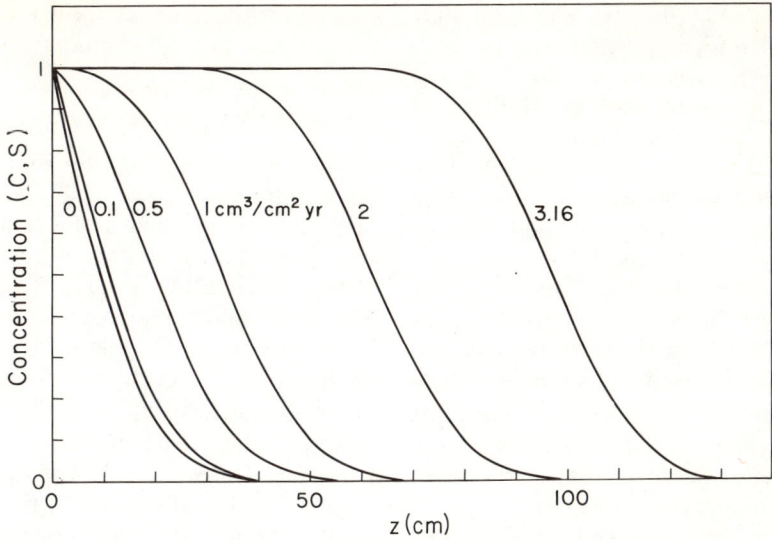

Fig. 6. Combined diffusion and infiltration for the linear isotherm $S = C$. Concentration profiles are shown as a function of distance for $t = 500$ years and for different infiltration fluxes $\beta\bar{v} = 0$ (pure diffusion), 0.1, 0.5, 1, 2, and 3.16 cm^3/cm^2 yr. The diffusivity for all profiles is as in Fig. 5.

the extent to which diffusion has flattened the initial sharp front is the same, these profiles may be roughly superposed by translating them so as to remove the effect of differences in infiltration.

In general, we can say that the shape of the front for a linear isotherm is a function of the amount of diffusion, and the distance of penetration depends on the amount of infiltration, provided a concentration plateau separating the front from the reservoir has developed.

S-Shaped Isotherm and Isotherm with a Solubility Gap

A particularly interesting feature of infiltration (Korzhinskii, 1970; Hofmann, 1972) is the development of self-sharpening and "diffuse" fronts in the concentration profiles for the less simple isotherms characteristic of major-element exchange between a crystal and a pore fluid. For the present boundary-value problem, chromatographic theory (Hofmann, 1972, and included references) would predict the preservation of the sharp front at the reservoir-rock interface in the initial state if the isotherm is linear. We have used this result and have described the effects of superposed diffusion. An arbitrary initial profile, decreasing monotonically in the direction of transport of a fluid entering with a higher concentration, will tend to develop a sharp front if the isotherm is concave upward and a "diffuse" front if the isotherm is convex upward. This is so because, applying (10) to pure infiltration, it can be shown (Hofmann, 1972) that the rate at which a particular fluid concentration propagates in the Z direction is proportional to the slope of the isotherm. Thus, over a concave upward section, higher concentrations will propagate faster than lower concentrations, and a self-sharpening front will develop. Another feature of infiltration under the present conditions is that, in isotherms involving two or more solubility gaps with discontinuities in solid phases $S_I \mid S_{II} \mid S_{III}$, if the rate of propagation of the higher-concentration front is faster than the lower front, the zone containing the intermediate phase S_{II} will be eventually removed, leaving the sharp front

S_I | S_{III} (see Fig. 4). As has been discussed by Hofmann (1972), the two assemblages separated by an ideally sharp infiltration front are not in equilibrium with one another. Local equilibrium prevails on either side of the replacement surface, but assemblages at any two points that are separated by this surface will not be in equilibrium.

We ask, then, what modifications must be introduced in the more complex situation when diffusion is superposed on infiltration, as it must to some extent in any real process. Qualitatively the most important effect is the smoothing out of fluid concentrations so that surfaces of disequilibrium no longer exist.

The effects of combined diffusion and infiltration on such a sharp infiltration front is illustrated in Fig. 7 for an S-shaped isotherm. Pure infiltration (not shown in Fig. 7) would form a sharp front at concentrations greater than $C^* = 0.38$, and a "diffuse" front at lower concentrations. The value of C^* is given by the point of tangency of the line connecting the highest and lowest concentrations represented in the sharp front. The rate of propagation of the concentration C^* common to both fronts is

$$dz^*/dt = (dC/dS)_{C^*} \beta \bar{v}$$
$$= (\Delta C/\Delta S)_{C^*} \beta \bar{v} \qquad (15)$$

where ΔC and ΔS are the concentration discontinuities across the sharp front. (For a more detailed derivation of these results and an illustration for the case of pure infiltration, see Hofmann, 1972.) The fluid concentration profiles in Fig. 7 for the case of "moderate" infiltration (fluid flux = 1 cm³/cm²yr) superposed on diffusion show somewhat similar fronts. Diffusion modifies the sharp front into an S-shaped profile, and contributes to the diffuseness of the "diffuse" front. Here, the S-shaped "diffuse" front is characteristic of infiltration; the corresponding fluid profile for diffusion alone is everywhere concave upward. These effects are masked in the solid-concentration profiles. All solid concentration profiles, including the one for diffusion alone, are approximately S-shaped, because of the form of the isotherm. Only the concentration plateau which extends from the reservoir clearly

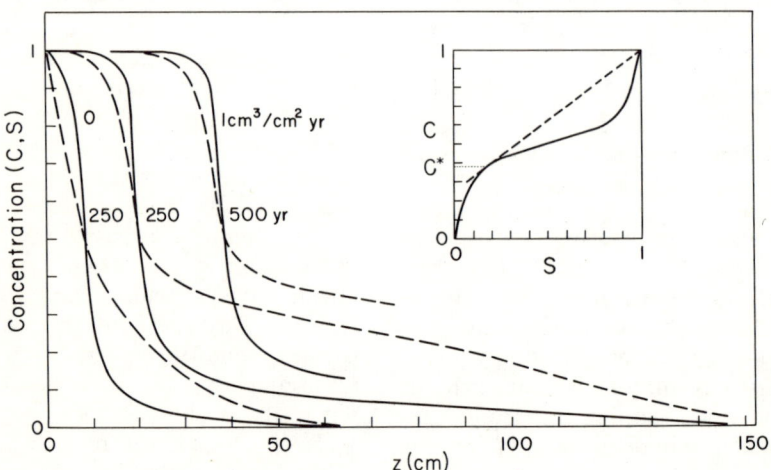

Fig. 7. Combined diffusion and infiltration for S-shaped isotherm (see inset). The fluid concentrations (dashed lines) and solid concentrations (solid lines) are shown for $\beta \bar{v} = 0$ (pure diffusion) at 250 years, and for $\beta \bar{v} = 1$ cm³/cm² yr at 250 and 500 years. The diffusivity for all profiles is as in Fig. 5.

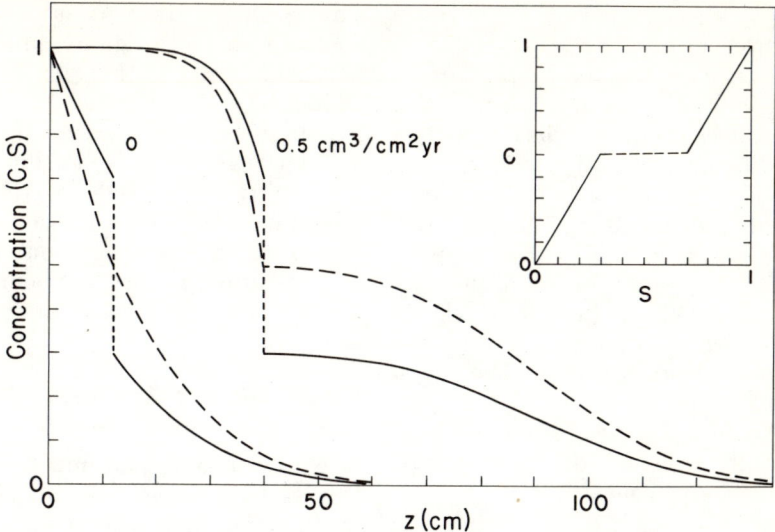

Fig. 8. Pure diffusion ($\beta\bar{v} = 0$) and combined diffusion and infiltration ($\beta\bar{v} = 0.5$ cm³/cm² yr) for isotherm with one solubility gap. The dotted lines in the profiles mark the solid-concentration discontinuity. For other symbols see previous figures.

indicates that infiltration contributed significantly to the transport. As transport continues, the shape of the sharp front tends towards a steady state. The stable shape represents a balance between the self-sharpening effect of infiltration and the tendency of diffusion to flatten the gradient. The rate of propagation of the front is the same as that in the case of infiltration alone, given in Equation 15.

The isotherm showing a symmetric solubility gap (Fig. 8) is a limiting case of an S-shaped isotherm as the minimum slope, dC/dS, tends to zero. The isotherm in Fig. 8 is simplified in that the continuous portions are straight rather than respectively convex upward and concave upward. Both the S-shaped isotherm and the isotherm with a solubility gap are typical of isotherms for ion exchange between alkali feldspars and chloride solutions (Orville, 1963). The solubility gap isotherm would correspond to a lower temperature than that of the S-shaped isotherm.

A solubility gap means that two solid compositions, S^* and S^{**}, will be in equilibrium across an interface at which the fluid composition is, for example, C^*. Mathematically, such a surface is an internal propagating boundary at which certain conditions (conservation of mass and constancy of fluid and solid compositions) must be met. Physically, it is useful to think of this situation as involving a sink for the transported component. This representation is used to derive some approximate concentration profiles for solubility-gap isotherms as merely a special case of the numerical computation for continuous isotherms (see Appendix). Likewise, except for the jump in solid concentration across the surface at which the fluid concentration is C^*, the profiles are on the whole similar to those for the continuous S-shaped isotherm. Figure 8 shows fluid and solid concentration profiles for the solubility-gap isotherm for diffusion alone and with an added infiltration flux of 0.5 cm³/cm²yr at $t = 500$ yr. Both fluid profiles have discontinuities in slope at the solubility gap, $C^* = 0.5$, although this is not easily seen in the pure diffusion profile at this depth of penetration. In the case of diffusion alone, the fluid profile is nearly linear above the solubility gap because, with the fixed value of C at the gap, and the slow movement of the surface of discontinuity relative to the rate of diffusion, diffusion within the slab to the left of the gap behaves like steady-state diffusion in a slab with fixed surface concentrations. The linear portion of the solid concentration profile depends on the special and not particularly realistic choice of a linear segment for the isotherm above the gap. The high-concentration plateau extending from the reservoir toward the sharp front is joined by an equally pronounced plateau at the base of the sharp front.

Discussion

Qualitative Features of Transport Models

Some observations that may be derived from our model calculations are:

1. Diffusion profiles of the fluid concentrations are always concave upward, but the corresponding solid-concentrations may be concave or convex, depending on the shape of the isotherm.

2. In pure diffusion metasomatism, sharp fronts caused by solubility gaps are connected by concentration profiles that closely approximate a steady-state shape for diffusion through a slab of fixed thickness and fixed boundary concentrations.

3. In combined diffusion and infiltration, the overall transport and shape of the concentration profiles is initially dominated by diffusion. As time increases, infiltration will effect virtually all transport through the origin of the column, and S-shaped concentration profiles and concentration plateaus become the characteristic features.

4. When the shape of an isotherm with continuous solid solution is such (concave upward) that infiltration alone would produce a sharp replacement front, the effect of added diffusion is to smooth the perfectly sharp front into a steady-state S shape. In contrast, the initially similar S shape caused by diffusion alone will continually flatten out as the front advances.

Using these and other qualitatively distinct features it is possible to assess the relative magnitude of the different modes of transport. In order to do this, one needs information about the actual concentration profiles, the boundary conditions, and the shape of the equilibrium isotherm.

Application of the Present Boundary-Value Problem

The boundary- and initial-value problem we have discussed provides a simple introduction to some of the basic features of combined infiltration and diffusion transport with local equilibrium described by realistically complex isotherms. However, it is likely to be directly applicable to only a very few special cases of natural metasomatism. One of these is wall-rock alteration adjoining a fluid-filled fissure or permeable fault gouge. The fissure concentration would be equivalent to the reservoir concentration in the model. An obvious modification of the model is the replacement of the constant boundary concentration with a time-dependent boundary composition—for example, one which represents a "pulse" of transport along the fissure. Such a modification could be easily handled within the framework of the numerical procedure used in the present paper.

When infiltration-generated sharp fronts propagate away from the reservoir, they take on steady-state shapes which are likely to be the characteristic signature of the respective isotherm and relative strengths of infiltration and diffusion under a more general range of one-dimensional transport situations.

Two extensions of the present treatment would involve (1) the discussion of transport in geometries more general than the one-dimensional case, and (2) the treatment of transport and reaction in multi-component systems. Such extensions would require a large amount of additional information, such as the relative diffusion coefficients and cross coefficients of the components and the specific boundary conditions. In addition, spatial variation of the parameters N_1 and N_2 might be included. These models are beyond the scope of this paper, which is intended to show some of the simple features of metasomatism.

Time Dependence of Diffusion and Infiltration Transport

The dependence of the effective distance of diffusive transport on the square root of time, also referred to as the parabolic rate law, has been widely discussed in the literature, especially with reference to oxidation rates of metals (see, for example, Tammann, 1920, and Kingery, 1959). It has also been found to apply to diffusion-controlled dissolution reactions of feldspar (Wollast, 1967; Helgeson, 1971) and it was used by Korzhinskii (1970) to describe diffusion in metasomatic columns. This rate law requires that both the distance of penetration of any given concentration (or moving boundary) and the amount of diffusing substance entering the medium through its surface are proportional to the square root of time (Crank, 1956, p. 36). This relationship is implicit in all solutions of the diffusion equation where transport distance z and time t occur only in the combination $z/(t)^{1/2}$ (see, for example, Equation 14). It is valid for diffusion in a semi-infinite medium with uniform initial concentration and constant boundary condition. The diffusion coefficient may vary as a function of concentration but not as a function of time or distance. For the purpose of approximate calculations, the parabolic rate law is useful even when it is not rigorously applicable. For example, diffusion into a single grain may be described in this way until the concentration in the center of the grain changes significantly.

In contrast with diffusion, the infiltration rate is constant with time, but only if the pressure gradient is constant. The pressure gradient will often be independent of the amount of transport. It may, for example, be controlled by tectonic forces, by the rate of metamorphic dehydration reactions, or by the presence of an intrusive body which acts as a heat pump. Thus, the flow of the pore solution does not necessarily result in a decrease in the driving forces. This may be thought of as the reason why, given enough time, the infiltration process is more efficient than diffusion, because the diffusion process will always tend to reduce its driving force, the chemical potential gradient.

In the concentration profiles for diffusion alone, the dimensionless concentrations are plotted against the dimensionless "depth of penetration" $Z/2(N_1 T)^{1/2} = z/2(\overline{D}t)^{1/2}$, where $\overline{D} = (c'/s')D$. For the choice of D and c'/s', the unit of this parameter corresponds to a depth of 1 cm for a time of 0.8 yr, or a depth of 100 cm for a time of 8000 yr.

For $N_1 = 1$, the characteristic time t' and the characteristic length z' are related by

$$t' = (c'/s') D (z')^2, \qquad (16)$$

or for the transport parameters chosen,

$$t' \text{ (yr)} = 3.16 \, [z'(\text{cm})]^2.$$

In the discussion of combined diffusion and infiltration the further choice of $z' = 1$ cm implies a characteristic time $t' = 3.16$ yr. A particular value of the fluid flux, $\beta \bar{v}$, then fixes the value of the dimensionless parameter N_2. We can generalize the application of the results by noting that a profile derived for a particular value of N_2 corresponds to a continuum of alternative values of distance, duration of transport, and fluid flux. These values can be found directly from Equation 16 and the relationship

$$\frac{N_2}{N_1} = (\beta \, \bar{v}/D) z' \qquad (17)$$

which is obtained from Equations 5 and 9. For example, in order to convert the distances in Figs. 5 through 8 from centimeters to meters, the flux $\beta \bar{v}$ must be decreased by a factor of 10^2 and the time must be increased by 10^4.

Absolute Rates of Transport

To assess quantitative aspects of the theory and to justify the choice of parameters used in computing the concentration profiles, it is necessary to know or to make assumptions about the effective diffusivity, the permeability, the viscosity of the fluid phase, and the pressure gradient acting on the fluid phase.

Diffusion coefficients in aqueous electrolyte solutions at 25°C are on the order of $D_i = 10^{-5}$ cm^{2}sec^{-1}. The temperature dependence of the diffusivity in liquids and gases is much smaller than in solids. Using Nigrini's (1970) method of calculating diffusivity from conductivity data, we obtain $D_i = 10^{-4}$ cm^{2}sec^{-1} at metamorphic temperatures of 500° to 700°C. This agrees with the value used by Korzhinskii (1970) and with measurements on intergranular diffusion in granodiorite by Kovalev (1971).

The flux of the fluid through the rock is given by Darcy's law

$$\beta \bar{v} = \frac{q}{A} = - \frac{k}{\eta} \frac{\partial p}{\partial z}$$

where q is the volume of fluid that passes through a cross section per second, A is the area of the cross section, β is the porosity, k is the permeability in darcys, η is the viscosity in centipoises (10^{-2} poise), and $\partial p / \partial z$ is the pressure gradient in bars per cm.

Estimates of the permeability of rocks may be obtained from the results of Brace *et al.* (1968) who measured the permeability of Westerly granite at room temperature and at effective pressures $\bar{P} = 0.1$ to 4 kbar. If the fluid pressure equals the rock pressure, this effective pressure equals zero and the extrapolated permeability will be 300 to 1000 nanodarcys. Brace *et al.* (1968) remark that "high temperatures will promote plastic flow of minerals, which could ultimately lead to a more or less complete closure of the tiny cavities that provide paths for flow of pore fluids. Thus, at high temperature or, in general, under conditions that promote flow or recrystallization, we might expect permeability to become vanishingly small." For the reader not acquainted with the paper of Brace *et al.* (1968) we note that the quoted statement is not based on high-temperature experiments. Following Orville (oral presentation at the Conference on Geochemical Transport and Kinetics, 1973), permeability of metamorphic rocks at high temperatures must be high enough to permit volatiles to escape during metamorphic dehydration and decarbonation reactions. At least during the time when such reactions take place, the pressures on the fluid and solid phases must be nearly equal *unless* the permeability is so large that the volatiles escape very rapidly. The equality of pressures will keep the pores open so that the only reduction in permeability is that caused by the sealing of pore connections. In any case, any reduction in permeability will reduce both the flow rate and the diffusion rate through the intergranular medium. For the purpose of our model calculations we use a permeability of 300 nanodarcys, which is equal to or lower than the values of Brace *et al.* (1968) at zero effective pressure and at room temperature.

The fluid viscosity of aqueous solutions at metamorphic temperatures is estimated to be 10^{-3} poise. This value is taken from measurements on water at temperatures up to 560°C and pressures up to 3500 bars by Dudziak and Franck (1966). The pressure and temperature dependence of the fluid viscosity is also relatively small. For example, at 560°C the viscosity of water changes from 0.7 to 1.0×10^{-3} poise when the pressure is increased from 1000 to 3400 bars; or, at 2000 bars the viscosity ranges from 1.2 to 0.8×10^{-3} poise between temperatures of 300° and 560°C.

A typical value of the infiltration flux used in the previous chapter is $\beta \bar{v} = 1$ cm³/cm²yr. Given our estimates of $\eta = 0.1$ centipoise and $k = 3 \times 10^{-7}$ darcy ($= 300$ nanodarcys), the pressure gradient required by the above flux is about 1 bar/meter. This relatively large gradient would cause the fluid to move through rock at a rate of 10 meters per year if the porosity is 10^{-3}. For a comparatively small gradient of 1 bar/km, the infiltration rate would still be 1 cm/year. To the extent that our values for the permeability and the pressure gradient are based on "high" estimates, our computed depths of penetration are probably near the maximum to be expected under natural conditions.

Our choice of the ratio of characteristic concentrations $c'/s' = 0.1$ corresponds approximately to the values obtained for the partitioning of potassium between K-feldspar and 1 molar KCl solution or sodium between albite and 1 molar NaCl solution. It is arbitrary in the sense that this ratio varies directly with the concentration of the fluid. For example, if one chooses a 0.1 molar alkali chloride solution, one obtains $c'/s' = 0.01$, and all the computed penetration distances must be reduced by a factor of 10.

To set limits on the possible transport mechanisms in large-scale metasomatic processes, we shall assume that the transported component does not react with the solid phases; this is to say, the solid phases are inert. In this way we obtain the largest conceivable, though perhaps unrealistic, transport distances. With an intrinsic diffusion coefficient of $D_i = 10^{-4}$ cm²sec⁻¹, we then obtain a characteristic diffusion distance of $(Dt)^{1/2} = 1.8$ km during a metamorphic event lasting 10^7 years. This distance is reduced by β $c'/s' = 10^{-4}$ if the solids do react with the diffusing component and if the values for the porosity and the characteristic concentrations are those used in our examples.

We conclude that intergranular diffusion is not a realistic model for metasomatism on a *regional* scale during a metamorphic event. This argument applies to such processes as granitization, change of bulk-oxygen isotopic composition in metamorphic rocks, and regional redistribution and equilibration of strontium isotopes. The infiltration process, on the other hand, is potentially efficient on a large scale. Thus the distance of penetration for the above choice of parameters is 100 km in 10^7 years, if the pressure gradient is maintained at 1 bar/km and if the solid phases are inert.

Essentially the same arguments may be made with respect to the Earth's upper mantle, where a silicate melt may be present as an intergranular medium and where the metasomatic processes involving migration of the melt and diffusion through the melt may be important in differentiation and homogenization processes. Diffusion coefficients in basaltic melts at solidus temperatures are on the order of 10^{-6} cm²sec⁻¹ so that for a duration of 100 m.y., $(Dt)^{1/2} = 560$ meters. Infiltration rates are difficult to estimate because of the lack of permeability data for partially molten rocks. The viscosity of the melt near the solidus temperatures is higher than that of metamorphic, aqueous fluids by a factor of 10^4 to 10^5 (Shaw, 1972), but this may be offset by a much greater permeability, so that it seems likely that magmatic liquids can migrate by infiltration through considerable distances of the upper mantle.

Acknowledgments

We thank S. R. Hart, H. R. Shaw, J. B. Thompson, Jr., and R. A. Yund for reviewing the manuscript.

Appendix

Concentration profiles for transport into a semi-infinite medium were evaluated by a straightforward finite-difference computation. (See Carslaw and Jaeger, 1959, Chapter 18, for additional discussion of the finite-difference method).

A slab is divided into $N-1$ laminas of uniform thickness ΔZ by the planes Z_n, $n = 1, \ldots, N$. Alternatively, with the exception of the first and Nth plane, each plane may be considered to be centered in a lamina of thickness ΔZ. The first and Nth planes bound laminas of thickness $\frac{1}{2} \Delta Z$. The concentrations at the interior planes may be supposed to represent the mean concentrations within the laminas in which they are centered. The concentrations at the bounding planes are held fixed.

Transport into a semi-infinite medium, bounded by the plane Z_1, from a reservoir with constant fluid concentration $C_R = 1$ is initiated by setting

$$C_1 = C(Z_1) = 1$$

and

$$C_n = 0, n = 2, \ldots, N.$$

The total rock, or solid, concentrations are correspondingly $S_1 = 1$, $S_n = 0$, $n = 2, \ldots, N$.

At time T the derivatives at the mth interior plane are approximated by

$$(\partial C/\partial Z)_m \cong (C_{m+1} - C_{m-1})/(2\Delta Z)$$
$$(\partial^2 C/\partial Z^2)_m \cong (C_{m+1} + C_{m-1} - 2C_m)/(\Delta Z)^2$$

The mass increment added to the lamina centered at the plane Z_m in the finite time increment ΔT is then, from (10),

$$\begin{aligned}(\Delta S)_m &\cong (\partial S/\partial t)_m \Delta T \\ &= [N_1(\partial^2 C/\partial Z^2)_m \\ &\quad - N_2(\partial C/\partial Z)_m]\Delta T \\ &\cong N_1 \Delta T/(\Delta Z)^2 [(C_{m+1} + C_{m-1} \\ &\quad - 2C_m) - \phi \Delta Z (C_{m+1} \\ &\quad - C_{m-1})/2]\end{aligned}$$

where $\phi = N_2/N_1$. The new solid (total rock) concentrations at time $T + \Delta T$ are then

$$S_m(t + \Delta t) = S_m(t) + (\Delta S)_m$$

The concentrations in the pore fluid may be determined directly from the isotherm (11).

A procedure used by Eyres *et al.* (1946) to treat a moving interface between melt and solid in one-dimensional solidification is used here to treat the moving sink of transported component represented by the presence of a solubility gap. This approximate procedure ignores the details of the concentration profile near the solubility gap. Increments to S_m are computed uniformly by the above procedure. If S_m enters the range S^* to S^{**} of the solubility gap, C_m is held fixed at C^* until S_m reaches S^{**}. This is equivalent to supposing that the interface is held fixed at the center of lamina Z_m until the concentration of the entire lamina is incremented by the amount $S^{**} - S^*$. As a consequence, the propagation of the interface is abrupt at a time scale corresponding to its passage through a thickness ΔZ, whereas in reality it propagates smoothly. Associated with the sudden jump of the nominal position of the interface to the next lamina are sudden fluctuations of the concentration profiles in the vicinity of the interface; profiles for such time steps are not representative.

Initially, the concentrations at all interior planes and the bounding plane Z_N are 0. As time passes, non-zero values of concentration propagate inward. During the time that the concentration at the plane Z_{N-1} remains zero, the computation gives an approximate solution for transport into a half-space. After the non-zero concentrations migrate to this plane, the numerical solution approximates that for a slab of finite thickness with zero concentration boundary condition imposed at Z_N. (We are not interested in these solutions in the present paper.)

References Cited

Brace, W. F., J. B. Walsh, and W. T. Frangos, Permeability of granite under high pressure, *J. Geophys. Res., 73,* 2225–2236, 1968.

Burt, D. M., Metasomatic zoning in Ca-Fe-Si exoskarns, this volume.

Carslaw, H. S., and J. C. Jaeger, *Conduction of Heat in Solids,* 2nd edition, Oxford University Press, London, 1959.

Cooper, A. R., Jr., Vector space treatment of multicomponent diffusion, this volume.

Crank, J., *The Mathematics of Diffusion,* Oxford University Press, London, 1956.

De Vault, D., The theory of chromatography, *J. Amer. Chem. Soc., 65,* 532–540, 1943.

Dudziak, K. H., and E. U. Franck, Messungen der Viskosität des Wassers bis 560°C und 3500 bar, *Ber. Bunsenges. Phys. Chem., 70,* 1120–1128, 1966.

Eyres, N. R., D. R. Hartree, J. Ingham, R. Jackson, R. J. Sarjant, and J. B. Wagstaff, The calculation of variable heat flow in solids, *Phil. Trans. Roy. Soc. London, Ser. A, 240,* 1–57, 1946.

Fisher, G. W., and D. Elliott, Criteria for quasi-steady diffusion and local equilibrium in metamorphism, this volume.

Foland, K. A., Alkali diffusion in orthoclase, this volume.

Fyfe, W. S., F. J. Turner, and J. Verhoogen, Metamorphic reactions and metamorphic facies, *Geol. Soc. Amer., Mem. 73,* 1958.

Glueckauf, E., Theory of chromatography. Pt. II. Chromatograms of a single solute, *J. Chem. Soc., London,* 1302–1308, 1947.

Glueckauf, E., Theory of chromatography. VII. The general theory of two solutes following non-linear isotherms, *Discussions Faraday Society, No. 7,* 12–25, 1949.

Helgeson, H. C., Kinetics of mass transfer among silicates and aqueous solutions, *Geochim. Cosmochim. Acta, 35,* 421–470, 1971.

Hofmann, A., Chromatographic theory of infiltration metasomatism and its application to feldspars, *Amer. J. Sci., 272,* 69–90, 1972.

Hofmann, A., Theory of metasomatic zoning. A reply to Dr. D. S. Korzhinskii, *Amer. J. Sci., 273,* 960–964, 1973.

Kingery, W. D., Introductory chapter in *Kinetics of High-temperature Processes,* W. D. Kingery, ed., Technology Press of MIT and John Wiley and Sons, New York, pp. 1–7, 1959.

Korzhinskii, D. S., Derivation of the equation of infiltration metasomatic zoning (in Russian), *Dokl. Akad. Nauk SSSR, 77,* 305–308, 1951.

Korzhinskii, D. S., Derivation of the equation of a simple diffusion metasomatic zoning (in Russian), *Dokl. Akad. Nauk SSSR, 84,* 761–764, 1952.

Korzhinskii, D. S., *Theory of Metasomatic Zoning* (translated by J. Agrell), Oxford University Press, London, 1970.

Korzhinskii, D. S., Theory of metasomatic zoning, a reply to Dr. A. Hofmann, *Amer. J. Sci., 273,* 958–959, 1973.

Kovalev, G. N., Diffusion along grain boundaries in rock in the presence of water, *Dokl. Akad. Nauk SSSR, 197,* 1410–1412, 1971.

Nigrini, A., Diffusion in rock alteration systems: I. Predictions of limiting equivalent ionic conductances at elevated temperatures, *Amer. J. Sci., 269,* 65–91, 1970.

Orville, P. M., Alkali ion exchange between vapor and feldspar phases, *Amer. J. Sci., 261,* 201–237, 1963.

Orville, P. M., Plagioclase cation exchange equilibria with aqueous chloride solutions: results at 700°C and 2000 bars in the presence of quartz, *Amer. J. Sci., 272,* 234–272, 1972.

Shaw, H. R., Viscosities of magmatic silicate liquids; an empirical method of prediction, *Amer. J. Sci., 272,* 870–893, 1972.

Tammann, G., Über Anlauffarben von Metallen: *Z. anorg. allg. Chem., 111,* 78–89, 1920.

Thompson, J. B., Jr., Local equilibrium in metasomatic processes, in *Researches in Geochemistry,* Vol. 1, P. H. Abelson, ed., Wiley, New York, pp. 427–457, 1959.

Wilson, J. N., A theory of chromatography, *J. Amer. Chem. Soc., 62,* 1583–1591, 1940.

Wollast, R., Kinetics of the alteration of K-feldspar in buffered solutions at low temperature, *Geochim. Cosmochim. Acta, 31,* 635–658, 1967.

INFILTRATION METASOMATISM IN THE SYSTEM K_2O-SiO_2-Al_2O_3-H_2O-HCl

John D. Frantz
Geophysical Laboratory
Carnegie Institution of Washington
2801 Upton Street, N.W.
Washington, D.C. 20008

and

Alain Weisbrod
École Nationale Supérieure de Géologie
B. P. 452
54-Nancy, France

ABSTRACT

A quantitative infiltration metasomatism model has been calculated for the alteration of a pyrophyllite-quartz rock by potassium-rich solutions at 500°C, 1 kbar, using data of Hemley (1959) and Weill and Fyfe (1964). Aluminum was considered to be conserved, and chloride assumed not to be a constituent of the solid phases. The relative thicknesses, sequence, and character of the reaction zones were determined for four different original quartz/porosity ratios (Fig. 1). Considering an original aluminum concentration equal to 0.28 mol/(cal/bar) and porosities less than 0.16, the volume of the pores is insufficient to accommodate the volume expansion required in the conversion of the quartz-pyrophyllite assemblage to the final quartz-orthoclase assemblage. Three possibilities are discussed: (1) The rock may expand. (2) Infiltration, and thus reaction, may cease with loss of porosity. (3) Solid pressures may exceed fluid pressures with maintenance of sufficient porosity to allow infiltration and reaction to continue. The third possibility is shown to be highly unlikely, as it requires maintenance of 6000 bars nonhydrostatic pressure differences on the solid phases.

INTRODUCTION

The use of quantitative mass transport models can be of great assistance in understanding the important processes and controls of the evolution of nonisochemical rock systems. Models involving diffusion of solutes through a static solvent have successfully been used to explain mineral textures that occur at metamorphic isograds (Carmichael, 1969). Fisher (1970) and Vidale (this volume) used similar theories in explaining mineral segregations and metamorphic differentiation layering. A quite different model, infiltration metasomatism (Korzhinskii, 1936), involves transport of material by the flow of the solvent in response to gradients in fluid pressure. Such a process may provide a mechanism explaining the transfer of material over distances greater than one might expect from diffusion. In spite of its potential importance, however, the model has received little attention in Western literature, possibly because of some confusion concerning its fundamental principles and their application to rock systems. This paper is an attempt to illustrate the use of infiltration metasomatism and investigate quantitatively its potentials and weaknesses in explaining long-range transport. The mineralogy and relative thickness of the reaction zones resulting from the alteration of a pyrophyllite-quartz rock by the infiltration of a potassium-rich solution were calculated at 500°C and 1000 bars fluid pressure using data presented by Hemley (1959) for the system K_2O-SiO_2-Al_2O_3-H_2O-HCl.

Infiltration metasomatism, as explained by Korzhinskii (1965, 1970) and Hofmann (1972), involves reactions between rock-forming minerals and aqueous solutes supplied by the movement of an inflowing fluid. In many cases several alteration zones of varying thickness and mineralogy are formed. Assuming local equilibrium and neglecting diffusion (i.e., dealing with pure infiltration as a theoretical "end member" of the transport processes), it can be demonstrated that when dealing with stoichiometric minerals (1) the boundaries—or "fronts"—of the zones are sharp (compositional gradients cannot occur in the rock), and (2) progressive variations of composition cannot occur in the solution (gradients of concentrations take place at the fronts and are infinite).

It is possible to identify these assemblages of the zones and calculate the relative rates of progression of their fronts using Hofmann's (1972) Equation 8:

$$dz/dv = \frac{1}{A}(\Delta C_i^f/\Delta C_i^s) = \frac{1}{A}(\Delta C_j^f/\Delta C_j^s), \quad (1)$$

where dz/dv refers to the rate of progression of a front as a function of the amount of fluid, by volume, passing through a unit area of rock, A. The term $\Delta C_i^s = (C_{ib}^s - C_{ia}^s)$ refers to the change in mass of component i in the solid per unit volume of solid at the front between the zones a and b. The term ΔC_i^f refers to the corresponding concentration change in the fluid. The terms ΔC_j^f and C_j^s refer to changes in concentration of component j at the same front. Implicit in this expression is the assumption that the pore volume is negligible in comparison with the rock volume. In systems with significant porosities, however, a better approximation derived in a manner similar to Equation 1 may be

$$dz/dv = \frac{1}{A}(\Delta C_i^f/\Delta C_i^r) = \frac{1}{A}(\Delta C_j^f/\Delta C_j^r) \quad (2)$$

where C_i^r is the mass of i and C_j^r is the mass of j contained in the solid phases per unit volume of rock (including pores). In this expression the masses of the components in the pore volume and the effects of changes in porosity on flow rates have been neglected.

CONCENTRATION UNITS

The choice of units for the concentrations in the fluid and in the rock is somewhat arbitrary because ratios of differences rather than absolute amounts are important. The present calculations use molality for fluid concentrations and the number of moles per 1 cal/bar of rock for rock concentrations. For example, a rock formed only of muscovite with no porosity would have silicon and potassium concentrations calculated as follows:

$$C_{Si}^r = 3/V_{Mu} \quad (3)$$
$$C_K^r = 1/V_{Mu}$$

where V_{Mu} refers to the molar volume of muscovite, $KAl_3Si_3O_{10}(OH)_2$, in cal/bar (1 cal/bar = 41.8 cm³).

ROCK CONCENTRATION DIAGRAMS

The concentrations of the three components in a rock composed of one or more minerals can be represented on a C_i^r-C_j^r-C_k^r concentration diagram of three dimensions. A diagram for the solid phases in the K_2O-SiO_2-Al_2O_3-H_2O-HCl system is given in Fig. 1. The axes $C_{H_2O}^r$ and C_{HCl}^r are not present because (1) H_2O is an excess component and (2) chloride is assumed not to be present as a component of any of the solid phases. Points Q, Pyr, Or, and Mu, representing rocks composed entirely of quartz, pyrophyllite, orthoclase, or muscovite, respec-

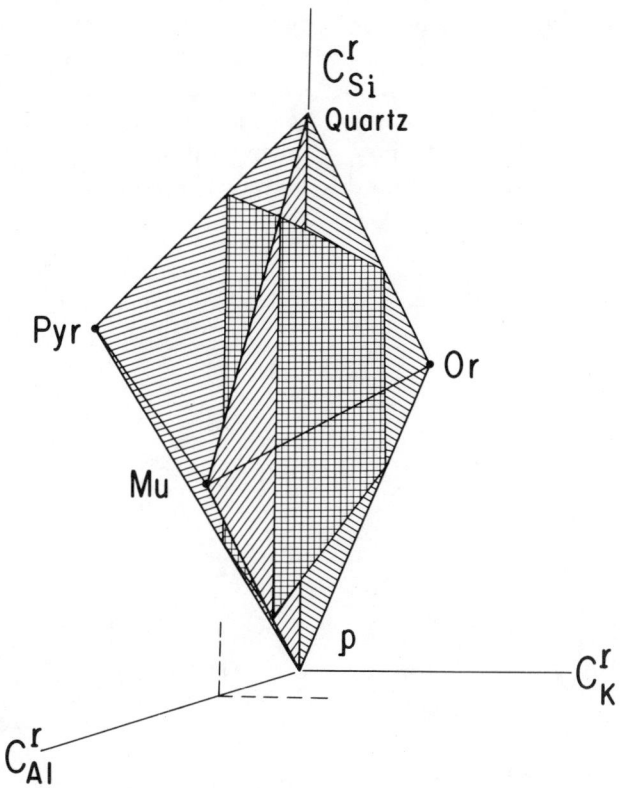

Fig. 1. Diagram showing concentrations of silica, potassium, and aluminum in a rock composed of one or more of the following minerals: orthoclase (Or), quartz (Q), muscovite (Mu), and pyrophyllite (Pyr). See text for discussion of the units. p (origin) = 100% porosity. A plane of constant Al concentration (shown by the dashed lines on the C_{Al}^r–C_K^r and C_{Al}^r–C_{Si}^r planes) cuts the polyhedron along a polygon, shown by the crosshatched area.

tively, were calculated using equations similar to Equation 3.

The origin p in Fig. 1 represents 100 percent porosity. The composition of a rock of any porosity comprising one or more of these minerals can be represented by a point on or within the polyhedron Q-Pyr-Mu-Or-p. For example, a porous rock containing quartz and pyrophyllite lies on the p-Q-Pyr plane; a nonporous rock of muscovite and orthoclase lies on the Mu-Or line.

Because of the low solubility of aluminum in many supercritical aqueous fluids (Morey and Hesselgesser, 1951), the aluminum content in the rock can be expected to remain approximately constant during metasomatic processes. The concentration of aluminum in a given rock volume, including the pores, (C_{Al}^r) could also be assumed to remain constant. Considering the alteration of a quartz-pyrophyllite rock, changes in C_{Si}^r and C_K^r can be represented on constant C_{Al}^r sections as shown in Fig. 1. Values of C_{Al}^r can be calculated from the composition of the starting rock using the initial porosity and an initial ratio of quartz to pyrophyllite. Figure 2 shows two sections: $C_{Al}^r = 0.28$ and $C_{Al}^r = 0.52$. The lines in Fig. 2 represent the intersections of the planes (Q-Pyr-p, Q-Mu-p, Q-Or-p, Q-Mu-Or, Q-Pyr-Mu, Mu-Or-p, and Pyr-Mu-p) in Fig. 1 with a plane perpendicular to the C_{Al}^r axis. In Fig. 2a the aluminum concentration ($C_{Al}^r = 0.28$) is low enough so that all planes are intersected. In Fig. 2b ($C_{Al}^r = 0.52$), however, the plane Q-Or-p is not intersected, indicating insufficient silica in the original rock for a quartz-orthoclase alteration zone to appear. Using constant-aluminum sections, the original porosity, and the quartz/pyrophyllite ratio, one is able to follow changes in the rock composition during infiltration in terms of the remaining parameters: C_K^r and C_{Si}^r.

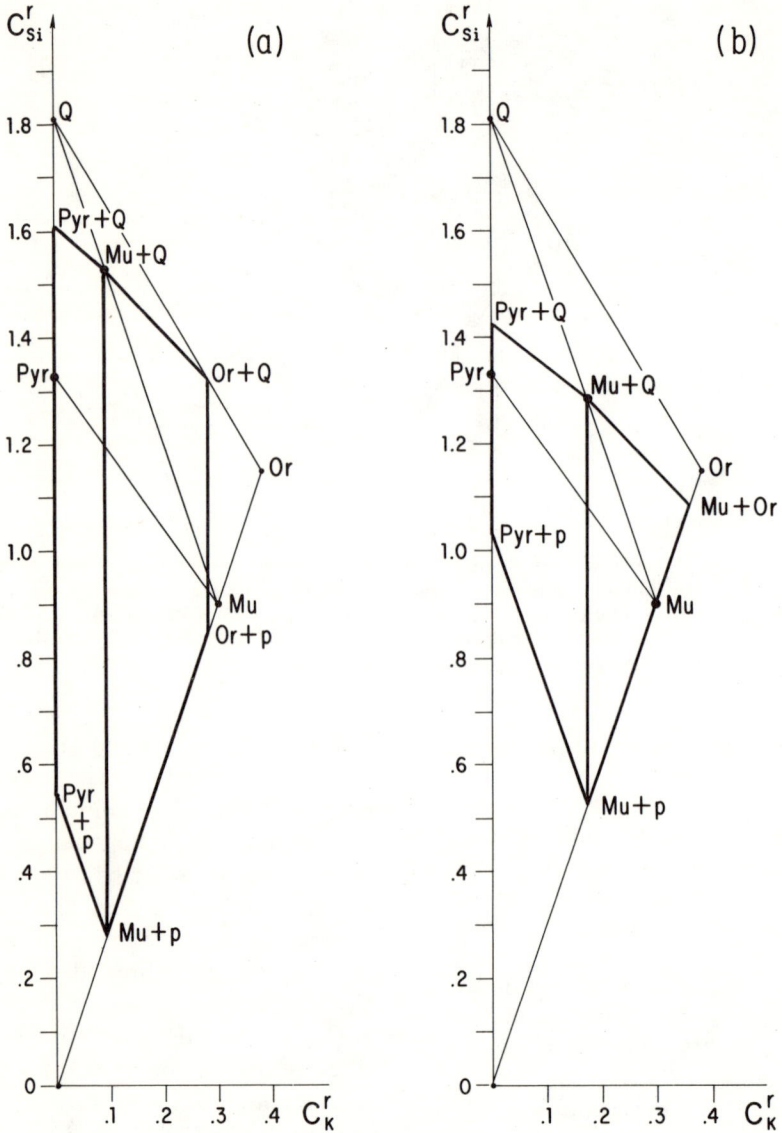

Fig. 2. C_{Si}^r–C_K^r constant-aluminum sections. In section a, $C_{Al}^r = 0.28$; section b, $C_{Al}^r = 0.52$. See Fig. 1 for identification of mineral labels. Heavier lines represent section lines; light lines, projected in the sections. See text for explanation of units.

Solution Concentration Diagrams

Because mass in the fluid-rock system is conserved, the relative changes in the C_K^r-C_{Si}^r sections must be identical with the relative KCl and SiO$_2$ changes in the fluid at each front. This can be seen by rewriting Equation 2:

$$\Delta C_K^r / \Delta C_{Si}^r = \Delta C_{KCl}^f / \Delta C_{Si}^f. \quad (4)$$

Thus, information concerning the relative fluid concentrations is necessary to predict reaction paths on the constant aluminum sections of Fig. 2. The silica and potassium concentrations in equilibrium with the minerals pyrophyllite, muscovite, quartz, and orthoclase are represented in Fig. 3. A constant 1 molal chloride concentration is assumed. Points A and C represent the following two sets of reactions:

$$3 \text{ Pyr} + 2 \text{ KCl} = 2 \text{ Mu} + 6 \text{ Q} + 2 \text{ HCl} \quad (5)$$

$$\text{Quartz} = \text{SiO}_2 \text{ (aqueous)} \quad (6)$$

$$\text{Mu} + 2 \text{ KCl} + 6 \text{ Q} = 3 \text{ Or} + 2 \text{ HCl} \quad (7)$$

$$\text{Quartz} = \text{SiO}_2 \text{ (aqueous)} \quad (8)$$

The curve Cn represents the equilibria:

$$\text{Mu} + 2 \text{ KCl} + 6 \text{ SiO}_2\text{(aqueous)} = 3 \text{ Or} + 2 \text{ HCl} \quad (9)$$

The molality of silica (0.04) for quartz-saturated solutions at 500°C, 1000 bars pressure (Equations 6 and 8) has been measured by Weill and Fyfe (1964) and Anderson and Burnham (1965). The KCl/HCl ratios (8 and 148) for Equations 4 and 6, respectively, are from Hemley (1959). The equation of the curve Cn (8) has been calculated, assuming ideal solution.

Predicted Reaction Paths

Using the data presented in Figs. 2 and 3, it is possible to predict the reaction paths by which a porous quartz-pyrophyllite rock may proceed if infiltrated by a quartz-saturated, 0.998 m KCl solution (point E, Fig. 3). For a rock with $C_{Al}^r = 0.28$, four reaction paths are possible, depending on the initial porosity (Fig. 4).

Path 1 ($p = 0.5$; $C_{Si}^r = 0.7$). The path represents the alteration of a rock with ample porosity but insufficient quartz for conversion of the quartz-pyrophyllite assemblage to quartz-orthoclase. The first front (Pyr-Q-p → Mu-Q-p) has quartz-bearing assemblages on either side, indicating no change in C_{Si}^f and thus C_{Si}^r. It is represented by a horizontal line (A_1C_1, Fig. 4). Because insufficient quartz exists to convert all the muscovite in zone C_1 to orthoclase, two additional fronts occur: C_1F_1, the partial conversion of the muscovite in zone C_1 to orthoclase; and F_1E_1, conversion of the remaining muscovite in this zone (F_1) to orthoclase using silica from the infiltrating solution (as per Equation 8).

The values of C_{Si}^r and C_K^r shown for zone F_1 were calculated as follows. In accordance with Equation 4, expressing conservation of mass, the slope of line FE (Fig. 3) must equal that of F_1E_1 in Fig. 4. The slope of F_1E_1 and the initial solution (point E) are known, and as zone F_1 contains the assemblage Mu + Or, the solution composition of F_1 must lie on curve Cn at point F (Fig. 3). Again, owing to Equation 4, the slope of C_1F_1 (Fig. 4) must equal that of the straight line CF (Fig. 3), thus defining the silica and potassium concentrations in zone F_1. With this information, summarized in Fig. 5 (1), it is possible to calculate the relative velocities of the three fronts using Equation 2; see Fig. 6 (1):

$$(dz/dv)_{A_1C_1} = (C - A)/(C_1 - A_1)_K \quad (10)$$

$$(dz/dv)_{C_1F_1} = (F - C)/(F_1 - C_1)_K = (F - C)/(F_1 - C_1)_{Si} \quad (11)$$

$$(dz/dv)_{F_1E_1} = (E - F)/(E_1 - F_1)_K = (E - F)/(E_1 - F_1)_{Si} \quad (12)$$

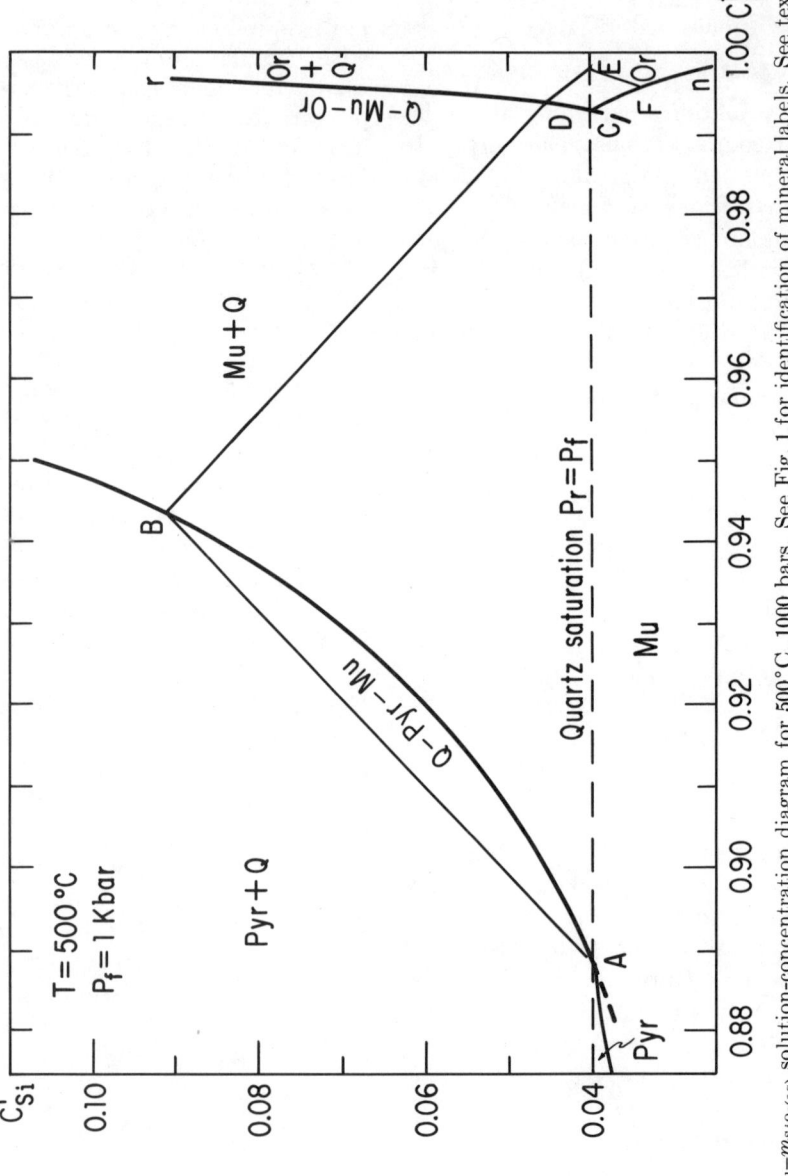

Fig. 3. $m_{KCl}-m_{SiO_2(aq)}$ solution-concentration diagram for 500°C, 1000 bars. See Fig. 1 for identification of mineral labels. See text for identification of letter symbols. Values of $C_{Si}{}^f$ above the quartz-solution line refer to fluids in which $P_s > P_r$.

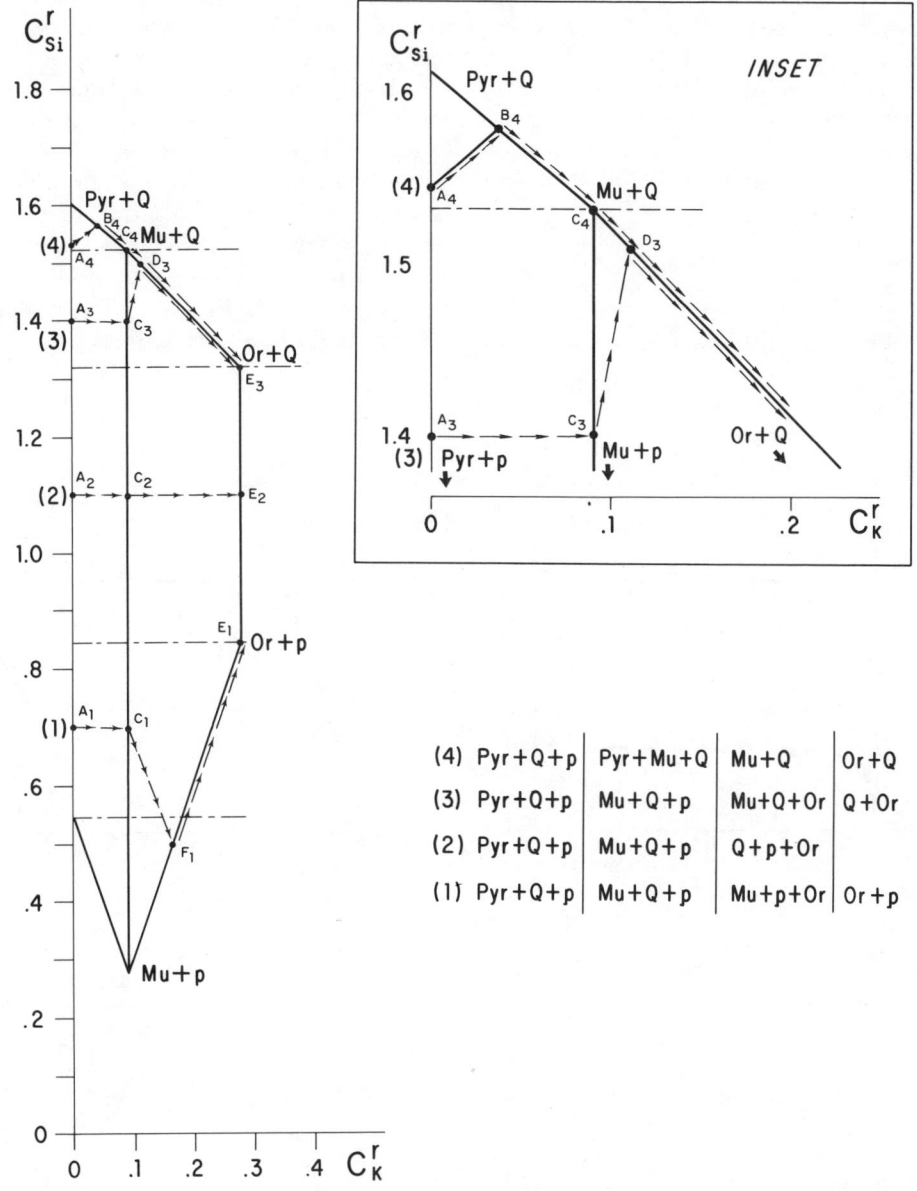

Fig. 4. Reaction paths depicted on constant-aluminum section a of Fig. 2 ($C_{Al}{}^r = 0.28$). Dashed lines with arrows indicate four possible reaction paths; subscript letters indicate reaction zones (see text). The inset is an enlargement of the upper part of the figure.

thus providing the following metasomatic sequence:

Pyr-Q-p | Mu-Q-p | Mu-Or-p | Or-p.

The relative concentrations of the minerals and the porosity can easily be calculated for each zone using the concentrations of elements in the rock.

Path 2 ($p = 0.28$; $C_{Si}{}^r = 1.10$). This path represents the alteration of a rock with both ample porosity and quartz for complete conversion of the porous pyrophyllite-quartz rock to a porous orthoclase-quartz rock. The fronts A_2C_2 (Equation 4) and C_2E_2 (Equation 6) were calculated in a manner similar to that of front A_1C_1 because quartz is present in all assemblages. The concentration of silica and potassium in the rock and fluid are represented in Fig. 5 (2). Calculation of the relative velocities (Fig. 6 (2)) indicates the following sequence of reaction zones:

$$\text{Pyr-Q-}p \mid \text{Mu-Q-}p \mid \text{Or-Q-}p.$$

Path 3 ($p = 0.12$; $C_{\text{Si}}^r = 1.40$). In this reaction path, the starting rock has enough quartz and porosity for the quartz-pyrophyllite assemblage to give muscovite-quartz as in fronts A_1C_1 and A_2C_2. There is, however, insufficient porosity to balance the volume increase associated with the alteration of muscovite + quartz to orthoclase + quartz. (Line A_3C_3 would intersect the Or-Mu-Q line if extended, Fig. 4). Three possibilities regarding further reaction exist.

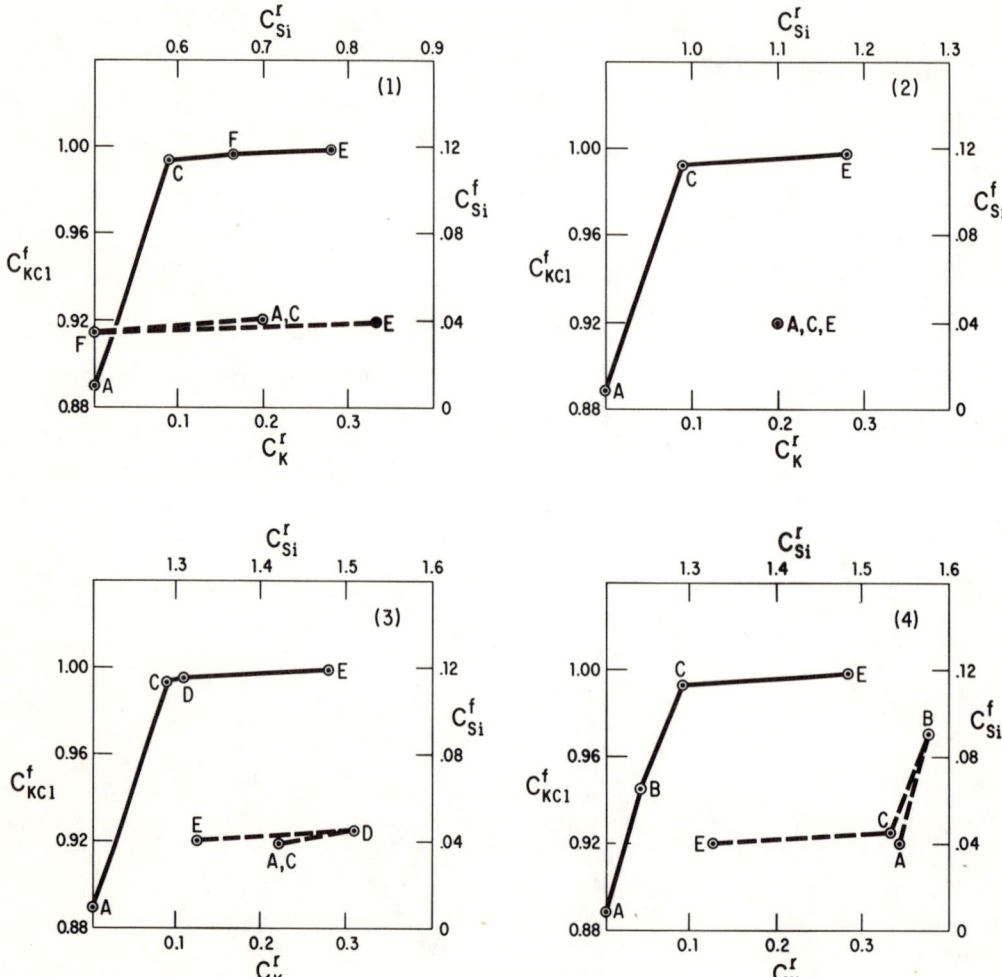

Fig. 5. Changes in silicon and potassium concentrations in the solution (C_{KCl}', C_{Si}') and in the rock (C_{K}^r, C_{Si}^r) at the fronts for the four reaction paths shown in Fig. 4. Letters indicate the reaction zones. Dashed lines, silicon; solid lines, potassium.

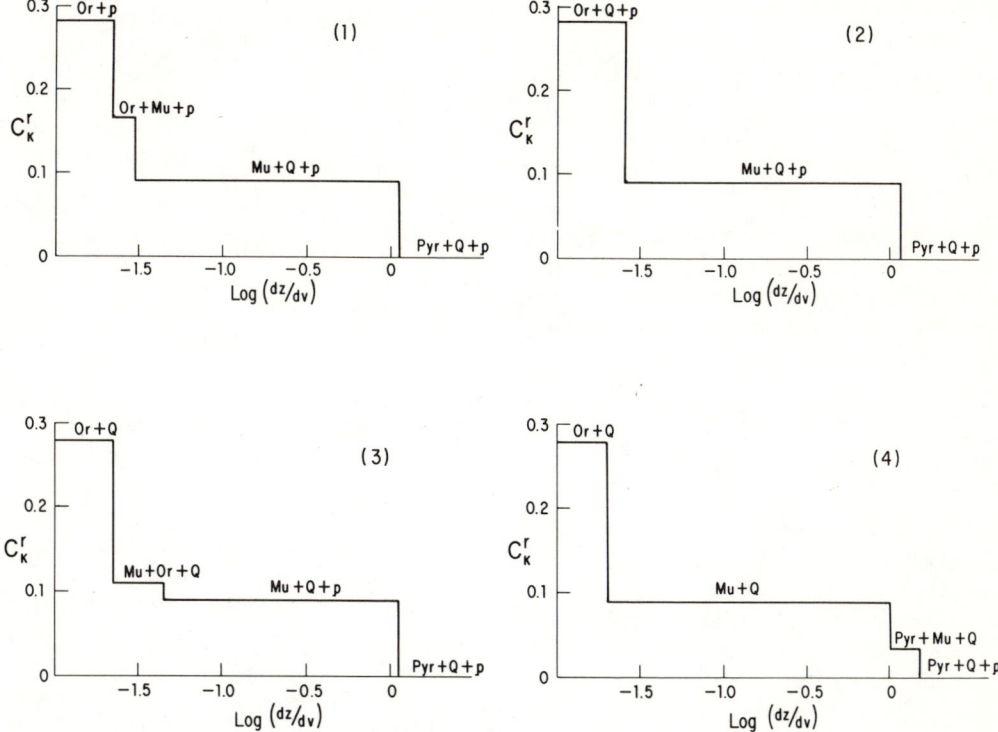

Fig. 6. Plot of the relative front velocities (log dz/dv) versus concentration of potassium in the rock (C_K^r) See Fig. 1 for mineral labels.

The entire rock may expand. On these grounds, the assumption of constant volume is no longer valid, and thus the sections in Figs. 2 and 4 are inappropriate. This situation could be approximated by considering a section with a larger initial porosity but the same C_{Al}^r/C_{Si}^r ratio, thereby giving a reaction path similar to that of path 1 or path 2.

A second possibility is that the volume remains constant and the pores are filled during the formation of part of the orthoclase. Then the flow of the infiltrating fluid would stop, thus prohibiting the formation of any macroscopic reaction zones (including A_3C_3).

The third possibility assumes constant volume during the formation of orthoclase with the maintenance of sufficient porosity for continued infiltration. The existence of two fronts is implied: C_3D_3, the partial conversion of muscovite to orthoclase filling up the major volume of pores; D_3E_3, the equal volume replacement of the remaining muscovite with a change in the quartz concentration. But zone D_3 would have too many phases (three instead of two), a situation that is impossible unless the system is subject to one less constraint. Thompson (1955) and Korzhinskii (1970) suggested that situations may exist where the pressure on the solids could be greater than that on the fluid. Such an eventuality provides the extra state parameter for the existence of assemblage D_3 and the equal volume transition of muscovite + orthoclase + quartz to orthoclase + quartz as indicated by front D_3E_3.

Quantitative calculations of this model involve the determination of the equilibrium concentration of potassium and silica in the solution for solid pressure greater than the fluid pressure. Considering reactions 7 and 8, and differentiating their free energy expressions with respect to solid pressure:

$$\left(\partial \ln(a_{KCl}/a_{HCl})/\partial P_s\right)_{P_f,T} = \Delta V_s/2RT \quad (13)$$

$$\left(\partial \ln a_{SiO_2}(aq)/\partial P_s\right)_{P_f,T} = V_Q/RT \quad (14)$$

where V_Q is the molar volume of quartz and ΔV_s is the volume change for the solids in reaction 6. Integrating and combining expressions 12 and 13 (assuming equal pressure on all solids):

$$a_{SiO_2}(aq)/a^0{}_{SiO_2}(aq) = \quad (15)$$
$$\left[(a_{KCl}/a_{HCl})/(a^0{}_{KCl}/a^0{}_{HCl})\right]^{2V_Q/\Delta V_s}$$

where $a^0{}_{SiO_2}$ (aqueous), $a^0{}_{KCl}$ and $a^0{}_{HCl}$ refer to activities of silica, KCl and HCl at $P_s = P_f$.[1] This equation expresses the relative fluid concentrations of silica and KCl/HCl in equilibrium with the Mu-Q-Or assemblage as a function of the pressure on the solids. Line CD (Fig. 3) illustrates this relation, assuming a 1 M chloride concentration and ideal mixing.

The relative fluid and solid concentrations of silica and potassium are calculated as follows. Owing to Equation 3 and the constant volume restriction, line DE in Fig. 3 must have a slope equal to that of D_3E_3. Because point E is fixed by the infiltrating solution and zone D contains the assemblage Q-Or-Mu, the solution composition in zone D_3 must lie at the intersection of lines DE and Cr on the solution concentration diagram. Similarly, line C_3D_3 (Fig. 4) must have a slope equal to that of CD (Fig. 3). Thus the $C_{Si}{}^r$ and $C_K{}^r$ for zone D_3 are determined, and one now has the necessary

[1] Equations 12, 13, and 14 do not take into account the effects of nonhydrostatic stress on the solids. Therefore, these equations are used only as approximations.

information (Fig. 5 (3)) for the calculation of the relative front rates (Fig. 6 (3)), giving the following sequence of reaction zones:

Pyr-Q-p | Mu-Q-p | Mu-Or-Q | Or-Q.

Path 4 ($p = 0.04$; $C_{Si}{}^r = 1.54$). In this case, the porosity is not sufficient to even convert the pyrophyllite-quartz assemblage to muscovite-quartz (a horizontal line from A_4 intersects the Pyr-Q-Mu line in Fig. 4), and the three possibilities discussed for path 3 exist. Considering the possibility of a solid pressure different from that on the fluid, calculation of the relative reaction-front velocities was attempted in a manner similar to that for path 3. This entails determination of the equilibrium concentrations of the dissolved components (SiO$_2$ and KCl) at various values of P_s for the assemblage Pyr-Mu-Q and Mu-Or-Q. These calculations were made using equations similar to Equation 14. As in path 3, the volume constraints for the transition of a Mu-Q rock to an Or-Q rock (Fig. 4) require solution concentrations of D and E (Fig. 3) in zones C_4 and E_4, respectively (Fig. 4). Further, the transition of a Mu-Pyr-Q assemblage to Mu-Q requires concentrations B and D in zones B_4 and C_4 (slope ED = slope E_4C_4; slope DB = slope C_4B_4). At the front A_4B_4, the slope A_4B_4 in Fig. 4 must equal that of AB in Fig. 3, but since AB and A_4 are known, B_4 may be located, thus providing the necessary information to calculate the relative front velocities using Equation 1 (Figs. 5 (4) and 6 (4)). The following is the predicted sequence of reaction zones:

Pyr-Q-p | Pyr-Mu-Q | Mu-Q | Or-Q.

Conclusions

Diffusion metasomatism occurs as a response to concentration gradients. Because these gradients decrease as the thickness of the reaction zone (or zones) increases, the process stops rapidly.

Thus, infiltration metasomatism seems to provide an adequate mechanism for long-distance transport, for it depends on fluid pressure gradients, which may remain unchanged regardless of the thickness of the reaction zones. However, as has been shown here, the occurrence during the metasomatic evolution of reactions involving positive volume changes puts serious limitations on the applicability of the infiltration model, unless one of the following assumptions is accepted: (1) The rock can expand. (2) The initial porosity is large enough to balance the expansion of the solids (if this expansion is important, the necessary minimum value of the initial porosity would be larger than expected in metamorphic rocks). (3) The solid pressures can exceed the fluid pressure. It seems unlikely that the situation described in the third assumption could exist without material precipitation, thereby filling the pores, releasing the nonhydrostatic pressure differences, and stopping the flow of infiltrating fluids. For instance, the solid pressure on the zone B_4 should be, at 500°C and 1 kbar fluid pressure, around 7 kbar! This value is, indeed, not realistic in rocks.

The example chosen may help others to assess the applicability of the infiltration metasomatism model to the various situations where metasomatic activity is suspected to have occurred.

Acknowledgments

The authors benefited greatly from helpful criticisms by Drs. Fisher, Hofmann, Vidale, Yund, and Yoder. This work was undertaken while both authors were in residence at The Geophysical Laboratory, Carnegie Institution of Washington.

References Cited

Anderson, G. M., and C. W. Burnham, The solubility of quartz in supercritical water, *Amer. J. Sci.*, *263*, 494–511, 1965.

Carmichael, D. M., On the mechanism of prograde metamorphic reactions in quartz-bearing pelitic rocks, *Contrib. Mineral. Petrol.*, *20*, 266–267, 1969.

Fisher, G. W., The application of ionic equilibria to metamorphic differentiation: an example, *Contrib. Mineral. Petrol.*, *29*, 91–103, 1970.

Hemley, J. J., Some mineralogical equilibria in the system K_2O-Al_2O_3-SiO_2-H_2O, *Amer. J. Sci.*, *257*, 241–270, 1959.

Hofmann, Albrecht, Chromatographic theory of infiltration metasomatism and its application to feldspars, *Amer. J. Sci.*, *272*, 69–90, 1972.

Korzhinskii, D. S., Mobility and inertness of components in metasomatism (in Russian), *Izv. Akad. Nauk SSSR Ser. Geol.*, *1*, 36–60, 1936.

Korzhinskii, D. S., The theory of systems with perfectly mobile components and processes of mineral formation, *Amer. J. Sci.*, *263*, 193–205, 1965.

Korzhinskii, D. S., *Theory of Metasomatic Zoning*, Clarendon Press, Oxford, 1970.

Morey, G. W., and J. M. Hesselgesser, The solubility of some minerals in superheated steam at high pressurees, *Econ. Geol.*, *46*, 821–835, 1951.

Thompson, J. B., The thermodynamic basis for the mineral facies concept, *Amer. J. Sci.*, *253*, 65–101, 1955.

Vidale, Rosemary, Metamorphic differentiation layering in pelitic rocks of Dutchess County, New York, this volume.

Weill, D. F., and W. S. Fyfe, The solubility of quartz in H_2O in the range 1000–4000 bars and 400°–550°C, *Geochim. Cosmochim. Acta*, *28*, 1243–1256, 1964.

METAMORPHIC DIFFERENTIATION LAYERING IN PELITIC ROCKS OF DUTCHESS COUNTY, NEW YORK

Rosemary Vidale
Geophysical Laboratory
Carnegie Institution of Washington
Washington, D.C. 20008

ABSTRACT

A distinct and nearly ubiquitous layering is observed in all pelitic rocks at metamorphic grades above the sillimanite isograd in and near Dutchess County, New York. Between the sillimanite and sillimanite-orthoclase isograds, most layers are 2–10 mm thick and alternately rich and poor in quartz and plagioclase. Above the sillimanite-orthoclase isograd, the layers are generally 5–10 mm thick and alternately rich and poor in quartz, plagioclase, and potassium feldspar. There is no significant change in the bulk chemistries of individual rock units with increasing metamorphic grade; the layering is therefore considered to be the product of a metamorphic differentiation process.

The layering always conforms to folding; thus it differs in structural orientation from an axial-plane cleavage layering seen at low metamorphic grade and from an axial-plane schistosity seen in mica-rich rocks at high grade.

The variation in mineral proportions that constitutes the layering can be achieved by recrystallization of the same mineral phases that appear locally in fissure veins, en-echelon gash fractures, garnet and staurolite shadows, boudin necks, and fold-nose segregations. One possible mechanism for the differentiation process is movement of components by diffusion within small local pressure gradients generated by the application of tectonic stress to a slightly layered rock. The early layering may be original bedding or axial-plane cleavage.

Experimental data on pore solution composition as a function of total pressure support a diffusion mechanism for transport of quartz and feldspar components in a pressure gradient. Local transport of this kind may play a significant part in rock deformation at relatively high temperature, pressure, and water fugacity.

Field Area

Field observations were made in a block of twelve 7½ minute quadrangles (1700 square kilometers) in and near Dutchess County, New York (Fig. 1). The major geologic mapping of the area was done by Barth (1936) and by Balk (1936) with significant additions by Fisher, Isachsen, and Rickard (1970).

Metamorphic grade increases to the east as shown by the appearance in pelitic rocks of biotite, garnet, staurolite, sillimanite, and orthoclase with sillimanite (Fig. 2; Barth, 1936; Vidale, 1974). Pelitic rock units can be traced almost continuously from the lower-greenschist to the upper amphibolite metamorphic facies. Bulk chemical compositions of the more massive units change little with increasing grade except for loss of H_2O and CO_2. Table 1 lists bulk chemical compositions and metamorphic grades for sequences of samples of three units, Walloomsac, Everett, and Manhattan. The samples chosen for analysis came from relatively homogeneous outcrops where carbonate

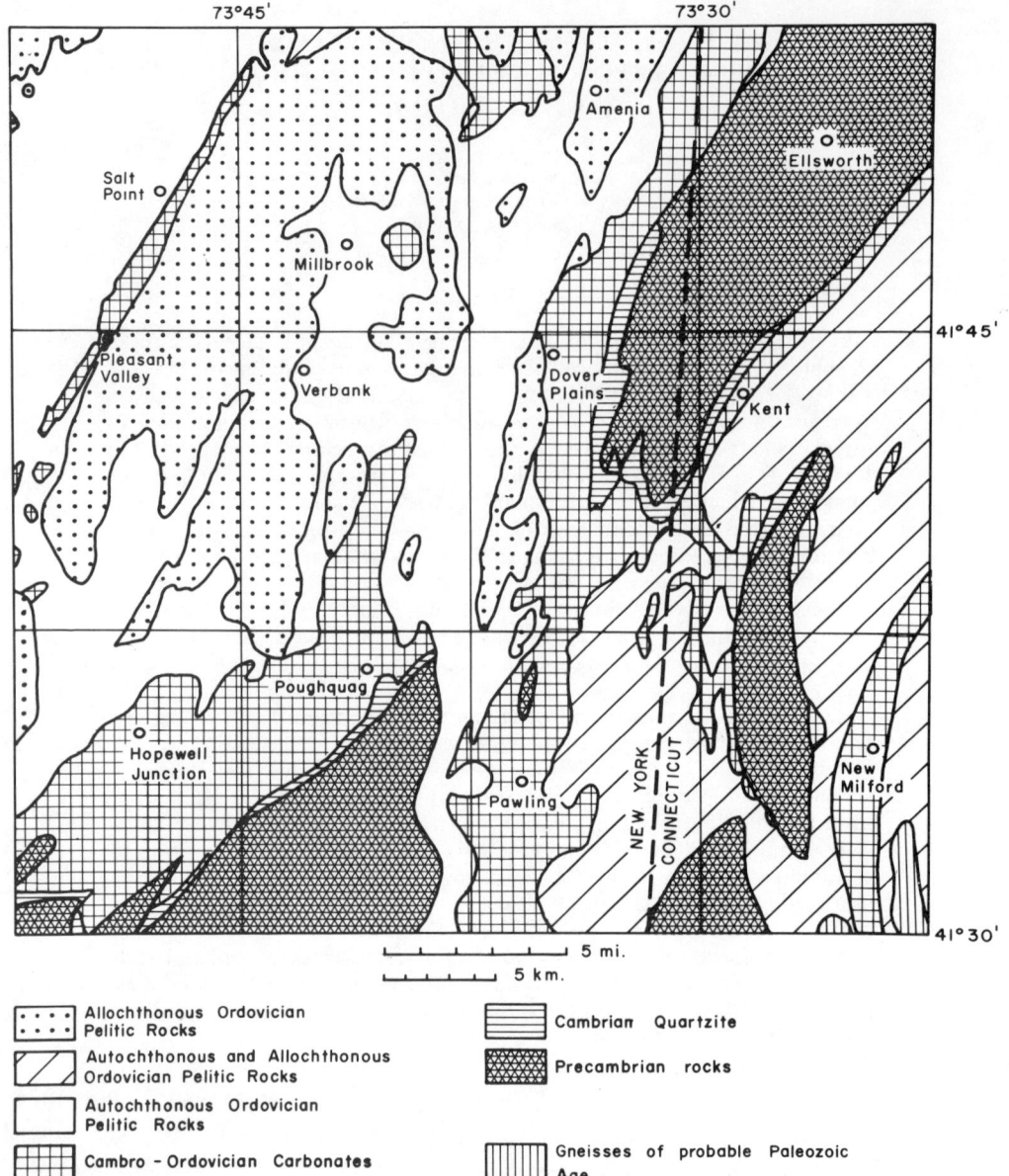

Fig. 1. Geologic map of the region studied (modified from Fisher, Isachsen, and Rickard, 1970). Quadrangle names are indicated by locality designations.

minerals could not be detected in the matrix by use of acid.

Vein mineral assemblages in the pelitic units consist of a very few phases and correlate strongly with metamorphic grade (Fig. 2, Vidale, 1974). The following vein assemblages are observed with increasing grade: quartz and quartz-calcite in all outcrops up to just above the staurolite isograd, with limited occurrence of quartz-albite just above the breakdown of detrital plagioclase and of quartz-albite-potassium feldspar in rare potassium feldspar bearing rocks; quartz and quartz-plagioclase (An_{20}-An_{50}) from the staurolite isograd up to the sillimanite-orthoclase isograd; and quartz, quartz-plagioclase, and quartz-plagioclase-orthoclase above the sillimanite-orthoclase isograd. The same limited assemblages are found in gash fractures, garnet and staurolite shadows,

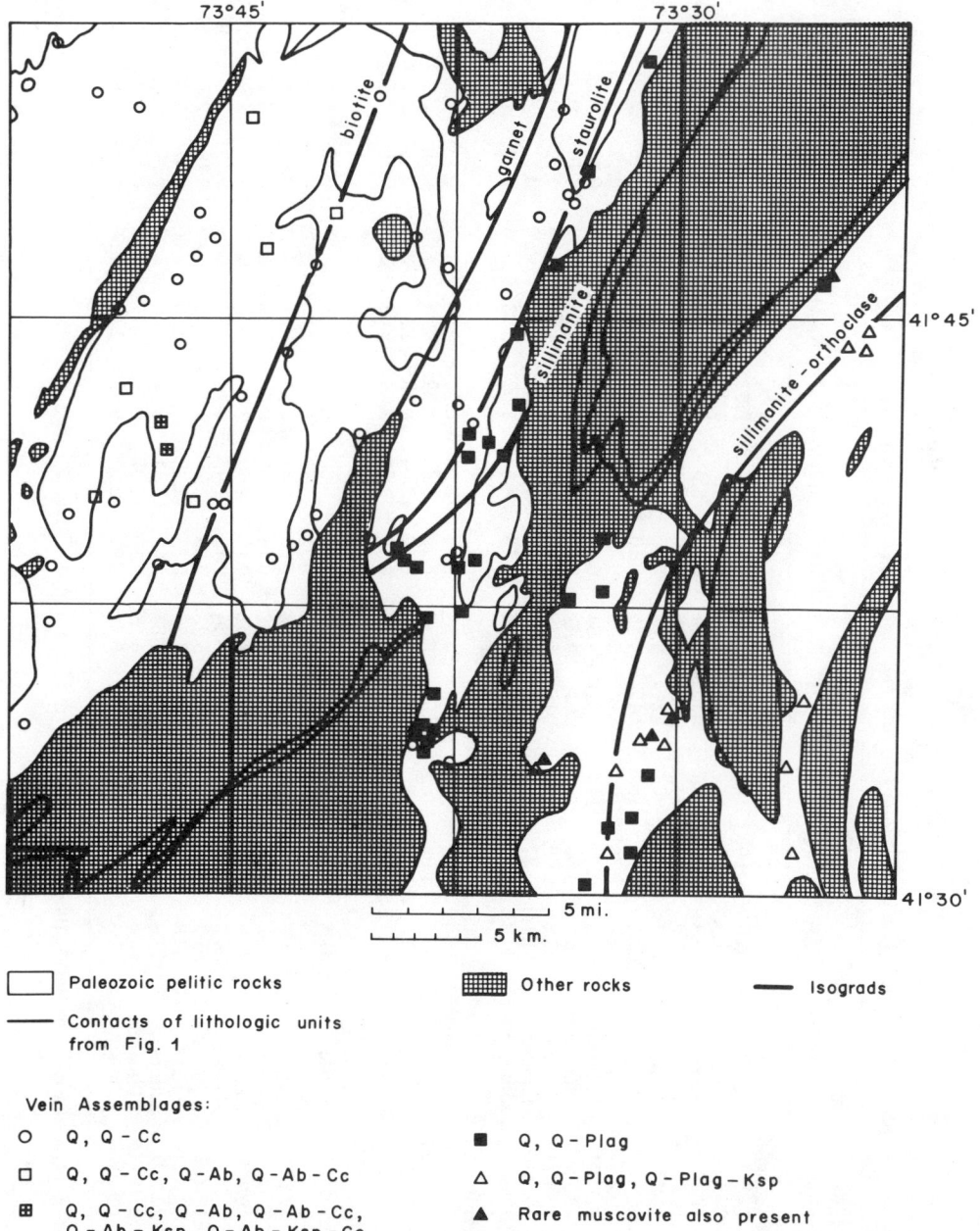

Fig. 2. Metamorphic isograds and vein mineral assemblages. Abbreviations: Q, quartz; Cc, calcite; Ab, albite; Ksp, potassium feldspar; Plag, plagioclase in the An_{20} to An_{50} composition range.

TABLE 1. Chemical analyses and metamorphic grade of sequences of samples from three units

Unit and Sample No.	Metamorphic Grade	SiO_2	TiO_2	Al_2O_3	Fe_2O_3	FeO	MnO	MgO	CaO	Na_2O	K_2O	P_2O_5	H_2O	CO_2	Total
Walloomsac															
B-2*	chlorite	65.0	0.89	16.2	1.13	5.80	0.04	1.74	0.02	2.63	2.80	0.12	3.23	...	99.6
202-72	chlorite	72.9	0.49	8.7	0.48	2.19	0.30	1.65	4.96	0.94	1.89	0.18	1.39	...	96.1
13-71	chlorite	70.3	0.69	13.2	2.32	4.04	0.09	3.04	0.18	0.04	2.76	0.10	2.13	0.00	99.2
B-4*	biotite	66.9	0.65	14.3	1.06	5.16	0.10	3.45	0.28	0.90	3.09	0.20	3.32	...	99.4
15-71	biotite	77.4	0.48	9.9	1.31	2.48	0.20	2.09	0.04	0.12	2.79	0.08	1.51	0.00	98.4
95-69	garnet	62.5	0.91	16.3	2.45	4.36	0.10	3.34	0.91	0.99	5.05	0.16	2.29	...	99.4
B-5*	garnet	62.6	0.83	16.5	0.54	5.45	0.12	3.47	1.98	1.35	4.56	0.14	1.61	0.15	99.4
21A-71	sillimanite	64.5	0.73	14.6	1.61	4.84	0.82	5.00	1.15	1.51	3.41	0.15	0.92	0.00	99.3
19-71	sillimanite	66.5	0.69	15.8	1.92	4.44	0.74	5.06	0.55	0.98	2.67	0.10	0.69	0.00	100.2
20B-71	sillimanite	66.2	0.69	14.8	1.18	5.48	0.16	4.48	0.77	1.41	3.79	0.12	0.87	0.00	99.9
106A-72	sillimanite	61.1	0.81	17.6	1.39	6.77	0.64	4.53	0.89	0.93	4.20	0.13	1.45	...	100.4
136-69	sillimanite	60.8	0.86	17.9	0.86	6.45	0.73	4.09	1.08	3.23	3.23	0.13	1.24	...	100.6
Manhattan															
25-71	sillimanite	64.9	0.98	18.4	2.43	4.88	0.07	2.00	0.29	0.18	3.77	0.15	1.66	0.00	99.8
142A-69	sill.-or.	66.1	1.06	16.8	1.52	5.17	0.22	1.88	0.32	0.28	4.35	0.16	3.23	0.00	101.1
23A-71	sill.-or.	62.3	1.07	18.1	2.03	5.60	0.12	2.06	0.79	1.11	3.46	0.14	2.71	0.00	99.4
24-71	sill.-or.	62.4	0.81	17.0	2.24	5.28	0.26	2.38	2.06	2.34	3.33	0.16	1.12	0.00	99.4
Everett															
212-72	biotite	51.8	1.20	24.2	1.80	7.70	0.14	2.30	0.10	1.52	4.35	0.12	4.74	...	100.0
B-19*	garnet	52.9	1.20	25.3	1.22	7.83	0.13	1.67	0.25	2.25	3.62	0.17	3.94	0.00	100.5
17A-71	garnet	58.3	0.98	22.2	3.16	4.76	0.07	1.41	0.20	1.31	3.34	0.17	2.43	0.00	98.3
B-6*	garnet	55.4	0.99	22.3	1.30	6.82	0.14	2.35	0.38	1.02	5.45	0.20	3.55	0.00	100.0
127-72	garnet	44.2	1.51	28.3	2.29	8.02	0.16	2.60	0.69	1.33	5.73	0.12	3.89	...	98.8
B-7*	staurolite	56.5	0.99	22.5	1.11	7.26	0.19	1.97	0.49	1.66	5.00	0.12	2.80	0.00	100.6
137-72	staurolite	55.2	1.27	22.9	1.73	5.95	0.25	2.28	1.05	2.64	4.49	0.15	2.74	...	100.6
155-72	staurolite	58.8	0.87	20.4	1.29	7.59	0.35	2.38	1.45	3.05	2.06	0.04	2.24	...	100.5
22A-71	sillimanite	57.8	1.16	22.7	3.86	3.64	0.07	1.44	0.81	1.38	5.37	0.23	1.33	...	99.8
22B-71	sillimanite	56.9	1.23	22.0	5.24	3.78	0.15	2.07	0.58	1.77	4.72	0.15	2.35	0.00	100.9
138-69	sillimanite	57.8	1.23	21.0	2.28	6.48	0.27	2.28	0.90	2.40	4.58	0.08	1.82	...	101.2

*Analyses taken from Barth, 1936. Other analyses by Max Budd, R. Vidale.

boudin necks, and fold-nose segregations.

The material in the veins and other segregations in these rocks is believed to be derived from the local matrix for the following reasons: Veins at lower metamorphic grades commonly consist of greatly elongated crystals reaching completely across the vein, as if these crystals had continued to grow at one or both ends from matrix-derived components as the vein slowly opened. At higher grades, fragments of matrix material extend perpendicularly into the veins with no evidence of displacement by flow of material through the veins. The vein assemblages always consist of a few of the matrix phases and are a consistent function of the metamorphic grade of the matrix. Bulk chemical composition of samples including networks of veins does not change significantly with increasing grade except for loss of H_2O and CO_2. (H_2O and CO_2 may have moved through a microfissure system whose only remaining traces are planes of secondary fluid inclusions as suggested by Orville, oral presentation at the Conference on Geochemical Transport and Kinetics).

Layering

Figure 3 illustrates types of layering commonly observed in these pelitic rocks by line tracings of typical micrographs. Figure 3a shows part of a fold in a muscovite-chlorite-quartz-albite-graphite phyllite. A concentration of graphite and micas is observed between folded layers (original bedding) and in closely spaced cleavage layers parallel to the axial plane of the fold. Bedding layering is usually faint where present and varies greatly in thickness. Cleavage layering is always present in the more micaceous rock units in the greenschist metamorphic facies. It is closely spaced (each layer is usually less than 1 mm thick) and consists of layers rich in mica and graphite (where graphite is present) alternating with layers rich in quartz and feldspar. Similar cleavage layering is described by Turner (1941), Brown (1963), Williams (1972), Schamel (1973), and by many others.

Figure 3b shows discrete quartz segregations near the nose of a gentle fold in a rock containing the same mineral phases as the sample shown in Fig. 3a but with a lower proportion of mica. These quartz segregations parallel the folded bedding. Figure 3c shows similar quartz segregations parallel to bedding in a relatively massive layer in a quartz-plagioclase-biotite-muscovite-garnet-ilmenite rock that is just above the garnet isograd. Note that the quartz layers cut across biotite and garnet grains, ruling out the possibility that they are depositional features. These segregations range up to a few millimeters in thickness and are spaced from a few millimeters to a few centimeters apart. Many individual segregations may be traced for tens of centimeters around the gently curving noses of concentric folds.

Figure 3d shows typical layering in a quartz-plagioclase-biotite-muscovite-garnet-kyanite-ilmenite schist above the sillimanite isograd. Abundant fibrolitic sillimanite is present in the biotite but is not shown. The light layers contain a high proportion of plagioclase and especially quartz, and the dark layers contain a high proportion of all other mineral phases. Quartz and quartz-plagioclase vein assemblages are seen in this rock.

Figures 4a and 4b are photographs of typical layering in outcrop. The rock in Fig. 4a has the same mineral assemblage as that of Fig. 3d but is only slightly above the sillimanite isograd. There is a faint but definite layering on a rather small scale (The coin is 24 mm in diameter.) The rock in 4b is a quartz-plagioclase-biotite-orthoclase-sillimanite-muscovite-ilmenite gneiss with a very high proportion of quartz, plagioclase, and orthoclase in the light layers. Veins in the outcrop shown in Fig. 4b contain quartz, quartz-plagioclase, and quartz-plagioclase-orthoclase assemblages.

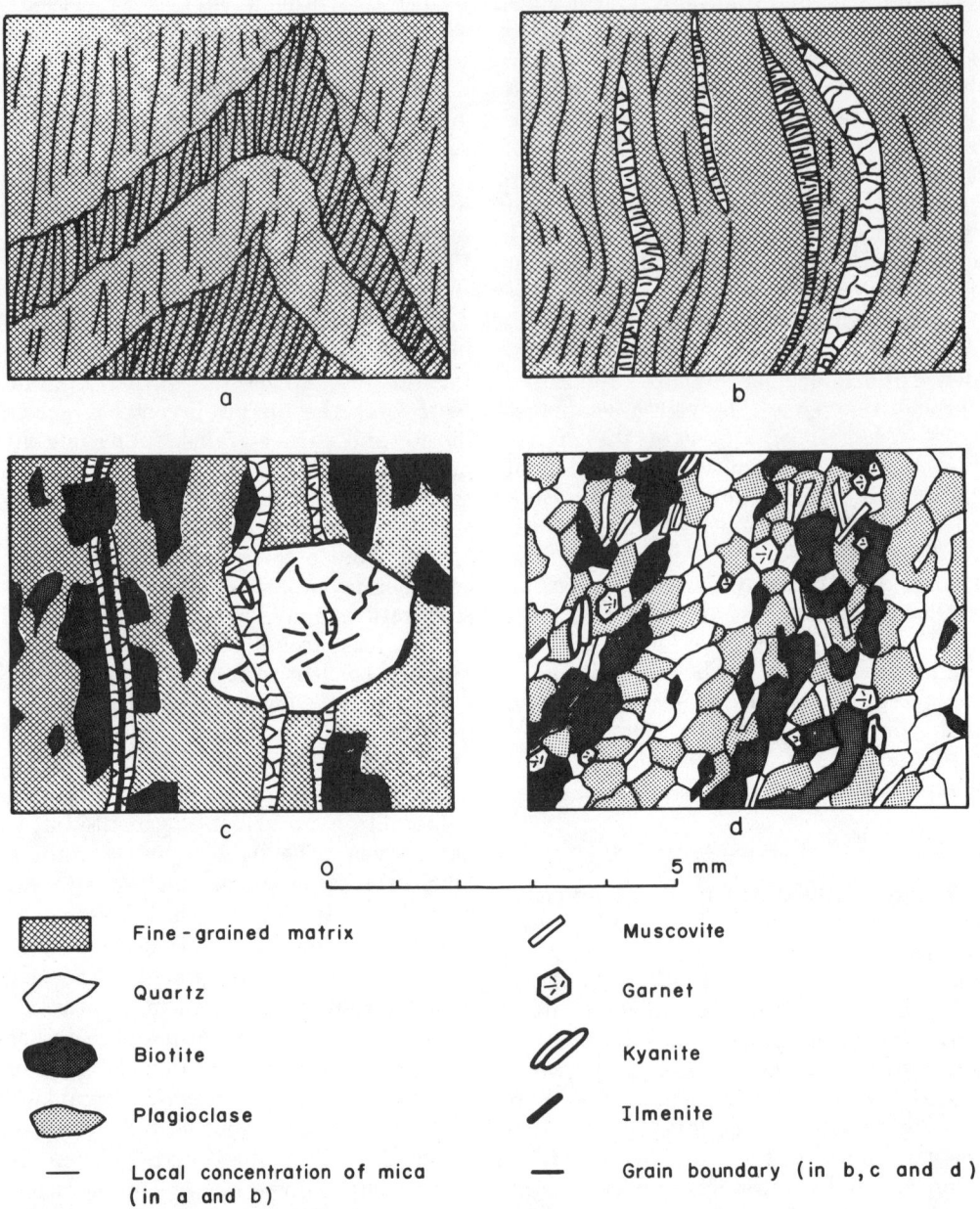

Fig. 3. Typical layering seen in thin section, in order of increasing metamorphic grade. (See text for descriptions.)

The type of layering shown in Fig. 3d, 4a, and 4b is the principal subject of this paper. It consists of alternating layers with high and low proportions of quartz and feldspar to other mineral phases in the rock. Similar layering has been described by Ramberg (1952), Dietrich (1960, 1963), Kretz (1961), Bowes and Park (1966), and by many others. In the Dutchess County area, this type of layer-

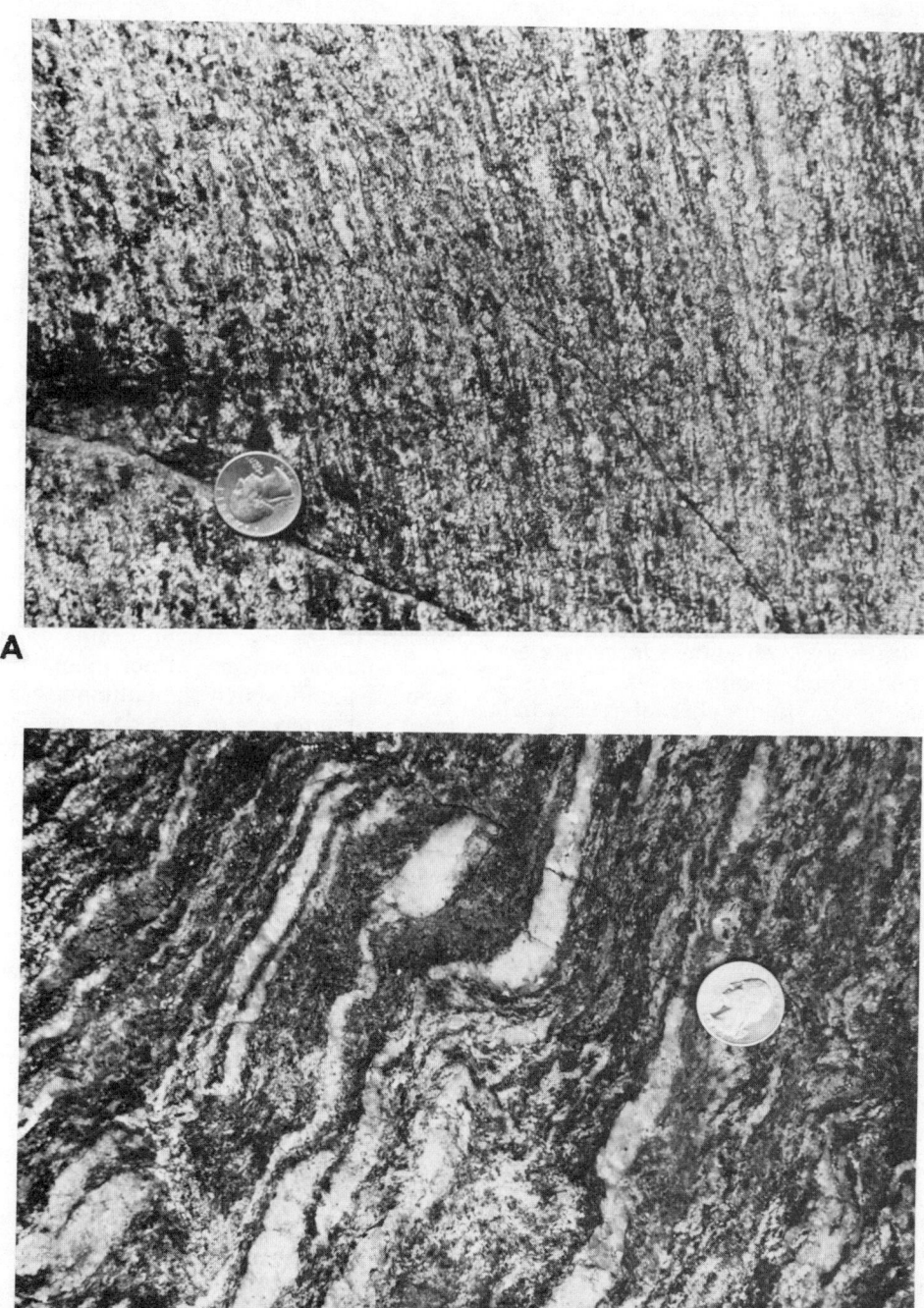

Fig. 4. Photographs of layering in outcrop. The coin is 24 mm in diameter. (a) Rock just above the sillimanite isograd. (b) Rock just above the sillimanite-orthoclase isograd. See text for descriptions of mineralogy.

ing usually occurs on a scale of 2–10 mm and always curves around the noses of folds. The layers are commonly slightly thicker on fold noses. This layering is never observed in an axial plane orientation except on the limbs of isoclinal folds. Axial plane schistosity, on the other hand, is common in the more mica-rich rocks. All of these higher grade rocks have been kneaded tectonically to the extent that original bedding cannot be identified.

The type of layering shown in Figures 3d, 4a, and 4b appears to have formed by metamorphic differentiation (Eskola, 1932) for the following reasons. The layering is present in all pelitic units above the sillimanite isograd, including units known to contain large homogeneous sections at lower grades. Thus the same units that are commonly homogeneous phyllites with only closely spaced cleavage layering in the greenschist facies are penetratively layered on a larger scale in the middle amphibolite facies. These sections retain their characteristic bulk chemical compositions (Table 1). There would be no reason to postulate a change in original sedimentary texture of all pelitic units (allochthonous and autochthonous, Fisher, Isachsen, and Rickard, 1970) near the present sillimanite isograd. Nor would such regular small-scale layering be a common result of sedimentary processes. Varves in periglacial lakes or other cyclic lake sedimentation might give this kind of texture, but the depositional environment for the Dutchess County pelitic rocks was most likely fairly deep-water marine, with pelitic sedimentation overstepping a carbonate shelf in the Walloomsac and pelitic sedimentation well beyond the shelf and below the carbonate compensation depth in the Everett and other Eastern facies. Such marine sediments are not strongly layered on a small scale and in a uniform manner. It is concluded that the observed layering formed during metamorphism by a differentiation process.

DIFFERENTIATION PROCESS

Differentiation of a rock into layers a few millimeters thick suggests the local operation of a diffusion process. Diffusion must be driven by gradients in the activities (or chemical potentials) of some of the species in the system, and these gradients in turn must be generated by some kind of heterogeneity in the system. Small, repeated temperature differences are unlikely in the Dutchess County rocks. Chemical gradients, inherited from original layering, can drive local diffusion processes and produce segregation layering (Orville, 1962, 1969; Vidale, 1969; Vidale and Hewitt, 1973). However, only one coexisting phase assemblage is found in each of the banded Dutchess County samples. Electron microprobe analyses of plagioclase and biotite show no detectable composition change between or within grains of the solid solution phases. Minor amounts of kyanite found above the sillimanite isograd and interiors of zoned garnets are considered unreactive (Hollister, 1969). Local variations in bulk chemical composition will not impose chemical potential gradients as long as the phase assemblages and phase compositions are uniform and only the proportion of phases changes. The existence of former chemical heterogeneities that are now completely homogenized must remain a possibility.

Two other kinds of heterogeneities may have provided driving mechanisms for differentiation of these rocks. The first consists of small, local shear zones generated during deformation. Components may move down their activity gradients from unstable strained crystals in shear zones into unsheared regions or into veins where they form unstrained crystals. This mechanism and similar ones have been suggested in the literature, especially for the formation of cleavage layering (see, for example, Dewey, 1965; Williams, 1972). The observed offset of earlier layering across cleavage planes makes a mechanism involving differential movement within these planes highly probable. A shear mechanism may also accentuate folded layering. Concentrations of mica and graphite are nearly always observed between folded layers in the greenschist facies pelites in Dutchess County (Fig. 3a, 3b). Shear parallel to earlier layering may also accentuate layering at higher metamorphic grades.

Another possible mechanism for differentiation is diffusion driven by small, local pressure gradients generated by the application of stress to rocks that are initially slightly layered. The initial layering may be an original sedimentary texture or it may be previously developed axial plane cleavage or other deformational layering. (The new layering might have a larger spacing than the initial layering, much like the layering produced by bending an old deck of cards.) The common appearance in Dutchess County of discrete segregations of the type shown in Figs. 3b and 3c in relatively massive layers of upper greenschist facies rocks strongly suggests the generation of low-pressure regions between original layers as the rock is folded. Mineral grains in the segregations are commonly elongate parallel to the axial plane of the fold. This texture is often seen in fissure veins and in pressure shadows in greenschist facies rocks (Vidale, 1974; Elliott, 1973). At higher temperature and pressure where rock behavior is more ductile and less brittle, segregations formed in pressure shadows parallel to deformed bedding might be gradational rather than discrete. Theoretical and experimental studies on the folding of layered rock (for example, Chapple, 1970; Dietrich, 1970; Hudleston, 1973a; Ramberg, 1964) deal with homogeneous layers of different competency, showing relative strain geometries of these layers. Analogous studies could provide estimates of internal pressure-gradients that might drive diffusion of constituents between heterogeneous layers of different competency.

Diffusion transport within small local pressure gradients appears to be the most likely mechanism for formation of the layering observed above the sillimanite isograd in the Dutchess County pelitic rocks on the basis of the following field observations. The mineral phases dominating the lighter bands always correspond to the mineral phases found at the same metamorphic grades in fissure veins, gash fractures, garnet and staurolite shadows, boudin necks, and fold-nose segregations, all of which constitute probable low pressure regions. The layering always curves around the noses of folds; it is never found parallel to the axial plane except in the case of isoclinal folding (axial plane layering would favor a shear mechanism). The layering is commonly more distinct near fold noses, and the lighter layers often thicken slightly near noses. These observations suggest generation of coarser layering by local pressure gradients as a slightly layered rock is repeatedly deformed at high temperatures and pressures.

It becomes important to distinguish between tectosilicate-rich layers that exhibit relatively competent behavior during folding (more buckling and less shortening, Hudleston, 1973b) at lower metamorphic grades, and tectosilicate-rich layers that may be thickening by transport of tectosilicate components within local pressure gradients at higher metamorphic grades. Experimental data on pressure solubility sheds considerable light on what components might be expected to move within a pressure gradient in a pelitic rock and on what conditions favor such movement.

EXPERIMENTAL DATA RELATING TO
PRESSURE GRADIENT TRANSPORT

Diffusion transport in a rock probably takes place predominantly through the pore solution if pore solution is present (Vidale, 1969, p. 866). If pore solution is not present as a separate phase, a somewhat more structured "dispersed phase" existing in and near grain boundaries may provide the highest diffusivity path (Elliott, 1973; Fisher and Elliott, this volume; Manning, this volume). A pressure gradient will generate activity gradients in species present in the solution or "dispersed phase" and thus provide the driving potential for their diffusion (Gresens, 1966).

Experimental data on solution composition as a function of total pressure may give a rough estimate of the kinds of activity gradients to be expected in a pressure gradient. Table 2 lists observed mineral solubilities and chloride salt compositions, each determined separately. The compositions of 2-normal chloride salt solutions in equilibrium with one pelitic assemblage are included in order to evaluate the effect of the presence of salt in a pore solution on transport in a pressure gradient.

Limited experimental data on the solubility of other silicates suggest that the tectosilicates are the most soluble (Morey and Hesselgesser, 1951; Ellis and Mahon, 1963). The tectosilicates should also exhibit the greatest pressure sensitivity to solution because of their relatively low density (Gresens, 1966). It is assumed here, therefore, that the composition of 2-normal chloride solution in equilibrium with the iron-free pelitic assemblage, muscovite-phlogopite-albite-microcline-quartz can be roughly approximated by summing chloride composition (buffered by reaction with the pelite), quartz solubility, albite solubility, and microcline solubility. (Common ion effects have been neglected here and may be significant for solubility of albite and microcline in the presence of NaCl and KCl.)

Component concentrations in the solution, except for those of silica and alumina, are dominated by the chloride salts (Table 2). Changes in solution composition with pressure, on the other hand, are dominated by mineral solubility. Thus the concentration gradients generated by pressure gradients in deforming pelitic rocks may be mostly a function of mineral solubility differences.

Solubilities of many compounds increase exponentially with increasing pressure in a moderately high pressure range because increased concentration of ionized solute decreases the partial molal volume of the solution (Mao and Bell, 1971). This strong effect of pressure will be modified somewhat by the presence of dissolved salts in the solution and by the decreasing ionization of electrolytes with increasing temperature (Barnes and Helgeson, 1966); however, there are probably pressure-temperature ranges over which mineral solubilities in an aqueous phase are extremely pressure sensitive and in which segregation caused by pressure gradients should be most pronounced.

It seems reasonable to conclude that tectosilicates present in a pelitic matrix would tend to move into any relatively low pressure region in the rock. The quartz and feldspar vein assemblages in Dutchess County pelitic rocks give evidence that this does happen, and the development of layering above the sillimanite isograd suggests that similar transport takes place within the matrix of folded layered rocks.

Acknowledgments

This study was funded by National Science Foundation Grant GA-30862. The author would like to thank David Miller for stimulating discussions of this work, David A. Hewitt, and Hatten S. Yoder, Jr., for helpful suggestions during the preparation of the manuscript, Max Budd for chemical analyses, and Frank Sedlak for thin section preparation. The mistakes are Ms. Vidale's.

TABLE 2. Solution composition as a function of pressure

				Normality of:						
Composition of Chloride Salts[1]	T,°C	P,kbar	Solution	Na	K	Ca	Mg	Al	Si
muscovite-phlogopite-microcline-albite-quartz	550	2	2N Cl⁻	1.56	0.41	0.011	0.021	<0.02	<0.02
muscovite-phlogopite-microcline-albite-quartz	550	4	2N Cl⁻	1.56	0.41	0.011	0.017	<0.02	<0.02
muscovite-phlogopite-microcline-oligoclase-quartz	550	2	2N Cl⁻	1.56	0.42	0.023	0.012	<0.02	<0.02
muscovite-phlogopite-microcline-oligoclase-quartz	550	4	2N Cl⁻	1.56	0.41	0.026	0.007	<0.02	<0.02
Quartz solubility[2]	500	2	H_2O						0.27
	500	4	H_2O						0.44
	600	2	H_2O						0.48
	600	4	H_2O						0.79
Albite solubility[3]	500	2	H_2O	0.010				0.023	0.107
	500	3.5	H_2O	0.024				0.043	0.213
Microcline solubility[4]	500	1	H_2O		0.0013			0.008	0.034
	500	2	H_2O		0.0042			0.023	0.011

[1] Vidale, see appendix.
[2] Anderson and Burnham, 1965, 1967.
[3] Currie, 1968.
[4] Morey and Hesselgesser, 1951.

Starting materials for hydrothermal runs

Mineral	Locality	SiO_2	TiO_2	Al_2O_3	Fe_2O_3	FeO	MnO	MgO	CaO	Na_2O	K_2O	Total
Muscovite	Stoneham, Maine	45.7	0.04	32.2	2.07	2.09	0.46	0.00	0.00	0.54	9.78	92.9
Phlogopite	North Burgess, Ontario	40.9	1.28	14.0	0.83	1.17	0.03	24.3	0.11	0.30	10.0	92.9
Albite	Amelia Courthouse, Va.	67.5	0.00	19.6	...	...	0.00	0.05	0.30	11.98	0.26	99.7
Microcline	Perry Sound, Ontario	65.2	0.00	18.1	...	...	0.00	0.02	0.09	2.46	12.1	98.0
Oligoclase	Mitchell Co., N.C.	63.4	0.04	22.8	...	...	0.00	0.01	4.38	7.88	0.55	99.1
Quartz	Minas Gerais, Brazil	...	0.00	0.16	...	...	0.00	0.02	0.05	0.08	0.02	

Analyst, Max Budd.

Appendix: Experimental procedure used for determination of chloride compositions

Samples consisting of 220 mg of muscovite-phlogopite-microcline-albite (or oligoclase)-quartz mix (see table of starting materials p. 284) and 100 μl of 2 N NaCl solution (made from Fisher Reagent Grade NaCl and double distilled water, tested for conductivity) were placed in platinum or gold capsules 4 mm i.d. and 35 mm long. (No difference in final data can be detected between runs in gold and runs in platinum for these assemblages.) The capsules were placed in 24-inch cold-seal tipping bombs (modified from Wellman, 1970) and run for 25 days. The runs were quenched by tipping the bombs so that the sample-bearing capsules slid down into the cold, water-cooled front end of the bomb. Estimated quench time is 2–3 seconds to 50°C.

The capsules were kept at room temperature for at least 24 hours before they were opened; this permits precipitation of metastable silicate species from the solution but does not seem to affect the quenched chloride salts. Then the capsules were placed in liquid nitrogen to freeze the solutions. (The initial spurt of air released on opening the slightly pressurized capsule may disperse significant amounts of solution.) The capsule contents were emptied into a small Teflon beaker, ground slightly under distilled water, and the solution was passed through a Millipore filter. Solution volume was brought up to 40 ml. Then 1.00 g of concentrated HCl was added to the filtrate; empirically, this gives improved precision to the data.

X-ray diffraction patterns (Guinier camera) of the solids were made both before and after the runs to confirm the presence of the proper phases. Solutions were analyzed for Na^+, K^+, Ca^{2+}, and Mg^{2+} by atomic absorption and for Si^{4+} and Al^{3+} by Beckman DU spectrometer. Recovery of chloride salts was good but not complete (75–95%, usually 90–95%). Their sum was normalized to 2.000 N.

These exchange reactions are not reversed in any manner. Similar exchange reactions with the assemblage muscovite-phlogopite-sanidine-anorthite-quartz did converge satisfactorily starting from pure KCl, $CaCl_2$, $AlCl_3$, $MgCl_2$, and HCl solutions (Vidale, 1969); however, the Na-bearing system of this study should include significant solid solution of muscovite-paragonite, alkali feldspar, and plagioclase and probably equilibrates less readily (Orville, 1972, plagioclase). The data, therefore, are considered only approximate.

References Cited

Anderson, G. M., and C. W. Burnham, The solubility of quartz in supercritical water, *Amer. J. Sci.*, *263*, 494–511, 1965.

Anderson, G. M., and C. W. Burnham, Reactions of quartz and corundum with aqueous chloride and hydroxide solutions at high temperatures and pressures, *Amer. J. Sci.*, *265*, 12–27, 1967.

Balk, R., Structural and petrologic studies in Dutchess Co., N.Y.: Part I: geologic structure of sedimentary rocks, *Geol. Soc. Amer. Bull.*, *47*, 685–774, 1936.

Barnes, H. L., H. C. Helgeson, and A. J. Ellis, Ionization constants in aqueous solutions, in *Handbook of Physical Constants*, S. P. Clark, ed., *Geol. Soc. Amer. Mem. 97*, 401–413, 1966.

Barth, T. F. W., Structural and petrologic studies in Dutchess Co., N.Y.: Part II: petrology and metamorphism of the Paleozoic rocks, *Geol. Soc. Amer. Bull.*, *47*, 775–850, 1936.

Bowes, D. R., and R. G. Park, Metamorphic segregation banding in the Loch Kerry basite sheet from the Lewisian of Gairloch, Ross-Shire, Scotland, *J. Petrology*, *7*, 306–330, 1966.

Brown, E. H., The geology of the Mt. Stoker area, eastern Otago, *N.Z. J. Geol. Geophys.*, *6*, 847–871, 1963.

Chapple, W. M., The finite-amplitude instability in the folding of layered rocks, *Can. J. Earth Sci.*, *7*, 457–466, 1970.

Currie, K. L., On the solubility of albite in supercritical water in the range 400–600°C and 750–3500 bars, *Amer. J. Sci.*, *266*, 321–341, 1968.

Dewey, J. F., Crenulation differentiation, *Tectonophysics*, *1*, 459–486, 1965.

Dietrich, J. H., Computer experiments on mechanics of finite amplitude folds, *Can. J. Earth Sci., 7*, 467–476, 1970.

Dietrich, R. V., Banded gneisses, *J. Petrology, 1*, p. 99–120, 1960.

Dietrich, R. V., Banded gneisses of eight localities, *Norsk Geologisk Tidsskrift, 43*, 89–121, 1963.

Elliott, D., Diffusion flow laws in metamorphic rocks, *Geol. Soc. Amer. Bull., 84*, 2645–2664, 1973.

Ellis, A. J., and W. A. J. Mahon, Natural hydrothermal systems and experimental hot-water/rock interactions, *Geochim. Cosmochim. Acta, 28*, 1323–1357, 1963.

Eskola, P., On the principles of metamorphic differentiation, *Bull. Comm. Geol. Finlande, 97*, 68–77, 1932.

Fisher, D., Y. W. Isachsen, and L. V. Rickard, *1970 Geologic Map of New York, New York State Museum and Science Service, Map and Chart Series No. 15*, 1972.

Gresens, R. L., The effect of structurally produced pressure gradients on diffusion in rocks, *J. Geol., 74*, 307–321, 1966.

Hollister, L. S., Contact metamorphism in the Kwoiek area of British Columbia: an end member of the metamorphic process, *Geol. Soc. Amer. Bull., 80*, 2465–2494, 1969.

Hudleston, P. J., An analysis of 'single-layer' folds developed experimentally in viscous media, *Tectonophysics, 16*, 189–214, 1973a.

Hudleston, P. J., The analysis and interpretation of minor folds developed in the Moine rocks of Monar, Scotland, *Tectonophysics, 17*, 89–132, 1973b.

Kretz, R., Preliminary examination of quartz-plagioclase layers and veins in amphibolite facies gneisses, southwestern Quebec, *Proc. Geol. Ass. Can., 13*, 23–44, 1961.

Mao, H. K., and P. M. Bell, Theory of the solubility of minerals at high water pressures, Annual Report of the Director of the Geophysical Laboratory 1971–1972, *Carnegie Inst. Washington Yearb. 71*, 457–459, 1972.

Morey, G. W., and J. M. Hesselgesser, Solubilities of some minerals in superheated steam at high pressures, *Econ. Geol., 46*, 821–835, 1951.

Orville, P. M., Alkali metasomatism in feldspars, *Norsk Geol. Tidsskr. 42* (Feldspar volume), 283–315, 1962.

Orville, P. M., A model for metamorphic differentiation origin of thin-layered amphibolites, *Amer. J. Sci., 267*, 64–86, 1969.

Orville, P. M., Plagioclase cation exchange equilibrium with aqueous chloride solution: results at 700°C and 2000 bars in the presence of quartz, *Amer. J. Sci., 272*, 234–272, 1972.

Ramberg, Hans, *The Origin of Metamorphic and Metasomatic Rocks,* Univ. of Chicago Press, 317 pp., 1952.

Ramberg, H., Selective buckling of composite layers with contrasted rheological properties, a theory for simultaneous formation of several orders of folds, *Tectonophysics, 1*, 307–341, 1964.

Schamel, S., Eocene subduction in central Liguria, Italy, Ph.D. thesis, Yale Univ., 1973.

Turner, F. J., The development of pseudo-stratification by metamorphic differentiation in the schists of Otago, New Zealand, *Amer. J. Sci., 239*, 1–16, 1941.

Vidale, R., Metasomatism in a chemical gradient and the formation of calc-silicate bands, *Amer. J. Sci., 267*, 857–874, 1969.

Vidale, R., Vein assemblages and metamorphism in Dutchess County, New York, *Geol. Soc. Amer. Bull., 85*, 303–306, 1974.

Vidale, R., and D. Hewitt, "Mobile" components in the formation of calc-silicate bands, *Amer. Mineral., 58*, 1973.

Wellman, T., Fugacities and activity coefficients of NaCl in NaCl-H_2O fluids at elevated temperatures and pressures, *Amer. J. Sci., 269*, 402–413, 1970.

Williams, P. F., Development of metamorphic layering and cleavage in low grade metamorphic rocks at Bermaqui, Australia, *Amer. J. Sci., 272*, 1–47, 1972.

METASOMATIC ZONING IN Ca-Fe-Si EXOSKARNS

D. M. Burt
Department of Geology and Geophysics
Yale University
New Haven, Connecticut 06520

ABSTRACT

Metasomatic zoning sequences commonly observed in Ca-Fe-Si exoskarns can be explained by simple diffusion models that assume simultaneous development of all major zones, as the result of chemical potential gradients set up between iron- and silica-rich solutions and calcium carbonate host rocks. The diffusion models can be represented graphically as saturation surfaces on chemical potential diagrams, made popular by D. S. Korzhinskii. These diagrams suggest that in some skarns at least, certain species in solution could diffuse "uphill," that is, against their concentration gradients. A given zoned skarn represents only a single curved path across the saturation-surface model referred to above. Therefore, the zoning sequence that is observed usually does not contain all of the mineral assemblages needed to specify the facies of the skarn.

INTRODUCTION

Skarns (also called tactites) are coarse-grained, metasomatically zoned, generally dark-weathering silicate rocks that can develop between carbonate-rich rocks (limestones, dolomites, Mn-rich carbonates) and Al- and Si-rich rocks (granitic or other intrusives, shales, schists, etc.). Skarns can be classified according to a number of criteria, some of which are given below.

Reaction skarns are small silicate reaction rims developed as the result of local diffusion between incompatible rock types, one of which is carbonate-rich. An interesting example is provided by the successive reaction rims of tephroite, Mn_2SiO_4, and rhodonite, $MnSiO_3$, formed between layers of rhodochrosite, $MnCO_3$, and chert, SiO_2, during contact metamorphism in Japan (Watanabe et al., 1970). Replacement skarns (ore skarns and other names) are extensive silicate replacements of carbonate and adjacent rocks due to the massive inflow of solutions out of equilibrium with the host rocks. These skarns are distinguished by their size and extent and by the fact that they commonly are mined (for ores of Fe, Cu, Zn, W, and other metals), although they also grade into the reaction type (cf. the bimetasomatic skarns of Korzhinskii, 1959; Zharikov, 1970).

Endoskarns are Al-rich skarns (rich in epidote or grossularitic garnet), generally rather limited in extent, that replace non-carbonate rocks, while exoskarns replace carbonate rocks. Exoskarns replacing dolomites tend to be very rich in Mg, whereas exoskarns replacing pure limestones tend to be rich in Ca, Fe, and occasionally Mn (hence the name "Ca-Fe-Si skarns," here applied).

The most common silicate minerals in Ca-Fe-Si exoskarns are andradite, $Ca_3Fe^{3+}_2Si_3O_{12}$, and hedenbergite, $CaFe^{2+}Si_2O_6$. In hydrous, low-temperature Ca-Fe-Si skarns ilvaite, $CaFe^{2+}_2Fe^{3+}Si_2O_7O(OH)$, is typical.

Facies diagrams for Ca-Fe-Si skarns (mineral stability diagrams for the model system Ca-Fe-Si-C-O-H as determined by P, T, μO_2, and μCO_2) have been presented elsewhere (Burt, 1971a, b, c), and need not be discussed further here, except to note that instability of the assemblage hedenbergite-calcite in some skarns can be indicated by enrichment of the clinopyroxene in Mn toward the contact with marble.

Metasomatic Zoning

Metasomatism in Ca-Fe-Si exoskarns results in a general tendency for the minerals to be segregated into more or less monomineralic bands around the presumed passageways for skarn-forming solutions (igneous contacts, sedimentary contacts, cross-cutting fractures and faults). The sequence of metasomatic bands or zones in a given deposit can range in total width from a few centimeters to several tens of meters or more. These variations in width presumably reflect local variations in the porosity and permeability ("plumbing") of the rock. The same sequence of zones tends to be found throughout a given deposit or group of deposits, although the relative thickness of individual zones may vary considerably from place to place, and some zones can be missing entirely in places.

Metasomatic zoning sequences commonly found in Ca-Fe-Si exoskarns are listed in Table 1. This summary table is based on observations by the author at more than 50 individual deposits in the U.S.A., Mexico, Peru, Japan, and Italy, as well as on the geologic literature.

Some superposition of different periods of skarn formation must have occurred in the deposits visited, and the skarn minerals rarely were pure end-member andradite or hedenbergite, but rather were continuously varying iron-rich solid solutions. These complications will generally be ignored in the discussion that follows.

Diffusion Models

Reaction and replacement skarn zoning sequences such as those given in Table 1 generally can be explained by simple diffusion models that assume simultaneous development of all major zones, as the result of chemical potential gradients set up between dissimilar host rocks, under the influence of skarn-forming solutions (*cf.*, Thompson, 1959; Korzhinskii, 1959). In the deposits on which Table 1 is based, at least, fluid motion along channelways and through the rock itself seems merely to spread the zones out over larger distances. The relevance to skarns of the end-member

TABLE 1. Zoning sequences in Ca-Fe-Si exoskarns

1. Vein or intrusive or epidote endoskarn or magnetite	andradite	calcite		
2. Vein or intrusive or garnet endoskarn or magnetite	andradite	wollastonite	calcite	
3. Vein or grossular endoskarn or magnetite	hedenbergite	calcite		
4. Magnetite or intrusive	ilvaite	hedenbergite	calcite	
5. Vein or grossular endoskarn	hedenbergite	wollastonite	calcite	
6. Vein or intrusive or endoskarn or magnetite	andradite	hedenbergite	calcite	
7. Magnetite or endoskarn	andradite	hedenbergite	wollastonite	calcite

model of infiltration metasomatism (cf., Korzhinskii, 1970; Hofmann, 1972) therefore is difficult to assess on the basis of present evidence, and will not be discussed further here.

Diffusion models for some of the sequences in Table 1 are represented graphically on Figs. 1 to 3. Figure 1 is for an andradite exoskarn in marble. Part A is the relevant facies diagram (Burt, 1971a), and part B shows the result if cut prisms of calcite, quartz, and magnetite are left in contact under conditions such that andradite is stable (cf., Thompson, 1959). Part C shows that andradite reaction rims can form between chert nodules and limestone under the influence of a chemical potential gradient in iron, as at the Iron King and Iron Queen Mines, New Mexico (Burt, 1972).

Lastly, part D depicts a μCa-μFe-μSi saturation surface for the assumption that Ca, Fe, and Si all are free to diffuse through a pore fluid. Inside this saturation surface, the pore fluid is undersaturated with Ca, Fe, and Si, and no crystalline phases in the end-member system Ca-Fe-Si-C-O are stable. The saturation surface flattened is topologically the same as the reacted cut prisms in Fig. 1B. This last type of representation was introduced to geologists by D. S. Korzhinskii (1959 and earlier publications).

Development of a wollastonite zone between andradite and marble is possible on Fig. 2. On part B note that the wollastonite zone can be missing if the solutions reacting with calcite are undersaturated with SiO_2. The typical zoning sequence for this facies is as follows: Ca, Fe, and Si undersaturated phases in the endoskarn (interior of the diagram)/andradite/wollastonite/calcite, but the wollastonite zone can be absent, as mentioned above, or a zone of magnetite or quartz can separate andradite from the endoskarn.

This example shows that more than one zoning sequence is possible within the confines of a single skarn facies and, conversely, that the same zoning sequence seen at different localities can form within different facies. The skarn zoning sequence at a given locality represents only one curved path across the saturation surface of the μCa-μFe-μSi diagram. It thus does not contain enough mineral assemblages, in general, to specify the facies of the skarn.

In Figs. 1D and 2B, μFe and μSi could be allowed to decrease away from the endoskarn or source of the solutions, and μCa could be allowed to decrease away from the limestone, but for the zoning sequence endoskarn (or magnetite)/andradite/hedenbergite/calcite in Table 1, it would appear that μSi must increase through the andradite and into the hedenbergite zone, as shown by the curve drawn in Fig. 3B.

This apparent increase in μSi towards the limestone leads to the speculation that SiO_2 was diffusing "uphill," that is, against its initial concentration gradient, as discussed by Cooper (this volume). A similar example of a local maximum in μAl_2O_3 inside a zoned skarn is given by Korzhinskii (1959, p. 96).

Unfortunately for the above argument, the present example can also be explained both by (1) an increase in μO_2 away from the limestone, which would tend to stabilize andradite at the expense of hedenbergite, and by (2) an increase in μAl_2O_3 towards the endoskarn, which would likewise tend to stabilize (aluminous) garnet. Less ambiguous examples of "uphill diffusion" should therefore be sought in nature.

CONCLUSIONS

Other saturation surfaces could be depicted to explain the remaining zoning sequences in Table 1 (see Burt, 1972), without changing the conclusions already reached—namely, that the diffusion models appear to work and that "uphill diffusion" could have occurred in some cases. That the diffusion models here

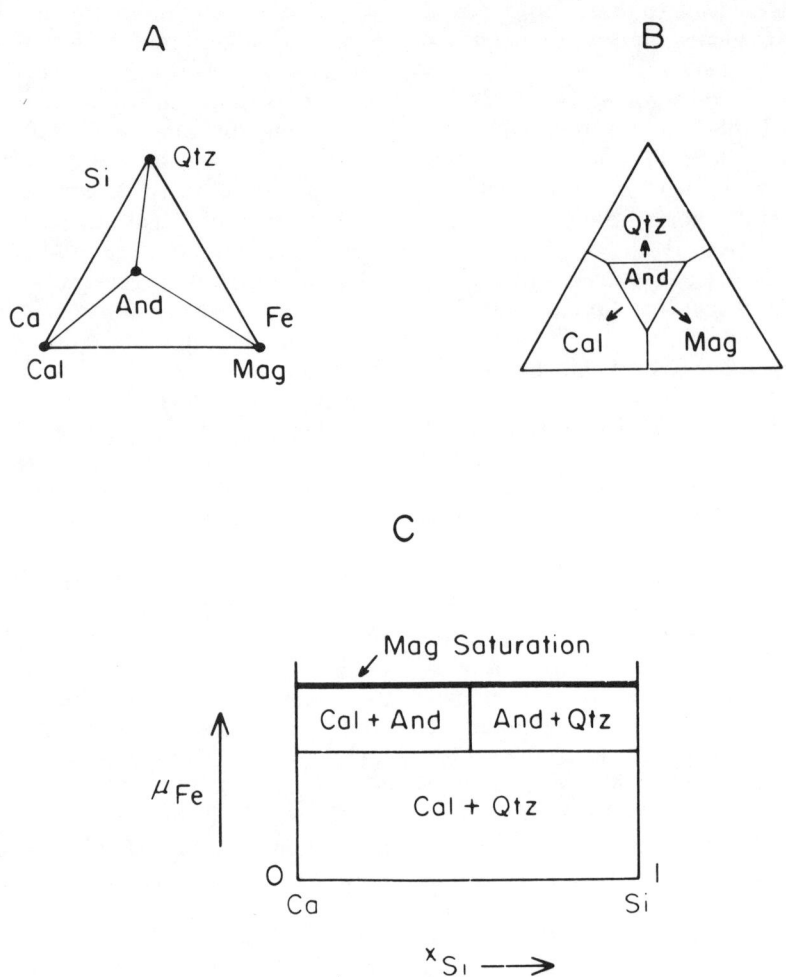

Fig. 1. Chemical potential and related diagrams for the exoskarn zoning sequence (magnetite)/andradite/calcite. For this and the following figures, the system is Ca-Fe-Si-C-O, and the variables T, P, μO_2, and μCO_2 are considered to be externally controlled. (A) Facies diagram, assuming that wollastonite is unstable. (B) Configuration of phases if cut prisms of calcite, magnetite, and quartz undergo prolonged reaction under the assumed conditions. (C) μFe-X diagram for the system, assuming that μCa and μSi are dependent variables. The maximum possible value of μFe is that of saturation with magnetite. Intermediate values of μFe are indicated by the assemblages calcite +

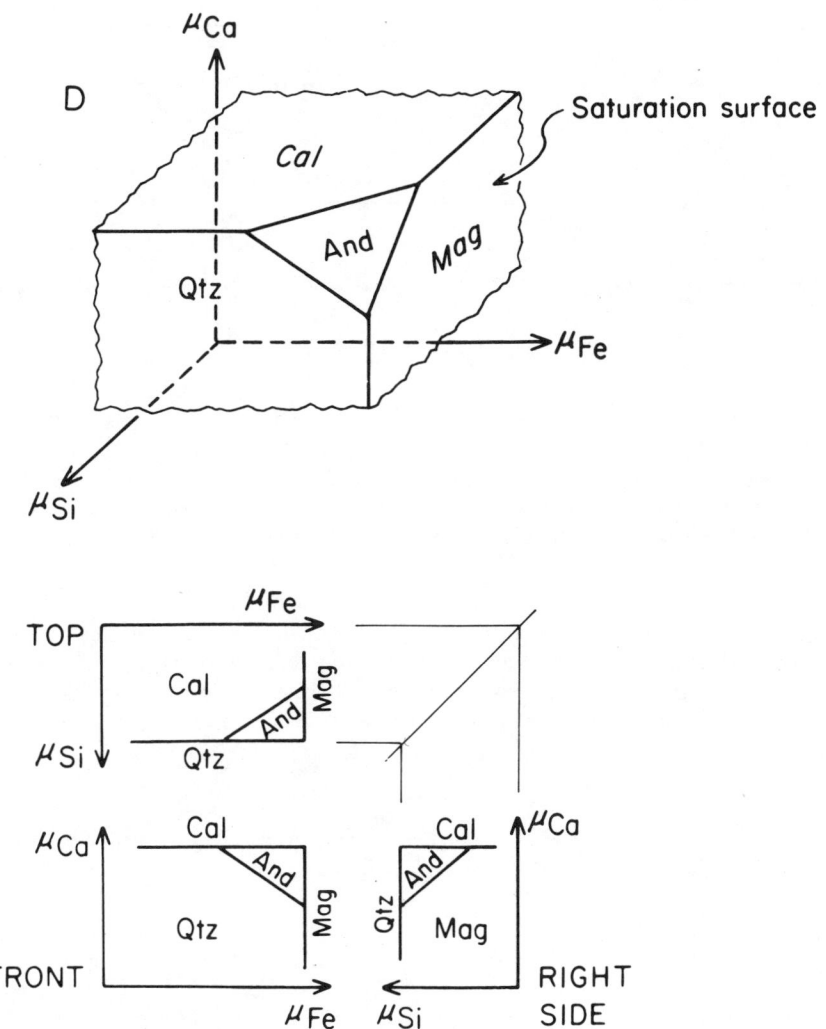

andradite or andradite + quartz. Low values of μFe are indicated by the assemblage calcite + quartz. (D) μCa-μFe-μSi diagram for the system and possible projections (μCa-μFe, μSi-μFe, μCa-μSi diagrams). The saturation surface has the appearance of the corner of a cube with an oblique face removed for andradite. The calcite, magnetite, and quartz faces extend infinitely far away from this single corner, as values of μCa, μFe, and μSi become infinitely small. Inside the saturation surface, the "pore fluid" is undersaturated with Ca, Fe, and Si, and no phases in the end-member system are stable.

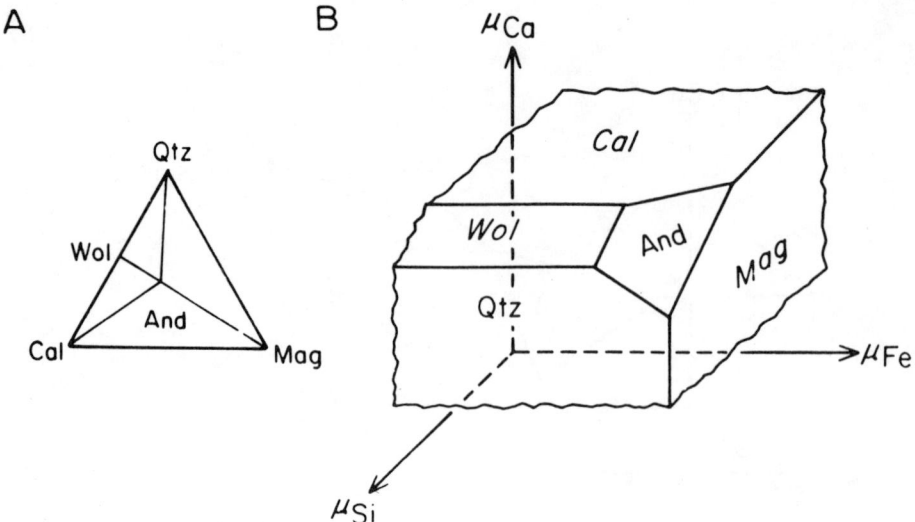

Fig. 2. Facies and chemical potential diagrams for the exoskarn zoning sequence (magnetite)/andradite/wollastonite/calcite. (A) Facies diagram. (B) μCa-μFe-μSi diagram. The saturation surface looks like that of Figure 1-D, except that a beveled wollastonite face separates the calcite and quartz faces.

presented are extremely simple in comparison to the skarn deposits being modeled perhaps explains why such models have as yet received little attention among economic geologists in the U.S. Nevertheless, the models would appear to be useful, especially for predicting where a given ore-bearing skarn zone will occur with respect to the other zones present.

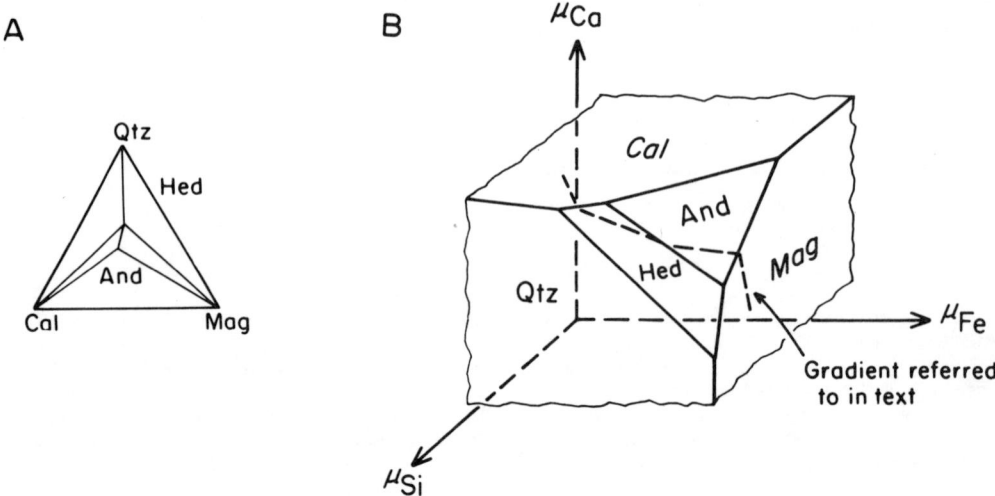

Fig. 3. Facies and chemical potential diagrams for the exoskarn zoning sequence (magnetite)/andradite/hedenbergite/calcite. (A) Facies diagram. (B) μCa-μFe-μSi diagram. The dashed line on the saturation surface shows the diffusion path referred to in the text. Note that going from the andradite to the hedenbergite face in a direction of increasing μCa and decreasing μFe implies an increase in μSi.

References Cited

Burt, D. M., Some phase equilibria in the system Ca-Fe-Si-C-O, *Carnegie Inst. Washington Yearb. 70,* 178–184, 1971a.

Burt, D. M., The facies of some Ca-Fe-Si skarns in Japan, *Carnegie Inst. Washington Yearb. 70,* 185–188, 1971b.

Burt, D. M., Multisystems analysis of the relative stabilities of babingtonite and ilvaite, *Carnegie Inst. Washington Yearb. 70,* 189–197, 1971c.

Burt, D. M., Mineralogy and geochemistry of Ca-Fe-Si skarn deposits, unpublished Ph.D. thesis, Harvard Univ., Cambridge, Mass., 1972.

Cooper, A. R., Vector space treatment of multicomponent diffusion, this volume.

Hofmann, A., Chromatographic theory of infiltration metasomatism and its application to feldspars, *Amer. J. Sci., 272,* 69–90, 1972.

Korzhinskii, D. S., *Physicochemical Basis of the Analysis of the Paragenesis of Minerals,* Consultant's Bureau, Inc., New York, 142 pp., 1959.

Korzhinskii, D. S., *Theory of Metasomatic Zoning* (trans. by Jean Agrell), Oxford Univ. Press, London, 162 pp., 1970.

Thompson, J. B., Jr., Local equilibrium in metasomatic processes, in *Researches in Geochemistry,* P. H. Abelson, ed., Wiley, New York, pp. 427–457, 1959.

Watanabe, T., S. Yui, and A. Kato, Bedded manganese deposits in Japan, a review, in *Volcanism and Ore Genesis,* T. Tatsumi, ed., Univ. of Tokyo Press, Tokyo, pp. 119–142, 1970.

Zharikov, V. A., Skarns, *Int. Geol. Rev., 12,* 541–559, 619, 647, 760–775, 1970.

A NEW APPROACH TO THE DETERMINATION OF TEMPERATURES OF INTRUSIVES FROM RADIOGENIC ARGON LOSS IN CONTACT AUREOLES

S. B. Brandt [1]
Institute of the Earth's Crust,
Siberian Division
Academy of Sciences, USSR

Concentration distributions of radiogenic substances in the neighborhood of contacts with intrusive bodies have been used to obtain information on physical, chemical, and time conditions of magma intrusion in wall rocks. In the paper of Brandt et al. (1967), a method of estimating the temperatures and durations of the action of a dike on the wall rocks was proposed based on quantitative comparisons of radiogenic argon losses in the wall rocks with kinetic parameters (argon retentivities) determined by laboratory means. On the other hand, in the investigations of Hart (1964) and Brandt et al. (1972) methods of obtaining migration parameters (activation energies of diffusion) of radiogenic argon and strontium were described which used argon concentrations in contact aureoles and hypothetical, computed thermal models for the intrusive (including the temperature of the intruded magma).

Both approaches are questionable, as their premises (the techniques of laboratory determinations of the activation energy, and the thermal properties of the magma and country rocks) are uncertain within a considerable range. It is doubtful whether the conditions of the emplacement of an intrusive can be imitated by means of laboratory heat treatment in a vacuum, under atmospheric conditions or in other media. Eutectic relations and the action of fluids are not taken into account. Equally indeterminate are estimates of heat flow from intrusives.

These considerations justify the investigation of the possibility of developing a potassium-argon geothermometer, which is independent of any hypotheses and external quantitative estimates, and is based on concentrations of radiogenic argon only. The purpose of the present paper is not to develop a precise and complete geothermometer, but rather to demonstrate the principles underlying a method which may be refined in the future, perhaps one involving complex computer programs.

Consider the argon-potassium ratio (Ar^{40}/K) for some mineral fraction (for instance, biotite) of a rock in a contact aureole, (see Fig. 1). Ar^{40}/K smoothly increases from a value $(Ar^{40}/K)_0$, which is close to the $(Ar^{40}/K)_{intr}$ of the intrusive at zero distance from the contact ($x = 0$), and trends towards an asymptotic value $(Ar^{40}/K)_\infty$ with increasing x.

Without restricting the generality, we shall base the arguments on a diffusional mechanism of radiogenic argon loss by minerals. Similar considerations may be carried out for arbitrary mechanisms. We suppose further that the duration of the thermal action is insignificant in geologic time.

It is known (Lykov, 1948) that, for diffusion models, the concentration of radiogenic argon is fully determined by the dimensionless criterion of similarity of Fourier

$$\text{Fo} = \frac{D\tau}{a^2} = \frac{D_o \tau}{a^2} \exp(-E/RT) \quad (1)$$

---

[1] Due to last-minute unforeseen difficulties, Dr. Brandt was unable to attend the Conference. The editors learned subsequently that Dr. Brandt was preparing a manuscript for inclusion in this volume. We were fortunate to be able to include our colleague's paper just as the volume went to press.——Eds.

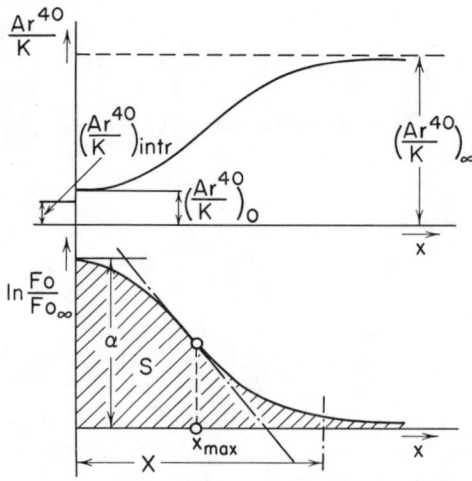

Fig. 1. Variation of the radiogenic argon to potassium ratio (Ar^{40}/K) as a function of the distance, x, from the intrusive in a contact aureole.

(D = the coefficient of diffusion, a = the characteristic grain size of the mineral, τ = time of interaction, E = the energy of activation of diffusion, T = the temperature).

Values of the Fourier number may be determined from the relative concentrations of radiogenic argon

$$f = \frac{(Ar^{40}/K)_x}{(Ar^{40}/K)_\infty}$$

using the nomograms given in the book of A. A. Lykov.

Suppose further that the thermal action on the minerals of the contact zone may be approximated by a thermal pulse of equal duration τ throughout (Fig. 2), but of a varying magnitude T, which may be expressed, for example, by the function

$$T = T_0 \exp(-kx) + T_\infty \quad (2)$$

(T_0 = the excess temperature at the contact ($x = 0$), k = the damping factor of the thermal pulse, T_∞ = the temperature of the surrounding medium).

Under these conditions the Fourier number (1) becomes

$$\text{Fo} = C \exp\left[-\frac{E}{R(T_0 e^{-kx} + T_\infty)}\right] \quad (3)$$
$$C = D_0 \tau/a^2$$

$$\ln \text{Fo} = \ln C - \frac{E}{R}\frac{1}{T_0 e^{-kx} + T_\infty}$$

Introducing the asymptotic value of the Fourier number, we have

$$\text{Fo}_\infty = \lim_{x \to \infty} C \exp\left[-\frac{E}{R(T_0 e^{-kx} + T_\infty)}\right] \quad (4)$$
$$= C \exp(-E/RT_\infty)$$

Then

$$\ln \frac{\text{Fo}}{\text{Fo}_\infty} = \frac{E}{R}\left[\frac{1}{T_\infty} - \frac{1}{T_0 e^{-kx} + T_\infty}\right] \quad (5)$$

Now we determine some parameters characterizing the model chosen.

First we determine the integral with an infinite upper limit which equals the area beneath the curve (5):

$$S = \int_0^\infty \ln \frac{\text{Fo}}{\text{Fo}_\infty} dx = \frac{E}{RT_\infty k} \ln(1 + y)$$
$$y = \frac{T_0}{T_\infty} \quad (6)$$

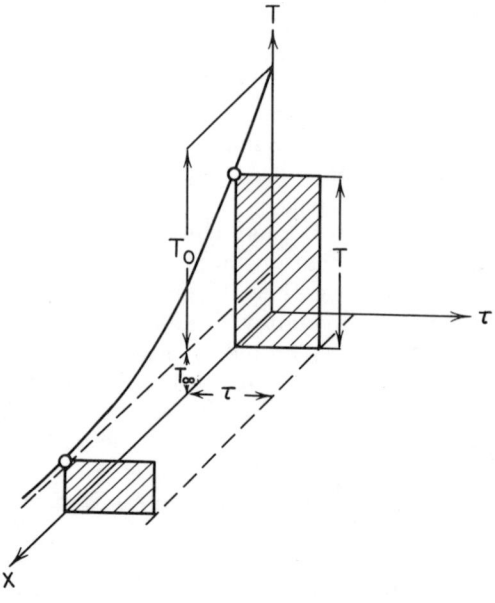

Fig. 2. The thermal action on the contact rocks is expressed by an equivalent thermal pulse of equal duration τ and of decreasing temperature T.

Further, we ascertain that the first derivative of (5) with respect to distance is

$$\frac{\partial(\ln \text{Fo}/\text{Fo}_\infty)}{\partial x} = -\frac{kT_0 E}{R}\frac{e^{-kx}}{T^2} \quad (7)$$

It has a maximum (or, in other words, the curve (5) has a point of inflection) at $X_{\max}$, where

$$e^k X_{\max} = \frac{T_0}{T_\infty} = y \quad (8)$$

It may be readily seen that $X_{\max}$ does not depend on Fo_∞.

Substituting (8) into (2) we find that at the point of inflection the temperature is twice the temperature of the medium

$$T_{X_{\max}} = 2\, T_\infty \quad (9)$$

We shall also make use of the expressions

$$\alpha = \ln\frac{\text{Fo}_0}{\text{Fo}_\infty} = \frac{E}{RT_\infty}\left(\frac{y}{1+y}\right) \quad (10)$$

$$\beta = \ln\frac{\text{Fo}_0}{\text{Fo}_{X_{\max}}} = \frac{E}{2RT_\infty}\left(\frac{y-1}{y+1}\right) \quad (11)$$

and of the value of the derivative (7) at the point of inflection

$$\gamma = \left(\frac{\partial(\ln \text{Fo}/\text{Fo}_\infty)}{\partial X}\right)_{X_{\max}}$$
$$= -\frac{E \ln y}{4R X_{\max} T_\infty} \quad (12)$$

These relations are adequate for determining the excess temperature of the contact with respect to the temperature of the medium $y = T_0/T_\infty$ as well as the activation energy E for the mineral considered. Here two approaches are possible.

First approach. Dividing (6) by (10) and accounting for (8), we obtain

$$\frac{S}{\alpha X_{\max}} = \frac{(1+y)\ln(1+y)}{y \ln y} = f(y) \quad (13)$$

On the left hand side of (13) all the quantities are determinable from a given halo (Fig. 1): S is the area beneath the curve $\ln \text{Fo}/\text{Fo}_\infty$, Alpha is the maximum ordinate of the curve, and $X_{\max}$ is the abscissa of the point of inflection. Hence, the left hand side provides a certain parametric number, say a, and the problem is reduced to the solution of the equation $f(y) = a$.

It can be readily shown that S/α is the active width of the halo X. Hence, (13) may be brought into an obvious form

$$\frac{X}{X_{\max}} = f(y) \quad (14)$$

The ratio of the active width of the aureole to the abscissa of the point of inflection is a function of y only, that is, it determines the excess temperature T_0.

Thereafter, the activation energy may be easily obtained from (10):

$$E = R\,\alpha\,T_\infty\left(\frac{1+y}{y}\right) \quad (15)$$

Although (14) and (15) completely solve the problem considered, in practice we encounter the need to determine the asymptotic value of the Fourier number Fo_∞—a rather small quantity that cannot be obtained by measurement. Therefore (14) is suitable for rough estimates of the temperatures and for comparisons of two halos. To obtain E another approach is advisable.

Second approach. Dividing (11) by (12), and after some transformations, we form the expression

$$\frac{\beta}{2\gamma X_{\max}} = \left(\frac{y-1}{y+1}\right)\frac{1}{\ln y} = \phi(y) \quad (16)$$

Here the left hand side of the equation again consists of measurable quantities $\beta, \gamma, X_{\max}$, whereas the right hand side is a function of excess temperature only. After determining the excess temperature, E may be obtained from the formula

$$E = 2\,R\,\beta\,T_\infty\left(\frac{y+1}{y-1}\right) \quad (17)$$

On Fig. 3, a curve of the function $\phi(y)$ is given.

Now we illustrate the above method on a numerical example of an aureole of

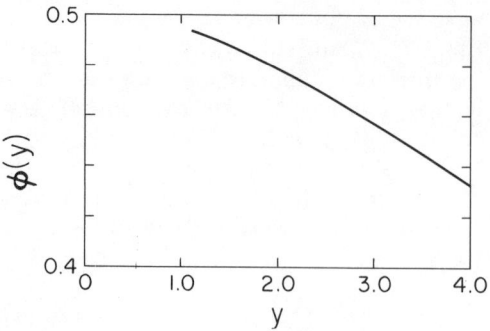

Fig. 3. Graph of the function $\phi(y)$ used in calculations of T_0 by means of formula (16).

argon concentrations in biotites from the Front Range (Colorado) described by S. R. Hart (1964). On Fig. 4, the curves for Ar^{40}/K, as well as for $\log Fo/Fo_\infty$ are given.

From the curve, it can be seen immediately that the point of inflection occurs at $X_{max} = 1.8$ km. Further, $\log Fo_0 - \log Fo_{x_{max}} = 3.7$ and the slope of the tangent at the point of inflection equals $\beta = 2.18$ km^{-1}.

Substituting these values in (16) we obtain $\phi(y) = 0.471$ and, from the curve in Fig. 3, determine $y \approx 2.5$. Hence, taking $T_\infty = 300°K$, the temperature of the zone near the contact would have been

$$T_{X=0} = 3.5 \times 300 = 1050°K \text{ or } \approx 780°C.$$

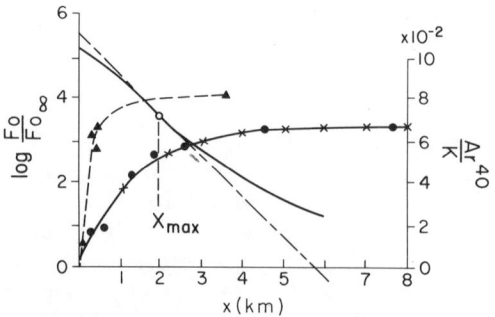

Fig. 4. The ratio of Ar^{40}/K in biotites in the contact aureole of a diorite intrusion in granite country rocks of the Front Range (after S. R. Hart, 1964) where X is the distance from the intrusive; solid circles, measured values for the biotites; crosses, averaged values; triangles, measured values for hornblendes; open circles, values of $\log Fo/Fo_\infty$.

This value approximates the temperature of solidification of a diorite magma under hypabyssal conditions (920°C). The activation energy from (17) has an unusually low value, $\sim$ 9.5 to 10.0 kcal/mole. Similar anomalously low values were obtained by us for the biotite of the same aureole in a paper by Brandt et al. (1972) using a completely different procedure. About the reasons for this phenomenon we can merely guess.

At the same time, we can compare the activation energies of biotite$_1$ and hornblende$_2$ from the same aureole, using Equation 11.

$$\frac{E_2}{E_1} = \frac{\beta_2}{\beta_1} \quad (18)$$

At a distance of $x = 1.8$ km from the contact the biotite has lost 30% of its argon, whereas the hornblende has lost 1% or less (Fig. 4). Hence we readily obtain

$$E_2 \geq (3 \text{ to } 4) E_1 = 30 \text{ to } 40 \cdot \text{kcal/mole}.$$

Therefore, the activation energy obtained for hornblende is similar to the values obtained by Hart (1964) and Brandt et al. (1972), namely, 40–52 kcal/mole.

A basic requirement for the application of the method developed here is the availability of detailed data on concentrations of radiogenic substances in contact aureoles.

REFERENCES CITED

Brandt, S. B., N. V. Volkova, and V. A. Utenkov, The estimate of mineral retentiveness of radiogenic argon (strontium) by contact concentration relation (without laboratory heat treatment), in *Year Book 1971* of the Siberian Institute of Geochemistry, Novosibirsk, 1972.

Brandt, S. B., V. I. Kovalenko, N. V. Volkova, and P. P. Kriventsov, Approach to potassium argon geothermometry and estimate of thermodynamic parameters of formation of intrusive bodies, *Izv. Akad. Nauk SSSR, Ser. Geol.*, No. 1, 1967.

Hart, S. R., The petrology and isotopic mineral age relations of a contact zone in the Front Range (Colorado), *J. Geol.*, 72, No. 5, 1964.

Lykov, A. A., *Theory of Thermal Conductivity*, Gostekhizdat, Moscow, 1948.

OXYGEN AND HYDROGEN ISOTOPE EVIDENCE FOR LARGE-SCALE CIRCULATION AND INTERACTION BETWEEN GROUND WATERS AND IGNEOUS INTRUSIONS, WITH PARTICULAR REFERENCE TO THE SAN JUAN VOLCANIC FIELD, COLORADO[1]

Hugh P. Taylor, Jr.
Division of Geological and Planetary Sciences
California Institute of Technology
Pasadena, California 91109

ABSTRACT

The 33 to 25 m.y.-old intrusive and volcanic rocks from western San Juan Mountains (Silverton-Ouray area of Colorado) are all abnormally low in O^{18} relative to most igneous rocks, particularly along the eastern edge of the Silverton Caldera where the average $\delta O^{18} = -5$. The only exceptions to this are certain plutons emplaced into Paleozoic and Mesozoic sedimentary rocks. Therefore, although in the volcanic rocks very large-scale convective circulation systems involving heated meteoric ground waters were established by the epizonal igneous intrusions, this did not occur in the less permeable sedimentary sections. The very low δD values of the hydrous minerals in the igneous rocks confirm these conclusions ($\delta D = -137$ to -150). The same processes also occurred in the vicinity of the 29 m.y.-old Alamosa stock in the eastern San Juan Mountains, but the δO^{18} depletion effects are not so extreme and the δD values are much higher (-101 to -118). This seems to imply that in the mid-Tertiary, the meteoric waters were about 35 per mil different in these two areas, even though they are only 70 miles apart. The meteoric-hydrothermal alteration processes that have affected these types of rocks can in certain circumstances produce almost complete oxygen exchange in plagioclase phenocrysts while preserving igneous textural features and delicate oscillatory zoning in the plagioclase.

With two exceptions all of the low-O^{18} igneous rocks throughout the world are of late Mesozoic to Tertiary age; these are now known to be very extensive and quite common in all volcanic fields where there are shallow igneous intrusions. In Precambrian rocks, however, the reverse is true. Alkali feldspars from the red-rock granophyres and granites in the Muskox, Bushveld, and Duluth Complexes, and in the St. Francois and Keweenawan volcanic terranes are commonly higher in O^{18} than coexisting quartz. Although this could have resulted from much higher-O^{18} meteoric waters in the Precambrian (which would imply that the oceans also were much higher in O^{18}), it seems more likely that this is due to very low-temperature ($\sim 150°C$) exchange with ground-water brines that circulated through these rocks for long periods of time. All these alkali feldspars are turbid and contain disseminated hematite dust; this high oxidation state is readily explained if the alteration occurs at low temperatures. It would also explain why Rb-Sr ages on these rocks are generally younger and much more variable than are the Pb-U zircon ages. It is because zircon is very resistant to exchange, and the turbid feldspars and altered mafic minerals are not.

[1] Contribution No. 2359, Publications of the Division of Geological and Planetary Sciences, California Institute of Technology, Pasadena, California 91109.

Introduction

In recent years it has become well established by means of oxygen isotope analyses that certain epizonal igneous intrusions have interacted on a very large scale with meteoric ground waters. In favorable terranes, that is, in highly jointed, permeable, flat-lying volcanic rocks, these intrusions act as gigantic "heat engines" that provide the energy necessary to promote a long-lived convective circulation of any mobile H_2O in the country rocks surrounding the igneous body. These systems may represent the "fossil" equivalents of the deep portions of modern geothermal water systems such as those in Steamboat Springs, Nevada, and Yellowstone Park, Wyoming (White, 1968).

The interaction and transport of large amounts of meteoric ground waters through hot igneous rocks produces a depletion of O^{18} in the igneous rocks and a corresponding O^{18} enrichment or "O^{18} shift" in the water. Fortunately, primary, unaltered igneous rocks throughout the world display a relatively narrow range of δO^{18}, only $+5.5$ to $+10.0$ (Taylor, 1968). This provides a datum by which we can discern the effects of any processes that drastically affect the δO^{18} value of an igneous rock. It is also fortunate that the only commonly occurring natural processes that are known to produce O^{18} depletions in igneous rocks are (1) interaction of the rocks (or magmas) with meteoric ground waters or ocean waters at high temperatures, or (2) strong decarbonation and loss of CO_2 from carbonate-bearing rocks.

It should be pointed out, however, that certain rare occurrences of low-O^{18} igneous rocks are not readily explained by either of the above processes. An example is the Roberts Victor eclogite pipe discussed by Garlick et al. (1971). In any event, as the deuterium contents of meteoric ground waters are commonly lower than ocean water and in many cases are much lower than the primary magmatic waters in igneous rocks, detailed geologic observations combined with both O^{18}/O^{16} and D/H analyses will usually allow a decision to be made between the possibilities mentioned above.

Low-O^{18} igneous rocks produced by interaction with meteoric ground waters have now been observed in the Skaergaard intrusion, the Stony Mountain ring-dike complex in Colorado, the Scottish Hebrides, in Iceland, in the Western Cascades, the Ag-Au deposits at Tonopah, Goldfield, and the Comstock Lode, Nevada, and in portions of the Boulder batholith and the Southern California batholith (Taylor and Epstein, 1963; Taylor, 1968, 1971, 1973; Taylor and Forester, 1971, 1973; Forester and Taylor, 1972; Sheppard and Taylor, 1974; Muehlenbachs et al., 1972). The purpose of the present paper is to present O^{18}/O^{16} and D/H data on another extensive area that shows these features, namely, the San Juan volcanic field, Colorado, and then to integrate all these data to try to understand the mechanisms that produce low-O^{18} igneous rocks. Finally, comparisons will be made between these low-O^{18} rocks and a somewhat analogous set of rocks of Precambrian age that also seem to have exchanged on a large scale with hydrothermal solutions. However, instead of being depleted in O^{18} by the hydrothermal fluids, these Precambrian rocks have instead become enriched in O^{18}.

San Juan Mountains, Colorado

Western San Juan Mountains (Silverton Caldera)

The δO^{18} and δD data obtained on volcanic rocks and epizonal igneous intrusions of the western San Juan Mountains are presented in Fig. 1 and Table 1. Analyses are principally from the area of the Silverton Caldera, but some samples were obtained from the Ouray area and from diorite porphyry intrusions em-

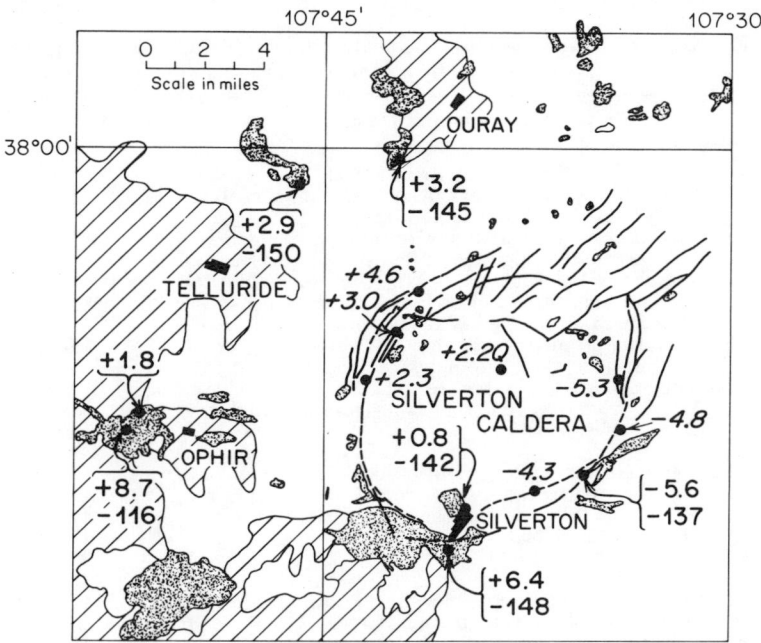

Fig. 1. Generalized geologic map of the Western San Juan Mtns., Colorado (after Luedke and Burbank, 1968), showing some of the oxygen and hydrogen isotope data obtained on igneous rocks in this area (see Table 1 for the complete set of isotopic data). The plotted δO^{18} analyses are on whole rock samples, except that Q = quartz. The analyses on volcanic rocks are given in italics, those on intrusive rocks in regular lettering. The large negative numbers (−142, −150, etc.) represent δD values on biotite or chlorite. The intrusions are shown in stippled pattern, the volcanic country rocks in blank pattern (this includes local outcrops of Precambrian basement), and the Paleozoic and Mesozoic sedimentary rocks in a diagonal-lined pattern.

placed into the Paleozoic and Mesozoic sedimentary rocks that outcrop to the west and to the south of the volcanic terrane. These various igneous rocks were largely emplaced in the late Oligocene to early Miocene about 33 to 25 m.y. ago (Luedke and Burbank, 1968; Lipman et. al., 1970).

The isotopic compositions of samples from the Tertiary volcanic terrane of the western San Juans are very depleted in both O^{18} and deuterium relative to "normal" igneous rocks. This is particularly true of samples collected in the deeply eroded Animas River Canyon along the eastern ring fracture of the Silverton Caldera, where four different samples have an average δO^{18} value of −5 per mil. These samples must have become depleted in O^{18} by at least 10 to 12 per mil during exchange with heated meteoric ground waters. These meteoric-hydrothermal solutions were set into convective circulation at a variety of igneous centers by the many epizonal intrusions along the ring fractures of the Silverton Caldera.

A detailed study of one of these intrusive centers, at Stony Mountain, six miles southwest of Ouray, was made by Forester and Taylor (1972). In this study it was shown that, even in a small geographical area, large δO^{18} variations are produced during the interaction of heated meteoric ground waters with a composite stock that has been emplaced by successive multiple intrusions of magma. In particular, although the bulk of the δO^{18} variations in the Stony Mountain complex are clearly due to exchange between

TABLE 1. Oxygen and hydrogen isotope analyses of whole-rock samples and minerals from the San Juan Mountains, Colorado

	Sample Location and Description	Mineral	$\delta O^{18}(^o/_{oo})$
	Western San Juan Mtns.		
Col-1-8	Propylitized diorite porphyry, on Telluride-Rico highway, 2.6 mi N of Trout Lake.	WR	+1.8
Col-2-8	Slightly chloritized quartz diorite porphyry, north side of Telluride-Rico highway, 2 mi W of Ophir, 2.2 mi N of Trout Lake. $\delta D = -116$ (chlorite + biotite).	WR F H(+C)	+8.7 +8.6 +6.9
Col-8-8	Diorite porphyry, just N of La Plata, 2.2 mi N of Kroeger campground and 8.4 mi N of junction with U.S. 160.	WR	+9.6
Col-11-8	Granodiorite porphyry at contact with Hermosa Formation, Silverton-Durango highway, 7 mi S of Col-13-8.	WR	+5.9
Col-13-8	Medium grained augite-biotite quartz monzonite, 1 mi SW of Silverton on Silverton-Durango highway. $\delta D = -148$ (biotite).	WR Q KF B	+6.4 +6.6 +8.8 +3.9
Col-13-8a	Fine grained leucocratic granitic dike cutting Col-13-8.	WR	+5.1
Col-14-8	Porphyritic quartz monzonite, somewhat propylitized, just NE of Silverton at mouth of Cement Creek. $\delta D = -142$ (chlorite).	WR Q KF F C	+0.8 +6.3 −0.2 −1.9 −4.1
Col-16-8	Chloritized granodiorite porphyry, Animas River canyon, mouth of Cunningham Creek. $\delta D = -137$ (chlorite).	WR F C	−5.6 −5.2 −7.6
SJ-25	Altered rhyolite, Burns Formation, Animas River canyon, mouth of Eureka Gulch, 7 mi NE of Silverton.	WR	−5.3
SJ-26	Altered rhyolite, Burns Formation, Animas River canyon, mouth of Maggie Gulch, 6 mi ENE of Silverton.	WR	−4.8
SJ-33	Altered rhyolite, Burns Formation, resistant ridge overlooking Animas River canyon, 2.5 mi NE of Silverton.	WR	−4.3
Col-3R (88316)	Vein quartz from an ore body in the Gold King mine, 6 mi N of Silverton.	Q	+2.2
SJ-14	Altered Treasure Mtn. rhyolite, roadcut on U.S. highway 550, 1.1 mi S of Red Mtn. Pass.	WR	+2.3
SJ-12	Silverton Volcanic Group, roadcut on U.S. highway 550, 0.9 mi N of Red Mtn. Pass.	WR	+3.0
SJ-10	Silverton Volcanic Group, roadcut on U.S. highway 550, 2.5 mi N of Red Mtn. Pass. (San Juan Co. line).	WR	+4.6
Col-17-8	Fresh, unaltered med. gr. pyroxene gabbro from Stony Mountain, 6 mi SW of Ouray (see Forester and Taylor, 1972, and Dings, 1941). $\delta D = -150$ (biotite).	WR F B M	+2.9 +3.3 +4.0 +2.1
Col-18-8	Highly propylitized diorite porphyry in Canyon Creek, 3 mi SW of Ouray. $\delta D = -145$ (chlorite).	WR	+3.2
Col-19-8	Slightly chloritized and uralitized gabbro dike, N-trending, 1.0 mi N of Cedar Hill Cemetery, N of Ouray.	WR	+3.8
	Eastern San Juan Mtns.		
Col-27-8	Granodiorite porphyry, just W of and above Annella Lake. Plagioclase shows well-developed, fine-scale oscillatory zoning (see Fig. 5).	Q F KF B Cpx	+6.8 +6.5 +5.8 +1.7 +4.5
Col-28-8	Andesite porphyry, Conejos formation, chloritized biotite phenocrysts, Alamosa River, 0.9 mi W of Lake Annella.	F B(+C)	+6.1 −0.9
Col-29-8	Biotite-augite granodiorite, Alamosa River stock, on road 2.5 mi E of Lake Annella, 1.4 mi W of junction to Stunner Campground.	F B	+4.8 +2.1
Col-32-8	Strongly kaolinized quartz monzonite porphyry, N side of Alamosa River stock, on road 0.7 mi W of junction to Stunner campground. $\delta D = -91$ (kaolinite).	Kaol	+4.2
Col-34-8	Strongly kaolinized quartz monzonite cut by quartz veins (3-4 mm thick), N side of Alamosa River stock, along road just W of bridge over the Alamosa River.	Q Kaol Q(vein)	+5.6 +3.9 +4.7

TABLE 1. Continued

	Sample Location and Description	Mineral	$\delta O^{18}(^o/_{oo})$
Col-36-8	Uralitized quartz monzonite with turbid K feldspar, center of Alamosa River stock, on road to Platoro, 0.9 mi S of bridge over the Alamosa River. $\delta D = -101$ (chlorite + biotite).	WR KF M	+2.3 +2.6 −2.8
Col-36-8a	Same locality as Col-36-8, quartz monzonite and associated fine gr. xenolith of altered andesite.	WR (qtz. monz.) WR (xenolith)	+1.5 +1.1
Col-37-8	Hornblende granodiorite with turbid K feldspar, Alamosa River stock, on road to Platoro, 0.3 mi S of bridge over the Alamosa River.	WR	+3.6
Col-39-8B	Augite-biotite granodiorite, turbid K feldspar, oscillatory zoned plagioclase, on road at mouth of Bitter Creek, 1.3 mi E of bridge over the Alamosa River.	F M	+3.8 0.0
Col-39-8A	Pink, fine gr. granitic dikelet (1 cm across) cutting Col-39-8B.	WR	+5.0
Col-41-8	Chloritized quartz monzonite with turbid alkali feldspar, along road at E edge of Alamosa River stock. $\delta D = -118$ (chlorite).	WR	+1.3

Abbreviations: WR = whole rock; F = plagioclase feldspar; KF = potassium feldspar Q = quartz; B = biotite; C = chlorite; H = hornblende; M = magnetite; Cpx = clinopyroxene; Kaol = kaolinite. δO^{18} values are relative to SMOW (standard mean ocean water); in many cases two or more separate determinations were made on each sample, with an analytical error of ±0.1 per mil. δD values are also relative to SMOW; the analytical error is ±3 per mil.

heated meteoric ground waters and *solidified* igneous rocks, Forester and Taylor (1972) concluded that a low-O^{18} magma must have been produced by some process of direct or indirect exchange with the meteoric ground waters that circulated to depths of several kilometers in this highly fractured, permeable, volcanic terrane.

We therefore cannot rule out the possibility that some of the low δO^{18} values shown in Fig. 1 are the result of crystallization of low-O^{18} magmas. However, this is clearly not the case for samples such as Col-14-8 and Col-17-8 which display marked isotopic disequilibrium within an assemblage of coexisting minerals. In these two rocks, quartz-feldspar fractionations of +5.5 to +8.2 per mil ("normal" values are +1.0 to +1.5 in igneous rocks) and a plagioclase-biotite fractionation of −0.7 per mil ("normal" values are +3 to +5) are clearly the result of preferential O^{18} depletion of the feldspar relative to coexisting quartz and biotite. Several other such examples of isotopic disequilibrium were cited by Forester and Taylor (1972) and this phenomenon has by now been well documented in several other localities (Taylor, 1968; Taylor and Forester, 1971, 1973). This implies that both plagioclase and alkali feldspar are much more susceptible to hydrothermal oxygen isotope exchange and/or recrystallization than any other common rock-forming mineral (with the probable exception of calcite).

The low-O^{18} rocks from the western San Juans all exhibit uniformly low δD values of −137 to −150, indicating that the meteoric ground waters in question were also relatively low in deuterium. If we assume that equilibration temperatures were on the order of 300°–600°C, these waters would have had δD values about 20 to 40 per mil higher than the associated chlorite or biotite with which they exchanged (Suzuoki and Epstein, 1973). This implies a δD of about −100 to −130 for these meteoric waters, and in subsequent discussion we shall assume that such waters had a uniform $\delta D \approx -115$. Applying the meteoric water equation ($\delta D = 8\delta O^{18} + 10$, Craig, 1961), this implies an initial $\delta O^{18}{}_{H_2O} \approx$

−16. Such waters would, of course, have undergone an O^{18} shift upward to δO^{18} values of about −5 to zero after having been involved in high-temperature exchange with most of the rock samples shown in Fig. 1. However, the extreme low-O^{18} rocks that occur along the eastern edge of the Silverton Caldera must have been in exchange equilibrium with H_2O having a δO^{18} value lower than −5. Depending on the temperature, the δO^{18} of this H_2O may have been as low as −10 to −15. These parameters *require* that the integrated water/rock ratios along the eastern ring fractures of the caldera were *at least* unity and were probably on the order of 5 to 10 (Fig. 10, Taylor, 1971). Note that these are minimum values because of the likelihood that an appreciable amount of water circulated upward along the fractures without completely equilibrating with the surrounding volcanic rocks.

The meteoric-hydrothermal solutions described above were not totally pervasive throughout the area shown in Fig. 1, as shown by the δO^{18} data from Col-13-8, a sample from the center of the large quartz monzonite stock just southwest of Silverton. The whole-rock δO^{18} value of +6.4 is only about 2 per mil lower than "normal" for a quartz monzonite (Taylor, 1968), and the quartz-alkali feldspar O^{18} fractionation of 2.2 per mil is just 1 per mil larger than "normal". Thus, only minor quantities of meteoric-hydrothermal solutions penetrated into the center of this stock (the integrated water/rock ratio may have been as low as 0.05 to 0.10). Nonetheless, such amounts of H_2O are more than enough to have overwhelmed the small amounts of primary magmatic water that may have been originally present in the biotite or hornblende of this quartz monzonite. This conclusion is of course also well documented by the very low δD value of the biotite in this rock (−145).

In addition to this example, which is a large stock emplaced along the boundary between the permeable volcanic terrane and a Paleozoic sedimentary rock section, certain other igneous rocks listed in Table 1 and Fig. 1 also show little or no O^{18} depletion. It is notable that *all of these* represent intrusions into the sedimentary-rock section, which apparently was much less permeable to ground water than was the volcanic section. Of such samples, only a single highly altered specimen collected at an intrusive contact near the boundary between the volcanic rocks and the sedimentary rocks shows any clear-cut O^{18} depletion whatsoever (Col-1-8, $\delta O^{18} = +1.8$, left side of Fig. 1). Other samples (Col-2-8, Col-11-8, and Col-8-8) lie well within the range of δO^{18} values of normal igneous rocks. The abnormally high δD value obtained on the stock west of Ophir (Col-2-8, −116) also indicates that only minor amounts of meteoric ground water probably exchanged with this intrusion during its crystallization and cooling. Although lower than normal, this δD value is the highest yet obtained from igneous rocks in the western San Juan Mountains.

In summary, the following picture emerges during the complex late-Oligocene to early-Miocene intrusive and volcanic history of the western San Juan Mountains. Repeated cauldron subsidence, volcanic eruption, and emplacement of epizonal intrusions provided the heat energy and fracture permeability necessary to establish widespread meteoric-hydrothermal convection systems throughout the volcanic terrane shown in Fig. 1. The hydrothermal fluids circulated to depths of at least several kilometers. Water/rock ratios were at a minimum on the order of 0.5 to 1.0 and were locally higher than 5.0. Circulation in the Paleozoic and Mesozoic sedimentary rocks (and probably the Precambrian basement rocks exposed to the south in the Needle Mountains) was, however, much more restricted, as a result of lower permeabilities in these rock types.

Very little isotopic data is available on the ore deposits of the western San Juans, but by analogy with data obtained on other epithermal mineral deposits in volcanic terranes (Taylor, 1973), it is practically certain that most of the ore deposits in the San Juan volcanic field were formed from hydrothermal fluids that contained a significant component of meteoric ground water. This statement is supported by the low-O^{18} quartz obtained from an ore body in the Gold King mine near the center of the Silverton Caldera (Fig. 1). The δO^{18} value of this quartz ($+2.2$) indicates formation from a hydrothermal fluid with $\delta O^{18} \approx -5$ to -10, if the temperature of deposition was in the range 200°–400°C (Clayton et al., 1972). Such a low-O^{18} fluid must be dominantly (or wholly) of ground water origin. Even more definite statements of this sort can be made concerning a vein dickite from the Ouray area ($\delta O^{18} = -6.2$, $\delta D = -141$, see Sheppard et al., 1969), and various low-O^{18} vein quartz samples near the Stony Mountain stock (Forester and Taylor, 1972).

Eastern San Juan Mountains (Alamosa River Stock)

Isotopic analyses were made on several samples from the composite granodiorite–quartz monzonite Alamosa River stock from the southeast part of the San Juan volcanic field (see Table 1 and Fig. 2). The K-Ar age of this stock is 29.1 m.y. (Lipman and others, 1970). The geology of this area, which lies within the Platoro Caldera, is described by Lipman and Steven (1970).

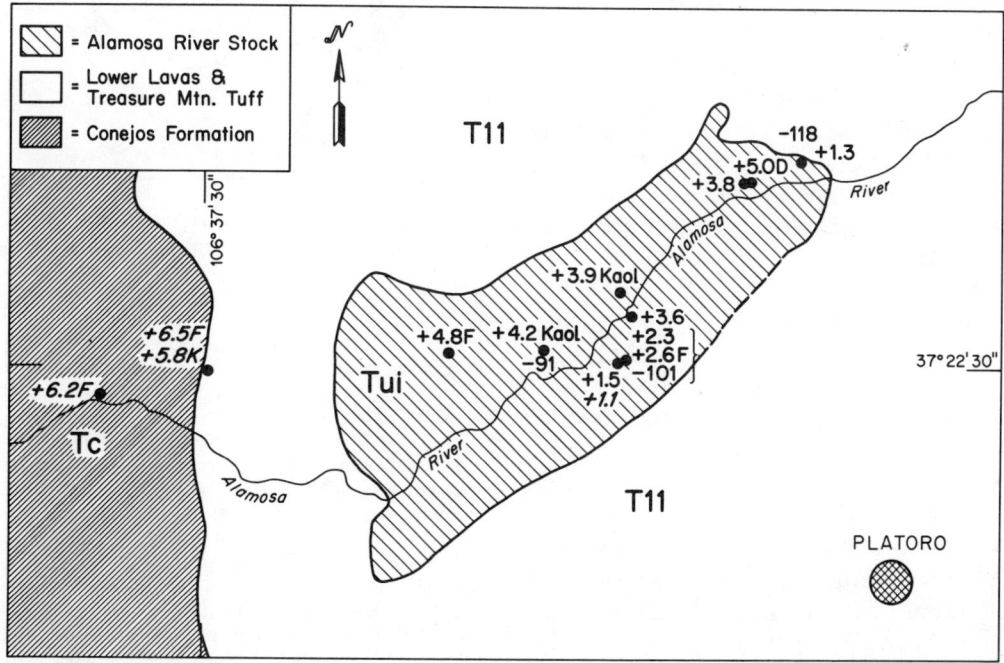

Fig. 2. Generalized geologic map of the Alamosa River stock and vicinity, San Juan Mtns., Colorado (after Lipman and Steven, 1970). The plotted numbers indicate some of the oxygen and hydrogen isotope data obtained on igneous rocks in this area (see Table 1 for a complete list of the isotopic data). The notation is identical to that given in Fig. 1, except that D = granitic dike, F = plagioclase, K = K feldspar, and Kaol = kaolinite.

The δO^{18} values of samples from the Alamosa River stock are abnormally low relative to "normal" igneous rocks (Taylor, 1968), although the O^{18} depletions are not quite so extreme as those described above from the western San Juans. The observed O^{18} depletions must have locally amounted to as much as 5 per mil (e.g., Col-36-8a, $\delta O^{18} = +1.5$). Therefore, the same general type of meteoric-hydrothermal convective system must have been set up after emplacement of this stock, but judging by the observed isotopic effects in the stock and in its country rocks, this system either involved meteoric water with a higher δO^{18} value or the amounts of water involved were less. Also, the convective circulation could have persisted to lower temperatures where the mineral-H_2O O^{18} fractionations are larger. One indication that the latter might have occurred is the very extensive kaolinitic alteration zone present on the north side of the Alamosa River stock. The δO^{18} values of two kaolinite samples from this part of the stock are 3.9 and 4.2. Assuming these formed in the temperature range 100°–200°C, they imply a δO^{18} of about —5 to zero in the low-temperature solfataric-type hydrothermal solutions that formed the kaolinite (Sheppard et al., 1969).

An interesting aspect of the hydrogen isotope data shown in Fig. 2 is that the samples from the eastern San Juans are markedly richer in deuterium than those from the Silverton area only 70 miles to the northwest (~ -110 vs ~ -145, respectively). It is clear from the δO^{18} data that meteoric-hydrothermal systems were important in both localities, and the time of formation of the calderas and associated epizonal intrusions was late Oligocene to early Miocene in both areas (Lipman and others, 1970). Therefore, the meteoric ground waters in the two areas apparently had markedly different δD values about 30 m.y. ago. This circumstance conceivably could be explained as follows: (1) The elevation of the volcanic plateau in the Silverton area may have been appreciably higher than in the Platoro area. This could have allowed greater snowfall in the Silverton area and thus a lower δD in the integrated annual local precipitation. This is actually the case today in the two

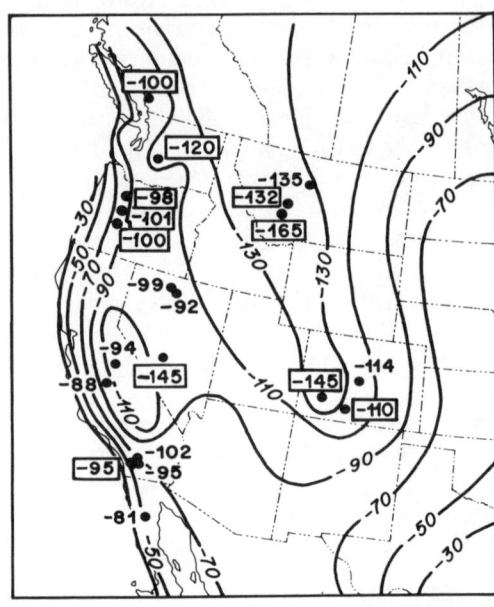

Fig. 3. Map of the western United States showing generalized contours of δD in present-day meteoric surface waters (rivers, lakes, etc.) after Friedman et al. (1964). The lettered numbers represent δD analyses (or averages for a given locality) of biotite, chlorite, or hornblende in a variety of Tertiary or late Mesozoic epizonal igneous intrusions (data from the present work; Taylor, 1973; Taylor and Epstein, 1968; Shieh and Taylor, 1969; Sheppard and Taylor, 1974). The numbers enclosed in rectangles are for samples that are depleted in O^{18} and are thus known to have interacted with abundant meteoric ground waters. For example, the $\delta D = -110$ and -145 in southwest Colorado represent the data from Table 1 on the Eastern and Western San Juan Mtns., respectively. Apparently, a number of the other intrusions with "normal" δO^{18} values also have exchanged with small quantities of meteoric ground waters. Note for comparison that most deep-seated igneous rocks from regional metamorphic or batholithic terranes have δD values of about -60 to -80 (Taylor and Epstein, 1966; Turi and Taylor, 1971; Godfrey, 1963).

areas. (2) The topographic barriers in the mid-Tertiary may have been such that the western San Juans obtained most of their rainfall and snowfall from Pacific storms, whereas the southeastern San Juans obtained most of their precipitation from air masses formed over the Gulf of Mexico. This pattern is also generally true today in the two areas, and it leads to higher deuterium values in the surface waters of northern New Mexico (see Friedman and others, 1964, and the map in Fig. 3). (3) The meteoric-hydrothermal systems may in fact have been terminated at different times in the two areas. The exact ages of such hydrothermal alteration zones are not easily determined, and the isotopic precipitation patterns could be expected to be different at different times in the Tertiary.

Again, as is the case in the Silverton district, several areas of ore mineralization are found in the vicinity of the Alamosa River stock, notably the Summitville district to the north and the Platoro district to the southeast. By analogy, these areas of hydrothermal alteration and ore deposition also very likely involve fluids that contain a dominant meteoric-water component. Note that, based on the D/H data in Fig. 2, the δD value of this meteoric water would have been about -80, implying an initial δO^{18} of about -11 rather than the -16 suggested in the Silverton area. Therefore, this might also in part explain why the δO^{18} values of the Alamosa River stock are higher than in the western San Juans.

Hydrogen Isotope Variations in Low-O^{18} Igneous Rocks in North America

The scale and the widespread nature of meteoric-hydrothermal convective circulation systems in volcanic-intrusive terranes are shown in Figs. 3 and 4. They are summary diagrams based upon data from a variety of sources, as well as upon the data given in Tables 1 and 2.

The map in Fig. 3 shows δD values of chlorite, biotite, and hornblende from a variety of igneous rocks throughout western North America. The δD values enclosed in rectangles represent samples (or averages for a locality) where the igneous rocks are abnormally low in O^{18} and are known to have exchanged at high temperatures with meteoric ground waters. The idea here is that if meteoric water effects are at all discernible in the O^{18}/O^{16} data, the D/H values of the hydroxyl-bearing minerals will totally reflect the D/H ratio of the local meteoric water, making some allowance for the variations introduced by differences in temperature and chemical composition of the minerals. In general, the lower the temperature or the higher the Fe/Mg ratio in the mineral, the larger is the water-mineral D/H fractionation (Suzuoki and Epstein, 1973), and the heavier would be the calculated δD of the H_2O.

Note in Fig. 3 that in spite of the above complications, there is a general correspondence between the δD of low-O^{18} igneous rocks and geographical position. Most of the samples shown on Fig. 3 are mid- to early-Tertiary. The samples from the Rocky Mountains–Great Basin area are consistently lower in δD than those from the areas near the Pacific Coast. As shown by the δD contours of present-day surface waters, this is also the general isotopic pattern produced in present-day rainfall and snowfall; this correlation is to be expected in view of the fact that the general shape of the North American continent in the Tertiary is known to have been roughly similar to the present-day configuration. The δD values calculated for waters that would have coexisted in equilibrium with the low-O^{18} igneous rocks from Fig. 3 at high temperatures are in general somewhat heavier than the values of present-day surface waters. This is, however, to

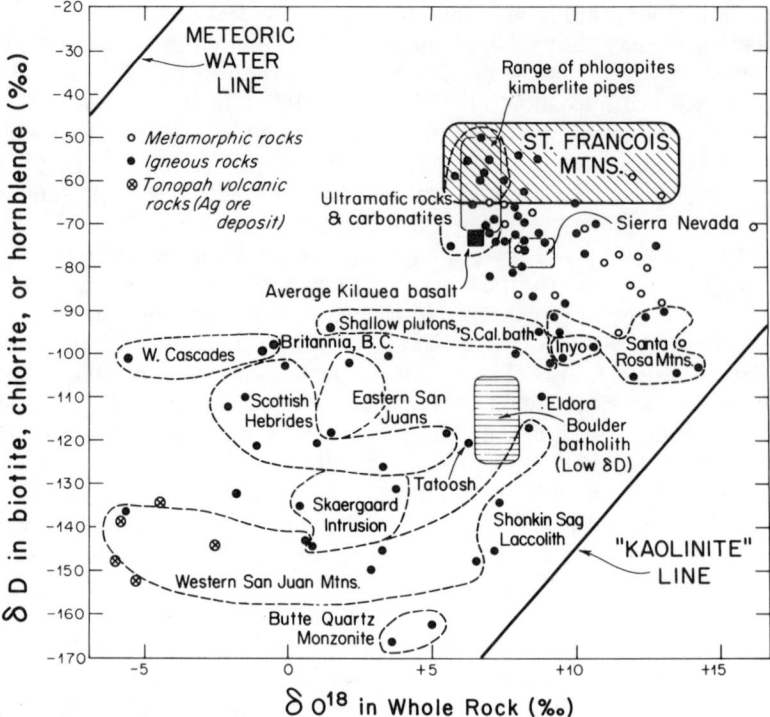

Fig. 4. Plot of δD in biotite, chlorite, or hornblende vs δO¹⁸ in whole rock for a variety of igneous and metamorphic rocks. The data are from the present work, and from Sheppard and Epstein (1970), Sheppard and Taylor (1974), Taylor (1973), Taylor and Epstein (1966, 1968), Shieh and Taylor (1969), Turi and Taylor (1971), Friedman (1967), and Godfrey (1963, Group II only). The samples labeled Butte quartz monzonite actually represent the early dark micaceous (biotite) alteration in the Butte ore deposit. Also shown is the general range of δO¹⁸ and δD from Precambrian igneous rocks of the St. Francois Mtns., Missouri (Wenner and Taylor, 1972). The meteoric water line (Craig, 1961) and sedimentary kaolinite line (Savin and Epstein, 1970) are shown for reference. Note that there is a rough correspondence between low δD values and low δO¹⁸ values but that at a given locality the δD is roughly constant while the δO¹⁸ may vary considerably.

be expected because the climate is known to have been generally warmer in the mid- to early-Tertiary than it is at present (Axelrod, 1964).

The δD values shown in Fig. 3 that are not enclosed in rectangles are from epizonal to mesozonal plutons which do not show any significant O¹⁸ depletion. In some instances even these plutons show a D/H correlation similar to that established for the low-O¹⁸ igneous rocks. This probably indicates that these epizonal igneous intrusions (all of which are emplaced into crystalline rocks or sedimentary rocks, not into permeable volcanic rocks) only exchanged with very tiny amounts of meteoric H_2O, enough to affect the D/H ratios of the hydroxyl-bearing minerals, because these contain only a small amount of hydrogens, but not enough to appreciably affect the much larger oxygen reservoir present in the whole-rock systems.

The Problem of Low-O¹⁸ Magmas

Reference was made above to the probable emplacement of the inner diorite in the Stony Mountain complex as a low-O¹⁸ magma with δO¹⁸ ≈ +2.5 (Forester and Taylor, 1972). Such low-O¹⁸ silicate liquids also were apparently produced at least in minor amounts during the very

TABLE 2. Oxygen isotope analyses of other igneous rock samples analyzed in the present work

	Location and Sample Description	Mineral	$\delta O^{18}(°/_{oo})$
IC.2	Obsidian, fresh, nonhydrated, postglacial, Hrafntinnuhryggur, near Myvatn, Iceland, donated by I. Carmichael (see Wright, 1915; Carmichael, 1962).	WR	+3.0
Brit-4	Quartz monzonite porphyry, 4000-ft level of Britannia Mine, British Columbia. $\delta D = -100$ (biotite).	WR	+3.8
		Q	+7.1
		F	+2.8
		KF	+2.8
		B	−1.9
O-23-9	Granodiorite, fine gr., center of Champion Creek stock, Bohemia Mining District, Oregon (see Fig. 7).	Q	+6.2
	All plagioclase grains show well-developed oscillatory zoning. Typical examples are shown in Fig. 5b and c.	F (phenocryst)	+2.6
		WR	+1.6
O-28-9	Hydrothermally altered andesite, very fine gr., porphyritic, 1.5 mi W of contact of Nimrod stock, McKenzie River, Western Cascade Range, Oregon (see Fig. 5 of Taylor, 1971). Many of the abundant andesine phenocrysts show moderately well-preserved oscillatory zoning, but recrystallization is much more apparent than in O-23-9. The quartz in this rock is present in aggregates and has clearly been recrystallized at some stage in the hydrothermal alteration process.	Q	+0.3
		F (phenocryst)	−4.1
		WR	−5.0

Abbreviations: Same as in Table 1.

latest stages of differentiation of the Skaergaard intrusion in east Greenland (Taylor, 1968; Taylor and Forester, 1973). However, the first definitive evidence for the existence of large volumes of low-O^{18} magmas was found in a number of recent volcanic rocks from Iceland (see Table 2 and Muehlenbachs et al., 1972; Muehlenbachs, 1973). Many such fresh, unaltered low-O^{18} lava flows in Iceland have now been identified by Muehlenbachs and his co-workers. They range in chemical composition from rhyolite to basalt. However, none of these magmas exhibits an O^{18} depletion anywhere near as large as that of many epizonal instrusions or altered volcanic rocks such as those from the Animas River Canyon shown in Fig. 1. The δO^{18} values of the low-O^{18} magmas are typically about +4 to +5 and none have been found to be lower than +2 per mil.

Thus, although low-O^{18} magmas certainly are generated in volcanic-intrusive terranes, it is clear that most of the O^{18} depletion observed in areas such as the San Juan volcanic field and the other localities noted in Fig. 4 is due to circulation of heated meteoric ground waters along joints and fractures in solidified igneous rocks. This effect is usually easily discerned by demonstrating a correlation between δO^{18} and grain size (e.g., Forester and Taylor, 1972), or by δO^{18} analyses of quartz-feldspar or feldspar-pyroxene mineral pairs. If the more easily exchanged feldspar is found to be abnormally depleted in O^{18} relative to the quartz or the pyroxene, it can be concluded that the abnormal depletion must have occurred after magmatic crystallization. However, if the δO^{18} values of the relatively resistant minerals such as quartz and augite are also lower than "normal," there will in general be some ambiguity about whether or not these minerals might have crystallized from a low-O^{18} magma. For example, the quartz from Col-14-8 and Col-34-8 (Table 1) might have formed from low-O^{18} silicate melts with $\delta O^{18} \approx +4$ to $+5$.

It is easy to envisage how convective circulation and migration of large amounts of meteoric ground water can occur in highly fractured rocks under essentially hydrostatic conditions. This represents the typical situation observed in modern hot spring systems (e.g., see White, 1968). However, it is much more difficult to envisage how significant amounts of water could migrate into a magma from an essentially hydrostatic fissure system lying outside an epizonal magma chamber. The magma must be under a lithostatic pressure that would be a factor of 2.5 to 3 higher than the hydrostatic pressure in the fissure system. The fractures that provide the major conduits for H_2O circulation outside the intrusion obviously cannot be present in the immediate contact zone at the edge of the magma body; otherwise they would be forcibly filled with magma. Therefore, if H_2O is to gain access directly to the magma, it must be by grain-boundary diffusion up a thermal gradient through the hot contact zone and into the magma (see Shaw, this volume).

Once the H_2O is inside the magma chamber, convective circulation of the silicate melt or successive injections of new magma can in principle aid the diffusion process in distributing this H_2O through the interior of the magma body. Nonetheless, this is likely to be a relatively slow process, even though most of these magmas are probably initially undersaturated with respect to H_2O. The well-studied Skaergaard intrusion (Wager and Brown, 1967) is a beautiful example of a magma body that definitely did not become depleted in O^{18} throughout 95 to 99% of its crystallization history, even though it is known that a large-scale meteoric-hydrothermal convection system was set up outside the pluton throughout its crystallization (Taylor and Forester, 1973). In principle this should have been a relatively favorable example, because of the very low H_2O content of the initial magma.

The difficulties outlined above are compounded by the fact that very large amounts of H_2O are required to produce any significant δO^{18} lowering of the magma. First of all, any H_2O finally able to diffuse into the magma would probably have already undergone a significant O^{18} shift to much higher δO^{18} values than those characteristic of the cool ground waters in the surrounding terrane. For example, in the San Juan volcanic area, it would be remarkable if such H_2O had a δO^{18} much lower than -5 at the time of influx into the silicate melt. To produce even a modest δO^{18} lowering of a magma from a "normal" value of about $+7$ to $+4$ with such H_2O would require a water/rock ratio of about 0.3. This is far more H_2O than can be dissolved in a magma at such shallow depths in the Earth's crust, thereby implying that most of this H_2O would have to diffuse or bubble through the magma chamber and then out again!

It is therefore doubtful that any model utilizing only direct influx of low-O^{18} meteoric H_2O into a magma chamber can by itself account for more than about a one per mil lowering of the δO^{18} of a large magma body. However, as pointed out by Taylor (1968), other things being equal, we would expect larger O^{18} effects to be observed in small magma bodies, such as the thin sheets of granophyric melt formed during the latest stages of crystallization of the Skaergaard magma. It is also possible that for a short time H_2O pressures in the vicinity of a pluton may be locally raised to values equal to or slightly greater than the lithostatic pressure when a catastrophic event like cauldron subsidence occurs, perhaps bringing water-rich rocks directly into contact with magma. However, this would be a transient effect and would in any case be more likely to cause violent steam explosions than to greatly enhance migration of H_2O directly into a magma.

Obviously, more data must be gathered on this problem, but it seems likely that some model or models involving indirect exchange between magmas and meteoric water will be necessary to account for most low-O^{18} magmas. The most logical ways in which this might occur are: (1) by direct melting, above a magma chamber, of water-rich country rocks that had already been hydrothermally altered and strongly depleted in O^{18} by the meteoric-hydrothermal circulation system set up above the intrusive body; (2) by large-scale assimilation and dissolution of such low-O^{18} rocks directly into the magma; (3) by sinking of low-O^{18} xenoliths through the magma chamber, because the H_2O in the abundant hydrous minerals would certainly be driven off into the magma, and because in hydrous magmas such xenoliths rapidly exchange O^{18} with their host magma (Shieh and Taylor, 1969) although they may show no evidence of dissolution by the silicate melt; and finally (4) by direct exchange between the liquid magma and O^{18} depleted country rock at the edge of the magma chamber or along a fissure through which the magma penetrates. The fourth mechanism is actually a variant of (3), and it is known to occur invariably in deep-seated plutonic or mesozonal environments whenever small bodies of magma are intruded into country rocks having a distinctly different O^{18}/O^{16} ratio (Taylor and Epstein, unpublished data; Shieh and Taylor, 1969; Turi and Taylor, 1971).

Direct melting of hydrothermally altered roof-rocks above a magma chamber would be favored by the large amounts of H_2O present in that environment. If these melted rocks remain separate from the underlying magma (because of lower density or low mixing rates), such melts conceivably could have δO^{18} values as low as any of the hydrothermally altered rocks from which they formed. However, if either of the assimilation mechanisms is involved, extreme O^{18} depletions in the magmas would not be expected.

Note that the more complicated the intrusive igneous history, the more likely it is that one or all of these four mechanisms will be actuated. In particular, multiple intrusion, ring-dike formation, repeated cauldron subsidence, and periodic explosive activity would all be expected to be accompanied by the developments outlined above. The lack of O^{18} depletion of the main mass of Skaergaard magma may be due to the fact that the Skaergaard intrusion represents almost an end-member example of a single, simple textbook intrusion of basaltic magma that subsequently underwent a relatively straightforward sequence of fractional crystallization.

Continued assimilation or melting of water-rich hydrothermally altered rocks at the top of a magma body emplaced into a permeable volcanic terrane would after a period of time undoubtedly lead to H_2O saturation in the upper portions of the magma chamber. Water pressure could then exceed lithostatic pressure and an explosion might ultimately result, causing eruption of low-O^{18} magma from the top of the chamber. Release of pressure on the underlying H_2O-undersaturated magma column would cause it to vesiculate as well, and the entire magma column might then erupt. Thus, meteoric water may be the underlying cause of many ash-flow tuff eruptions in volcanic fields such as the San Juan Mountains.

Evidence for Preservation of Magmatic Zoning in Certain Plagioclase Crystals That Have Undergone O^{18} Exchange

One of the interesting textural features of many of the low-O^{18} igneous rocks from the San Juan Mountains (and other areas) is the presence of delicate, oscillatory zoning in much of the plagioclase. This is clearly a magmatic phenomenon, and most explanations of this feature have emphasized the role of H_2O pressure in some way. Repeated build-up and sub-

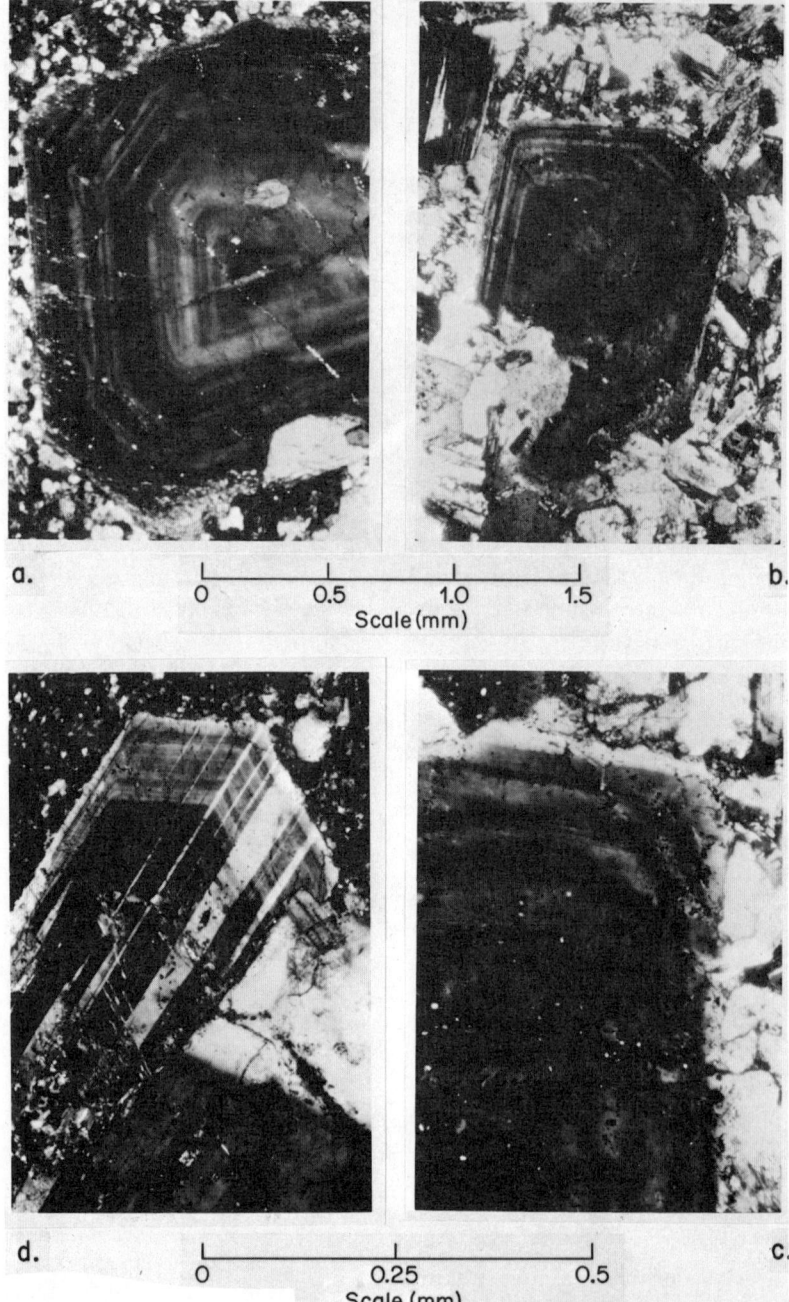

Fig. 5. (a) Photomicrograph (X-Nicols) of a typical plagioclase phenocryst ($\delta O^{18} = +6.1$) in Col-28-8 from the Eastern San Juan Mtns. (see Table 1). Although this rock has suffered some low-O^{18} alteration (δO^{18} of chloritized biotite $= -0.9$), the plagioclase has not undergone any appreciable O^{18} exchange, compatible with the well-preserved oscillatory zoning. (b) Photomicrograph (X-Nicols) of a typical plagioclase phenocryst ($\delta O^{18} = +2.6$) in sample O-23-9 from the Western Cascades (see Fig. 7 and Table 2). All plagioclase phenocrysts in this granodiorite sample from the Champion Creek stock show well-developed oscillatory zoning, even though the plagioclase has clearly undergone marked O^{18} depletion after its original crystallization from the granodiorite

sequent partial release of H_2O pressure in the magma, recurring on a fairly regular basis, seems to be a very adequate way to produce such oscillatory zoning. This is because of the sensitive response of the temperature of the liquidus in any silicate melting diagram to fluctuations in P_{H_2O}. In the light of the previous discussion involving incorporation of meteoric H_2O into magmas, it is not unreasonable to expect that this type of plagioclase would be very common in volcanic-intrusive terranes.

Many of these oscillatory-zoned plagioclases are strongly depleted in O^{18}. Photomicrographs of some examples are shown in Figs. 5 and 6. The question immediately arises as to whether this textural feature implies that the oxygen in such plagioclase grains was "frozen in" at the time of magmatic crystallization, and has not been subsequently disturbed by exchange with the meteoric-hydrothermal solutions that have permeated many of these rocks. This is an important concept because if it is true, it means that the presence of well-preserved oscillatory zoning in a low-O^{18} plagioclase would *demand* that the plagioclases have crystallized from a low-O^{18} magma.

Unfortunately, the answer to the above question is both yes and no. The ambiguities arise mainly because of difficulty in quantifying some measure of what is meant by "well-preserved oscillatory zoning." It is certainly true that if one takes a random collection of igneous rocks from an area like the San Juan volcanic field and arranges them into a sequence from "best preserved" to "worst preserved," subsequent O^{18} analyses will show that this sequence correlates roughly with the δO^{18} values (highest to lowest, respectively). An example of such a "well-preserved," high-O^{18} sample is given in Fig. 5a. However, the state of preservation of the magmatic zoning in the plagioclase also correlates with the degree of hydrothermal alteration of the rock in general, and it is certainly to be expected that the amounts of chlorite, epidote, calcite, actinolite, etc., in the rock should correlate roughly with O^{18}/O^{16} ratio, as in fact they do.

Nevertheless, sufficient O^{18} data now exist to state fairly positively that the oscillatory zoning in a plagioclase grain can survive almost complete oxygen isotope exchange with meteoric-hydrothermal fluids. Several examples of this effect are illustrated in Figs. 5 and 6. It must be emphasized that none of the O^{18} analyses of these plagioclases is on a single grain; they all represent analyses of 10–15 mg separates of a large number of grains, and it cannot be demonstrated that *all* grains show oscillatory zoning nor that the zoning is equally "well preserved" in all cases. However, one excellent example in which almost every plagioclase grain in the thin section *is* zoned in this fashion is sample O-23-9 (Table 2) from the Western Cascade Range, Oregon (Fig. 5b). The location of this sample is shown on the map in Fig. 7, modified after Fig. 4 of Taylor (1971).

Sample O-23-9 was chosen for O^{18} analysis in the present study because the plagioclase in thin-section shows such well-developed oscillatory zoning; it had not been analyzed previously by Taylor (1971). Note that this sample well illustrates the point made above about a correlation between δO^{18} value and "degree of preservation," because all other analyzed samples of the Champion Creek

magma. The whole-rock $\delta O^{18} = +1.6$, indicating that the fine-grained groundmass has undergone even more marked O^{18}-depletion than have the plagioclase phenocrysts. (c) Photomicrograph (X-Nicols) at higher magnification of another typical plagioclase phenocryst from O-23-9. (d) Photomicrograph (X-Nicols) of a reasonably typical plagioclase phenocryst in SM 147, an inclusion of San Juan Tuff in the Stony Mtn. ring-dike complex (Forester and Taylor, 1962). These plagioclase grains are strongly depleted in O^{18} ($\delta = +2.3$) relative to values in "normal" igneous rocks, but it is possible that this effect may in part be due to crystallization from a low-O^{18} magma (see text).

Fig. 6. (a) Photomicrograph (X-Nicols) of some of the better preserved oscillatory zoning shown by plagioclase phenocrysts in O-28-9, a propylitically altered andesite from the Western Cascade Range, Oregon (see Table 2). Many of the plagioclase phenocrysts in this rock show evidence of recrystallization in thin section, but remnant oscillatory zoning is common. The rock has undergone intense O^{18} depletion (δO^{18} whole rock $= -5.0$), and so have the plagioclase phenocrysts ($\delta O^{18} = -4.1$). (b) Photomicrograph (X-Nicols) of G-260, a gabbro from the Upper Border Group (α) of the Skaergaard intrusion. The labradorite crystals have preserved a normal magmatic zoning (calcic cores) even though their oxygen has been totally exchanged (δO^{18} plagioclase $= +2.0$;

stock shown in Fig. 7 are considerably more depleted in O^{18}. The O-23-9 plagioclase is drastically out of equilibrium with its coexisting quartz and after crystallization must have been depleted in O^{18} by at least 2 per mil. Some of the cores of the O-23-9 plagioclase grains are recrystallized, but not enough to explain the entire O^{18} effect. It thus seems clear that although the oscillatory zoning in the plagioclase of O-23-9 is somewhat "fuzzy" and not as sharp as, for example, that shown by the high-O^{18} sample in Fig. 5a, the basic zoning pattern has been essentially preserved during the O^{18} recrystallization process. Another grain shows this at higher magnification in Fig. 5c.

Analyses were also made on an altered andesite (O-28-9) from the most O^{18}-depleted locality in the Western Cascades (Table 2, see Fig. 5 of Taylor, 1971). Most of the abundant fine-grained, plagioclase phenocrysts in this rock are oscillatory-zoned, but the zoning is less well preserved than in O-23-9 (Fig. 6a). This plagioclase ($\delta O^{18} = -4.1$) has been depleted in O^{18} by about 10 per mil during hydrothermal alteration, based on comparisons with analogous volcanic rocks outside the area of propylitic alteration (Taylor, 1971).

Other samples which suggest the same conclusion are SM 147 (Fig. 5d), from an inclusion of San Juan tuff in the Stony Mountain ring-dike complex (see Forester and Taylor, 1972). The zoning in the plagioclase of this sample is exceptionally well preserved, even though the plagioclase has a relatively low $\delta O^{18} = +2.3$. Unfortunately, we do not know how much O^{18} depletion may have occurred in the San Juan tuff magma prior to eruption, so we do not know exactly how much δO^{18} lowering of this sample is due to later rock-water interaction.

Some of the best examples of preservation of zoning during O^{18} exchange are rocks of the Upper Border Group of the Skaergaard intrusion (Taylor and Forester, 1973). Some of these are coarse-grained (Fig. 6). Although they do not show the type of delicate oscillatory zoning discussed above, they do show pronounced magmatic zoning and here there is no ambiguity whatsoever; all the oxygen in these Skaergaard plagioclases has been totally exchanged with meteoric-hydrothermal solutions, as clearly shown by the strong isotopic reversal in the plagioclase-clinopyroxene O^{18} fractionations. The plagioclase has been depleted in O^{18} by at least 3 to 4 per mil, while the coexisting clinopyroxenes are virtually unaffected.

In summary, virtually all the plagioclase grains in the O^{18}-depleted igneous rocks shown in Tables 1 and 2 and Fig. 4 are in fact "pseudomorphs" after the original igneous plagioclase grains. In many cases, the magmatic textures in the rocks are almost perfectly preserved (a good example is the Skaergaard intrusion); and the internal zoning in the feldspars is also often found to be preserved. However, in areas of progressively more intense meteoric-hydrothermal alteration, the internal plagioclase zoning is progressively destroyed, just as are the other primary igneous textural features in the most strongly altered rocks. In general, the calcic cores of the oscillatory-zoned plagioclase grains are destroyed first, and the oscillatory zoning itself tends to become less and less distinct.

δO^{18} clinopyroxene $= +3.8$, Taylor and Forester, 1973). (c) Photomicrograph (plane light) of G-238, a granophyric ferrodiorite from the Upper Border Group (β) of the Skaergaard intrusion. The tiny prismatic crystals surrounding the plagioclase are quartz that has inverted from tridymite. (d) Photomicrograph (X-Nicols) of G-238. The andesine grains in this rock characteristically show a few well-developed oscillatory zones, even though the oxygen in the feldspar crystals has been totally exchanged (δO^{18} plagioclase $= +1.7$, δO^{18} clinopyroxene $= +4.6$).

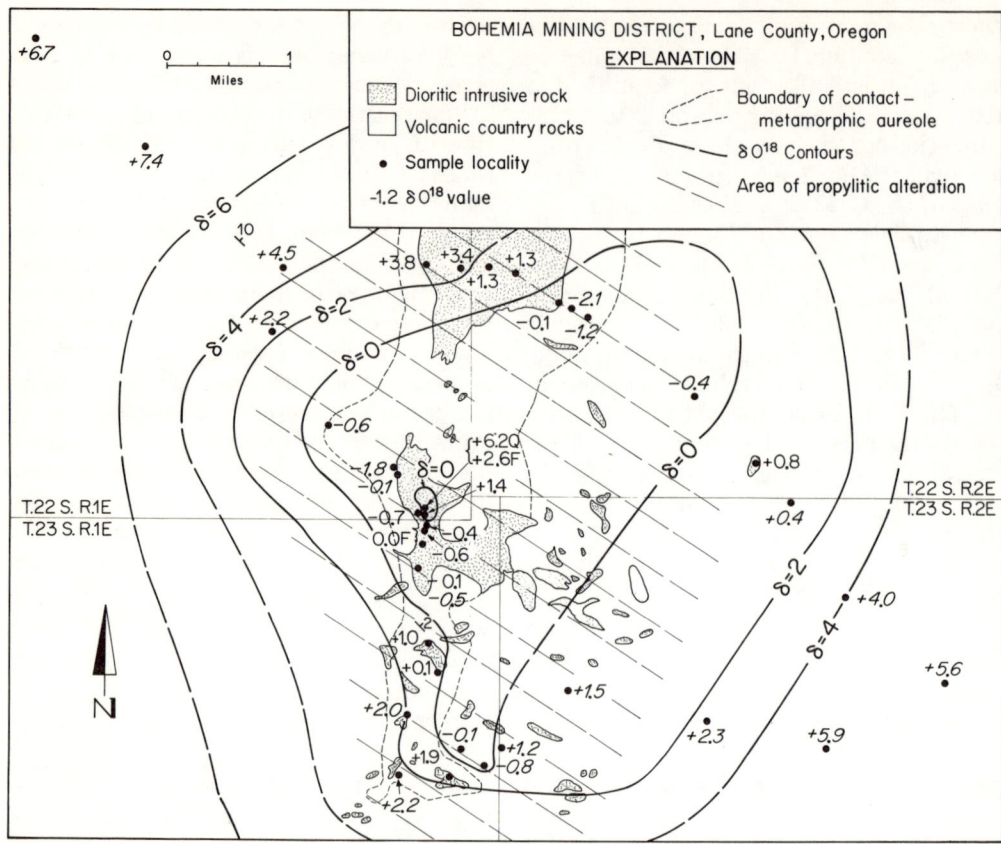

Fig. 7. Generalized geologic map of the Bohemia Mining District, Western Cascade Range, Oregon, showing δO^{18} values on whole-rock samples of epizonal intrusions (regular lettering) and volcanic country rocks (italic lettering), modified after Fig. 4 of Taylor (1971). Notation: Q = quartz, F = plagioclase. New δO^{18} data on granodiorite sample O-23-9, discussed in the text, are shown on the map (δO^{18} quartz = +6.2, δO^{18} plagioclase = +2.6). This sample is located near the center of the Champion Creek stock; it has the highest δO^{18} value of any granodiorite sample analyzed from the stock, and its plagioclase also shows the best-preserved oscillatory zoning (see Fig. 5b).

Solutions clearly penetrated the plagioclase crystals along cracks and imperfections, and it is reasonable to believe that they preferentially recrystallized the more calcic zones, thus exchanging oxygen with the feldspar but retaining as a relict texture the gross features of the original feldspar grains. If this occurs by a fine-scale solution and redeposition process in nature as it seems to do in some cases in the laboratory (O'Neil and Taylor, 1967), then the primary reason for partial preservation of the oscillatory zoning may well be the relative insolubility of aluminum in these solutions; the Al/Si ratio in the plagioclase grain will control the Na/Ca ratio regardless of how easily the Na and Ca exchange. It is also possible that under some hydrothermal conditions oxygen may diffuse through the feldspar lattice more easily than do the cations; this would also allow O^{18} exchange while preserving the oscillatory cation zonation. There is, in fact, some experimental evidence that this may be the case (see Yund and Anderson, this volume).

High-O^{18} Precambrian Igneous Rocks

A remarkable feature of the oxygen isotope investigations carried out on igneous rocks (Taylor, 1968; 1971; 1973; this work), is that practically without exception the occurrences of low-O^{18} igneous rocks are of Tertiary or late Mesozoic age (see Fig. 4). Only two such occurrences are known from Precambrian terranes, a 600 m.y. old granite from the Seychelles Islands and an isolated occurrence of granite in the 1450–1500 m.y. old volcanic-intrusive terrane in the St. Francois Mountains, Missouri. Neither of these two examples is strongly depleted in O^{18}, as the whole-rock δ values are no lower than $+3.0$ and $+5.0$, respectively (Taylor, 1968; Wenner and Taylor, 1972). In part, the lack of low-O^{18} rocks in older terranes is undoubtedly due to a lesser degree of sampling, as well as to the fact that later erosion and metamorphism may have removed many of the examples of such rocks. However, in spite of these qualifications, enough O^{18} data are on hand to indicate clearly that some peculiar process was involved in the petrologic history of some of these older igneous rocks.

Not only are low-O^{18} rocks largely absent from Precambrian terranes, many examples of these older igneous rocks throughout the world display exactly the opposite effects and are in fact abnormally high in O^{18}! In most instances these high whole-rock δO^{18} values are accompanied by strong disequilibrium in quartz-feldspar O^{18} fractionations, but instead of the feldspar being abnormally depleted in O^{18} as in the Tertiary examples, the feldspars consistently have higher δO^{18} values than the coexisting quartz. Localities where this phenomenon has been shown to exist are shown on a map of North America in Fig. 8, and the isotopic data are shown in Figs. 9 and 10. Such effects have now been observed in the granophyres and granitic rocks of the Muskox, Bushveld, and Duluth layered igneous complexes, and in various parts of the Keweenawan volcanic terrane, as well as throughout the sequence of ash-flow tuffs and epizonal granitic intrusions in the St. Francois Mountains, Missouri. Other examples occur in a pegmatitic vein in the Precambrian basement rocks brought up in the drill-core cuttings of a well at Sandhill, West Virginia, and in the hydrothermal alteration zones along fractures in the Precambrian Basement complex at Hopedale and Nain in Labrador.

All of the above examples have the following characteristics in common. The disturbed alkali feldspars are all turbid and brick-red in color due to disseminated hematite dust. The mafic minerals are partially or completely converted to chlorite. Several of the samples display granophyric textures. Except for the ubiquitous presence of hematite dust and their high-O^{18} character, these samples are in fact similar in character and to a certain extent in geologic setting to the low-O^{18} Tertiary igneous rocks described above. This is particularly true of the St. Francois Mountains terrane.

To what then should we attribute the reversed oxygen isotope behavior in these Precambrian rocks? There are basically only two possibilities: either (1) the hydrothermal solutions were much higher in O^{18} than those involved in the Tertiary occurrences; or (2) O^{18} exchange and alteration occurred at much lower temperatures ($\sim 150°C$) in the Precambrian examples. Either of these hypotheses presents difficulties, and sufficient data are not yet available to settle this problem. In either case, very large amounts of H_2O must have circulated through and exchanged with these Precambrian rocks.

Hypothesis (1) demands either that the meteoric ground waters in the Precambrian were much higher in O^{18} than Tertiary or present-day ground waters, or that these Precambrian granitic rocks contained far higher concentrations of primary magmatic water than did the younger igneous rocks. There is no in-

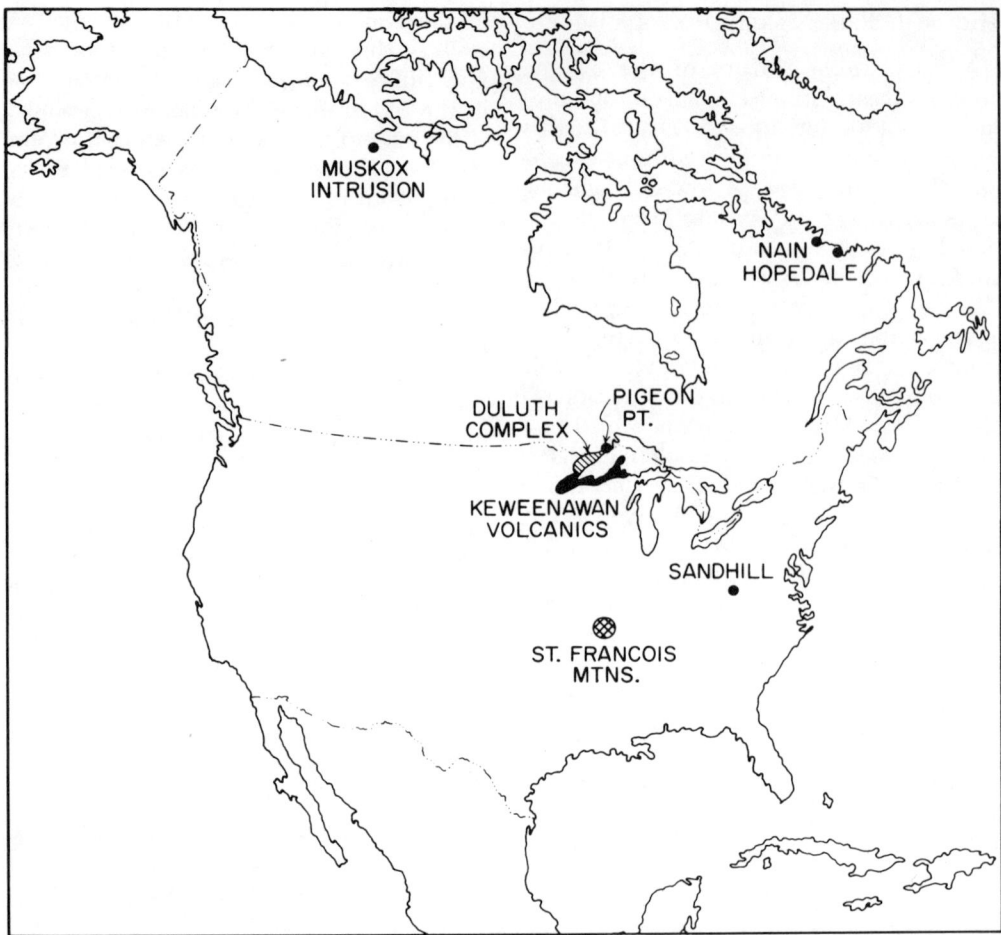

Fig. 8. Map of North America showing localities of Precambrian rocks discussed in the text, all of which show "reversed" quartz-feldspar O^{18} fractionations, abnormally high whole-rock δO^{18} values, and dissemination of hematite dust throughout the feldspar grains. To these North American localities we should also add the brick-red Bobbejaankop granite from the Bushveld Complex, South Africa.

dependent evidence that the latter was true, and if the former was true the only way it could logically come about would be if the Precambrian oceans were at least 5 to 10 per mil heavier in O^{18} than they are at present. The position of the meteoric water line on a δD-δO^{18} plot such as that shown in Fig. 4 is essentially fixed by the isotopic composition of the oceans existing at the time.

If the ocean was originally derived by degassing of the Earth's interior at high temperatures, its original δO^{18} value should have been about $+6$ rather than its present value of zero. This is because such H_2O would have been in exchange equilibrium with basalts and ultramafic rocks in the mantle, and these rocks have $\delta O^{18} = +5$ to $+6$. At about $1000°C$ the O^{18} fractionation between such rocks and H_2O is close to zero. Thus it is not unreasonable that at some time in the Earth's history the oceans may have been approximately 6 per mil richer in O^{18} than they are at present. However, as soon as sedimentation began, with consequent precipitation of low-temperature, high-O^{18} marine minerals (such as cherts, clay minerals, carbonates) the oceans should have become steadily lower in O^{18} with time. Material-balance calculations imply that the "locking-up" of so much O^{18} in these low-temperature sedimentary rocks must be balanced by loss of O^{18} in some other part of the system, and the only suitable candidate is the ocean water reservoir itself (see Silverman, 1951; Savin and Epstein, 1970).

Thus, a reasonable mechanism does exist by which the early Precambrian oceans may have been a few per mil high in O^{18}. However, many of the high-O^{18} igneous rock samples under discussion were formed in the interval 1100 to 1500 m.y. ago. At this relatively late stage in the Earth's history, the oceans had been in existence for at least 2000 m.y. and cycles of sedimentation had occurred many times. It therefore seems unlikely that these later Precambrian oceans would still have had a primordial δO^{18} value $\approx +6$. In fact, based upon data on the evolution of δO^{18} values of cherts with time, Perry (1967) has presented an entirely different model for the oceans, proposing that the oceans were *lower* in O^{18} during the Precambrian than they are at present.

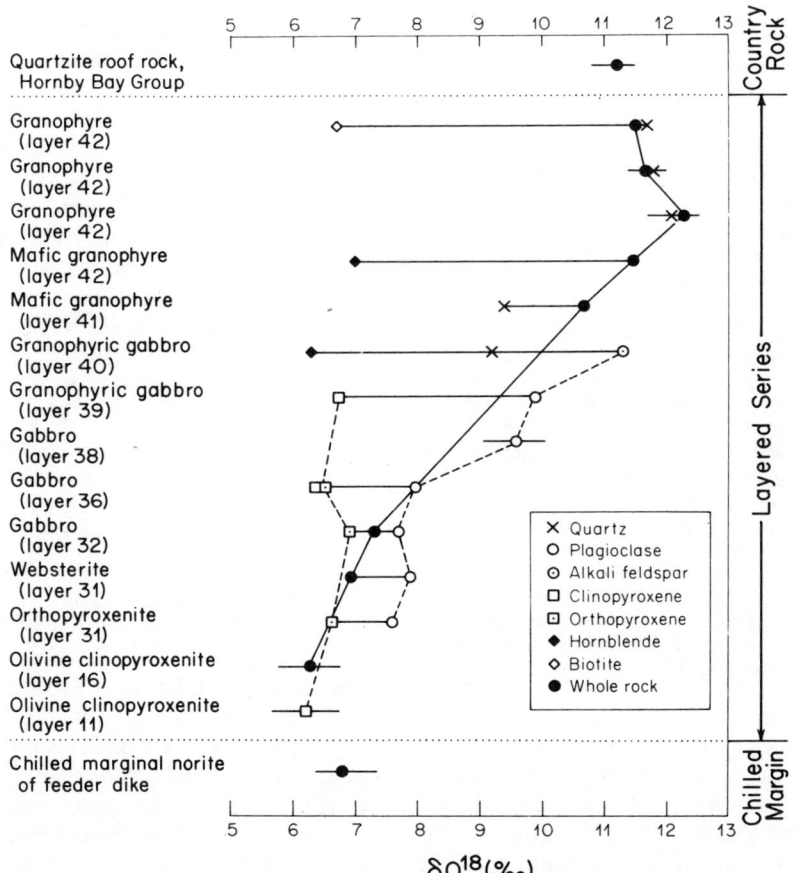

Fig. 9. Oxygen isotope data obtained on minerals and rocks from the Muskox intrusion, northern Canada, after Taylor (1968). Note the "reversed" quartz-feldspar fractionations in some of the late-stage, red-rock granophyres, and the strong O^{18}-enrichment in the feldspars at the top of the intrusion.

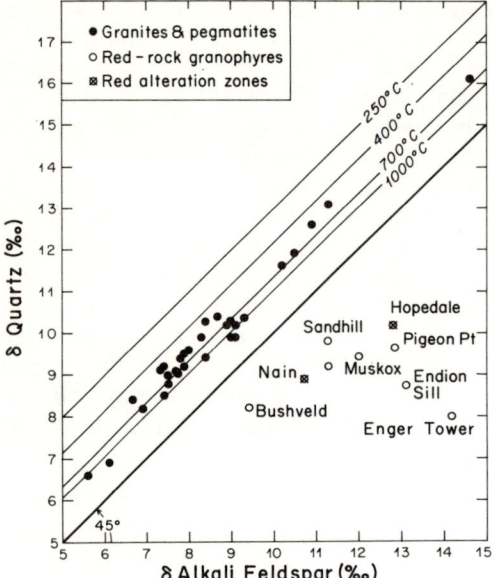

Fig. 10. Plot of δO^{18} in feldspar vs δO^{18} in quartz for various igneous rocks, including the Precambrian samples indicated in Fig. 8. The isotherms drawn parallel to the 45° line represent the equilibrium quartz-alkali feldspar O^{18} fractionations at various temperatures, based on the partial exchange quartz-H_2O experiments of Clayton et al. (1972) and the feldspar-H_2O experiments of O'Neil and Taylor (1967). Note that most of the "normal" igneous rock samples (black circles) exhibit fairly reasonable isotopic "temperatures," whereas all of the Precambrian red-rock samples represent drastic nonequilibrium. Data on the "normal" samples is from Taylor (1968) and Taylor and Epstein (1962 and unpublished data). Note that the nonequilibrium quartz-feldspar data-points for the low-O^{18} Tertiary examples discussed in the text would plot well to the left of this diagram. For the Nain and Hopedale data-points, which represent red hydrothermal alteration zones along fractures in gneiss, there is no change in δO^{18} quartz in going from the unaltered gneiss into the alteration zone (Taylor, 1967).

The history of oxygen isotopic evolution of the oceans with time remains a problem of prime geochemical importance, and further investigations of Precambrian igneous rocks that have exchanged with hydrothermal waters of surface origin may cast some light on this problem. However, an evaluation of the presently available evidence strongly suggests that the Precambrian oceans 1000 to 1500 m.y. ago were probably not significantly higher in O^{18} than they are at present. Well-preserved limestones and cherts formed at this time are not significantly different from analogous Paleozoic and Mesozoic samples (Knauth, 1973). Also, material-balance calculations by Savin and Epstein (1970) indicate that sedimentation has probably been responsible for no more than about a 1 per mil depletion of the oceans in the last billion years.

The extended discussion above leads us to the tentative conclusion that hypothesis (2) above is the most likely explanation of the high-O^{18}, brick-red feldspars in Precambrian igneous rocks. It is in fact likely that such rocks were in contact with circulating water at 100°–200°C for extended periods of time because they have all been deeply buried under younger sedimentary or volcanic rocks which would have contained connate waters or meteoric ground waters. For example, soon after the Muskox intrusion had completed crystallization, the entire area was buried under the very thick section of Coppermine River Basalt flows (Irvine and Smith, 1967). As another example, brines with δO^{18} values of -4 to $+4$, δD values of -40 to $+20$, and salinities of 1 to 6 gram-equivalents per liter are common throughout sedimentary basins in the mid-continent and Gulf Coast regions of the United States (Clayton et al., 1966). Brines with similar characteristics have been identified in fluid inclusions of minerals that make up the Mississippi Valley type Pb-Zn deposits in the Paleozoic sediments of the mid-continent region of the U.S. (Hall and Friedman, 1963). Ore deposits of this type are common throughout the world in such sedimentary sections, and various geothermometric techniques suggest they formed at about 100°–200°C. Also, the anomalous character of the Pb isotopic compositions in the galenas of these deposits strongly suggests that the hydrothermal solutions from which they were derived must have circulated through the underlying Precambrian basement (Heyl et al., 1966).

At 100° to 200°C, the equilibrium alkali feldspar–H_2O oxygen isotope fractionation varies from $+17.5$ to $+9.6$ (O'Neil and Taylor, 1967). Exchange between feldspars and one of the aforementioned brines at these temperatures would lead to feldspar δO^{18} values similar to those shown in Fig. 10. The relatively heavy δD values of such brines are also compatible with the high δD values obtained on Keweenawan and St. Francois igneous rocks, as shown in Fig. 4, and by Forester and Taylor (unpublished data). Oxygen isotope exchange would be much more likely with such a brine than with low-salinity meteoric ground waters because the brine would not be in cation equilibrium with the feldspar and would thus tend to attack the feldspar (Orville, 1963; O'Neil and Taylor, 1967). Exchange with such low-temperature hydrothermal fluids would be strongly oxidizing, thereby explaining the ubiquitous hematite dust disseminated through the feldspars (note that because of the shapes of oxygen buffer curves on a plot of P_{O_2} vs temperature, a simple lowering of temperature will move a system into steadily more oxidizing conditions and finally into the hematite field of stability (see Eugster, 1959).

It is conceivable that some of the iron necessary to form the ubiquitous hematite dust was originally present in solid solution in the alkali feldspars (e.g., see Ernst, 1960). However, it is even more likely that this Fe is released during the hydrous alteration of the coexisting mafic minerals in the rocks (D. R. Wones, personal communication). In any case, the constant association of high O^{18}/O^{16} ratios and hematite dust in these feldspars is clearly not a coincidence; this implies the hematite was definitely *not* formed by a simple exsolution process but must have been formed by the hydrothermal event that produced the high δO^{18} values in the feldspars (Taylor, 1967). Inasmuch as the hydrothermal O^{18} exchange process requires a complete breaking and re-forming of all Si-O and Al-O bonds throughout the feldspar crystal, it is perfectly reasonable that the Fe was derived from outside the original feldspar grain, as proposed by Wones.

Why, then, are these particular Precambrian alkali feldspars so susceptible to exchange by such low-temperature hydrothermal fluids? The answer probably lies in the fact that many of the feldspars that have been through a high-temperature hydrothermal exchange event (such as has affected all of the Tertiary examples discussed previously) are commonly turbid and full of imperfections and fluid inclusions. Such feldspars are probably much more susceptible to hydrothermal exchange during a later, lower-temperature event. One indication that this two-stage process might have occurred, at least in the St. Francois Mountains, is the apparent local preservation of some low-O^{18} rocks that may reflect an earlier episode of exchange with "normal" low-O^{18} meteoric waters (Wenner and Taylor, 1972). Certainly the general similarity between the geologic history of the Precambrian St. Francois Mountains volcanic terrane and the Tertiary San Juan volcanic field would imply that if low-O^{18} meteoric ground waters were present during the Precambrian, they would have interacted on a large scale with the St. Francois igneous rocks just as they have in the San Juans.

One other phenomenon that might have been produced by the postulated low-temperature hydrothermal event is a disturbance of Rb-Sr systematics in these rocks. Since these rocks were clearly "open" to a great deal of water that may have been very saline, any trace elements present in the highly altered feldspars and mafic minerals would be expected to have undergone redistribution. If this event occurred a significant time interval after primary crystallization of the red-rock granophyres and volcanic rocks, it could produce abnormally young Rb-Sr isochron ages. Just such a pattern is developed in the Keweenawan, Duluth, and St. Francois Mountains terranes because the Rb-Sr ages invariably show a greater spread and are slightly younger than the U-Pb concordia ages obtained from zircons (Faure et al., 1969; Chaudhuri and Faure, 1967; Silver and Green, 1963; Bickford and Mose, 1972). Bickford and Mose (1972) report that the Rb-Sr ages in the St. Francois Mountains tend to be "chaotic" and are up to 13 percent younger than the zircon ages; the zircon ages establish fairly clearly that the primary igneous events in this area occurred over a narrow time interval about 1475 m.y. ago. Note that recently formed zircons that have suffered little radiation damage or metamictization are not very susceptible to alteration during a hydrothermal alteration event (L. T. Silver, personal communication).

The envisaged hydrothermal event may have been responsible for such effects as the extensive serpentinization of the Muskox intrusion, and both there and in other areas such an event could have occurred intermittently over a very long time, perhaps extending into the Paleozoic Era or later. The postulated temperatures of 100°–200°C might either be due to another episode of magmatic activity in the various localities or simply to slight enhancement of the normal geothermal gradient (with perhaps just enough lateral temperature variation to produce some convective circulation of the H_2O). In any case, this second stage of the two-stage process might have little or nothing to do with the original magmatic heat in the intrusion. Therefore, although the "deuteric" explanation of these isotope data proposed previously (Taylor, 1967, 1968) may still have some local validity, it is very doubtful that it can serve as a general explanation (unless, of course, the oceans and hence the meteoric waters were in fact much higher in O^{18} during the mid- to late-Precambrian than they are at present).

Acknowledgments

The writer wishes to thank his colleagues, Samuel Epstein, Richard Forester, and Leon Silver, for fruitful discussions of this work. Arden Albee kindly assisted with the photomicrographs and Paul Yanagisawa aided in the laboratory work. This research was supported by the National Science Foundation, Grant 30997 X.

References Cited

Axelrod, D. J., The Miocene Trapper Creek flora of southern Idaho, *Calif. Univ. Publ. Geol. Sci.*, 51, 1–148, 1964.

Bickford, M. E., and D. G. Mose, Chronology of igneous events in the Precambrian of the St. Francois Mtns., S.E. Missouri: U-Pb ages of zircons and Rb-Sr ages of whole rocks and mineral separates (abstract), *Geol. Soc. Amer. Abstr. Programs*, 4, 451–452, 1972.

Carmichael, I., A note on the composition of some natural acid glasses, *Geol. Mag.*, 99, 253–264, 1962.

Chaudhuri, S., and G. Faure, Geochronology of the Keweenawan Rocks, White Pine, Michigan, *Econ. Geol.*, 62, 1011–1033, 1967.

Clayton, R. N., I. Friedman, D. L. Graf, T. K. Mayeda, W. F. Meents, and N. F. Shimp, The origin of saline formation waters: I. Isotopic composition, *J. Geophys. Res.*, 71, 3869–3882, 1966.

Clayton, R. N., J. R. O'Neil, and T. Mayeda, Oxygen isotope exchange between quartz and water, *J. Geophys. Res.*, 77, 3057–3067, 1972.

Craig, H., Isotopic variations in meteoric waters, *Science*, 133, 1702–1703, 1961.

Dings, M., Geology of the Stony Mountain stock, San Juan Mountains, Colorado, *Geol. Soc. Amer. Bull.*, 53, 695–720, 1941.

Ernst, W. G., Diabase-granophyre relations in the Endion sill, Duluth, Minnesota, *J. Petrology*, 1, 286–303, 1960.

Eugster, H. P., Reduction and oxidation in metamorphism, in *Researches in Geochemistry*, Vol. 1, P. H. Abelson, ed., Wiley, New York, pp. 397–426, 1959.

Faure, G., S. Chaudhuri, and M. O. Fenton, Ages of the Duluth gabbro complex and of the Endion sill, Duluth, Minnesota, *J. Geophys. Res.*, 74, 720–725, 1969.

Forester, R. W., and H. P. Taylor, Jr., Oxygen and hydrogen isotope data on the interaction of meteoric ground waters with a gabbro-diorite stock, San Juan Mtns., Colorado, *24th Int. Geol. Congr. Sec. 10*, 254–263, 1972.

Friedman, I., Water and deuterium in pumice from the 1959–60 eruption of Kilauea volcano, Hawaii, *U.S. Geol. Surv. Prof. Pap. 575-B,* B120–B127, 1967.

Friedman, I., A. C. Redfield, B. Schoen, and J. Harris, The variation of the deuterium content of the natural waters in the hydrologic cycle, *Rev. Geophys., 2,* 177–224, 1964.

Garlick, G. D., I. MacGregor, and D. E. Vogel, Oxygen isotope ratios in eclogites from kimberlites, *Science, 172,* 1025–1028, 1971.

Godfrey, J. D., The deuterium content of hydrous minerals from the East-Central Sierra Nevada and Yosemite National Park, *Geochim. Cosmochim. Acta, 26,* 1215–1245, 1963.

Hall, W., and I. Friedman, Composition of fluid inclusions, Cave-in-Rock fluorite district, Illinois, and Upper Mississippi Valley zinc-lead district, *Econ. Geol., 58,* 886–911, 1963.

Heyl, A. V., M. H. Delevaux, R. E. Zartman, and M. R. Brock, Isotopic study of galenas from the Upper Mississippi Valley, the Illinois-Kentucky, and some Appalachian Valley mineral districts, *Econ. Geol., 61,* 933–961, 1966.

Irvine, T. N., and C. H. Smith, The ultramafic rocks of the Muskox intrusion, Northwest Territories, Canada, in *Ultramafic and Related Rocks,* P. J. Wyllie, ed., Wiley, New York, pp. 38–49, 1967.

Knauth, L. P., Oxygen and hydrogen isotope rates in cherts and related rocks, unpublished Ph.D. thesis, California Institute of Technology, 369 pp., 1973.

Lipman, P. W., and T. A. Steven, Reconnaissance geology and economic significance of the Platoro Caldera, southeastern San Juan Mtns., Colorado, *U.S. Geol. Surv. Prof. Pap. 700-C,* C19–C29, 1970.

Lipman, P. W., T. A. Steven, and H. H. Mehnert, Volcanic history of the San Juan Mountains, Colorado, as indicated by potassium-argon dating, *Geol. Soc. Amer. Bull., 81,* 2329–2352, 1970.

Luedke, R. G., and W. S. Burbank, Volcanism and cauldron development in the Western San Juan Mountains, Colorado, in Cenozoic volcanism in the southern Rocky Mountains, R. C. Epis, ed., *Colo. Sch. Mines Quart., 63,* 175–208, 1968.

Muehlenbachs, K., The oxygen isotope geochemistry of acidic rocks from Iceland (abstract), *Trans. Amer. Geophys. Union, 54,* 499–500, 1973.

Muehlenbachs, K., A. T. Anderson, and G. E. Sigvaldason, The origins of O^{18}-poor volcanic rocks from Iceland (abstract), *Trans. Amer. Geophys. Union, 53,* 566, 1972.

O'Neil, J. R., and H. P. Taylor, Jr., The oxygen isotope and cation exchange chemistry of feldspars, *Amer. Mineral., 52,* 1414–1437, 1967.

Orville, P. M., Alkali ion exchange between vapor and feldspar phases, *Amer. J. Sci., 261,* 201–237, 1963.

Perry, E. C., The oxygen isotope chemistry of ancient cherts, *Earth Planet. Sci. Lett., 3,* 62–66, 1967.

Savin, S. M., and S. Epstein, The oxygen and hydrogen isotope geochemistry of ocean sediments and shales, *Geochim. Cosmochim. Acta, 34,* 43–64, 1970.

Shaw, H. R., Diffusion of H_2O in granitic liquids, this volume.

Sheppard, S. M. F., and S. Epstein, D/H and O^{18}/O^{16} ratios of minerals of possible mantle or lower crustal origin, *Earth Planet. Sci. Lett., 9,* 232–239, 1970.

Sheppard, S. M. F., R. L. Nielsen, and H. P. Taylor, Jr., Hydrogen and oxygen isotope ratios of clay minerals from porphyry copper deposits, *Econ. Geol., 64,* 755–777, 1969.

Sheppard, S. M. F., R. L. Nielsen, and H. P. Taylor, Jr., Hydrogen and oxygen isotope ratios in minerals from porphyry copper deposits, *Econ. Geol., 66,* 515–542, 1971.

Sheppard, S. M. F., and H. P. Taylor, Jr., Hydrogen and oxygen isotope evidence for the origin of water in the Butte ore deposits and the Boulder batholith, Montana, *Econ. Geol., 69,* in press, 1974.

Shieh, Y. N., and H. P. Taylor, Jr., Oxygen and hydrogen isotope studies of contact metamorphism in the Santa Rosa range, Nevada and other areas, *Contrib. Mineral. Petrol., 20,* 306–356, 1969.

Silverman, S. R., The isotope geology of oxygen, *Geochim. Cosmochim. Acta, 2,* 26–42, 1951.

Silver, L. T., and J. G. Green, Zircon ages for Middle Keweenawan rocks of the Lake Superior region (abstract), *Trans. Amer. Geophys. Union, 44,* 107, 1963.

Suzuoki, T., and S. Epstein, Hydrogen isotope fractionation between OH-bearing minerals and water, *Geochim. Cosmochim. Acta,* in press, 1974.

Taylor, H. P., Jr., Origin of red-rock granophyres, *Trans. Amer. Geophys. Union, 48,* 245–246, 1967.

Taylor, H. P., Jr., The oxygen isotope geochemistry of igneous rocks, *Contrib. Mineral. Petrol., 19,* 1–71, 1968.

Taylor, H. P., Jr., Oxygen isotope evidence for large-scale interaction between meteoric ground waters and Tertiary granodiorite intrusions, Western Cascade Range, Oregon, *J. Geophys. Res., 76,* 7855–7874, 1971.

Taylor, H. P., Jr., O^{18}/O^{16} evidence for meteoric-hydrothermal alteration and ore deposition in the Tonopah, Comstock Lode, and Goldfield mining districts, Nevada, *Econ. Geol., 68*, 1973.

Taylor, H. P., Jr., and S. Epstein, Relationship between O^{18}/O^{16} ratios in coexisting minerals of igneous and metamorphic rocks. Part 1: Principles and experimental results, *Geol. Soc. Amer. Bull., 73*, 461–480, 1962.

Taylor, H. P., Jr., and S. Epstein, O^{18}/O^{16} ratios in rocks and coexisting minerals of the Skaergaard intrusion, east Greenland, *J. Petrology, 4*, 51–74, 1963.

Taylor, H. P., Jr., and S. Epstein, Deuterium-hydrogen ratios in coexisting minerals of metamorphic and igneous rocks (abstract), *Trans. Amer. Geophys. Union, 47*, 213, 1966.

Taylor, H. P., Jr., and S. Epstein, Hydrogen isotope evidence for influx of meteoric ground water into shallow igneous intrusions, *Geol. Soc. Amer. Spec. Pap. 121*, 294, 1968.

Taylor, H. P., Jr., and R. W. Forester, Low-O^{18} igneous rocks from the intrusive complexes of Skye, Mull, and Ardnamurchan, Western Scotland, *J. Petrology, 12*, 465–497, 1971.

Taylor, H. P., Jr., and R. W. Forester, An oxygen and hydrogen isotope study of the Skaergaard intrusion and its country rocks (abstract), *Trans. Amer. Geophys. Union, 54*, 500, 1973.

Turi, B., and H. P. Taylor, Jr., An oxygen and hydrogen isotope study of a granodiorite pluton from the Southern California Batholith, *Geochim. Cosmochim. Acta, 35*, 383–406, 1971.

Wager, L. R., and G. M. Brown, *Layered Igneous Rocks*, W. H. Freeman, San Francisco, 588 pp., 1967.

Wenner, D. R., and H. P. Taylor, Jr., O^{18}/O^{16} and D/H studies of a Precambrian granite-rhyolite terrane in S.E. Missouri (abstract), *Trans. Amer. Geophys. Union, 53*, 534, 1972.

White, D. E., Hydrology, activity, and heat flow of the Steamboat Springs thermal system, Washoe County, Nevada, *U.S. Geol. Surv. Prof. Pap. 458-C*, 109 pp., 1968.

Wright, F. E., Obsidian from Hrafntinnuhryggur, Iceland: its lithophysae and surface markings, *Geol. Soc. Amer. Bull., 26*, 255–286, 1915.

Yund, R. A., and T. F. Anderson, Oxygen isotope exchange between potassium feldspar and KCl solutions, this volume.

MOBILITY OF OXYGEN ISOTOPES DURING METAMORPHISM

Yuch-Ning Shieh
Department of Geosciences
Purdue University
West Lafayette, Indiana 47907

ABSTRACT

The mobility of oxygen isotopes in rocks during metamorphism depends mainly on the availability of oxygen-bearing fluids and duration of heating. Studies on the O^{18}/O^{16} ratios in pelitic rocks from contact metamorphic aureoles indicate that oxygen isotopic exchange between intrusive and country rock occurs only to a distance of 1 or 2 feet from the main intrusive contacts. The exchange effects are much more extensive in xenoliths, in roof zones, and in re-entrants of country rocks projected into the intrusive. Inward migration of high-O^{18} metamorphic waters from the country rocks into the intrusive (to hundreds of feet) appears to have occurred in many granitic plutons. In regional metamorphism, the homogenization of O^{18}/O^{16} ratios among rocks of different lithologies over large distances (to several hundred feet or more) is observed in some areas. The δO^{18} values of regionally metamorphosed pelitic rocks tend to decrease with increasing metamorphic grade, implying that the rocks are open to oxygen isotope exchange with magmatic water ascending from depth during metamorphism. In the migmatite terrane of the Grenville province of Ontario (10,000 sq. mi.) the δO^{18} values of migmatites, granite gneisses, and paragneisses are all uniformly low ($\delta O^{18} = 5.0$–8.8 per mil). It is attributable to extensive synanatectic oxygen isotope exchange and homogenization with a mafic subcrustal reservoir through a granitizing fluid or silicate melt during migmatization.

INTRODUCTION

Oxygen, the most abundant element in the earth's crust, makes up more than 90% of its volume. As a consequence, the lithosphere may be regarded practically as a sphere of oxygen ions with interstices filled with cations. The problem of the extent and mechanism of the migration of oxygen in geologic processes, therefore, is one of the most interesting and important topics in geochemistry. Before the availability of isotopic tracer methods, study of the migration of oxygen in rocks and minerals was difficult because most replacement processes in rocks and minerals take place without appreciable change in volume (Lindgren's volume law of replacement). In other words, the number of oxygen ions has remained essentially constant in most replacement processes. Therefore, migration and exchange of oxygen ions between minerals and rocks normally cannot be detected by ordinary chemical means.

Fortunately, oxygen has three stable isotopes whose percent abundances in ocean water are (Garlick, 1969) $O^{16} = 99.763\%$, $O^{17} = 0.0375\%$, and $O^{18} = 0.1995\%$. The O^{18}/O^{16} ratios in natural materials exhibit a range of variation amounting to 10%, due to equilibrium and kinetic isotope fractionation effects (Urey, 1947; Bigeleisen, 1949). Therefore, we can utilize this natural variation of the abundance of oxygen isotopes as a tracer to study the migration of oxygen ions in many geochemical processes.

Figure 1 summarizes the general range of O^{18}/O^{16} ratios in some of the common rock types of the earth's crust. The oxygen isotope data presented in this paper are expressed in the usual δ notation:

$$\delta O^{18} = \left(\frac{O^{18}/O^{16} \text{ sample}}{O^{18}/O^{16} \text{ SMOW}} - 1 \right) \times 1000.$$

The δ values represent the per mil difference between the O^{18}/O^{16} ratio of the sample and that of Standard Mean Ocean Water (SMOW).

As can be seen, igneous rocks show the lowest and narrowest range of O^{18}/O^{16} ratios and sedimentary rocks show the highest. Metamorphic rocks have O^{18}/O^{16} ratios which overlap those of igneous rocks on the one end and sedimentary rocks on the other. A simple inspection of Fig. 1 suggests that sedimentary rocks exchange O^{18} for O^{16} in the process of metamorphism. The present paper attempts to describe and interpret some of the observations on the O^{18}/O^{16} distribution patterns in metamorphic rocks which bear on the problem of oxygen isotope exchange and transport during metamorphism.

Oxygen Isotope Exchange in Contact Metamorphic Aureoles

In view of the fact that the O^{18}/O^{16} ratios of igneous rocks are very different from those of sedimentary and low-grade regional metamorphic rocks (Fig. 1), there is always a *tendency* for the oxygen isotopes to undergo exchange when the two rock types are brought together because of the presence of an isotopic compositional gradient. Isotopic equilibrium is established in a system only if the isotopic compositions of a particular mineral are the same everywhere in that system, and provided that system is everywhere at a constant temperature. Therefore, studies of the variations of oxygen isotopes in the vicinity of an intrusive contact allow us to estimate the extent, and possibly the mechanism, of

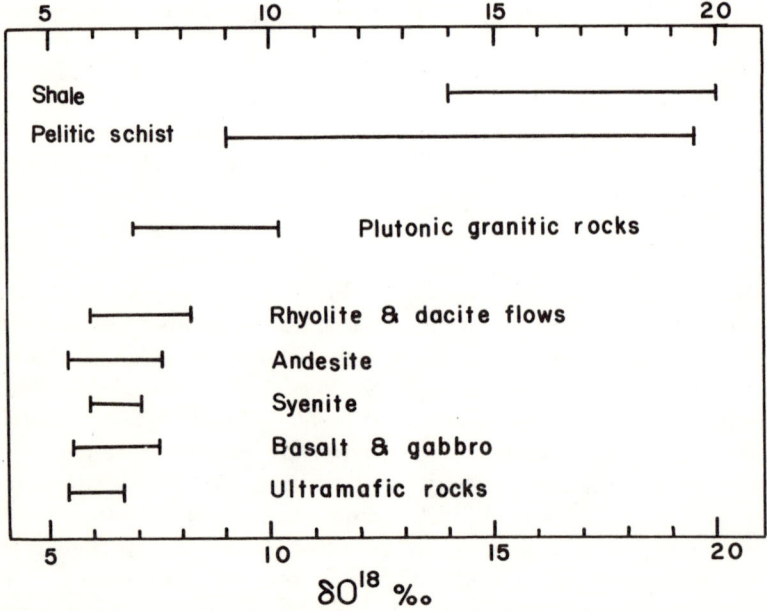

Fig. 1. Normal range of δO^{18} values in some of the common rock types of the earth's crust.

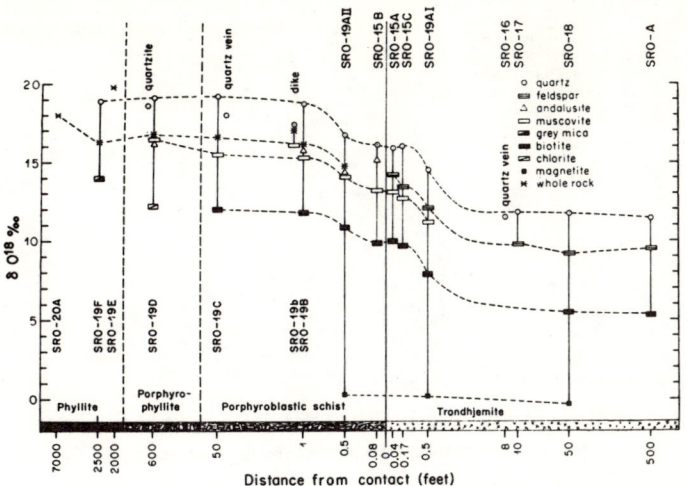

Fig. 2. Plot of δO^{18} values vs distance for samples from the contact metamorphic zone of Sawtooth stock, Santa Rosa Range, Nevada (Shieh and Taylor, 1969a).

isotope exchange and migration between the intrusive and its sedimentary country rock during contact metamorphism.

A typical δO^{18} profile for samples collected across an intrusive contact is shown in Fig. 2 (Shieh and Taylor, 1969a). Here we observe that a small-scale oxygen isotope exchange between the igneous intrusion and the adjacent pelitic country rock has taken place within a distance of about 1 foot of the intrusive contact. In the above example, the intrusive stock has a diameter of about 1.5 miles. Shieh and Taylor (1969a, b) and Turi and Taylor (1971a, b) have investigated the oxygen isotopic relationships on a variety of epizonal to mesozonal granitic contact aureoles and they found, in every case, that the exchanged zone is within several feet of the intrusive contacts. The width of the exchanged zone correlates well with the size of the intrusions, the presumed intrusive temperatures, the length of time of heating, and the availability of oxygen-bearing fluids. Figure 3 shows the percent oxygen isotope exchange between intrusive and country rock as a function of distance from the contact for several contact zones studied by Shieh and Taylor (1969a).

The narrowness and steepness of an isotopic gradient (about 3 per mil per ft) in the exchanged zones, on both the intrusive side and the country rock side of the contacts, suggest that such small-scale isotopic exchange occurred essentially in the solid state by a diffusion controlled recrystallization process. This presumably took place by grain-boundary diffusion in a static interstitial pore fluid. No massive horizontal *outward* movement of water from the intrusive into the contact metamorphic aureole seems to have occurred during the crystallization of the granitic magmas.

Although the oxygen isotope exchange between the intrusive and the country rock is very limited at the main contacts, the exchange effects in xenoliths, in roof zones and in re-entrants of country rock projected into the intrusive are much more extensive. In the Inyo batholith of California, a schist xenolith of 10 $\times$ 50 ft has undergone essentially 100% oxygen isotope exchange with the surrounding granites (Fig. 4). Upward movement of H_2O-rich volatile from the intrusive into the country rock is probably responsible for the extensive oxygen isotope exchange in these cases.

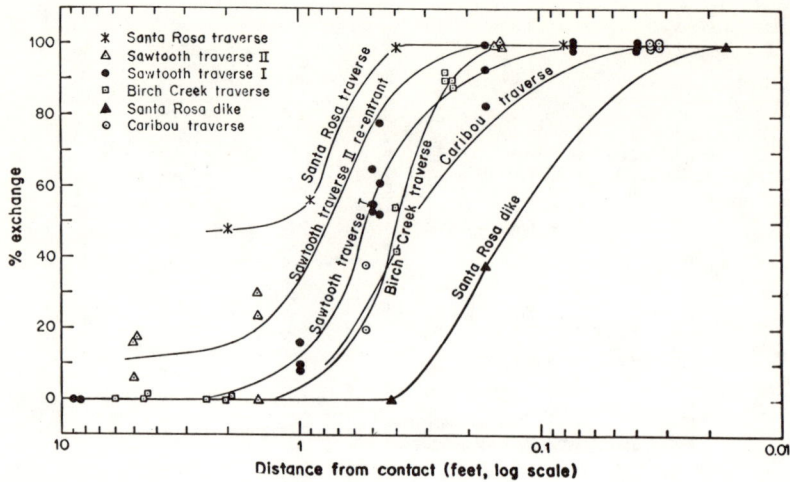

Fig. 3. Percent oxygen isotope exchange between intrusive and country rock as a function of distance from the intrusive contacts (Shieh and Taylor, 1969a).

It was first observed by Shieh and Taylor (1969a), and subsequently by Turi and Taylor (1971a), that in the marginal portions, up to several hundred feet from the intrusive contact, of most granitic plutons, the O^{18}/O^{16} ratios are abnormally high relative to "normal" igneous rocks from the central portions of the plutons (Table 1). These isotope effects are not accompanied by any significant chemical and mineralogical effects and thus cannot be explained by assimilation of the country rocks alone. They must in large part be brought about by exchange between the intrusive and the surrounding metasedimentary rocks by means of *inward* migration of high-O^{18} metamorphic waters through the marginal zones of the plutons. Thus, higher O^{18}/O^{16} ratios are usually observed in plutons that are relatively dry when emplaced into relatively wet sediments or metasedimentary rocks (e.g., Santa Rosa stock in Table 1). On the other hand, if the country rocks were regionally metamorphosed to very high grade (and were therefore almost completely dehydrated) prior to the intrusion of the stocks, no enrichment of O^{18} in the intrusive is usually observed (e.g., Eldora and Caribou stocks, also Johnny Lyon and Texas Canyon stocks; see Table 1).

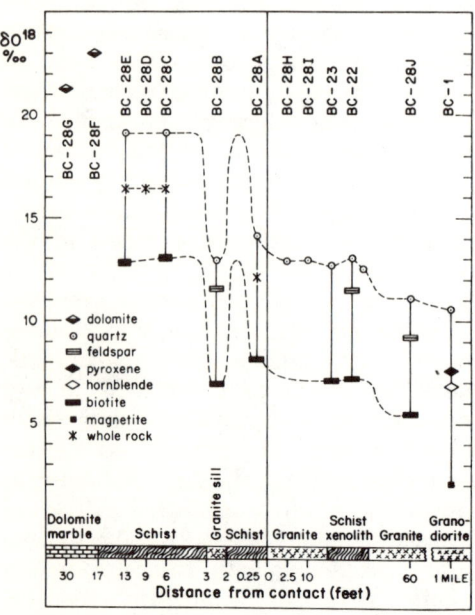

Fig. 4. Plot of δO^{18} values vs distance for samples from Birch Creek contact zone, Deep Springs Valley, Inyo County, California (Shieh and Taylor, 1969b). Note the nearly 100% oxygen isotope exchange in the schist xenolith.

In summary, under normal conditions of relatively dry intrusion into pelitic or carbonate country rock of the shallow

TABLE 1. Comparison of δO^{18} values of quartz from marginal zones (within 1000 ft. of contact) with those from central zones (greater than 1000 ft. of contact) of granitic plutons from various localities

	Marginal Zone	Central Zone
Santa Rosa Range, Nevada (Shieh and Taylor, 1969a)		
Santa Rosa stock	13.1 (1)*	9.5 ± 0.0 (2)
Small intrusive (300 × 600 ft)	14.0 (1)	9.5 (inferred)
Flynn stock	12.0 ± 1.4 (4)	9.5 (inferred)
Sawtooth stock	11.5 ± 0.1 (4)	9.5 (inferred)
Inyo batholith, California (Shieh and Taylor, 1969b)		
Birch Creek pluton	12.3 ± 0.6 (4)	10.6 (1)
Front Range, Colorado (Shieh and Taylor, 1969a)		
Eldora and Caribou stock	9.9 ± 0.0 (2)	9.9 (inferred)
Southern California batholith (Turi and Taylor, 1971a)		
Domenigoni Valley pluton	9.9 ± 0.5 (16)	8.7 ± 0.4 (6)
Small intrusive (200 × 600 ft)	13.2 ± 0.3 (3)	8.7 (inferred)
Cochise County, Arizona (Turi and Taylor, 1971b)		
Texas Canyon pluton	9.5 ± 0.1 (6)	9.4 ± 0.1 (2)
Johnny Lyon pluton	10.5 ± 0.1 (4)	10.3 (1)

*Number of samples.

crust, oxygen isotope exchange of the country rock is very limited, on a scale of several feet. This is in contrast with another isotopic interaction phenomenon observed by Taylor (1968, 1971), Taylor and Forester (1971), and Forester and Taylor (1972) who showed in the studies of the volcanic-plutonic complexes of the Skaergaard intrusion of Greenland, western Cascade Range of Oregon, Scottish Hebrides, and the San Juan Mountains of Colorado that shallow intrusions displaying evidence of interaction with hot, circulating water of meteoric origin are surrounded by aureoles of up to thousands of square miles in areal exposure, in which the country rocks as well as the intrusive have been exchanged with large amounts of O^{18}-poor meteoric waters. The water has evidently circulated through systems of open fractures in the volcanic country rock close to the earth's surface.

Oxygen Isotope Exchange in Regional Metamorphism

The oxygen isotopic relationships in regional metamorphic rocks are much more difficult to decipher than those in contact metamorphic rocks because in the former case the initial conditions (e.g., nature of protolith, configuration of heat source, etc.) are very difficult, if not impossible, to define. There is often ambiguity concerning the degree of oxygen isotope exchange in regional metamorphism because the original O^{18}/O^{16} ratios of rocks are seldom known with certainty.

Nevertheless, estimates on the dimensions of oxygen isotopic equilibrium attained during regional metamorphism have been attempted by many investigators. Anderson (1967) has described heterogeneous metamorphic rocks in which detectable isotopic disequilibrium occurs over distances of a few inches to a few feet during prograde metamorphism. Sheppard and Schwarcz (1970) also determined the size of the equilibrium exchange system for carbon and oxygen in marbles from southwestern Vermont to be less than a few inches across the inferred relict bedding. On the other hand, Taylor et al. (1963) and Garlick and Epstein (1967) have demonstrated in the pelitic schists of the chloritoid-kyanite zone from the Lincoln Mountain quadrangle in Vermont that isotopic equilibrium was established over several hundred feet because the O^{18}/O^{16} ratios of a particular mineral (e.g., quartz) are the same among the samples of different chemical composition collected several hundred feet apart. An example of oxygen isotope exchange and homogenization at a much larger scale was described by Taylor (1970) in the Adirondack Mountains, New York, covering an area of about 1200 square miles. The feldspars from metamorphosed anorthosites are on the average 3 to 4 per mil richer in O^{18} than feldspars from unmetamorphosed anorthosites and gabbros of world-wide occurrences. The feldspars from the associated metasyenites, metagabbros, and granite gneisses also displayed anomalously high O^{18}/O^{16} ratios. Taylor (1970) proposed that metamorphism in the presence of a pore fluid had permitted the anorthosites and some of the metaigneous rocks to exchange with the relatively O^{18}-rich paragneiss-carbonate metasedimentary rocks into which they had been earlier intruded. A similar relationship has been observed by Shieh et al. (1974) on rocks from the Grenville province of southeastern Ontario.

One striking feature of the O^{18}/O^{16} ratios of regionally metamorphosed pelitic rock is that the δO^{18} values tend to decrease with increasing metamorphic grade (Garlick and Epstein, 1967; Shieh and Taylor, 1969a; Dontsova, 1970). This is in contrast with the contact metamorphic rocks where the O^{18}/O^{16} ratios remain exceedingly constant throughout the entire contact aureole except for samples from the very narrow exchanged zone in the vicinity of the intrusive contact (Fig. 5). It implies that during regional metamorphism the pelitic rocks are open to oxygen isotope exchange with some external reservoir of isotopically light oxygen. Presumably this reservoir is ultimately derived from extensive deep-seated plutonic bodies of some type. At the highest grades of metamorphism, perhaps at the onset of migmatization, there is a tendency for metasediments to approach the igneous range of O^{18}/O^{16} ratios. One remarkable case has been observed from the migmatite terrane of the Grenville province of Ontario to be discussed below.

OXYGEN ISOTOPE EXCHANGE IN THE MIGMATITE TERRANE OF THE GRENVILLE PROVINCE OF ONTARIO

Figure 6 shows the part of the Grenville province that has been studied by Shieh and Schwarcz (1974). The northwestern region consists mainly of high grade (upper amphibolite to granulite facies) metamorphic rocks, hybrid granite gneisses, and migmatites. It is separated from the lower grade (middle amphibolite to greenschist facies) region in the southeast by a series of diapiric granitic plutons running approximately NE-SW (the Harvey-Cardiff arch). In the lower-grade region intrusive granites are abundant and they usually produce contact metamorphic aureoles in the metasedimentary country rocks.

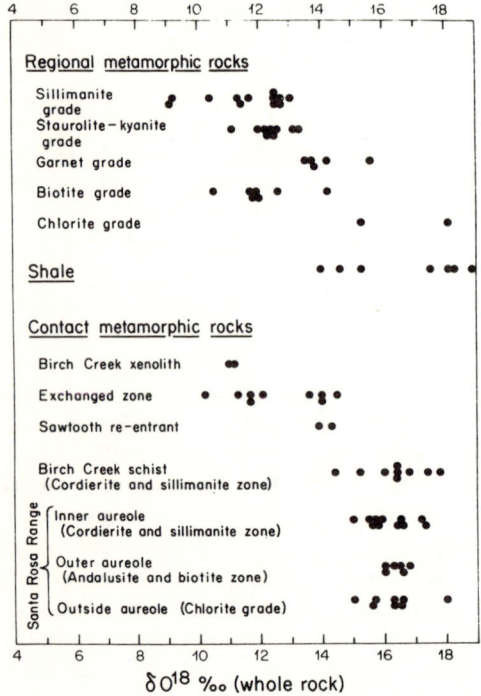

Fig. 5. δO^{18} values of pelitic rocks plotted against metamorphic grade for both regional and contact metamorphism (Shieh and Taylor, 1969a.)

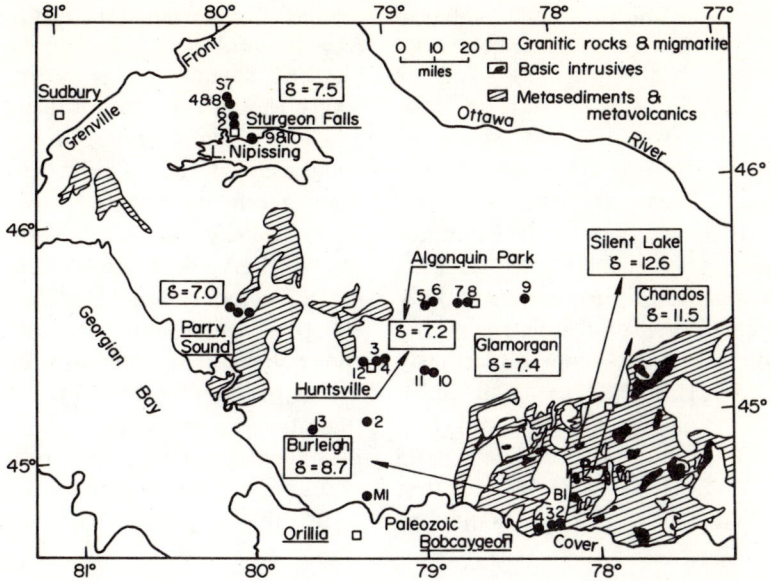

Fig. 6. Generalized geologic map of the Grenville province of Ontario, showing sample locations and average whole-rock δO^{18} values for samples from individual regions (Shieh and Schwarcz, 1973).

The isotopic analyses are summarized in Fig. 7. It can be observed that: (1) The granitic rocks from the northwestern part of the area studied (the migmatite terrane), including Glamorgan, Huntsville, Algonquin Park, Sturgeon Falls, and Parry Sound, form a group that is significantly lower in O^{18} than those of the Chandos Lake pluton and Silent Lake pluton to the southeast. This low-O^{18} group also shows a range that is clearly distinguishable from that shown by the "normal" plutonic granites of Taylor (1968); the range is, however, identical to that shown by the hypersolvus granites, rhyolites, and dacites. Note that the charnockitic granites and granulites from the Precambrian shield of central Australia (Wilson et al., 1970) and the Precambrian Grimstad granite in Norway (Friedrichsen, 1971) also fall within the low-O^{18} group. (2) In the Chandos Lake area, the paragneisses are distinctly higher in O^{18} than the granites intruded into them. In contrast, paragneisses associated with the low-O^{18} granites are isotopically indistinguishable from the granites, which in most cases form concordant laminae or subconcordant veins through the paragneisses.

To explain the low and uniform O^{18}/O^{16} ratios observed in the granite gneisses and the paragneisses from the migmatite terrane, Shieh and Schwarcz (1974) have considered the following three hypotheses: (A) The granitic rocks were formed by anatectic melting of fresh felsic volcanic rocks. (B) The granitic rocks were formed by magmatic differentiation of a basaltic magma and subsequently injected into the paragneiss. (C) The granitic rocks were formed by anatectic melting of ordinary metasediments having $\delta O^{18} = 10$–20 per mil but during or after the melting process the oxygen of both the granite (or granitic melt) and the metasedimentary precursor (the paragneiss) was isotopically exchanged with a mafic or ultramafic subcrustal reservoir through a H_2O-rich pore fluid or silicate melt.

Hypothesis A should be seriously considered if we can demonstrate that the protolith of "paragneiss" was fresh, unaltered felsic volcanic rock such as dacite or rhyolite. We have seen that such rocks possess δ values in the range observed for the low-O^{18} granites. Partial or total melting of such rocks would of course yield magmas of about the same isotopic composition. However, Shieh *et al.* (1974) have shown that in the Chandos Lake–Apsley area, most of the gneisses of possible igneous origin were very likely hydrated or altered volcanic rocks because many of them show abnormally high O^{18}/O^{16} ratios (δ = 12.0–13.5). Taylor (1968, p. 36), Garlick and Dymond (1970), and Muehlenbachs and Clayton (1972) have also shown that submarine volcanic rocks are easily hydrated and enriched in O^{18} even before any detectable change in major or trace element chemistry. It is therefore unlikely that all of the rocks in the northwestern region, underlying an area of thousands of square miles, were derived from fresh unhydrated silicic volcanic rocks before metamorphism. Furthermore, there is abundant field, petrographic, and chemical evidence, such as the presence of interlaminated marbles, the well-bedded character of the paragneiss, frequent subtle textural and compositional changes from layer to layer, and local occurrence of aluminous silicates in gneiss, to suggest that at least some of the paragneisses were derived

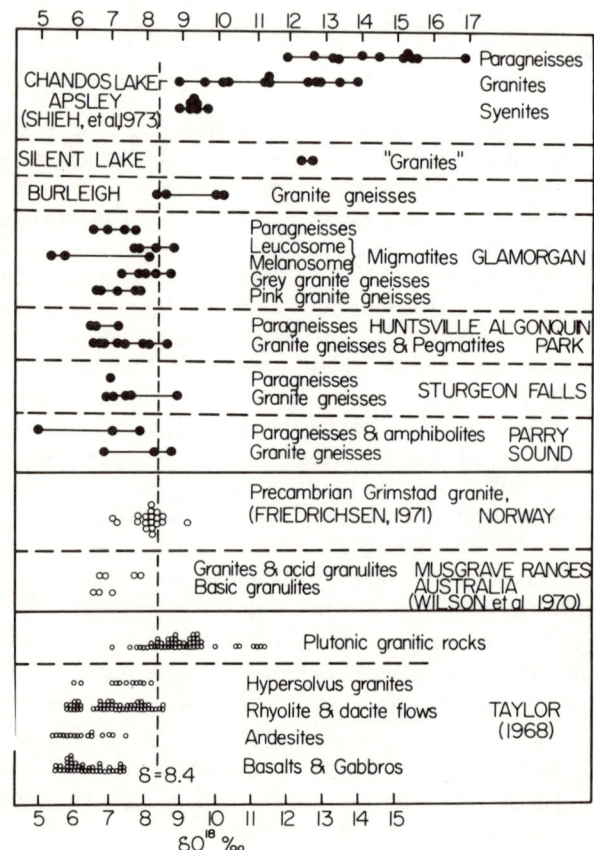

Fig. 7. Comparison of whole-rock δO^{18} values of granites, migmatites, and paragneisses from the Grenville province and from other shield areas with those of "normal" plutonic granites and other common igneous rocks of worldwide occurrences (Shieh and Schwarcz, 1973).

from marine sediments such as graywacke, sandstone, and siltstone (Lumbers, 1971; Shaw, 1962, 1972). All of these rock types have δO^{18} values much higher than fresh volcanic rocks. Thus, even if large parts of the paragneiss were derived from silicic volcanic rocks, it is probable that much of this material was deposited in the sea, an environment in which its oxygen isotopic composition would have been rapidly changed to heavier values than that of fresh volcanic rocks.

Hypothesis B is able to explain the low O^{18}/O^{16} ratios of the granites because during magmatic differentiation very little oxygen isotope fractionation is expected to occur (Garlick, 1966). Therefore, a granite differentiated from a basaltic magma would be expected to have an O^{18}/O^{16} ratio very similar to basalt. However, this hypothesis cannot explain the equally low O^{18}/O^{16} ratios observed in the paragneisses if the latter originally had δ values of 12 to 17 per mil as in the Apsley gneiss of Chandos township (Shieh et al., 1974). It is possible that after the basalt-derived granite was injected, it exchanged oxygen isotopes with the enclosing metasediments, causing them to attain low O^{18} values, but unless the volume of granite available for exchange was much larger than that of metasedimentary material, such exchange would result in an assemblage (granite + paragneiss) having O^{18}/O^{16} intermediate between those of the two end-members. This, however, is not observed. Furthermore, field evidence suggests that the volumes of leucosome and melanosome are roughly equal. If much larger volumes of granite, hidden unexposed at depth in the crust are involved as an oxygen reservoir, we then require oxygen exchange on a large scale in an open system, equivalent to hypothesis C.

By way of elimination, hypothesis C, namely open-system large-scale oxygen isotope exchange with a mafic reservoir in the lower crust or upper mantle during the anatectic melting of metasediments, remains as the most likely origin of the low-O^{18} granites, although the mechanism of exchange is highly speculative at this stage.

The exchange hypothesis explains not only the oxygen isotopic composition of the low-O^{18} granites but also that of the paragneiss-migmatite complex as well, since the exchange process would induce, at a constant temperature, the same isotopic composition on a given phase (e.g., quartz) regardless of what rock type it occurred in. If we assume a small thermal gradient between the subcrustal reservoir and the site of anatexis, then the isotopic composition that would be produced is that which would be in approximate isotopic equilibrium with a subcrustal reservoir at temperatures recorded by oxygen isotope fractionations among the coexisting minerals in the anatectic rocks themselves. Shieh and Schwarcz (1974) have observed $\delta = 7.5-9.0$ per mil for quartz from the granite gneiss, while the oxygen isotope temperatures indicated that isotopic exchange ceased at about 500°–550°C. The quartz δ values are exactly those one would expect if they were in isotopic equilibrium at 500°–550°C with plagioclase (An_{60}) having $\delta = 5.7-7.2$, corresponding to analyses of feldspars commonly observed in fresh unaltered basalts and gabbros believed to have originated from the lower crust or upper mantle (Taylor, 1968).

Since all the low-O^{18} granites and paragneisses that have been studied are associated with the large-scale development of migmatite, we infer that extensive oxygen isotope exchange is associated with the process of migmatization. Perhaps the presence of a silicate melt or granitizing fluid would greatly enhance the isotopic exchange process. Supporting such a contention is the observation that although the granite gneiss, the migmatite, and the paragneiss are uniformly low in O^{18}/O^{16} ratios, the associated non-migmatitic units, including the marbles ($\delta = 17.8-20.8$) and the metagabbros and metanorthosites ($\delta = 7.5-9.7$) are all relatively enriched in O^{18}, indicating that they are rather resistant to isotopic exchange. In addition, a paragneiss xenolith (5 × 15 ft) enclosed in metagabbro, both nonmigmatitic, has $\delta = 10.7$ per mil; it is about 3–4 per mil higher than migmatized paragneiss occurring in the same area. The paragneiss xenolith was able to retain its relatively high O^{18} mainly because it was protected from exchange by being embedded in the relatively infusable metagabbro.

SUMMARY AND CONCLUSIONS

The mobility of oxygen isotopes in rocks and minerals during metamorphism depends critically on the following factors: (1) temperature, (2) duration of heating, (3) availability of an exchange medium (oxygen-bearing fluids), and (4) permeability and degree of structural deformation in rocks. Thus, in the essentially static geometrical relationships between intrusive and country rock during contact metamorphism, the oxygen isotope exchange effects at the main intrusive contacts are confined to a zone of 1 or 2 feet, but the exchange effects in xenoliths, in roof zones, and in re-entrants of country rock projected into the intrusive are much more extensive (to tens or hundreds of feet). The intrusive bodies themselves also appear to have been isotopically contaminated, in favorable cases, by the enclosing country rocks for hundreds of feet into the intrusives through the influx of high-O^{18} metamorphic waters.

In regional metamorphism, the scale of oxygen isotope exchange varies from area to area and even from rock to rock, depending on the factors mentioned above, but there is definitely a trend of decreasing O^{18}/O^{16} with increasing metamorphic grade. One possible explanation is that there is a continuous supply of extraneous oxygen to the metamorphic system. This oxygen, in the form of aqueous fluids ultimately derived from depth, undergoes isotopic exchange with the system. The oxygen isotope exchange culminates at the stage of migmatization of metasedimentary rocks, as witnessed by the large-scale isotopic homogenization and lowering of δO^{18} in rocks from the migmatite terrane of the Grenville province of Ontario. The observation that large-scale isotopic homogenization and lowering of O^{18}/O^{16} ratios is associated with migmatization while nonmigmatitic units do not display such isotopic alterations suggests that the fluid mobilisate present during anatexis acts as a medium for oxygen transport, to permit oxygen isotopic exchange over large distances. While such mobilisate is absent, no extensive isotopic exchange occurs.

REFERENCES CITED

Anderson, A. T., The dimensions of oxygen isotopic equilibrium attainment during prograde metamorphism, *J. Geol.*, *75*, 323–332, 1967.

Bigeleisen, J., The relative reaction velocities of isotopic molecules, *J. Chem. Phys.*, *17*, 675, 1949.

Dontsova, Ye. I., Oxygen isotope exchange in rock-forming processes, *Geochem. Int.*, *7*, 624–636, 1970.

Forester, R. W., and H. P. Taylor, Jr., Oxygen and hydrogen isotope data on the interaction of meteoric ground waters with a gabbro-diorite stock, San Juan Mountains, Colorado, *Int. Geol. Congr.*, *24th Session*, Section 10, 254–263, Montreal, 1972.

Friedrichsen, H., Oxygen isotope fractionation between coexisting minerals of the Grimstad granite, *Neues Jahrb. Mineral., Monatsh.*, 26–33, 1971.

Garlick, G. D., Oxygen isotope fractionation in igneous rocks, *Earth Planet. Sci. Lett.*, *1*, 361–368, 1966.

Garlick, G. D., The stable isotopes of oxygen, in *Handbook of Geochemistry*, Vol. 2, K. H. Wedepohl, ed., Section 8B, 1969.

Garlick, G. D., and J. R. Dymond, Oxygen isotope exchange between volcanic materials and ocean water, *Geol. Soc. Amer. Bull.*, *81*, 2137–2142, 1970.

Garlick, G. D., and S. Epstein, Oxygen isotope ratios in coexisting minerals of regionally metamorphosed rocks, *Geochim. Cosmochim. Acta*, *31*, 181–214, 1967.

Lumbers, S. B., Geology of the North Bay area, Districts of Nipissing and Parry Sound, *Ont.*

Dep. Mines N. Affairs Geol. Rep. 94, 104 pp., 1971.

Muehlenbachs, K., and R. N. Clayton, Oxygen isotope studies of fresh and weathered submarine basalts, *Can. J. Earth Sci., 9,* 172–184, 1972.

Shaw, D. M., Geology of Chandos Township, Peterborough County, *Ont. Dep. Mines Geol. Rep. 11,* 1–28, 1962.

Shaw, D. M., The origin of the Apsley gneiss, Ontario, *Can. J. Earth Sci., 5,* 561–583, 1972.

Sheppard, S. M. F., and H. P. Schwarcz, Fractionation of carbon and oxygen isotopes and magnesium between coexisting metamorphic calcite and dolomite, *Contrib. Mineral. Petrol., 26,* 161–198, 1970.

Shieh, Y. N., and H. P. Taylor, Jr., Oxygen and hydrogen isotope studies of contact metamorphism in the Santa Rosa Range, Nevada and other areas, *Contrib. Mineral. Petrol., 20,* 306–356, 1969a.

Shieh, Y. N., and H. P. Taylor, Jr., Oxygen and carbon isotope studies of contact metamorphism of carbonate rocks, *J. Petrology, 10,* 307–331, 1969b.

Shieh, Y. N., and H. P. Schwarcz, Oxygen isotope studies of granite and migmatite, Grenville province of Ontario, Canada, *Geochim. Cosmochim. Acta, 38,* 21–45, 1974.

Shieh, Y. N., H. P. Schwarcz, and D. M. Shaw, An oxygen isotope study of the Loon Lake pluton and the Apsley gneiss, Grenville province of Ontario (in preparation), 1974.

Taylor, H. P., Jr., The oxygen isotope geochemistry of igneous rocks, *Contrib. Mineral. Petrol., 19,* 1–71, 1968.

Taylor, H. P., Jr., Oxygen isotope studies of anorthosites with special reference to the origin of bodies in the Adirondack Mountains, New York, in *Origin of Anorthosites, N.Y. State Museum and Sci. Serv. Mem. 18,* 111–134, 1970.

Taylor, H. P., Jr., Oxygen isotope evidence for large-scale interaction between meteoric ground waters and Tertiary granodiorite intrusions, Western Cascade Range, Oregon, *J. Geophys. Res., 76,* 7855–7874, 1971.

Taylor, H. P., Jr., and R. W. Forester, Low-O^{18} igneous rocks from the intrusive complexes of Skye, Muil, and Ardnamurchan, Western Scotland, *J. Petrology, 12,* 465–497, 1971.

Taylor, H. P., Jr., A. L. Albee, and S. Epstein, O^{18}/O^{16} ratios of coexisting minerals in three assemblages of kyanite-zone pelitic schist, *J. Geol., 71,* 513–522, 1963.

Turi, B., and H. P. Taylor, Jr., An oxygen and hydrogen isotope study of a granodiorite pluton from the Southern California batholith, *Geochim. Cosmochim. Acta, 35,* 383–406, 1971a.

Turi, B., and H. P. Taylor, Jr., O^{18}/O^{16} ratios of the Johnny Lyon granodiorite and Texas Canyon quartz monzonite plutons, Arizona, and their contact aureoles, *Contrib. Mineral. Petrol., 32,* 138–146, 1971b.

Urey, H. C., The thermodynamic properties of isotopic substances, *J. Chem. Soc.,* London, 562–581, 1947.

Wilson, A. F., D. C. Green, and L. R. Davidson, The use of oxygen isotope geothermometry on the granulites and related intrusives, Musgrave Ranges, Central Australia, *Contrib. Mineral. Petrol., 27,* 166–178, 1970.

SUBJECT INDEX

Absorption bands, infrared quartz, 131, 132
Acoustic loss peak, 131, 132
Actinolite, 313
Activation energy, 1, 4, 5, 10, 35, 39, 46, 53–57, 64–74, 77, 90–96, 101, 111–115, 131–137, 151–153, 171, 185, 189–192, 195, 200–201, 239, 295–298
Activation enthalpy, 117, 119, 125
Activation volume for diffusion, 117, 119, 122
Activity gradients, 280, 282
Adatoms, 5, 6
Adirondack Mountains, New York, 330
Adularia, 78, 90, 91, 92, 99–105, 175, 177, 180
Affinity of reaction, 233, 234, 240
Age, discordances, 61–76, 78, 94, 96, 107–115, 299, 321, 322
  resetting, 62, 69, 107, 114
Ages, radiometric, 61–76
Ag-Zn-Cd system, 20, 21
Alamosa River stock, 299, 302–307
Albite, 74, 78, 79, 91, 94, 99, 102, 104, 173–183, 185–193, 257, 274, 277, 282, 283, 284
Albite transformation, low-high, 185
Albite twinning, 177
Algonquin Park, 331, 332
Alkali Diffusion in Orthoclase, 77–98
Alkali equilibrium partitioning determination, 77
Alkali exchange, 70, 99, 102, 103, 173–183, 185–193
Alkali feldspar, 273–286, 299, 304, 317, 320, 321
  diffusion in, 39, 43, 77–98, 99–105, 180, 192, 243
  disordering in, 175, 180, 185–193
  exsolution in, 78, 173–183
  see also Feldspar; Potassium feldspar
Alkali metasomatism in feldspars, 244
Allochthonous pelitic rocks, 280
Alloy, multicomponent, 15
Almandine, 34
AlO_4 tetrahedra, 44
Alpha-alumina, 104
Alpha-beta transformation, quartz, 132, 133
Alpha-quartz, 131, 132
Alpine fault, New Zealand, 63, 64
Al-Si disordering, 95, 185–193
Al-Si ordering in feldspars, 74, 99, 102, 103, 177, 180–182
Alteration zones, 262
Alumina, 93, 104, 282
Aluminum, 35, 43, 94, 261–271, 285, 287, 289, 316, 321
Amelia (Virginia) albite, 185–193, 282–285
Amelia (Virginia) microcline, 99–105, 175
Amphibolite facies, 273–286, 330, 332
Analbite, 173–183
Andalusite, vi, 327, 330

Andesite, 159, 160, 302, 303, 309, 314, 315, 326, 332
Andradite, 287–293
Anelasticity in alpha-quartz, 131
Animas River Canyon, 301, 302, 309
Anisotropy, diffusional, 61, 109, 132
Annite, 73, 74, 107–115
Anorthite, 285
Anorthosite, 330
Apollo mission, lunar pyroxenes, 195
Apophyses, 156
Apsley, 332, 333
Ar^{40} diffusion, 89–95, 107–115
Argon diffusion, 62–70, 74, 78, 89–92, 107–115
  Arrhenius plot, 68, 69, 107, 108, 111–115
  data, 67–69, 90, 94, 110
  in biotite, 74, 114
  in muscovite, 74
  in phlogopite, 107–115
Argon diffusivities at moderately high temperatures, 64
Argon loss, 62, 63, 68, 69, 94, 96, 107–115, 295–298
Argon in Phlogopite Mica, Studies in Diffusion, 107–115
Argon-potassium ratio, 295–298
Arrhenius plot, Alpine fault, New Zealand, 64
  argon diffusion, 68, 69, 107, 108, 111–115
  K diffusion, 85, 87
  Na diffusion, 83
  oxygen diffusion, 102
  Rb diffusion, 89
  water diffusion, 136
Arrhenius relations and plots, 1, 2, 77, 90, 91, 94, 96, 112, 117, 131, 135, 171, 181, 190, 200, 238
Atom drift velocity, 3, 7–12
Atomic driving forces, 8–12
Atomic mobility model, 31
Atom jumps, 3–11, 46–50, 57
Atoms, nonrandom motion of, 11
Augite, 302, 303, 309
Aureoles, contact, 214, 295–298, 325–335
Autochthonous pelitic rocks, 280
Axial-plane layering, 273–286

Barometer, geological, 53
Basalt, 139–170, 257, 309, 311, 318, 326, 331–333
Batholith, 300, 329
Benson Mines, orthoclase, 77–98
Berman balance, 145
Beryllium, 45
Beta-alumina, 93
Beta-Quartz, Diffusion of Tritiated Water in, 131–138
Binary diffusion, 15–30, 147
Binary system, 15
Biotite, 33, 34, 69–74, 107, 114, 273–286, 295, 298, 302–308, 312, 328, 330

Biotite-garnet transition, 34, 35
Biotite, Rb-Sr vs K-Ar age data, 62, 63
Biotite zone rocks, 33, 34
Birch Creek, 328–330
Black Hills, South Dakota, 186
Bobbejaankop granite, 318
Boltzmann constant, 9, 118, 154, 206
Boudin necks, 273–286
Boulder batholith, 300
Boundary layers, 139–170
Bourdon gauge, 141
Bromide, ionic radius, 53
Bromine pentafluoride, 100
Buoyancy, 157, 161, 165, 167
Burgess County, Ontario, phlogopite, 107–115, 282–285
Burleigh, 331, 332
Burns Formation, 302
Bushveld Complex, 299, 317, 318

$CaCO_3$-CO_2 isotope exchange, 219
$CaCO_3$-H_2O system, 219–227
Cadmium chalcogenide, 53
Cadmium, ionic radii, 53
Cadmium sulfide, 55–57
Calcite, 171, 274, 288–292, 303, 313
Calcite, Mass Transfer during Hydrothermal Recrystallization, 205–218
Calcite-water system, isotope exchange, 219–227
Calcium, 43, 47–49, 140, 164, 165, 201, 207, 246, 285, 287–293, 316
Calcium carbonate, 219, 287
Ca-Mg exchange in diopside, 201
CaO-Al_2O_3-SiO_2 system, 27
CaO-K_2O-SiO_2 system, 19, 20
Carbon, 219
Carbonate minerals, 213, 219, 273, 280, 287, 300, 319, 330
Cardiff, Harvey—arch, 330
Caribou stock, 328, 329
Cascades, Western, 300, 309, 313–316, 329
$CaSiO_3$, 201
Cation-anion vacancy pairs, 95
Cation diffusion, 53, 54, 152
Cationic Diffusion in Olivine to 1400°C and 35 kbar, 117–129
Cave-in-Rock district, 215
Chalcocite, 55
Chalcogenides, 53
Champion Creek stock, 309, 312, 313, 316
Chandos Lake, 331–333
Chemical diffusion coefficients, 39, 140, 147
Chemical diffusion in magma, 140, 156–168
Chemical gradient, 147, 156, 163
Chemically closed system, 205–218
Chemical potential, 33, 42, 44, 131–141, 165, 231–241
  gradient, 3, 9, 12, 42, 44, 49, 118, 154, 168, 231–241, 257, 280, 287–293
Chemical solvus, 173–183
Chemical spinodal, 173, 174

Chert, 205, 213, 287, 289, 319, 320
Chloride, ionic radius, 53
Chlorides, 207, 244, 246, 257, 261–271, 282–285
Chlorite, 276, 277, 302–308, 312, 313, 317, 327, 329, 330
Chromatographic theory, 243, 251
Circulation, large-scale, 299–324
Classical solvus, 173–183
Clay minerals, 319
Cleavage, axial-plane, 273–286
Cleavage layering, 273–286
Clinopyroxene, 43, 47, 69, 70, 201, 288, 303, 314, 315
Cochise County, Arizona, 329
Coherent Exsolution in the Alkali Feldspars, 173–183
Coherent phase relations, 173–183
Coherent solvus, 173–183
Coherent spinodal, 173, 174
Cold-seal pressure vessels, 140
Cold-seal tipping bombs, 285
Colorado, Eldora stock intrusion into Precambrian, 63
Comminution of minerals, 73
Composition, effect on diffusion, 15–30, 117–129, 131–136, 139–170
Comstock Lode, Nevada, 300
Concentration gradients and profiles, 2, 3, 7, 11, 12, 42, 43, 49, 78, 139–170, 232–235, 243–259, 261–271
Concordia plots, 62, 72
Conduction, ionic, 118
Conductivity, electrical, 61, 73, 117–129
Conductivity, thermal, of magma, 140, 157
Convection, diffusion with, 22, 23
Convective circulation systems, 139–170, 299–324
Copper, diffusion coefficients, 55
  ionic radii, 53
Coppermine River basalt, 320
Copper sulfide, 54–56
Cordierite, vi, 330
Correlated-jump effects, 48
Correlation-factor effects, 3, 9–11
Critical radius of crystal, 206–210
Crowdion mechanism, 6
Cryptoperthite, 181
Crystallites, 140, 141, 145
Crystallization, 15–30, 156, 157, 161
  see also Recrystallization
Crystallization of melts, partitioning during, 31
Crystallographic orientation, effect on diffusion, 117–129
Crystals, diffusion in, 54
  dissolution of, 62
  structure, 54–57
Crystals, Simple, Diffusion Kinetics and Mechanisms in, 3–13
Cu-Ag-Au system, 28, 29
Cubic silver sulfide, 54, 55
Cu-Zn-Ni system, 27, 28
Cylindrical diffusion model, 77–98, 107–115

SUBJECT INDEX

Dacite, 326, 331, 332
Darcy's law, 256
Darken's equation, 16, 39, 40
Dating, isotopic, 61
   K-Ar, 62, 67–70, 94, 107–115, 305
   Rb-Sr, 62, 70–72, 299, 321, 322
   U-Pb, 62, 72, 299, 322
   U-Th-Pb, 62, 72
Decomposition, spinodal, 171, 173, 176, 197, 198, 201
Defects, Frenkel, 94
   lattice, 73, 131
   point, 54
   Schottky, 95
Defect structure, influence on diffusion, 53
Deformation, plastic, 62
Dehydration, 139–170
Density, magma, 140, 157
   measurements, 145
Determination of Temperatures of Intrusives from Radiogenic Argon Loss in Contact Aureoles, 295–298
Deuterium, 300–308
Diagenesis, 205–218
Diapiric granitic plutons, 330
Dickite, 305
Differentiation layering, metamorphic, 261, 273–286
Diffusion, atomic driving forces in, 8, 9
   effect of composition on, 15–30, 117–129, 131–136, 139–170
   effect of crystallographic orientation on, 117–129
   effect of pressure on, 117–129
   effect of temperature on, 63–65, 71, 115, 117–129, 139, 151, 152
   effective grain size for, 107–115
   grain-boundary, 2, 4, 5, 231–241, 243, 327
   isothermal, 15–30, 40, 157, 158
   kinetic-atomic theory of, 3–13
   low-temperature, 64
   multicomponent, 15–30, 31, 39–48, 140, 152, 244
   nonisothermal, 15–30
   quasi-steady, 231–241
   surface, 4–6
   uphill, 15–30, 229, 287–293
   volume, 3–12
Diffusion: Argon in Phlogopite Mica, 107–115
Diffusion, Multicomponent, Vector Space Treatment, 15–30
Diffusional anisotropy, 61, 109, 132
Diffusional kinetics in solid and melt, 78
Diffusion coefficient, 1, 2, 3–10, 15, 31–50, 55, 57, 61, 62, 70, 73, 79, 80, 85–88, 91, 94, 95, 99–104, 107, 110, 131–136, 139–170, 219, 231–241, 243–247, 255–257, 295–296
   interdiffusion, 1, 8, 92, 93, 117–129, 202
   intrinsic, 1, 7, 8, 15, 93, 257
   self-diffusion, 1, 15, 47, 77, 79, 82, 84, 124, 219
   tracer, 1, 3, 7–11, 48, 79, 90, 92
   see also Diffusivity, Diffusion constants
Diffusion and Combined Diffusion-Infiltration Metasomatism, Simple Models of, 243–259
Diffusion constants, 61, 62, 70, 73, 147
Diffusion-Controlled Properties of Silicates, Modeling of, 31–52
Diffusion-controlled and reaction-controlled growth processes, criteria for distinguishing between, 236–239
Diffusion with convection, 22, 23
Diffusion couple interface, effect on diffusion, 122
Diffusion couples, 15–30, 117–129
Diffusion equation for a well-stirred reservoir, 100
Diffusion equations, flux, 31–50, 245
   kinetic, 45–50
   kinetic-atomic, 3, 7–12
   random-walk, 3, 7–11
   thermodynamic-continuum, 3, 12
   vector, 15–30
Diffusion gradient, vi, 147, 248
Diffusion and growth, partitioning by, 31
Diffusion of H_2O in Granitic Liquids, 139–170
Diffusion-infiltration metasomatism, 243–259
Diffusion Kinetics and Mechanisms in Simple Crystals, 3–13
Diffusion matrix, 15–30, 232
Diffusion mechanisms, crowdion, 6
   divacancy, 6
   exchange, 5
   indirect interstitial, 6, 94
   interstitial, 5–7, 54, 57, 94–96
   interstitialcy, 6, 9, 54, 77, 94–96
   ring, 5, 46
   short-circuit, 3, 4
   vacancy, 3, 6, 9–11, 39–50, 54, 66, 67, 77, 94–96
   volume, 3–7
Diffusion mechanisms in crystals, 3–13
Diffusion models, 31–52, 77–98, 107–115, 205–218, 231–241, 243–259, 261, 287–293, 295
Diffusion paths, 3, 4, 15–30, 45, 46, 245
Diffusion in polycrystals, 5
Diffusion in a pressure gradient, 273–286
Diffusion processes, linear, 15–30
Diffusion rate, 139–170, 231–241, 254, 256
Diffusion Related to Geochronology, 61–76
Diffusion in Sulfides, 53–59
Diffusion of Tritiated Water in beta-Quartz, 131–138
Diffusion up a chemical potential gradient, 24, 25
Diffusion up a concentration gradient, 23, 24
Diffusivity, *see also* Diffusion, Diffusion constant
   concentration dependence, 139–170
   effective, 245, 248, 256
   intrinsic, 245
   net, 147–152, 155, 164, 166
   relative, 61
   temperature dependence, 63, 65, 139
   thermal, 155, 157
Digenite, 55

Diopsides, enstatite exsolution from, 195–203
Diorite, 298, 300, 302, 308, 314
Disaggregation, 224, 226, 227
Discordances, age, 61–76, 78, 94, 96, 107–115, 299, 321, 322
Disequilibrium partitioning between phenocrysts and host lava, 78
Disordering rate, alkali feldspar, 185–193
   effect of composition, 188, 189
   effect of grain size, 187, 188
Dissociation equilibrium, 244
Dissolution and reprecipitation, 15–30, 62, 70, 79, 92, 99, 102, 103, 105, 185, 192, 193, 219–227, 238, 316
   see also Solution
Divacancy mechanism, 6
Dolomite, 287, 328
Domenigoni Valley pluton, 329
Drift velocity, 3, 7–11
Driving forces, atomic, 8–12
   thermodynamic, 12
Driving potential gradient, 147
Duhem, Gibbs—equation, 41, 44
Duluth Complex, 299, 317, 321
Dutchess County, New York, 273–286

Eclogite, 300
Effective grain size for diffusion, 107–115
Eigenvalue, 15–30
Eigenvector, 15–30
Einstein, Nernst—equation, 11, 118
Einstein, Stokes—equation, 139, 154
Egypt, 117
Elastic strain, 171, 173–176
Eldora stock, 328, 329
   intrusion into Precambrian, 63
Electrical conductivity, 61, 73
   in olivine, 117–129
   pressure dependence, 117, 126
   temperature dependence, 117
Electron microprobe, vi, 2, 15, 31, 42, 119–121, 147, 201, 229, 280
Endoskarn, 287
En-echelon gash fractures, 273–283
Everett samples, 273–286
Enstatite exsolution from supersaturated diopsides, 195–203
Enthalpy of migration, 95
Entropy, 232, 241
Epidote, 287, 288, 313
Epizonal igneous intrusions, 299–324
Equilibrium, dissociation, 244
   local, vi, 229, 231–241, 243–245, 252, 254
   pyrite-pyrrhotite, 53
   solid-fluid, 78
   trace-element, 205–218, 246
Equilibrium fractionation, 219, 221
Equilibrium during metamorphism, domains of local, 31
Equilibrium partitioning, 79, 81, 245
Exchange, ion, 70, 79, 85–87, 92, 99, 102, 244

Exchange, isotope, 77–98, 99–105, 205–218, 219–227
Exchange between crystal and inhomogeneous reservoir, 31–33
Exchange mechanism, 5
Exoskarns, metasomatic zoning, 287–293
Exsolution, coherent, 173–183
   diopside-enstatite, 195–203
   kalsilite-nepheline, 199
   pyrite-pyrrhotite, 199, 200
Exsolution lamellae, 67, 68, 174, 176, 180
Eyring, Ree—theory, 139, 154

Fayalite, 43, 47–49, 117–129
Fayalite-tephroite join, 47
Feldspar, vii, 67–70, 74, 77, 165, 243, 309, 314–321, 327, 328, 330
Feldspar, *see also* Alkali feldspar, Potassium feldspar
Feldspar argon-diffusion data, 67–69, 90, 94
Feldspar-water fractionation correction, 100
Fe-Ni system, 78
Ferrodiorite, 314
Ferrous orthosilicate, 42, 47
Ferrous oxide, 54
Ferrous sulfide, 53, 54, 57
Fick's laws, 1, 15, 17, 38, 41, 42, 147, 245, 246
Finite-difference computation method, 246, 258
Fission tracks, spontaneous, 61, 73
Fissure veins, 273–286
Fluoride ion radius, 53
Flux equations, 31–50, 245
Flynn stock, 329
Fold-nose segregation, 273–286
Forsterite, 117–129, 196
Forsterite-fayalite, join, 47–49
   system, 43, 117–129
Fourier number, 295–298
Fourier's law of heat conduction, 147
Frank-Griggs hypothesis, 105, 131
Free energy, Gibbs, 34, 46–49
   surface, 206
Frenkel defects, 94
Frequencies, jump, 3, 7–11, 94
Freundlich, Thomson—equation, 206, 209
Front Range, Colorado, 298, 329
Fugacity, oxygen, 33, 195
   water, 33, 95, 143, 148–150, 165–167

Gabbro, 302, 314, 326, 330, 332
Galena, 320
Garnet, vii, 31–38, 43, 47, 48, 273–286, 287–289, 330
Garnet isograd, 34, 279
Garnet shadows, 273–286
Garnets, metamorphic, zoning in, 31–33
Gash fractures, en-echelon, 273–283
Geochronology, 78, 107
Geochronology, Diffusion Related to, 61–76
Geological barometer, 53
   thermometer, 32, 53, 295–298, 320

SUBJECT INDEX

Geotherm, 118, 126, 322
Geothermal water systems, 300, 310
Geothermometer, see Geological thermometer
Gibbs-Duhem equation, 41, 44
Gibbs energy, 174, 180
    free energy, 34, 46–49
Gibbs-Thomson equation, 206
Gibbs triangle, 16, 19, 20
Glamorgan, 331, 332
Glauconite, 69, 112
Gneiss, 32, 277, 320, 325–335
Goldfield, Nevada, 300
Gold, ionic radii, 53
Grade, metamorphic, vi, 205, 273–286, 325–335
Grain-boundary diffusion, 2, 4, 5, 231–241, 243, 327
    network, 231–241
Granite, vii, 139–170, 230, 245, 287, 298, 299, 305, 317, 325–335
Granitic Liquids, Diffusion of H_2O in, 139–170
Granodiorite, 256, 302, 303, 305, 309, 312, 316, 328
Granophyre, 299, 310, 314, 317, 319, 321
Granulite facies, 330
Graphite, 277, 281
Grashof number, 157–159
Greenschist facies, 273–286, 330
Grenville Province gneisses, 32
Grenville province, Ontario, migmatite terrane, 325–335
Grimstad, Norway, granite, 331, 332
Ground water, 299–324, 329
Growth, see Nucleation

Harvey-Cardiff arch, 330
Heat capacity of magma, 140
Heat flow, 139–170
Heat of fusion of magma, 140
Hebrides, Scottish, 300, 329
Hedenbergite, 287–293
Heise bourdon gauge, 141
Hematite, 299, 317, 318, 321
Henry's law, 244, 246
Hermosa Formation, 302
High-temperature diffusion, 71
Hill Mine, 215
Homogeneous transformation, 187, 188, 191, 193
Hopedale, Labrador, 317, 320
Hornblende, 298, 303, 306–308, 328
Hrafntinnuhryggur, Iceland, 309
Hugo microcline (Black Hills, South Dakota), 185–193
Hugo pegmatite (Black Hills, South Dakota), 186
Huntsville, 331, 332
Hydration of magma, 139–170
Hydration rates, 139
Hydrogen diffusion, in quartz, 132
    in silica glass, 139, 152–154
Hydrogen isotopes, 299–324
Hydrolytic weakening, 131
Hydrothermal annealing experiments, 191–193

Hydrothermal conditions, isotope exchange under, 219–227
Hydrothermal diffusion experiments, 77–98, 107–115
Hydrothermal recrystallization, 205–218, 220
Hydrothermal self-diffusion experiment, 65
Hydroxyl loss, 108
Hydroxyl in quartz, 131
Hydroxyl reactions in silica glass, 152, 154

Iceland, 300, 309
Iceland spar, 220
Igneous intrusions, 139–170, 299–324
    rocks, vi, vii, 299–324, 325–335
Ilmenite, 277
Ilvaite, 287, 288
Impurity conduction, 118
Impurity diffusion by vacancy mechanism, 11
Indirect interstitial mechanism, 6, 94
Infiltration metasomatism, 78, 243–259, 261–271, 289
Infiltration Metasomatism in the System K_2O-SiO_2-Al_2O_3-H_2O-HCl, 261–271
Infrared absorption bands, quartz, 131, 132
Interaction between ground water and igneous intrusions, 299–324
Interdiffusion, 20, 65, 66, 73, 78, 90–92, 192, 200, 201
    Fe and Mg in olivine, 117–129
Interdiffusion coefficient, 1, 8, 92, 93, 117–129, 202
    equation for, 118
Interstitialcy mechanism, 6, 9, 54, 77, 94–96
Interstitial mechanism, 5–7, 54, 57, 77, 94–96
Interstitial mechanism, indirect, 6, 94
Intrinsic diffusion coefficient, 1, 7, 8, 15, 93, 257
Intrinsic semiconduction, 118
Intrusions, 139–170, 295–298, 299–324
Intrusive rocks, 299–324, 325–335
Intrusives, temperature of, 295–298
Inyo batholith, California, 327, 329
Iodide, ion radius, 53
Ion activity product, 206
Ion diffusivities in Fe-Ni system, 78
Ion exchange, 70, 79, 85–87, 92, 99, 102, 244
Ionic conduction, 118
    diffusion, 118
    electrical conductivity, 117, 118, 127
    migration, 118, 125
    radii, 53, 54
Ion jumps, 54, 57
Ion microprobe, vi, 2, 31, 65
Ion mobility, oxide, 53, 99
    sulfide, 53
Iron, 33–49, 53, 54, 57, 73, 78, 287
    ionic radii, 53
Isochrons, strontium, 62
Isograd, metamorphic, vi, 34, 231, 261, 273–286
Isotherm, 141, 142, 243–259, 320
Isothermal boundary, 157
    diffusion, 15–30, 40, 157, 158

heating experiments, 107–115, 171, 187, 196
Isotope dilution analysis, 109
Isotope exchange, between calcite and water, 219–227
    between crystal and fluid, 205–218
    between feldspar and alkali chloride, 77–98, 99–105
    in equilibrium system, 78, 83–85
Isotope mobility, oxygen, 325–335
Isotopic dating methods, 61
Isotopic equilibrium, 71, 83, 205–218, 221
Isotropic diffusion model, 77–98, 100
Itatiaia K albite, 191, 193

Jemez Mountains, New Mexico, 145
Johnny Lyon stock, 328, 329
Jump frequencies, 3, 7–11, 94
Jumps, atom, 3–11, 46–50, 57
    ion, 54, 57

$KAlSi_3O_8$, 179, 180
$KAlSi_3O_8$-$NaAlSi_3O_8$-$NaCl$-KCl-H_2O system, 244
$KAl_3Si_3O_{10}(OH)_2$, 262
Kaolinite, 303–308
Kalsilite-nepheline exsolution, 199
K-Ar dating, 62, 67–70, 94, 107–115, 305
Keweenawan volcanic terrane, 299, 317, 321, 322
Kinetic-atomic diffusion, equations, 3, 7–12
    theory, 3–13
Kinetic equations, 45–50
Kinetics, nonequilibrium, 31
Kinetics of Al/Si Disordering in Alkali Feldspars, 185–193
Kinetics of Enstatite Exsolution from Supersaturated Diopsides, 195–203
Kirkendall effect, 41
Kirkendall shift, 40
K_2O-SiO_2-Al_2O_3-H_2O-HCl system, 261–271
K_2O-SrO-SiO_2 system, 20, 24
Kristallina (Switzerland) adularia, 99–105, 175
Kyanite, vii, 277, 280, 329, 330

Labrador, 317, 320
Labradorite, 314
Lake Toxaway quartz, 197
Lamellae, exsolution, 67, 68, 174, 176, 180
Lattice defects, 73, 131
Lattice diffusion, 232
Lead, 53, 62, 72, 320
Lead chalcogenide, 53
Lead loss, 72
Leadville limestone (Mississippian, Colorado), 215
Leucosome, 332, 333
Lifshitz-Slyozov-Wagner theory, 208, 209
Limestone, 165, 205, 215, 287, 289, 320
Lincoln Mountain, Vermont, 329
Lindgren's law, vii, 229, 325
Linear rate law, 244, 245
Lithostatic pressure, 310, 311

Low-temperature diffusion, 64, 115
Lunar pyroxene, 195

Mafic dikes, age studies near, 63
Mafic minerals, 299, 317, 321
Mafic reservoir, 325–335
Magma chambers, 139–170
Magma intrusion, 295–298, 299–324
Magnesia, 104, 117–129
Magnesium, 42–49, 207, 285, 287
Magnesium orthosilicate, 42, 44, 47
Magnetite, 288–292, 303, 327, 328
    formation from biotite, 71
Manganese, 33–49, 287
Manganous orthosilicate, 42, 47
Manhattan samples, 273–286
Mantle minerals, 131
Marble, 215, 289, 328–333
Marine sediments, 325–335
Mass balance equation, 147, 245, 248
Mass flow, 41, 139–170, 245
Mass Transfer of Calcite during Hydrothermal Recrystallization, 205–218
Mass transfer, in magma chambers, 155–168
    model, 205–218, 231–241
    rates, silicates and aqueous solutions, 78
Mass transport models, 261
Matano-Boltzmann analysis, 134
Matano interface, 118
Matrix, 15–30, 232
Melanosome, 332, 333
Mesozoic, 299–306, 317, 320
Metagabbro, 330, 333, 334
Metaigneous rocks, 330
Metamict Ceylon zircon, dating, 72
Metamorphic differentiation layering, 261
Metamorphic Differentiation Layering in Pelitic Rocks, 273–286
Metamorphic facies concept, 205
Metamorphic garnets, zoning in, 31–33
Metamorphic grade, vi, 205, 273–286, 325–335
Metamorphic rocks, vi, 205–218, 231–241, 243, 256, 325–335
Metamorphism, 31, 34, 62, 63, 107, 205–218, 229, 231–241, 243, 287, 325–335
    prograde, 33, 205, 329
    retrograde, 33, 34
Meta-anorthosite, 333
Metasedimentary rocks, 328, 330
Metasomatic processes, 31, 229, 261, 263
Metasomatic Zoning in Ca-Fe-Si Exoskarns, 287–293
Metasomatism, 70, 78, 289
    diffusion-infiltration, 243–259
    Infiltration, in the System K_2O-SiO_2-Al_2O_3-H_2O-HCl, 261–271
Metasyenite, 330
Meteoric ground water, 299–324, 329
Meyer-Darken equation, 16
$MgAl_2O_4$, 104
MgO-Cr_2O_3 system, 118

SUBJECT INDEX

MgO-MgAl$_2$O$_4$ system, 118
MgSiO$_3$-CaMgSi$_2$O$_6$ system, 196, 197
Mg$_2$SiO$_4$-Fe$_2$SiO$_4$-Mn$_2$SiO$_4$ system, 42, 47
Mica, 277, 280, 281, 327
   argon loss, 63
   water loss, 69
   see also Biotite, Phlogopite, Muscovite
Microcline, 74, 78, 91, 94, 99–105, 173–183, 185–193, 282–285
   perthitic, 186, 192
Microcline–low-albite series, 173–183, 186
Microcline-sanidine transformation, 185–193
Microprobe, 31, 32, 42, 65, 78, 92
   see also Electron microprobe, Ion microprobe
Migmatite, 325–335
Minas Gerais, Brazil, quartz, 282–285
Mineral-specimen characterization, 67
Mineral segregation, 231–241, 261, 273–286
Minerals, stability to age resetting, 62, 69
Miocene, 301–306
Mississippian, Colorado, 215
Mississippi Valley, 320
Mitchell Co., N.C., oligoclase, 283–285
Mobility, oxide ion, 53, 99
   sulfide ion, 53
   matrix, 17
Mobility of Oxygen Isotopes during Metamorphism, 325–335
Modeling of Diffusion-Controlled Properties of Silicates, 31–52
Models, diffusion, 31–52, 77–98, 107–115, 205–218, 231–241, 243–259, 261, 287–293, 295
Monoclinic silver sulfide, 54, 55
Monzonite, 71, 302–309
Moon rocks, 195
Muffle furnace, 71
Mugearite, 164
Multicomponent diffusion, 15–30, 31, 39–48, 140, 152, 244
   Vector Space Treatment, 15–30
Muscovite, 33, 71, 74, 107, 261–271, 277, 282–285
Musgrave Ranges, Australia, 332
Muskox Complex, 299, 317–322
Myvatn, Iceland, 309

Nain, Labrador, 317, 320
Needle Mountains, 304
Nepheline-kalsilite system, 199
Nernst-Einstein equation, 11, 118
Nernst-Planck equation, 15
New Zealand, Alpine fault, 63, 64
Nickel, 78
Nickel sulfide, 57
Nimrod stock, 309
Nonequilibrium kinetics, 31
Nonequilibrium thermodynamics, 31, 229, 231–241
Nonisothermal diffusion, 15–30
Nonisochemical rock systems, 261
Nonrandom motion of atoms, 11

Nucleation and growth, 171, 176, 180–182, 187, 191, 195–203, 205–218
Nusselt number, 158–161, 165, 166

Obsidian, 139–170, 309
Octahedra, 45, 46
Octahedral sites, 53, 54, 57, 73
Oligocene, 301–306
Oligoclase, 283–285
Olivine, 39, 42–49, 117–129
Onsager reciprocity theorem, 40–42
Onsager's Reciprocal Relations, ORR, 15, 29
Orthoclase, 68, 69, 70, 72, 74, 77–98, 261–271, 273–286
   Alkali Diffusion in, 77–98
Orthoclase-quartz, 261–271
Orthopyroxene, 69, 70, 201
   in metamorphic rocks, 47
Orthorhombic enstatite, 200
Orthosilicates, 42–47
Ostwald ripening, 205–218, 219–227
Ouray, Colorado, 299–305
Oxide, ferrous, 54
Oxide ion, radius, 53
   mobility, 53, 99
   polarizability, 53
Oxides, 43, 46
   diffusion in, 53, 54
Oxide-spinel system, 118
Oxygen, 43, 45, 46, 78, 105, 131, 134
   diffusion, in quartz, 132
      in silica glass, 152–154
   exchange, between feldspar and air, 99, 100, 102–104
      between feldspar and CO$_2$, 99, 100, 102–104
   fugacity, 33, 195
Oxygen and Hydrogen Isotope Evidence for Large-Scale Circulation and Interaction, 299–324
Oxygen Isotope Exchange between Calcite and Water under Hydrothermal Conditions, 219–227
Oxygen Isotope Exchange between Potassium Feldspar and KCl Solutions, 99–105
Oxygen isotopes, 99–105, 165, 206, 215, 219–227, 299–324, 325–335
Oxygen vacancies, 95

Paleozoic, 299–304, 320, 322
Parabolic rate law, 150, 244, 245, 255
Paragneiss, 325–335
Paragonite, 285
Parry Sound, 331, 332
Partitioning, 31, 32, 79, 81, 85, 96, 245, 257
Partition coefficients, 32, 33, 81, 89
Pegmatite, vii, 186, 332
Pelite, 34, 35
Pelitic rocks, 273–286, 325–335
Penetration distance, 155, 156, 167, 247, 255
Pericline twinning, 177
Permeability, 256, 257, 288

Perry Sound, Ontario, microcline, 282–285
Perthite, 67, 70, 74, 78, 91, 181
　coherent submicroscopic, 181
　formation of, 78
　two-phase lamellar, 74
Phase equilibria, pyroxene systems, 195, 201
Phase relations in Fe-Ni system, 78
Phenocrysts, vii, 299, 302, 309, 312, 314, 315
Phlogopite, 69–74, 93, 107–115, 282–285
Phlogopite-annite series, 73, 74, 107–115
Photoconductive properties, cadmium sulfide, 55
Photovoltaic devices, CdS-Cu$_2$S, 55, 56
Phyllite, 277, 280, 327
Pigeonite, 195, 196
Plagioclase, 244, 246, 248, 273–286, 299, 302–305, 309–316, 333
Planar crystal growth, 23
Planck, Nernst—equation, 15
Plastic deformation, 62
Plate diffusion model, 107–115
Platoro, 306, 307
Platoro Caldera, 305
Pluton, 74, 139–170, 299, 301, 308, 310, 325–335
Point defects, motion of, 54
Polarizability, oxide ion, 53
　sulfide ion, 53
Polycrystals, diffusion in, 5
Polymeric systems, 150
Pore fluid, diffusion and infiltration, vi, 2, 243–259, 289, 291
Porosity, 214, 215, 229, 245, 256, 257, 261–271, 288
Porphyry, 300, 302, 309
Potassium, 39, 62, 67–74, 77–98, 99–105, 107, 180, 243, 257, 261–271, 273–286, 295–298, 303, 305
Potassium-argon geothermometer, 295–298
Potassium diffusion, in alkali feldspar, 39, 77–98, 180, 243
　in orthoclase, 77–98
Potassium feldspar, 99–105, 185–193, 257, 273–286, 303, 305
Potential, *see also* Chemical potential
　driving, gradient, 147
　saturation, 146
Prandtl number, 157–160, 165, 166
Precambrian, 299, 300, 304, 308, 317–322, 331
　Eldora stock intrusion into, 63
Precambrian Basement, 317
Precambrian shield, Australia, 331, 332
Precession photographs, 119, 173–183, 185–193
Precipitation, *see* Dissolution, Recrystallization, Crystallization
Pressure, effect on diffusion, 117–129
Pressure gradient, 229, 255–257
Prograde metamorphism, 33, 205, 329
Protoenstatite, 196, 200
Providencia, Mexico, limestone, 215
Pyrite-pyrrhotite equilibrium, 53
Pyrite-pyrrhotite exsolution, 199, 200
Pyrophyllite-quartz, 261–271
Pyroxene, 43, 47, 69, 70, 195–203, 302, 309, 328
　lunar, 195
Pyrrhotite, pyrite—equilibrium, 53

Quartz, vii, 33, 35, 71, 95, 105, 131–138, 165, 261–271, 273–286, 289–292, 299–324, 327–329
　alpha-beta transformation, 132, 133
　infrared absorption bands, 131, 132
Quartzite, 327
Quartzofeldspathic constituents in basalt, 165
Quartz-orthoclase assemblage, 261–271
Quartz-pyrophyllite assemblage, 261–271
Quasi-steady Diffusion and Local Equilibrium in Metamorphism, Criteria for, 231–241
Quaternary alloy, 15

Radiogenic argon, 69, 107–115, 295–298
Radiogenic strontium 87, 72, 295
Radiometric ages, 61–76
Radon leakage, 72
Random-walk diffusion equations, 3, 7–11
Rate, diffusion, 139–170, 231–241, 254, 256
　reaction, vi, vii, 231–241
Rate laws, 150, 244, 245, 255
Rayleigh number, 157–161, 165, 166
Rb-Sr dating, 62, 70–72, 299, 321, 322
Reaction rate, vi, vii, 231–241
Reaction zone, 231–241
Recrystallization, 62, 65, 74, 99–103, 107, 171, 205–218, 219–227, 256, 273, 303, 309, 314–316, 327
Recrystallized monomineralic rocks, 205–218
Red Sea olivine, 117–129
Ree-Eyring theory, 139, 154
Reentrants, 325–335
Relaxation, volume, 143–146
Reprecipitation, *see* Dissolution, Recrystallization
Reset ages, 62, 107, 114
Retrograde isotopic reequilibration, 215
Retrograde metamorphism, 33, 34, 168
Reynolds number, 163, 164
Rhodochrosite, 287
Rhodonite, 287
Rhombic enstatite, 196
Rhyolite, 302, 309, 326, 331, 332
Ring dike, 300, 311, 312, 315
Ring mechanism, 5, 46
Ripening, Ostwald, 205–218, 219–227
Rockport, Mass., fayalite, 117–129
Roof zones, 325–335
Rubidium, 62, 70–72, 77–98
　diffusion in orthoclase, 77–98

St. Francois Mountains, Missouri, 308, 317, 321
St. Francois volcanic terrane, 299, 321, 322
St. John Island, Red Sea, 117
Sandhill, West Virginia, 317
Sanidine, 69, 100, 102, 173–183, 185–193, 285
Sanidine-analbite series, 173–183
San Juan Mountains, Colorado, 299–324, 329
San Juan tuff, 312, 315
San Juan volcanic field, Colorado, 299–324
Santa Rosa Range, Nevada, 327–330
Santa Rosa stock, 328, 329

Saturation potential, 146
Saturation, surface, 148
Saturation surface, 287–293
Sawtooth stock, 327–330
Schist, 273–286, 287, 326–329
Schistosity, axial-plane, 273–286
Schmidt number, 163–166
Schottky defects, 95
Scottish Hebrides, 300, 329
Sedimentary rocks, vi, 299–324, 325–335
Segregation, 231–241, 261, 273–286
Selenide ion radius, 53
Self-diffusion, 65, 66, 73
   cadmium, 55, 56
   calcium, 201
   carbon, 219
   iron, 57
   oxygen, 99, 219
   potassium, 70, 83–85, 91, 92, 94
   rubidium, 93
   silver, 54
   sodium, 70, 82, 83, 90, 94
   sulfur, 55, 56
Self-diffusion coefficient, 1, 15, 47, 77, 79, 82, 84, 124, 219
Self-diffusion by vacancy mechanism, 11
Semiconduction, intrinsic, 118
Seychelles Islands, 317
Shadows, 273–286
Shale, 34, 287, 326, 330
Short-circuit diffusion mechanisms, 3, 4
Silent Lake, 331, 332
Silica, 261–271, 282
Silica glass, 139–170, 152–154
Silicate glass, 237
Silicates, framework, 147, 152, 154
Silicates, Modeling of Diffusion-Controlled Properties of, 31–52
Silicon, 35, 43, 48, 94, 105, 131, 134, 285, 287–293, 316, 321
Sillimanite, 273–286, 330
Silver, ionic radii, 53
Silver sulfide, 54–56
Silverton, Colorado, 299–307
Silverton Caldera, 299–305
SiO_4 tetrahedra, 44, 48
Skaergaard, 300, 309–315, 329
Skarns, 229, 243, 287–293
Sodium, 39, 70, 71, 72, 77–98, 99, 140, 152, 207, 246, 257, 282, 285, 316
   chemical equilibration 83, 87
   diffusion, in alkali feldspars, 39, 77–98, 180, 243
      in orthoclase, 77–98
      in quartz, 132
   isotopic equilibration, 83, 87
Solfataric hydrothermal solutions, 306
Solid-fluid equilibrium, 78
Solubility gap, 248–254, 258
Solubility product of crystal, 206
Solubility of water in quartz, 131, 135–137
Solution, *see also* Dissolution

Solution-precipitation, 70
Solution and redeposition, 316
Solvus, 173–183
Sorption, 140–150, 155
Southern California batholith, 300, 329
Spessartine, 34
Spherical diffusion model, 77–98, 100
Spinodal, chemical, 173–174
   coherent, 173, 174
   strain-free, 173, 174
Spinodal decomposition, 171, 173, 176, 197, 198, 201
Sr^{86}/Sr^{88} ratio, 72
Stability of minerals to age resetting, 62
Staurolite, 47, 273–286, 330
Staurolite shadows, 273–286
Steamboat Springs, Nevada, 300
Steel, ductility, 53
Stillwater Complex, 159
Stoichiometric minerals, 262
Stokes-Einstein equation, 139, 154
Stoneham, Maine, muscovite, 282–285
Stony Mountain, Colorado, 300–302, 305, 308, 312, 315
Strain, elastic, 171, 173–176
Strain-free spinodal, 173, 174
Strain-rate-dependent strength of mantle, 131
Strontium, 62, 70–72, 257
Strontium isochrons, 62
Sturgeon Falls, 331, 332
Sulfide, cadmium, 55–57
   copper, 54–56
   ferrous, 53, 54, 57
   nickel, 57
   silver, 54–56
Sulfide ion, mobility, 53
   polarizability, 53
   radius, 53
Sulfides, Diffusion in, 53–59
Summitville, 307
Supercell reflections in FeS at low temperatures, 57
Supercooled liquid, 15, 20
Supersaturated diopsides, enstatite exsolution, 195–203
Surface free energy, 206
Surface, saturation, 287–293
Surface saturation, 148
Syenite, 326, 332

Tactites, *see* Skarns
Tectosilicate, 281, 282
Telluride ion radius, 53
Temperature, effect on diffusion, 63, 65, 71, 117–129, 139, 151, 152
Temperatures of intrusives, 295–298
Tephroite, 47, 287
Ternary diffusion, 93
Ternary system, 15, 22, 40
Tertiary, 299, 301, 306–308, 317, 320, 321
Tetrahedra, 43–45, 48

Tetrahedral-oxygen (T-O) bonds, 189
Tetrahedral sites, 53, 54, 57, 105, 185, 191, 193
Texas Canyon stock, 328, 329
Thermal conductivity of magma, 140, 157
Thermal diffusivity, 155, 157
Thermodynamic-continuum diffusion equations, 3, 12
Thermodynamic driving forces, 12
   equilibrium, 206
   matrix, 17
Thermodynamics, nonequilibrium, 31, 229, 231–241
Thermometer, geological, 32, 53, 295–298, 320
Thermosyphon, 156
Tholeiite, 160
Thomson-Freundlich equation, 206, 209
Thomson, Gibbs—equation, 206
Thorium, 62, 72
TiO_2-Cr_2O_3 system, 118
Tonopah, Nevada, 300
Trace-element equilibrium, 205–218, 246
Trace-element exchange between crystal and fluid, 205–218
Trace elements, partitioning of, 31
Tracer diffusion coefficient, 1, 3, 7–11, 48, 79, 90, 92
Transport equations, 243, 244
   models, 243–259, 261
Treasure Mountain, 302
Tridymite, 314
Trioctahedral mica, Fe content, 114
Tritiated Water, Diffusion of in beta-Quartz, 131–138
Tritium, 131–138
Triton decay, 134
Trondhjemite, 327

Ultramafic rocks, 318, 326
U-Pb dating, 62, 72, 299, 322
Uphill diffusion, 15–30, 229, 287–293
Uranium, 62, 72
Uranium loss, 72
U-Th-Pb dating, 62, 72

Vacancy diffusion, 118
   mechanism, 3, 6, 9–11, 39–50, 54, 66, 67, 77, 94–96
   mobility, 54, 57
   structure, 15, 54, 57

Vacancy-wind effects, 3, 9–12, 48
Vacuum-fusion extraction of argon, 109
Valles Caldera, New Mexico, 140
Varves in periglacial lakes, 280
Vastervik segregations, 240
Vector Space Treatment of Multicomponent Diffusion, 15–30
Vermont marble, 329
Vesiculation, vii, 148, 156, 161, 165
Viscosity, relation with diffusion, 139–170
Viscosity of fluid phase, 256, 257
   of magma, 140, 157–168
Viscous flow, 139–170
Volcanic contact metamorphic effect, 71
Volcanic necks, 156
   rocks, 299–324, 329, 331–333
   terranes, 299, 317, 321, 322
Volume diffusion in solids, 61–76
Volume-diffusion mechanisms, 3–7
Volume relaxation, 143–146

Walloomsac samples, 273–286
Wall-rock alteration, 254
Water, diffusion, in granitic liquids, 139–170
   in quartz, 131–138
Water, ground, 299–324, 329
   isotope exchange with calcite, 219–227
Water fugacity, 33, 95, 143, 148–150, 165–167
Water systems, geothermal, 300, 310
Wollastonite ($CaSiO_3$), 201, 288–292

Xenoliths, 325–335

Yellowstone Park, Wyoming, 300
Yule marble, 215

Zeolites, 148, 154
Zinc chalcogenide, 53
Zinc, ionic radii, 53
Zircons, dating of, 72, 299, 322
Zoning, chemical, 139, 140
   magmatic, 311–316
   in metamorphic garnets, vi, vii, 31–33
   in skarns, 229, 243, 287–293

AUTHOR INDEX

Abelson, P. H., 52, 241, 259, 293, 322
Adams, J. A. S., 72, 75
Agrell, J., 259, 293
Ahrens, T. J., iv, v
Akimoto, S., 120, 126, 127
Albarede, F., 31, 50, 78, 97
Albee, A. L., 121, 127, 324, 335
Allen, R. L., 54, 58
Alves, G. E., 237, 241
Anderson, A. T., 323, 329, 334
Anderson, D. E., iv, v, 1, 31–52, 168
Anderson, G. M., 265, 271, 283, 285
Anderson, T. F., iv, v, 1, 78, 95, 98, 99–105, 171, 192, 193, 205, 206, 216, 219–227, 230, 316, 324
Ando, K., 104, 105
Ansel, G., 137
Appleman, D. E., 105, 182
Ardell, A. J., 209, 216, 217
Arkharov, V. I., 58
Ashbee, K. H. G., 131, 137
Austerman, S. B., 45, 46, 50
Avrami, M., 199, 202
Axelrod, D. J., 308, 322
Azároff, L. V., 45, 50

Baadsgaard, H., 69, 71, 73, 75
Bachinski, S. W., 177, 181, 182
Baëta, R. D., 131, 137
Bailey, A., 39, 50, 78, 79, 91, 97, 180, 182
Baldwin, R. L., 38, 41, 42, 44, 49, 51
Balk, R., 273, 285
Bardeen, J., 39, 44, 50
Barnes, H. L., 218, 282, 285
Barr, L. W., 13
Barrer, R. M., 148, 150, 169
Barrett, C. S., 205, 216
Barretto, P. M. C., 72, 75
Barth, T. F. W., 217, 273, 276, 285
Bartkowicz, I., 55, 58
Bartlett, R. W., 241
Bartnitsky, E. N., 112, 115
Baruch, P., 137
Baskin, Y., 185, 193
Bassett, W. H., 70, 75
Baynes, R. A., 30
Beck, P. A., 205, 216
Becke, F., 206, 216
Bell, P. M., iv, v, 127, 282, 286
Bell, T., 152, 169
Bence, A. E., 121, 127, 203
Berner, R. A., 214, 216
Bertaut, E. F., 54, 57, 58
Beswick, A. E., 81, 97
Bickford, M. E., 322
Bigeleisen, J., 325, 334
Bigelow, S. L., 206, 216
Birchenall, C. Ernest, iv, v, 1, 53–59
Bird, R. B., 154, 157, 163, 169, 237, 241

Blackburn, W. H., 31, 32, 50
Bollmann, W., 176, 182
Borchardt, V. G., 125, 127
Bottinga, Y., 31, 50, 78, 97, 140, 169
Botts, S., 108, 115
Boucher, D. R., 237, 241
Bowen, N. L., 140, 155, 169
Bowes, D. R., 278, 285
Bown, M. G., 195, 202
Boyd, F. R., 120, 127, 195, 197, 201, 202
Brace, W. F., 167, 169, 256, 258
Bradley, R. S., 126, 127
Brady, J. B., iv, v
Brandt, S. B., 69, 75, 112, 115, 230, 295–298
Brett, Robin, 203
Brigham, R. J., 31, 40, 51
Brindley, G. W., 193
Brock, M. R., 323
Brook, D. W., 148, 150, 169
Brown, E. H., 277, 285
Brown, G. E., 203
Brown, G. M., 195, 202, 310, 324
Brown, W. L., 182
Brunner, G., 131, 137
Buckley, G. R., 1, 31–52
Budd, Max, 276, 282, 284
Buening, D. K., 118, 127
Burbank, W. S., 301, 323
Burnham, C. W., 140, 143, 165, 169, 265, 271, 283, 285
Burt, D. M., iv, v, 216, 229, 243, 259, 287–293
Buseck, P. R., 118, 127

Cahn, J. W., 173, 174, 175, 182
Carlson, P. T., 20, 21, 29
Carman, P. C., 80, 97
Carmichael, D. M., 231, 240, 241, 261, 271
Carmichael, I., 309, 322
Carron, J., 140, 152, 169
Carslaw, H. S., 18, 29, 66, 75, 148, 155, 161, 169, 258, 259
Chai, B. H. T., iv, v, 171, 205–218, 219–227
Chamberlain, J. W., 202, 203
Champness, P. E., 195, 202
Chandrasekhar, S., 157, 169
Chapple, W. M., 281, 285
Chaudhuri, S., 322
Chernov, A. A., 34, 50
Choi, H. Y., 157, 158, 159, 169
Choudhury, A., 132, 137
Christian, J. W., 187, 193, 197, 199, 202
Christie, J. M., 195, 202, 203
Clark, A. M., 118, 125, 126, 127
Clark, J. R., 203
Clark, R. B., 72, 75
Clark, S. P., Jr., 127, 140, 169, 285
Clarke, R. L., 56, 58

Clayton, R. N., iv, v, 100, 105, 215, 217, 219, 221, 225, 227, 305, 320, 322, 332, 335
Cohen, J. B., 54, 58
Condit, R. H., 57, 58
Cooper, A. R., Jr., iv, v, 1, 15–30, 31, 39, 42, 44, 50, 52, 115, 131–138, 147, 169, 230, 238, 241, 244, 259, 289, 293
Correns, C. W., 217
Cos Garea, A., Jr., 39, 42, 52
Coslett, V. E., 128
Craig, H., 308, 322
Crank, J., 1, 23, 29, 38, 39, 42, 50, 51, 66, 75, 80, 97, 100, 105, 109, 110, 115, 140, 147–150, 155, 169, 247, 248, 255, 259
Curier, H., 137
Curran, P. F., 233, 237, 239, 241
Currie, K. L., 283, 285
Curtis, G. H., 69, 75, 115

Dahl, O., 33, 50, 51
Danckwerts, P. V., 33, 51
Darken, D. S., 39, 42, 51
Darken, L. S., 16, 29, 148, 169
Davidson, L. R., 335
Davis, B. T. C., 126, 127, 201, 202
Davis, N. F., 140, 165, 169
Dayanada, M. A., 29
Dayanandu, M. A., 15, 30
deBruyn, P. L., 241
deGroot, S. R., 17, 29, 30, 38, 41, 51, 231, 241
Dehoff, R. T., 27, 28, 30
Delevaux, M. H., 323
Demianiuk, M., 56, 59
Deuser, W. G., 71, 75
De Vault, D., 243, 259
Dewey, J. F., 280, 285
Dietrich, J. H., 281, 286
Dietrich, R. V., 278, 286
Dings, M., 302, 322
Dodd, D., 131, 132, 137
Donnay, G., 105
Dontsova, E. I., 104, 105, 330, 334
Douglass, D. L., 197, 202
Drury, T., 132, 138, 150–152, 169
Duba, A., 117, 118, 125–127
Dudziak, K. H., 256, 259
Dumbgen, G., 132, 138
Dunlop, P. J., 38, 41, 42, 44, 49, 51
Dwornik, E. J., 203
Dymond, J. R., 332, 334

Eckert, E. R., 23, 30
Eckhardt, D., 125, 126, 127
Eirich, F. R., 169
Eitel, W., 152, 169
Elliott, David, 229, 231–241, 243, 259, 281, 282, 286
Ellis, A. J., 225, 227, 282, 285, 286
Emleus, C. H., 202
Engel, A. E. J., 215, 217
Engelke, H., 19, 20, 30
Engell, H. J., 125, 128

Engels, J., 70, 75
England, J. L., 120, 126, 127
Engstrom, Arne, 128
Epstein, S., 100, 105, 215, 217, 220, 227, 300, 303, 306–308, 311, 319, 320, 322–324, 329, 330, 334, 335
Erga, O., 241
Ernst, W. G., 321, 322
Eskola, P., 206, 217, 280, 286
Etienne, A., 55, 58
Eugster, H. P., 112, 115, 321, 322
Evans, H. T., Jr., 100, 105, 175, 182
Evernden, J. F., 69, 75, 108, 112, 115
Eyres, N. R., 258, 259
Eyring, H., 154, 169, 217

Fairbairn, H. W., 63, 64, 75
Fasiska, E. J., 57, 58
Faul, H., 96
Faure, G., 322
Fenton, M. O., 322
Fern, F. H., 209, 217
Ficca, James, 56
Fine, M. E., 34, 51, 197, 202
Finger, L. W., 121, 127
Finn, D., 39, 42, 52
Fisher, D., 273, 274, 280, 286
Fisher, G. W., iv, v, 118, 127, 229, 231–241, 243, 259, 261, 271, 282
Fisher, R. M., 202, 203
Fitts, P. D., 41, 42, 51
Fleischer, R. L., 73, 75
Fletcher, R. C., iv, v, 229, 241, 243–259
Foland, K. A., iv, v, 1, 68–70, 72–75, 77–97, 101, 105, 180, 182, 243, 259
Folinsbee, R. E., 69, 75
Forester, R. W., 300–303, 305, 308–310, 312–315, 321, 322, 324, 329, 334, 335
Foster, W. R., 195, 203
Franck, E. U., 256, 259
Frangos, W. T., 169, 258
Franklin, A. D., 13
Frantz, J. D., iv, v, 229, 261–271
Fraser, D., 131, 132, 137
Frey, F. A., iv, v
Friedman, Irving, 143, 169, 306–308, 320, 322, 323
Friedrichsen, H., 331, 334
Frieman, S. W., 29
Frischat, G. H., 132, 138
Fry, D. L., 152, 169
Fryt, E., 55, 58
Fujikawa, Y., 124, 125, 128
Fujino, T., 55, 58
Fujisawa, H., 120, 126, 127
Fujita, H., 17, 30
Fullman, R. L., 214, 217
Fulrath, R. M., 218
Fyfe, W. S., 243, 259, 265, 271

Gallagher, K. J., 79, 97
Galwey, A. K., 35, 51, 205, 217
Garlick, G. D., 300, 323, 325, 329, 330, 332–334

AUTHOR INDEX

Garrels, Robert M., 233, 241
Gast, P. W., 63, 75
Gay, P., 195, 202
Geffken, W., 23, 30
Gel'fman, A. Y., 56, 59
Gerling, E. K., 70, 75
Ghose, Subrata, 195, 202
Gibbs, G. B., 238, 241
Gifkins, R. C., 241
Giletti, B. J., iii–v, 1, 2, 61–76, 78, 93, 96, 97, 107–115, 193
Girifalco, L. A., 12
Glueckauf, E., 243, 244, 259
Godfrey, J. D., 306, 308, 323
Goetze, C., 168
Goldsmith, J. R., iv, v, 185, 190–193
Gorbunova, K., 58
Gordon, R. S., 209, 217
Gorman, R. R., 15, 30, 152, 170
Gosting, L. J., 17, 30, 38, 41, 42, 44, 49, 51
Gould, E. S., 95, 97
Grace, R. E., 15, 29, 30
Graf, D. L., 322
Grant, J. A., 78, 97
Green, D. C., 335
Green, H. W., II, iv, v
Green, J. G., 322, 323
Greenwood, H. J., 127
Gresens, R. L., 282, 286
Greskovich, C., 118, 128
Griggs, D. T., 95, 97, 105, 131, 138, 202, 203
Grigorév, D. P., 205, 217
Guillemin, C., 182
Gupta, P. K., 16, 18–20, 22–24, 26, 27, 29–31
Gurry, R. W., 148, 169
Guy, A. C., 27, 28, 30
Guy, A. G., 15, 30

Haase, R., 31, 45, 51
Hadidiacos, C. G., 120, 121, 127, 128
Hafner, S. S., 201, 203
Hall, H. T., 199, 200, 203
Hall, L. D., 42, 51
Hall, W., 320, 323
Haller, W., 155, 169
Hamilton, D. L., 143, 169
Hamilton, R. M., 126, 128
Hamza, M. S., 219, 227
Handwerker, D. S., 105, 182
Hanitzsch, E., 207, 217
Hanson, G. N., 63, 69, 71, 73, 75, 112, 115
Harker, A., 205, 206, 217
Harris, J. E., 238, 241, 323
Hart, S. R., iv, v, 62, 63, 74, 75, 189, 190, 193, 257, 298
Hartley, G. S., 39, 51
Hartree, D. R., 259
Haul, R. A. W., 80, 97, 132, 138, 219, 227
Haven, H., 138
Heard, H. C., 127
Heasley, J. H., 39, 50

Heckel, R. W., 180, 183
Heckman, R. W., 152, 170
Helgeson, H. C., iv, v, 77, 97, 233, 237, 241, 244, 255, 259, 282, 285
Hemley, J. J., 261, 262, 265, 271
Hench, L. L., 29
Henderson, D., 217
Herring, C., 39, 44, 50
Hess, G. B., 159, 163, 169
Hess, H. H., 169, 195, 202, 203
Hess, P. C., 32, 51
Hesselgesser, J. M., 263, 271, 282, 283, 286
Hetherington, G., 152, 169
Heuer, A. H., iv, v, 1, 115, 131–138, 202, 203
Hewitt, D., 280, 282, 286
Heyl, A. V., 321, 323
Hobbins, R. R., Jr., 57
Hoffman, W. P., 203
Hofmann, A. W., iii–v, 1, 2, 70–72, 75, 78, 96, 97, 168, 229, 230, 243–259, 262, 271, 289, 293
Holland, J. G., 202
Hollister, L. S., 280, 286
Holloway, J. R., 169
Hooyman, G. J., 41, 51
Howard, R. E., 13
Hsu, L. C., 33, 51
Hudleston, P. J., 281, 286
Hughes, H., 63, 64, 75, 117, 118, 125, 126, 128
Hulett, G. A., 206, 217
Hurley, P. M., 63, 64, 75, 128

Iiyama, J. T., iv, v
Ingamells, C. O., 196, 203
Ingham, J., 259
Irvine, T. N., 320, 323
Isachsen, Y., 96, 273, 274, 280, 286
Ishiguro, M., 55, 58

Jack, K. H., 152, 169
Jackson, E. D., 69, 75
Jackson, R., 259
Jaeger, J. C., 18, 29, 66, 75, 148, 155, 156, 161, 162, 169, 258, 259
Jäger, E., 78, 97
Jagitsch, R., 78, 97
Jain, S. C., 79, 97, 100, 105
James, P. F., 209, 217
Jamil, A. K., 127
Jander, W., 124, 125, 128
Jensen, M., 78, 90, 91, 97
Jones, E. D., 55, 56, 58
Jones, J. B., 189, 193
Jones, K. A., 35, 51, 205, 217
Jost, W., 217

Kahlweit, M., 207, 217
Kalade, G. A., 215, 217
Kanamori, H., 125, 128
Kanukov, A. B., 225, 227
Katchalsky, A., 237, 239, 241
Kato, A., 293

Kats, A., 131, 132, 138
Kegeles, G., 38, 41, 42, 44, 49, 51
Kendrick, F. B., 206, 217
Kennedy, G. C., 225, 227
King, T. B., 152, 169
Kingery, W. D., 30, 255, 259
Kirkaldy, J. S., 15, 30, 31, 39, 40, 45–47, 50, 51
Kirkwood, John G., 38, 41, 42, 44, 49, 51
Kirsch, P. F., 30
Kistler, R. W., 69, 75, 115
Kline, Stephen J., 232, 233, 241
Klotsman, S. M., 57, 58
Knauth, L. P., 320, 323
Kobayashi, Y., 125, 128
Koch, F., 54, 58
Komado, E., 127
Korbel, A., 209, 217
Koros, P. J., 152, 169
Korzhinskii, D. S., 229, 243, 244, 251, 255, 256, 259, 261, 262, 269, 271, 287–289, 293
Kovalenko, V. I., 69, 75, 298
Kovalev, G. N., 237, 238, 241, 256, 259
Kovaleva, A. D., 56, 59
Kraichnan, R. H., 159, 169
Kravchenko, N. G., 56, 59
Kretz, R., 35, 51, 205, 217, 278, 286
Kriventsov, P. P., 298
Kröger, F. A., 56, 58
Krogh, T. E., iv, v
Krop, K., 209, 217
Kulp, J. L., 70, 72, 75
Kumar, V., 56, 58
Kummer, J. T., 93, 98
Kushiro, I., iv, v, 114, 115, 127, 195, 196, 201–203
Kwak, T. A. P., 31, 32, 51

Lally, J. S., 195, 202, 203
Lane, J. E., 31, 39, 45–47, 50, 51
Lapides, I. L., 69, 75
Larmer, K., 127
Laves, F., 137, 176, 181, 183, 185, 189–193
Lazarus, D., 119, 128
Lee, R. W., 152, 169
Leith, C. K., 206, 217
Letolle, R., 70, 75
Levich, V. G., 23, 30
Libby, W., 138
Lidiard, A. B., 13
Lifshitz, I. M., 206, 207, 217
Lightfoot, E. N., 169, 237, 241
Lin, T. H., 39, 51, 74, 75, 78, 79, 90–92, 94, 96, 97, 99, 101–103, 105, 175, 180, 183
Linder, R., 200, 203
Lipman, P. W., 301, 305, 306, 323
Lipson, J., 69, 75
Loberg, B., 233, 240, 241
Long, J. V. P., 118, 121, 125, 127, 128
Long, W., 169
Lorimer, G. W., 195, 202
Lu, Tien-Lien, 56
Luce, R. W., 237, 238, 241

Luedke, R. G., 301, 323
Lumbers, S. B., 333, 334
Luth, W. C., 196, 203
Lykov, A. A., 295, 296, 298

MacGregor, I., 323
MacKenzie, W. S., 97, 181, 183, 193
Madden, T., 127
Mahon, W. A. J., 282, 286
Malinin, S. D., 225, 227
Manheim, F. T., iv, v
Manning, J. R., iv, v, 1, 3–13, 16, 30, 31, 39, 44, 48, 51, 66, 75, 94, 97, 282
Manson, V., 108, 115
Mao, H. K., 282, 286
Marchant, D. O., 209, 217
Markworth, A. J., 209, 217
Maruyamo, H., 125, 128
Mason, G. R., 31, 51
Massalski, T. B., 205, 216
Matano, C., 118, 128
Mathieu, H. J., 55, 58
Maurette, M., 73, 75
Mayeda, T. K., 100, 105, 219–221, 225, 227, 322
Mazur, P., 17, 29, 30, 38, 41, 51, 231, 241
McBirney, A. R., 140, 156, 169
McCallister, R. H., iv, v, 78, 98, 171, 173, 183, 195–203, 216
McConnell, J. D. C., iv, v, 176, 182, 183, 185, 188, 190, 191, 193
McCrea, J. M., 220, 227
McDonald, K. L., 117, 125, 126, 128
McKie, D., 185, 188, 190, 191, 193
McNutt, R. H., 71, 75
Medaris, L. G., Jr., 118, 127
Medford, G. A., 140, 152, 164, 169
Meents, W. F., 322
Megaw, H. D., 193
Mehnert, H. H., 323
Mérigoux, H., 78, 97, 99, 101–103, 105
Meyer, O. A., 202
Meyer, O. E., 16, 30
Miller, David, 282
Miller, D. G., 15, 30
Miller, D. S., 73, 75
Misener, D. J., iv, v, 1, 2, 39, 51, 117–129
Mizia, J., 209, 217
Mizutani, H., 125, 128
Moore, W. J., 54, 58
Morey, G. W., 263, 271, 282, 283, 286
Morgan, G. J., 205, 217
Morse, J. W., 215, 216
Mose, D. G., 322
Moulson, A., 132, 138
Mrowec, S., 55, 58
Muehlenbachs, K., 300, 309, 323, 332, 335
Mueller, G., 32, 33, 51
Müller, G., 177, 181, 182, 185, 188, 189, 191, 193
Munro, D. C., 127
Murase, T., 140, 169
Mussett, A. E., 67–69, 75, 78, 96, 97, 115

AUTHOR INDEX 351

Nachtrieb, N. H., 119, 128
Narahari Achar, B. N., 193
Naughton, J. J., 124, 125, 128
Nernst, W., 15, 30
Newland, B. T., 69, 75
Ng, George, 202
Nichols, I. A., 117, 118, 125–127
Nicholson, R. B., 209, 216
Nicolaysen, L. O., 62, 76
Nielsen, R. L., 323
Niggli, R., 97
Nigrini, A., 256, 259
Nissen, H.-U., 176, 182
Nord, G. L., 195, 203
Northrup, D. A., 219, 225, 227
Norwood, Curtis B., 74, 76, 114, 115
Nye, J. F., 32, 51

Obradovich, J. D., 69, 75, 115
Oda, F., 55, 58
Ogilvie, R. E., 40, 52
Oishi, Y., 15, 16, 22, 23, 26, 27, 30, 104, 105
Okazaki, H., 54, 58
O'Keefe, M., 118, 128
Olsson, M. G., 78, 97
O'Neil, J. R., 70, 76, 99, 100, 102, 105, 219, 221, 225, 227, 316, 320–323
Onsager, L., 15, 30, 38, 40, 41, 51, 52
Orville, P. M., iv, v, 81, 97, 103, 105, 175, 177, 183, 186, 192, 193, 216, 227, 240, 241, 244, 246, 253, 256, 259, 277, 280, 285, 286, 321, 323
Osborn, E. F., 169
Ostwald, W., 206, 217, 221, 227
Ovchinnikova, G. V., 70, 75
Overbeek, J. Th. G., 241
Owen, D. C., 176, 182, 183

Palmer, D., 137
Papike, J. J., 195, 203
Park, R. G., 278, 285
Parks, G. A., 241
Pask, J. A., 218
Paul, W., 128
Pecket, A., 202
Pellas, P., 73, 75
Perry, E. C., 319, 323
Peschanski, D., 58
Petrović, R., 78, 79, 90–97, 180, 183
Philibert, J., 15, 30
Phillips, R., 202
Pidgeon, R. T., 72, 76
Pinckney, D. M., 215, 217
Pinson, W. H., Jr., 63, 64, 75
Planck, M., 15, 30
Pluschkell, W., 125, 128
Poldervaart, A., 169, 195, 203
Prewitt, C. T., 203
Price, P. B., 73, 75, 76
Pugh, E. M., 197, 203
Purdy, G. R., 31, 51
Purohit, R. K., 56, 58

Radcliffe, S. V., 202, 203
Rahlfs, P., 54, 58
Ramberg, H., 33, 52, 278, 281, 286
Rastogi, P. K., 217
Raw, G., 40, 52
Ray, L., 121, 127
Redfield, A. C., 323
Ree, T., 154, 169
Reed, S. J. B., 121, 128
Reid, A. M., 203
Ribbe, P. H., 186, 187, 189, 191, 193
Ribble, T. J., 118, 128
Richards, T., 127
Rickard, L. V., 273, 274, 280, 286
Rickert, H., 55, 58
Ridley, I., 203
Ringwood, A. E., 118, 125–128
Robbins, Gary A., 74, 76
Roberts, G., 132, 135, 138
Roberts, J. P., 132, 135, 138, 150–152, 169
Robertson, E. C., 168
Robinson, D., 215, 217
Robinson, R. A., 238, 241
Rohsenow, W. M., 157–159, 169
Roseboom, E. H., 145, 169
Ross, M., iv, v, 195, 202, 203
Roth, W. L., 54, 59
Runcorn, S. K., 128
Rye, D. M., iv, v
Rye, R. O., 215, 217

Sabatier, G., 105
Sachs, S., 115
Sahama, Th. G., 119, 128
Sang, J. S-L., 1, 131–138
Sarjant, R. J., 259
Sauer, F., 27, 30
Sauthoff, G., 209, 217
Savin, S. M., 308, 319, 320, 323
Sawyer, B., 131, 138
Schairer, J. F., 195, 197, 201, 202
Schamel, S., 277, 286
Schmalzried, H., 125, 127
Schneider, A., 32, 33, 51
Schneider, T. T., 185, 193
Schober, 126, 128
Schock, R. N., 127
Schoen, B., 323
Schönert, H., 31, 42, 43, 52
Schreiner, G. D. L., 70, 76
Schult, A., 126, 128
Schwarcz, H. P., 329–333, 335
Schwartzman, D. W., 69, 70, 76
Sedlak, Frank, 282
Selby, S. M., 95, 97
Schaffer, E. W., 1, 2, 131–138
Shaffer, W. S., 132, 138
Shankland, T. J., 117–119, 125, 127, 128
Shapiro, L., 108, 115
Sharma, B. L., 56, 58
Sharp, J. H., 191, 193
Sharp, W. E., 225, 227

Shaw, D., 56, 59
Shaw, D. M., 333, 334
Shaw, H. R., iv, v, 1, 139–170, 257, 259, 310, 323
Sheppard, S. M. F., 300, 305, 306, 308, 323, 329, 335
Shewmon, P. G., 13, 34, 52
Shieh, Y. N., iv, v, 230, 306, 308, 311, 323, 325–335
Shimp, N. F., 322
Sigvaldason, G. E., 323
Silver, L. T., 72, 76, 322, 323
Silverman, S. R., 319, 323
Simonsen, D., 17, 30
Sipling, P. J., iv, v, 171, 185–193
Sippel, R. F., 78, 90, 91, 97
Skinner, B. J., 216
Slyozov, V. V., 206, 207, 217
Smirnov, V. N., 69, 75
Smith, A. F., 209, 218
Smith, C. H., 320, 323
Smith, D. K., 195, 203
Smith, J. V., 176, 181, 183, 189, 193
Smith, R. L., 168, 169
Somorjai, G. A., 56, 59
Spray, A., 205, 218
Sreedhar, A. K., 56, 58
Stamm, W., 124, 125, 128
Steiger, R. H., 189, 190, 193
Stein, L. H., 219, 227
Stevels, J. M., 138
Steven, T. A., 305, 323
Stevenson, D. A., 53, 59
Stewart, D. B., 78, 97, 103, 105, 168, 175, 177, 180, 183, 186, 187, 191, 193
Stewart, W. E., 169, 237, 241
Stokes, R. M., 238, 241
Strens, R. G. J., 73, 76
Stubican, V. S., 118, 128
Sucov, E. W., 15, 30, 152, 170
Suggate, R. P., 64, 76
Sullivan, G. A., 56, 57, 59
Sundelhof, L. O., 26, 30
Sundheim, B. R., 42, 52
Suzuoki, T., 303, 307, 323
Symes, E. M., 73, 75
Sysoev, L. A., 56, 59
Szeto, W., 56, 59

Takeda, Hiroshi, 195, 203
Tammann, G., 255, 259
Tanzilli, R. A., 180, 183
Tatsumi, T., 293
Taylor, H. P., Jr., iv, v, 70, 75, 99, 100, 102, 105, 215, 218, 230, 299–324, 327–335
Taylor, W. H., 189, 193
Terjesen, S. G., 239, 241
Thompson, J. B., Jr., 44, 52, 231, 240, 241, 244, 257, 259, 269, 271, 288, 289, 293
Thrower, P. A., 203
Tilton, G. R., 62, 63, 76
Timofeev, A. N., 57, 58
Tolland, H. G., 73, 76
Tomisaka, T., 185, 192, 193

Toor, H. L., 17, 30
Trakhtenberg, I. S., 57, 58
Trimble, H. M., 206, 216
Trimble, L. E., 39, 42, 52
Tullis, J., 177, 178, 181–183, 193
Turi, B., 306, 308, 311, 324, 327–329, 335
Turner, F. J., 259, 277, 286

Underwood, E. E., 214, 218
Urey, H. C., 325, 335
Utenkov, V. A., 298

van Breemen, O., 71, 73, 75
Vanderberg, Ludewig, 56
van der Merwe, A., 40, 52
van Lier, J. A., 238, 241
Varshneya, A. K., 15, 20, 24, 29, 30, 238, 241
Ve, P., 241
Verbeek, A. A., 70, 76
Verhoogen, J., 259
Vidale, R., iv, v, 229, 231, 241, 261, 271, 273–286
Virgo, D., 127, 201, 203
Vogel, D. E., 323
Volkova, N. V., 69, 75, 298

Wachtman, J. B., Jr., 13
Wager, L. R., 310, 324
Wagner, C., 23, 30, 42, 52, 118, 128, 206, 207, 218
Wagner, J. W., 45, 46, 50
Wagstaff, J. B., 259
Walker, R. M., 73, 75, 76
Walsh, J. B., 169, 258
Walter, L. S., 202
Warschauer, D. M., 128
Watanabe, T., 287, 293
Watkins, C., 202, 203
Weast, R. C., 95, 97
Wedepohl, K. H., 334
Weiblen, P. W., 78, 97
Weichert, D. H., 31, 40, 51
Weill, D. F., 140, 169, 261, 265, 271
Weisbrod, A., iv, v, 229, 261–271
Welke, H.-J., 70, 76
Wellman, T., 285, 286
Wenk, E., 97
Wenner, D. R., 308, 317, 321, 324
Westcott, M. R., 63, 76
Wetherill, G. W., 62, 76
Whelan, R. C., 56, 58
White, D. E., 300, 310, 324
White, J., 216, 218
Whitney, W. P., 118, 128
Willaime, C., 181, 182
Williams, E. L., 152, 170
Williams, P. F., 277, 280, 286
Wilson, A. F., 331, 335
Wilson, J. N., 243, 259
Winchell, P., 140, 152, 170
Winslow, G. H., 197, 203
Wollast, R., 255, 259

AUTHOR INDEX

Wondratschck, H., 137
Wones, D. R., iv, v, 112, 115, 127, 168, 321
Wood, D. L., 131, 138
Wood, J. A., 78, 97
Woodbury, H. H., 56, 59
Wright, F. E., 309, 324
Wright, T. L., 78, 97, 103, 105, 168, 175, 177, 180, 183
Wu, C. H., 203
Wyart, J., 105
Wyllie, P. J., 323

Yanagisawa, Paul, 322
Yang, Houng-Yi, 195, 203
Yao, Y-F. Y., 93, 98

Yoder, H. S., Jr., iii–vii, 114, 115, 119, 127, 128, 202, 229, 230, 271, 282
York, D., 83, 98, 111, 115, 197, 203
Yourgrau, W., 40, 52
Yui, S., 293
Yund, R. A., iii–v, 1, 39, 51, 74, 75, 78, 79, 90–92, 94–96, 98, 99–105, 171, 172, 173–183, 185–193, 199, 200, 202, 203, 216, 230, 257, 271, 316, 324

Zartman, R. E., 323
Zen, E-an., 168, 233, 241
Zharikov, V. A., 287, 293
Ziebold, T. O., 16, 28, 30, 31, 39, 40, 52
Zmija, J., 56, 57, 59
Zussman, J., 97

J.R. Beckett

$38.00